# Handbook of
# THIN FILM DEPOSITION

# Handbook of THIN FILM DEPOSITION

## THEORY, TECHNOLOGY AND SEMICONDUCTOR APPLICATIONS

**Fifth Edition**

Edited by

**DOMINIC SCHEPIS**
IBM Microelectronics, East Fishkill, NY, United States
GLOBALFOUNDRIES, Austin, TX, United States

**KRISHNA SESHAN**[†]
Materials Science, University of Arizona, Tucson, Arizona

ELSEVIER

[†] Deceased

Elsevier
Radarweg 29, PO Box 211, 1000 AE Amsterdam, Netherlands
125 London Wall, London EC2Y 5AS, United Kingdom
50 Hampshire Street, 5th Floor, Cambridge, MA 02139, United States

Copyright © 2025 Elsevier Inc. All rights are reserved, including those for text and data mining, AI training, and similar technologies.

Publisher's note: Elsevier takes a neutral position with respect to territorial disputes or jurisdictional claims in its published content, including in maps and institutional affiliations.

No part of this publication may be reproduced or transmitted in any form or by any means, electronic or mechanical, including photocopying, recording, or any information storage and retrieval system, without permission in writing from the publisher. Details on how to seek permission, further information about the Publisher's permissions policies and our arrangements with organizations such as the Copyright Clearance Center and the Copyright Licensing Agency, can be found at our website: www.elsevier.com/permissions.

This book and the individual contributions contained in it are protected under copyright by the Publisher (other than as may be noted herein).

**Notices**

Knowledge and best practice in this field are constantly changing. As new research and experience broaden our understanding, changes in research methods, professional practices, or medical treatment may become necessary.

Practitioners and researchers must always rely on their own experience and knowledge in evaluating and using any information, methods, compounds, or experiments described herein. In using such information or methods they should be mindful of their own safety and the safety of others, including parties for whom they have a professional responsibility.

To the fullest extent of the law, neither the Publisher nor the authors, contributors, or editors, assume any liability for any injury and/or damage to persons or property as a matter of products liability, negligence or otherwise, or from any use or operation of any methods, products, instructions, or ideas contained in the material herein.

ISBN: 978-0-443-13523-1

> For information on all Elsevier publications visit our website at
> https://www.elsevier.com/books-and-journals

*Publisher:* Matthew Deans
*Editorial Project Manager:* Toni Louise Jackson
*Production Project Manager:* Fizza Fathima
*Cover Designer:* Mark Rogers

Typeset by TNQ Technologies

# Dedication

To my wonderful wife and three amazing children, whose constant love and support have been my guiding stars, this book is lovingly dedicated.

To my good friend, scientist, and mentor, Dr. Krishna Seshan, who left too soon, may your legacy live on through these writings.

# Contents

*Contributors*   *xiii*
*Foreword*   *xv*
*Preface*   *xvii*
*Acknowledgments*   *xix*

## PART I   Introduction

**1. The role of thin films in nanotechnology**   **3**
Dominic J. Schepis

   1. Introduction   3
   2. Device scaling becomes the driver   4
   3. Thin film challenges   10
   4. Handbook organization   11
   References   14
   Further reading   14

## PART II   Reduction to practice

**2. Process integration for on-chip interconnects**   **17**
Jeff Gambino

   1. Introduction   17
   2. Device scaling   18
   3. Copper interconnect processing   21
   4. Reliability   46
   5. Future directions   66
   References   72

**3. Sputter processing**   **93**
Andrew H. Simon

   1. Introduction   93
   2. Energy and kinematics of sputtered atoms   94
   3. Energy dependence of sputtering   95
   4. Plasmas and sputtering systems   98
   5. Reactive sputter deposition   103
   6. Sputter-tool design and applications for semiconductor technology   105

7. Contamination and metrology — 126
8. Future directions — 132
References — 134

## 4. Epitaxial growth processes for high performance advanced CMOS devices — 141
Shogo Mochizuki

1. Introduction — 141
2. Strained Si technology with epitaxial process — 143
3. Strain engineering for nanoscale strained SiGe FinFET — 148
4. Advanced source drain extension formation for scalded devices — 165
5. Performance enhancement techniques for gate-all-around (GAA) pFET device — 173
6. Conclusions — 182
References — 183

## 5. Equipment and manufacturability issues in chemical vapor deposition processes — 193
Loren A. Chow

1. Introduction — 193
2. Basic principles of CVD — 194
3. A brief history of CVD equipment — 198
4. CVD applications and their impact on scaling — 205
5. Contamination and metrology — 211
6. Summary of CVD technologies — 218
7. CVD tool selection for research and manufacturing — 233
8. CVD trends and projection — 236
References — 242

## 6. CMP: Scaling down and stacking up: How the trends in semiconductors are affecting chemical-mechanical planarization — 257
Wei-Tsu Tseng

1. Introduction — 257
2. CMP challenges — 259
3. The path forward — 264
4. Conclusions — 282
Acknowledgments — 282
References — 283

## 7. Limits of gate dielectrics scaling — 289
Shahab Siddiqui, Takashi Ando, Rajan Kumar Pandey and Dominic J. Schepis

1. Introduction — 289
2. Dennard scaling theory — 290
3. Gate oxide and EOT scaling — 292
4. Hafnium based ternary, quaternary and bilayer oxides for EOT scaling — 300
5. EOT scaling through interfacial layer — 307
6. Ab-initio modeling — 311
7. Gate oxides in FinFET era — 320
8. High voltage (HV) input/output (I/O) gate oxides with HiK/MG for advanced SOC (FinFET and FDSOI) — 322
9. SiGe as a pFET channel (cSiGe) to enable gate oxide scaling — 327
10. Nano-sheet (NS) gate-all-around (GAA) transistor technology and its implication on gate-oxide for logic and I/O transistors — 329
11. Si/SiGe heterostructure-based I/O devices with low temperature ALD oxide and densification — 336
12. High mobility (high atomic % Ge, SiGe) channel: Logic IL and I/O gate oxide research results — 347
13. Conclusion — 352

References — 353
Further reading — 355

# PART III  Applications and limitations

## 8. Semiconductor reliability overview — 359
Fernando Guarin and Ed Hostetter

1. Introduction — 359
2. Definition of reliability — 360
3. Semiconductor reliability balancing act — 360
4. Challenges and principal degradation mechanisms in semiconductor reliability — 361
5. Front end of line reliability, back end of line reliability, and middle of the line reliability — 362
6. Reliability assessment methodologies — 364
7. Mitigation strategies — 364
8. Test structures and methodologies for reliability assessment — 365
9. Conclusion — 366

References — 366
Further reading — 367

## 9. Thin film development for LED technologies — 369
J. Lee, Y.C. Chiu, J.-P. Leburton and C. Bayram

| | |
|---|---|
| 1. Introduction | 369 |
| 2. Development of green-emitting hexagonal InGaN/GaN LEDs | 373 |
| 3. State-of-the art of bulk cubic GaN and InGaN/GaN LEDs | 379 |
| 4. Computation-based design of cubic InGaN/GaN LED | 382 |
| 5. Experimental growth of cubic GaN on U-grooved Si (100) for green LEDs | 392 |
| 6. Future work | 398 |
| Acknowledgments | 399 |
| References | 399 |

## 10. Emerging ferroelectric thin films: Applications and processing — 405
Santosh K. Kurinec, Uwe Schroeder, Guru Subramanyam and Roy H. Olsson III

| | |
|---|---|
| 1. Introduction | 405 |
| 2. History | 405 |
| 3. Principle | 408 |
| 4. Thin films | 412 |
| 5. Thin film deposition processes | 414 |
| 6. Patterning of ferroelectric thin films | 420 |
| 7. Characterization of ferroelectric films | 420 |
| 8. Ferroelectric thin film applications | 428 |
| 9. Three exemplary ferroelectric films | 434 |
| 10. Reliability of ferroelectric films | 440 |
| 11. Conclusions | 444 |
| Acknowledgments | 445 |
| References | 445 |

## 11. Thin films in semiconductor memory — 455
S.B. Herner

| | |
|---|---|
| 1. Introduction | 455 |
| 2. DRAM | 458 |
| 3. NAND | 465 |
| 4. Other semiconductor memories | 475 |
| 5. Conclusion | 479 |
| References | 480 |

## 12. Yield impact of defects from thin films and other processing steps    **485**
Ishtiaq Ahsan

  1. Introduction    485
  2. Examples of different fail modes    487
  3. Defect density and its impact on yield    488
  4. Various yield assessment structures    493
  5. SRAM yield learning methodology    497
  6. How to cheat defect density by adding redundency    500
  7. Conclusion    505
  Acknowledgments    506
  References    506

*Summary*    *507*
*Index*    *509*

# Contributors

**Ishtiaq Ahsan**
Semiconductor Technology Research & Development, IBM Research, Albany, NY, United States

**Takashi Ando**
IBM T. J. Watson Research Center, Yorktown Heights, NY, United States

**C. Bayram**
Department of Electrical and Computer Engineering, University of Illinois at Urbana-Champaign, Champaign, IL, United States; Micro and Nanotechnology Laboratory, University of Illinois at Urbana-Champaign, Champaign, IL, United States

**Y.C. Chiu**
Department of Electrical and Computer Engineering, University of Illinois at Urbana-Champaign, Champaign, IL, United States; Micro and Nanotechnology Laboratory, University of Illinois at Urbana-Champaign, Champaign, IL, United States

**Loren A. Chow**
SVT Associates, Los Altos, CA, United States

**Jeff Gambino**
Onsemi, Gresham, OR, United States

**Fernando Guarin**
GlobalFoundries, East Fishkill, NY, United States; IBM, Reliability Engineering, East Fishkill, NY, United States

**S.B. Herner**
ASM International, Portland, OR, United States

**Ed Hostetter, Jr.**
GlobalFoundries, East Fishkill, NY, United States; IBM, Reliability Engineering, East Fishkill, NY, United States

**Santosh K. Kurinec**
Electrical & Microelectronic Engineering, Rochester Institute of Technology, Rochester, NY, United States

**J.-P. Leburton**
Department of Electrical and Computer Engineering, University of Illinois at Urbana-Champaign, Champaign, IL, United States; Micro and Nanotechnology Laboratory, University of Illinois at Urbana-Champaign, Champaign, IL, United States; Department of Physics, University of Illinois at Urbana-Champaign, Champaign, IL, United States

**J. Lee**
Department of Electrical and Computer Engineering, University of Illinois at Urbana-Champaign, Champaign, IL, United States; Micro and Nanotechnology Laboratory, University of Illinois at Urbana-Champaign, Champaign, IL, United States

**Shogo Mochizuki**
IBM Research, Albany, NY, United States

**Roy H. Olsson III**
Electrical and Systems Engineering, University of Pennsylvania, Philadelphia, PA, United States

**Rajan Kumar Pandey**
Vellore Institute of Technology, Vellore, Tamil Nadu, India

**Dominic J. Schepis**
IBM Microelectronics, East Fishkill, NY, United States; GLOBALFOUNDRIES, Austin, TX, United States

**Uwe Schroeder**
NaMLab gGmbH, Dresden, Germany

**Shahab Siddiqui**
IBM Research Albany, Albany, NY, United States

**Andrew H. Simon**
IBM Research, Albany, NY, United States

**Guru Subramanyam**
Electrical and Computer Engineering, University of Dayton, Dayton, OH, United States

**Wei-Tsu Tseng**
IBM Semiconductor Technology Research, Albany, NY, United States

# Foreword

Semiconductors are at the heart of every aspect of modern technology. From smartphones to electric vehicles, every aspect of modern technology is underpinned by smaller, better and cheaper semiconductors. Miniaturization of semiconductor structures and devices to fuel such growth is made possible due to innovation in process technology especially patterning and thin film deposition. As the demand for higher performing and sustainable semiconductors is growing, innovations in fundamental materials and controls at the atomic dimensions are required.

This book is a collection of the progress in the application of thin films in modern semiconductor technology and its outlook for the future. It was designed to include topics from many varied groups to give a cross-section of thin film technology and applications. As semiconductors get even more complex, the reliance on the quality and repeatability of thin films becomes paramount. Future thin film technology needs to address continued scaling at atomic dimensions, increased topography, and defectivity challenges. Today's semiconductor technologies not only employ silicon but new materials such as new high-k dielectrics, ferroelectrics, hetero-epitaxial films, and newer deposition techniques to push the boundaries of tighter integration and materials compatibility to address the broad semiconductor application space.

The demand for innovation in semiconductor technology is further accelerated by new workloads such as artificial intelligence (AI) and new methods of computing such as Quantum Computing (QC). These new computing paradigms are also changing our thinking around chip architecture and material systems currently deployed in the semiconductor industry. The industry expects to produce exciting new technologies such as chiplet-based system designs that increase the speed of deployment, reuse of older technology nodes, and brings new start-up companies to contribute to a growing ecosystem. Acceleration of 3D stacking and heterogeneous integration technologies are required to meet the demand of new workloads and computing methods. In summary, we expect the continued progress to be more than conventional scaling but entirely new methods for design, build, and deposition.

We are in a very exciting era where for the first time in the history, high precision computing, artificial intelligence computing, and quantum computing are converging. Semiconductors will drive growth in each of these computing areas as well as their convergence for a better and sustainable future. This handbook on thin film deposition for semiconductors will provide a solid foundation on the state of the art in thin film technology and a glimpse into the bright future.

**Mukesh Khare**
IBM Research, Albany, NY, United States

# Preface

In this new fifth edition, the editor's goal was to continue to bring forth new material and processing innovations that are required to allow microelectronic devices to continue to scale in density and performance. As traditional scaling of the past decades began to challenge the technological capabilities of thin films at the atomic level, more recent changes have focused on design, new materials, and transistor architectures.

The approach in each chapter is to once again examine the applications and asking how nanometer microelectronics continues to evolve. In the last edition when planar CMOS devices were still plentiful in many application spaces, the last several years have seen changeovers to FinFET technology and more recently, architectures such as nanowires or nanosheet or gate-all-around devices. These too shifted the requirements of thin films to ensure they could meet both the morphology and reliability requirements to bring these semiconductors to market. In addition, a new chapter on reliability fundamentals has been added to this edition.

After the introduction, the next section of this book provides six chapters on thin film processing and some information on their reduction to practice. While it would be difficult to cover all of the processes involved in state-of-the-art semiconductor fabrication, these six are key disciplines for advancing semiconductor form and function. Here the reader will find updates to the latest advances in several key areas of semiconductor manufacturing and their challenges.

The next major section covers existing and innovative new applications of thin films including some new materials such as Ferroelectrics and heteroepitaxial materials or compound semiconductors used in new high-efficacy LED technology. Subsequent chapters also cover critical updates to specific applications of thin films for semiconductor memories and a discussion of yield issues and defect densities. There are many topics that cannot be covered in detail in such a broad field of nanoelectronics. Our goal will be to provide periodic updates to these and other films and applications.

Rather than this being an instructional textbook, this work is better served as a reference book where practicing engineers, students, and managers may get overviews of different aspects of this rapidly developing field. It is important to note that the examples and methods described are based on our best current understanding on how these materials and preferred methods are reduced to practice. These processes and methods in this document are for informational purposes only and are not intended to be a substitute for professional guidance or instruction. The editors and authors of these chapters do not guarantee the safety, completeness, or suitability of the processes for any particular purpose or situation. The editors and authors of this document are not liable for any damages, injuries, or losses that may result from following or attempting to follow the processes described in this publication. Anyone who chooses to follow or reproduce these processes does so at their own risk and responsibility.

Finally, to prolong the relevance of this book in changing times, both a print and online version will be offered, providing continuing updates as well as access to classic chapters from previous editions.

# Acknowledgments

First, I would like to thank all the exceptional authors who took time from their busy lives to contribute to this book. Their expertise made this handbook possible. I would also like to thank the incredible editorial staff at Elsevier, especially Kayla Dos Santos, Toni Louise Jackson, and Stephen Jones, for their invaluable support and guidance. I would also like to thank my friends and colleagues, Dr. Devendra Sadana and Dr. Jack Fitzsimmons, for their many helpful comments and suggestions. Finally, I thank my wife for her patience and support during this project.

# PART I

# Introduction

# CHAPTER 1

# The role of thin films in nanotechnology

Dominic J. Schepis[1,2]
[1]IBM Microelectronics, East Fishkill, NY, United States; [2]GLOBALFOUNDRIES, Austin, TX, United States

## 1. Introduction

The Handbook of Thin Film Deposition is now in its fifth edition, and in this edition, the editors devote most of the book's content on the intricacies of thin films for semiconductor fabrication. The chapters describe many of the challenges that emerge with each succeeding generation. While no publication could examine all of the workings of such a broad topic, the editor chose specific areas which are critical in chip fabrication and push the limits of thin film formation.

This book will also serve the reader by including chapters that were updated versions of previously discussed topics from earlier editions. We expect the reader will get both an introduction to the topic as well as some deep understanding of dielectric scaling, new materials and some of the challenges in the preparation of these films. Since many readers may be from industry or academia, we have included a wealth of references for further reading and background on these and related materials. We hope it will support you on your journey to understand the state-of-the-art fabrication of thin films used in today's semiconductor fabricators. As we learned from earlier editions, some students also found this book to be a resource for those learning the art of semiconductor fabrication in their college clean rooms, and expanded some of the sections to aid in this utilization.

In just the few years since the last edition of this book (*Handbook of Thin Film Deposition*), in 2018, the semiconductor industry has undergone a sea of change. The landscape of typical semiconductor manufacturing during the previous period was primarily planar complementary metal oxide semiconductor (CMOS) devices, either on bulk or silicon-on-insulator

(SOI) substrates, or FinFet transistors for advanced microprocessor designs. Today, many scaled FinFet and gate-all-around (GAA) devices have been fabricated and are in late stages of development or manufacturing. These designs or the upcoming complementary FETs may be the architecture of choice for the near future. Many of these changes would not have been possible without the innovation in materials and the methods designed to produce them with just the precise stoichiometry, size and composition required for the new applications.

One could say that the progress in thin film technology was developed primarily for the silicon integrated circuit field, and has made its way into other technologies. Armed with these thin films, it became possible to build large structures with multifunctional capabilities that range from the nanotechnology advances to the large macro structures composed of composite thin films.

The ability to pattern and deposit multiple thin films has opened up a large industry, not only in semiconductors, but in the microminiaturization of complete systems such as analytical laboratories on a single chip or micro-sized biological applications. This deposit, etch, and deposit capability have impacted many related fields as they can be applied to a plurality electronic and mechanical systems.

## 2. Device scaling becomes the driver

Before we discuss some of the new developments in thin films, let us briefly review some of the historical background providing the driving force for the thin film advances. One of the key drivers in all of the thin film development is the continued scaling of semiconductor devices. This driving force is the motivation of many of the breakthroughs in semiconductor materials and microarchitectures required to perpetuate the ever-shrinking design rules. Scaling has occurred in all 3 dimensions and architecture had to be reimagined due to the constraints that physics imposed on the designs of yesterday. Throughout the history of semiconductor development, we see traditional methods of scaling applied consistently over time. However, in order to keep the progress in development that drives continued scaling, there had to be novel innovative breakthroughs through either materials science, advanced design, or manufacturing capabilities.

Of course, the scaling provided a doubling of transistor density approximately every 2 years, which was known as Moore's law [1]. This device scaling allowed commercial processes to drop in cost as the economy of scale provided benefits of increased performance and density. Beginning with the history of scaling, one of the founding papers in this field was published by R.H. Dennard, et al. [2]. These design rules of scaling are the rules obeyed as the transistor shrinks.

Although much has changed since this groundbreaking paper, it still describes many of the present challenges fundamental to the continuation of device scaling. Updated reviews of CMOS silicon scaling have been made by multiple authors; however, each new change in device architecture requires new modeling to determining the critical device parameters for that generation of devices [3]. Although many transistor architectures have evolved during recent years, many of the basic challenges remain the same.

While device and wafer scaling have been remarkably successful, a new set of challenges occurred during each successive generation, including an increase in leakage current, more RC delays, increased numbers of I/O pins, thermal power output, and new film development, to name a few [4].

As engineers scaled these new films to thinner dimensions, it became apparent that as devices scaled, the variability of the device performance was critically tied to the variability of the thin dielectrics. For example, gate oxide variability is a critical layer that is necessary to control device Vt and on and off currents for the devices. The initial use of thermal silicon dioxide was an excellent candidate, until the scaling reached levels where the films were too thin and fundamental atomic level material defectivity limited these films precluding reliable leakage levels. The heat cycles of thermal oxides also were quite substantial, and this meant exposing the sensitive devices to excessive diffusion of the dopants required to keep the device parameters within specifications. Although novel lower temperature silicon dioxide films were developed, the industry moved on to new series of high-k dielectrics which provided the equivalent oxide thickness of the gate oxide enabling lower electrical leakages [5].

Along with the effective gate oxide scaling, it was also necessary to improve the device performance for future nodes. One method was to employ techniques to strain the silicon, in order to increase electron/hole mobility. It was found that compressive stress of the PFET (P-type FET) channels would effectively improve hole mobility and hence device performance. Some of the new thin films required to add this strain were silicon germanium, where the films were grown epitaxially on the source-drain regions of the transistor, to provide channel strain [6]. Other methods to produce NFET (N-type FET) strain were also developed including compressive silicon nitride films [7] and embedded SiC [8].

Eventually semiconductor performance began to increase with the incorporation of multiple cores and additional functionality which then allowed the progression of Moore's law by integration of multiprocessors, additional memory, and other discrete electronics.

Future devices will continue to evolve to enable more speed and density as well as lower power consumption to provide the needs of upcoming applications. Some of the many options being pursued in the industry are horizontal nanosheets and nanowires. Also, since silicon mobility becomes limiting at monolayer thicknesses, 2D materials such as transition metal dichalcogenides such as WSe or other films such as graphene have been suggested as highly conductive ultrathin films for advanced device channels. These materials may usher in devices as we approach the next decade [9].

The present trend to scale technology for high-performance processes to smaller and smaller dimensions without also scaling power supply is unsustainable due to increasing power and current density. The lower power device design offered solutions to some of these problems; however, device threshold voltage (Vt) variability became a new challenge at these reduced dimensions. Likely, future silicon devices will incorporate a design that minimizes external resistances and parasitic capacitance, as well as processes to control doping precisely to allow a minimum of variability in a fully depleted device. The addition of designs that employ a complete wrap-around gate or GAA seems promising for low voltage operations and excellent Vt control.

## 2.1 Device roadmaps (Fig. 1.1)

Device or semiconductor roadmaps show various innovations and device types used in the manufacture of semiconductors over the years. Roadmaps can show trends, for example, such as dimensions, nodes, device structures, etc., with the x-axis often shown in years and the y-axis the expected dimensions or innovations over the years. One such example is the IEEE International Roadmap for Devices and Systems that was formed as a continuation of the International Technology Roadmap for Semiconductors (ITRS). The ITRS has been publishing this roadmap since 1992 and it provides a history of past events as well as the expected trajectory for semiconductor devices in the future [10]. Most academic and corporate functions doing semiconductor research and development will also generate such a roadmap which drives their research goals. An example of a semiconductor roadmap showing the potential roadmap extension according to imec is shown in Fig. 1.1 [13].

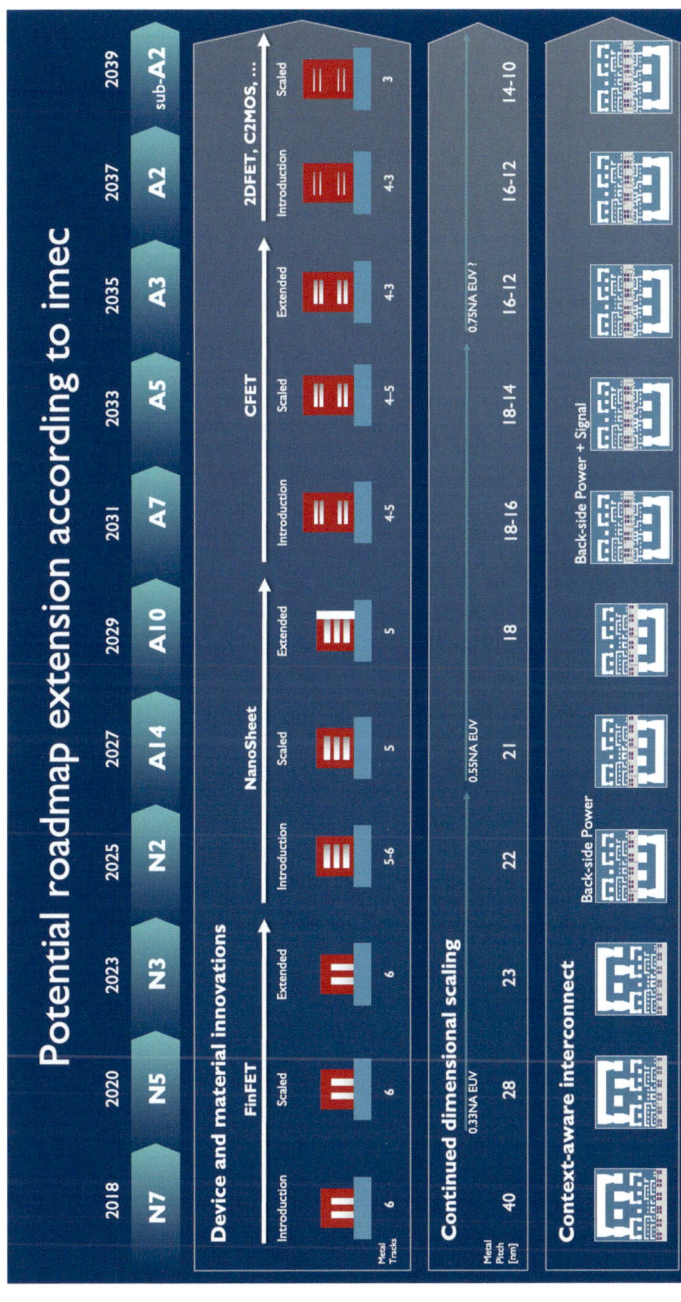

**Figure 1.1** Example of a semiconductor roadmap, highlighting some of the intricate structures that will challenge thin film requirements at each node. Copyright IMEC [13].

With the increase in device frequencies, as well as increased number of transistors on a chip, circuits also became less expensive per unit to manufacture. This led to a decrease in the prices of devices like those used in semiconductor memories over the years, which helped proliferate the use of computers in nearly every industry over the past decades. The sharp decrease in cost for memory over time is shown in Fig. 1.2.

## 2.2 Thin film characterization

At the same time that devices were scaling and thin films were being designed to very stringent requirements, new techniques to measure these films and characterize their homogeneous structure were required. For decades, optical methods of measurements were employed for film thickness and refractive index. This worked very well, especially for blanket nonpatterned films where the size of the laser or light source was not limited by the device dimensions. This also allowed for automated measurements in specified areas of the wafer kerf, often where measurement sites were located to preserve the useful space of device silicon. The measurements were very rapid and uniformity maps and other analysis were

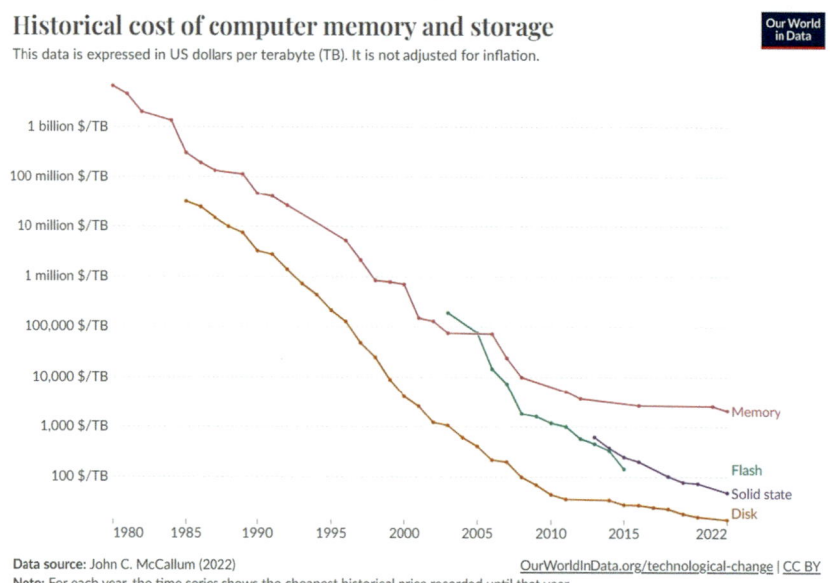

**Figure 1.2** The decrease in price of memory is one of the more dramatic successes of scaling. By 2008, it was possible to buy a Tb of memory for $100. Today that much memory is only half as much. *(Estimates from John C. McCallum (2002).)*

quickly generated, which were useful for feedback of films deposited in the steps prior to the measurement. These techniques are still in use today in many of the critical paths of the manufacturing sequence.

As time progressed, additional methods were required to get accurate determination of not only film thickness and uniformity, but also stoichiometry and interfaces between layers. This gave rise to the use of more advanced methods of analysis. These improvements in analytical techniques allowed more accurate measurements of thin films. While traditional optical methods are still used in several applications, the beam size often became a limiting factor. Optical methods often required large areas where the films could be measured which was often limited by chip layout and kerf requirements. In addition, methods such as cross-sectional scanning electron micrographs (SEM) and transmission electron microscopy (TEM) became key enablers, to quantify film thickness and morphology. Although effective, these techniques were destructive to the semiconductor devices and required analysis time. For determination of dopant concentration, secondary ion mass spectroscopy (SIMS) was employed, but this also was a destructive technique.

For more advanced thin films, especially those in a multilayer form, new methods such as focused ion beam and energy dispersive X-ray spectroscopy (EDS) measurements have evolved. This would allow a beam of ions to cut into a stack of thin films and either prepare a precise cut for examination, or to do analysis. A technique called EDS was developed to get scattered X-rays to emit their characteristic signal which provided elemental analysis of that sample. As is shown in Fig. 1.3, a thin film sandwich of titanium/vanadium/titanium was measured by multiple EDS measurements. The cross-sectional EDS is able to quantify the thickness of the vanadium layer even sandwiched in between metal layers of titanium.

Additional methods that improve on the already useful SIMS techniques for dopant profiles are the new time-of-flight SIMS (time-of-flight mass spectroscopy) instruments. ToF-SIMS is a technique that allows isotopic and elemental analysis, as well as molecular data from solid sources. The ion beam in this case causes secondary ions to be measured by a time-of-flight mass analyzer for detection [11]. For film roughness measurements, atomic force microscopy came into use on the manufacturing lines as well. All these methods as well as many others have helped the materials engineers to gain insight into the quality and cleanliness of the films during deposition.

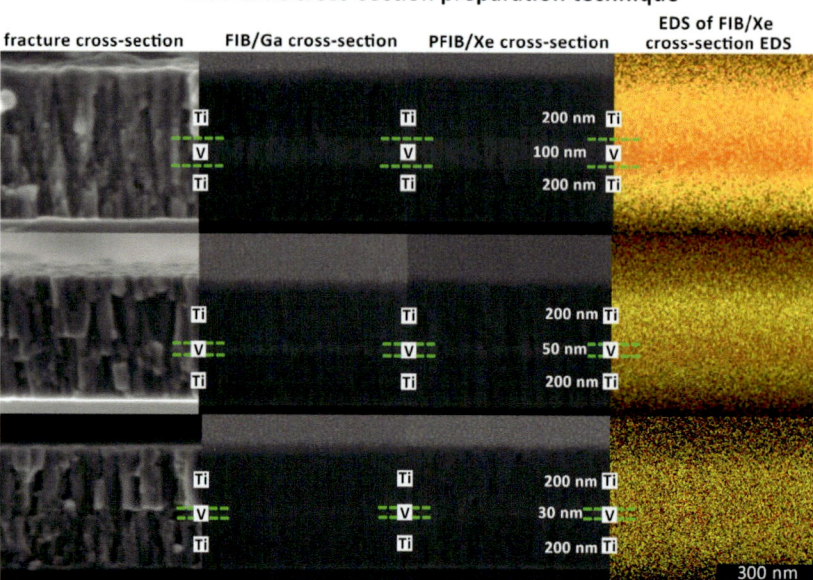

**Figure 1.3** Cross-sections of Ti/V/Ti multilayers with various thicknesses of middle V layer (100, 30, and 10 nm) based on SEM and EDS measurements, prepared by fracture technique, FIB/Ga, and PFIB/Xe. Note: The SEM images were recorded at high resolution and their original version can be found in the supplementary materials. *(Improved methodology of cross-sectional SEM analysis of thin-film multilayers prepared by magnetron sputtering. Coatings 2023, 13(2), 316; https://doi.org/10.3390/coatings13020316.)*

## 3. Thin film challenges

It will become evident from the chapters that follow that the thin film requirements for today's semiconductor are extremely rigorous. Some of the films are deposited by methods known as atomic layer deposition (ALD), which basically deposit films one atomic layer at a time. This control makes it possible to create films that were not possible in earlier days of chip development.

As we enter this new era of artificial intelligence (AI) and the potential of Quantum Computing, we expect to have new tools for modeling and predicting properties of films that have not yet been created. Generative AI will be used to design accelerator chips and semiconductor companies will use AI for more of the tasks where many variables are changing at once. These new capabilities should bring new insights into the optimum conditions for films in which we can dial in the stress, mobility, and other properties for the new semiconductors to come [12].

## 4. Handbook organization

One cannot summarize decades of progress of semiconductor development without an extensive treatise on the many contributions from companies in the industry and academia. However, this brief discussion was to just give a perspective on the many factors impacting the exacting requirements demanded by this industry over succeeding generations. In the world of semiconductor process development, we see it is heavily dependent on the optimization of thin films.

Now that we have briefly examined the rationale behind the continued scaling of thin films and new materials, and their analysis, let us discuss the layout of this handbook.

As we proceed through each of the chapters of this book, we hope to see how dimensional scaling has had its impact on the integrity, defectivity, and reliability of these films.

This fifth edition of this Handbook takes a wide swath of new developments, ideas, and applications of nanotechnology defined by thin films. It brings together a collection of research and development data that has been gathered by the many authors contributing to this edition. These researchers are on the forefront of nanotechnology and process engineering and their work will have important ramifications for the future of thin films and their applications. It will be, hopefully, something that a contemplative engineer will return to during the course of their work. No book can begin to contain all the major developments by the countless companies and universities providing this research, and this edition is designed to give a representation or sampling of some of the recent developments in thin films.

The approach in each chapter is to once again examine the applications and demonstrate how nanometer microelectronics continues to evolve. While in the last edition when planar CMOS devices were still plentiful in many application spaces, these last several years have seen changeovers to FinFET technology and more recently, architectures such as nanowires or nanosheet or GAA devices have been also introduced. These too shifted the requirements of thin films to ensure they could meet both the morphology and reliability requirements to bring these semiconductors to market. A new chapter on reliability has been added to this edition to emphasize its importance.

After this introduction, the next section of this book provides 6 chapters on thin film reduction to practice. Here, the reader will find updates to the latest state-of-the-art innovations in several key areas of semiconductor manufacturing and their challenges.

Chapter 2 begins a thorough review of past and future challenges of thin films for on-chip interconnects. This process of interconnecting the many transistors on each chip is sometimes referred to as the Back-end-of-Line, since it occurs directly after the transistors have been fully fabricated. The interconnects must not only be reliable, but they need to have the right combination of low resistance and capacitance while allowing power and signals to be transmitted across the chip. This development has increased in sophistication with each succeeding generation and will need to evolve into new materials and processes as we transition to future nodes.

In Chapter 3, we get introduced into the world of sputter processing. With sputtering, ionized atoms are accelerated on a surface which eject atoms off of that surface. This technique can be used to both etch a surface or to deposit material onto a surface. Within the semiconductor realm, both sputter etching and deposition are often employed to give a specific film or to remove a film. The author does a deep-dive on the intricacies of these processes which are extremely important in today's semiconductors.

Chapter 4 provides the latest understanding in the epitaxial film growth. Here, the films grown are actually a continuation of the crystal structure underneath. This provides a highly controllable profile of the materials and doping required to make today's high performance devices. Different heterogenous materials can be grown epitaxially, proving more flexibility with regards to stress and electrical properties.

In Chapter 5, we get a thorough coverage of the topic of Chemical Vapor Deposition (CVD). This method of film deposition is used extensively in most semiconductor processes and may be used to deposit both insulators and conducting materials. Atomic layer accuracy is now enabled due to processes such as ALD. Abrupt junction formation is needed for modern devices which can be enabled by CVD. Plasma-enhanced CVD is also very important today due to the low temperatures that they can be deposited.

Chapter 6 discusses the remarkable progress in chemical mechanical polishing (CMP). Here, the author highlights the many challenges of doing CMP on the tall stacks of materials that make up modern semiconductor devices. The methods require sophisticated end point detection to ensure only the intended layers are removed, with high selectivity in some cases. This technique allows a planar structure to be built which is necessary for making reliable contacts among other factors.

Chapter 7 concludes the discussion of pure semiconductor materials processing. In this chapter, the authors review the continued scaling of gate dielectrics and the new challenges in ensuring the gate materials can fill the topography and has the conformality required to meet the most demanding aspect ratios of next-generation devices.

The next section of the handbook covers existing and innovative new applications of thin films including some new materials such as ferroelectronics and hetero-epitaxial materials or compound semiconductors used in new high efficiency LED technology. These recent developments are a result of many years of research at prominent universities. Most well-known ferroelectrics had difficulties with integration into CMOS process flows, but new materials and new ways of depositing them have recently been discovered with high potential for integration into modern device flows. For LED technology, these materials are very important toward future devices, many of which may become parts of integrated semiconductor packages as well as optoelectronic applications. These are highly likely to be a part of future data centers, where high speed connectivity will be paramount.

The following chapter also covers specific applications of thin films for semiconductor memories. This chapter reviews the latest developments in the continued growing requirements for semiconductor memories used in countless devices today.

Finally, since defect density at these nanometer dimensions are critical, we would be remiss if we left out a critical discussion of yield issues and defect densities. This chapter gives an introduction to some of the current methods of yield learning and applications.

There are many topics that cannot be covered in detail in such a broad field of nanoelectronics. Our goal will be to provide periodic updates to these and other films and applications in future publications.

Rather than this being an instructional textbook, this work is better served as a reference book where practicing engineers may get overviews of different aspects of this rapidly developing field. It is important to note that the examples and methods described are based on our best current understanding on how these materials and preferred methods are reduced to practice. These processes and methods in this document are for informational purposes only and are not intended to be a substitute for professional guidance or instruction. The editors and authors of these chapters do not guarantee the safety, completeness, or suitability of the processes for any particular purpose or situation. The editors and authors of this document are not liable for any damages, injuries, or losses that may result from following or attempting to follow the processes described in this publication. Anyone who chooses to follow or reproduce these processes does so at their own risk and responsibility.

Finally, to prolong the relevance of this book in changing times, both a print and online version will be offered, providing continuing updates as well as access to classic chapters from previous editions.

## Supplementary materials

https://www.mdpi.com/2079-6412/13/2/316#app1-coatings-13-00316

## References

[1] G. E More, Cramming more components onto integrated circuits, Proc. IEEE 86 (1) (1998) 82–84.
[2] R.H. Dennard, F. H Gaensslen, H.N. Yu, V. Leo Rideout, E. Bassous, A.R. Leblanc, Design of ion-implanted MOSFETS's with very small physical dimensions, J. Solid State Circuits 9 (5) (1974) 256.
[3] B. Davari, R.H. Dennard, G.G. Shahidi, CMOS scaling for high performance and low power-the next ten years, Proc. IEEE 83 (4) (1995) 595–606.
[4] Overcoming research challenges for CMOS scaling: industry directions, T.C. Chen, Solid state and integrated circuit technology, in: 8th ICSICT Conference Proceedings, 2006, pp. 4–7.
[5] K. Mistry, et al., Tech. Dig, Int. Electron Devices Meet (2007) 247.
[6] S. Bedell, A. Khakifirooz, D. Sadana, Strain scaling for CMOS, MRS Bull. 39 (2014) 131–137, https://doi.org/10.1557/mrs.2014.5.
[7] M. Belyansky, Thin Film Deposition for the Front End of Line: The Effect of the Semiconductor Scaling, Strain Engineering and Pattern Effects, (see science direct reference).
[8] Epitaxial Growth of Si:C Alloys: Process Development and Challenges, A. Dube 2010 the Electrochemical Society, ECS Transactions, Volume vol 28, Number 1 Abhishek Dube et al, 2010 ECS Trans. 28 63.
[9] Semiconductor Digest (August/September issue, p12-p17).
[10] The International Roadmap for Devices and Systems: 2022 Copyright ©, IEEE, 2022 (All Rights Reserved).
[11] National Institute for Standards and Technology. https://www.nist.gov/programs-projects/time-flight-secondary-ion-mass-spectrometry.
[12] 2024 Semiconductor Industry Outlook, https://www2.deloitte.com/us/en/pages/technology-media-and-telecommunications/articles/semiconductor-industry-outlook.html.
[13] 2024 IMEC Semiconductor Roadmap. Copyright IMEC. https://www.imec-int.com/en/articles/view-logic-technology-roadmap.

## Further reading

[1] TechInsights—IEDM 2023—2D Materials—Intel and TSMC, Scotten Jones, 2023.

# PART II
# Reduction to practice

# CHAPTER 2

# Process integration for on-chip interconnects

**Jeff Gambino**
Onsemi, Gresham, OR, United States

## 1. Introduction

There has been tremendous progress in the manufacturing of integrated circuits over the past 40 years. The minimum feature size has gone from 10 μm down to 30 nm (Fig. 2.1), the cost per transistor has decreased by seven orders of magnitude, the maximum number of transistors per chip has increased by nine orders of magnitude [1]. Nanotechnology has been defined as "*structures, devices, and systems .... at a critical length scale of matter typically under 100 nm*" [2]. By this definition, the manufacturing of advanced silicon integrated circuits has been in the nanotechnology era since the year 2000.

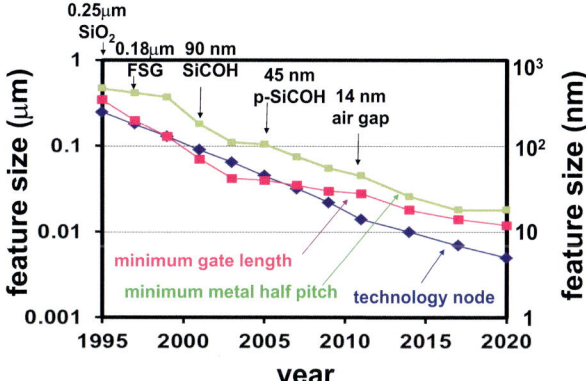

**Figure 2.1** Minimum feature size on silicon integrated circuits as a function of time [1]. Copper interconnects with $SiO_2$ dielectrics were introduced at the 0.25 μm technology node. The BEOL insulator has changed over time to reduce interconnect capacitance.

There are a variety of process technologies used for fabricating silicon integrated circuits. Bipolar transistors can achieve higher switching speeds than metal oxide semiconductor field effect transistors (MOSFETs). However, the great majority of silicon devices are manufactured using CMOS (complementary metal oxide semiconductor) circuits, where the devices are n-type or p-type (i.e., nMOS or pMOS). CMOS technology has a number of advantages compared to bipolar technology, especially lower power consumption and higher circuit density [3]. Because of the higher circuit density, the system performance is generally better for CMOS technology compared to bipolar technology, despite the slower switching speed of MOSFETs.

## 2. Device scaling

The remarkable progress in the microelectronics industry has been largely due to the scaling properties of MOSFET devices [4,5]. Device scaling theory states that if the transistor physical dimensions (both horizontal and vertical) and the operating voltage are decreased by a factor $f$ (where $f < 1$), then transistor area is reduced by a factor $f$ [2], gate delay is reduced by a factor $f$, and power per gate is reduced by a factor $f$ [2]. Unfortunately, interconnects delay does not decrease with scaling. For local wiring, the delay is constant with ideal scaling (i.e., all dimensions are scaled by a factor $f$) [6]. The resistance of the wire increases by a factor $1/f$ (because of the smaller cross-sectional area of the conductor, the capacitance of the wire decreases because of the smaller surface are of the wire (Fig. 2.2). Hence, the delay $(R_{int}C_T)$ is constant.

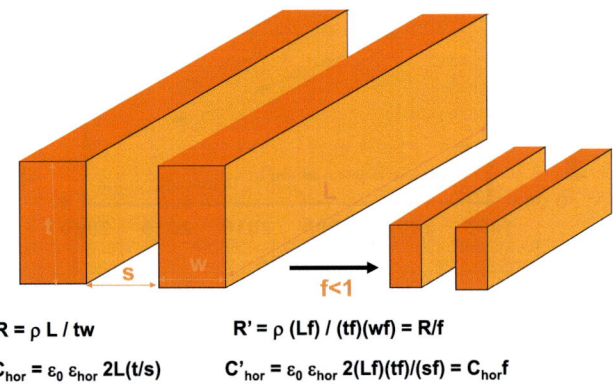

$R = \rho L / tw$         $R' = \rho (Lf) / (tf)(wf) = R/f$

$C_{hor} = \varepsilon_0 \varepsilon_{hor} 2L(t/s)$         $C'_{hor} = \varepsilon_0 \varepsilon_{hor} 2(Lf)(tf)/(sf) = C_{hor}f$

**Figure 2.2** Effect of scaling on wire resistance and capacitance [6].

For global wires, scaling is even more difficult, because the wire length generally does not decrease at smaller technology nodes. In fact, the wire length for global wires tends to increase with each technology generation [7], corresponding to the increase in chip size (Fig. 2.3).

Historically, the circuit delay was limited by the device delay and interconnect delay was not a concern (Fig. 2.4) [9].

However, at feature sizes below 1 μm, the delay from the interconnects becomes significant, and can dominate the total delay unless the process and design are optimized. At the 14 nm node and below, the delay from interconnects can be comparable to that of the devices [8,10]. The total delay in a circuit has contributions from both the device delay and the interconnect delay. The important parameters for determining the delay of a circuit, are the on-resistance of the driver transistor, $R_D$, the resistance and capacitance of the interconnect, $R_{int}$ and $C_{int}$, and the input capacitance of the transistors that form the load, $C_L$ (Fig. 2.5) [11].

As shown in Fig. 2.5, for a transition at the input of the inverter from a high voltage, $V_{dd}$, to a low voltage (i.e., from logic 1 level to logic 0 level), there is a delay associated with the voltage transition at the input of the load transistor, for the voltage to go from 0% to 90% of the final value (i.e., the

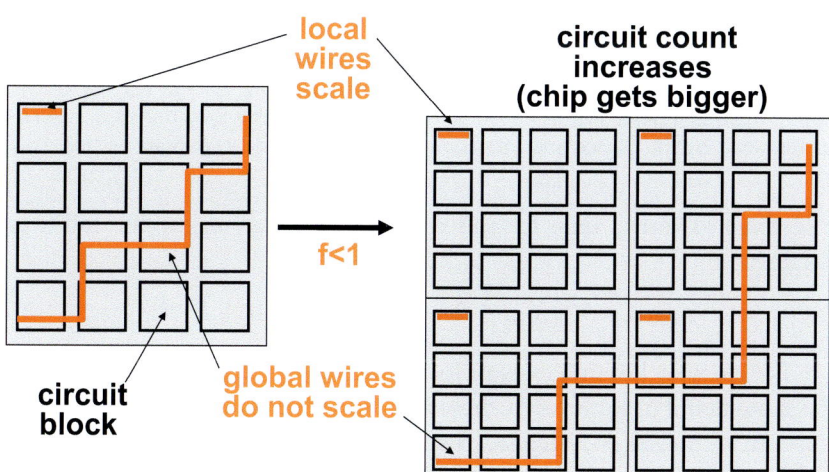

**Figure 2.3** The effect of scaling on the length of local wires and global wires. The length of local wires tends to scale at each technology node, because individual circuit blocks shrink. However, the length of global wires tends not scale at each technology node if more functionality is added to the chip [7].

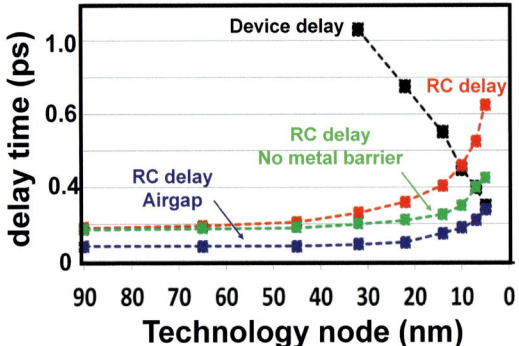

**Figure 2.4** Device delay and interconnect (RC) delay as a function of feature size for local interconnects [8]. The interconnect delay can be reduced by thinning the metal barrier layer or by using air gap dielectrics.

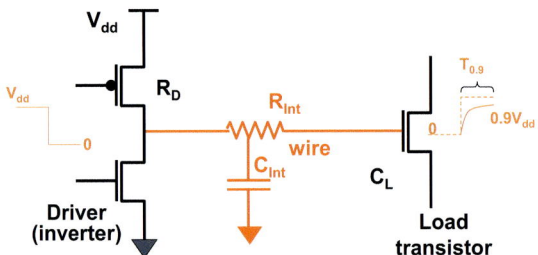

**Figure 2.5** Schematic of an inverter circuit showing resistance and capacitances from wires and devices. For a transition from the power supply voltage ($V_{dd}$) to 0 at the input, there is a delay in the transition from 0 to $V_{dd}$ at the load transistor, which is determined by the device and interconnect resistances and capacitances [11].

rise time). The rise time for this circuit, $\tau_{90\%}$, is giving by the following expression:

$$\tau_{90\%} = 1.0 R_{int} C_{int} + 2.3(R_D C_{int} + R_D C_L + R_{int} C_L) \qquad (2.1)$$

For the two limiting cases of local wires and global wires, Eq. (2.1) can be simplified. For local wires, the wire resistance is generally much less than the driver transistor resistance. For this case, Eq. (2.1) simplifies to the following:

$$\tau_{90\%} \approx 2.3 R_D (C_{int} + C_L) \text{ for local wires, } R_{int} \ll R_D \qquad (2.2)$$

Hence, for circuits with the local wires, the circuit performance is limited by the devices and by the interconnect capacitance. The other limiting case is for global wires, where the wire resistance and capacitance are generally much greater than the transistor resistances and capacitances:

$$\tau_{90\%} \approx 1.0 R_{int} C_{int} \text{ for global wires, } C_{int}, R_{int} \gg C_L, R_D \quad (2.3)$$

Hence, for circuits with global wires, the circuit performance is limited by the resistance and capacitance of the wires.

Two major changes have been made in on-chip interconnect processing, to address the performance limitations of wiring with scaling. The first change was from Al metalization to Cu metalization, which was first introduced at the 0.25 μm technology node [12]. On-chip copper interconnects have gained wide acceptance in the microelectronics industry due to improved resistivity and reliability compared to Al interconnects. Initially, copper interconnects were only used for high performance logic circuits. However, Cu interconnects are now used in a wide variety of integrated circuits, including dynamic random access memories (DRAM) [13], RF circuits [14], CMOS image sensors [15], power semiconductors ([16,16a], and 3D structures with through-silicon vias (TSVs) [17]. Copper interconnects will continue to be used for the advanced nodes for metal pitches above 20 nm (i.e., upper wiring levels), though alternate metals such as Co and Ru may replace Cu for metal pitches below 20 nm (i.e., 3 nm node and below) [17a]. The second major process change was in the dielectric, going from $SiO_2$ to materials with a lower dielectric constant ("low K dielectrics"), such as SiCOH [18,19]. However, there are many challenges with integration of Cu interconnects and low K dielectrics at these nodes, including increased resistivity, difficult patterning, dielectric damage, and reliability problems [20–22]. In this chapter, each of these topics is addressed.

## 3. Copper interconnect processing
### 3.1 Process flow

Copper cannot be easily patterned by reactive ion etching (RIE), due to the low volatility of Cu chlorides and Cu fluorides [22,23] Hence, Cu interconnects are formed using the "dual damascene" process (Fig. 2.6) [12,20–22].

After processing of M1, the V1/M2 dielectric is deposited (SiCOH, for example) and V1 vias are patterned (Fig. 2.6a), stopping on the SiCN layer that protects the Cu from oxidation. Next, the M2 trenches are patterned

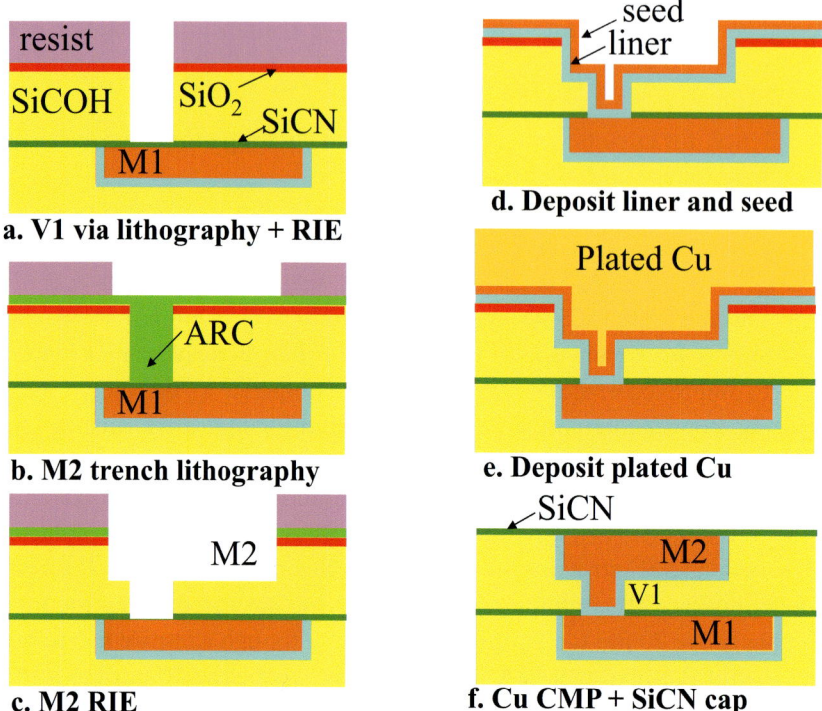

**Figure 2.6** Schematic of process flow for via-first dual damascene process [21].

(Fig. 2.6b,c), the final step being the removal of the SiCN etch stop from the bottom of the via. The first part of the metalization is sputter deposition of a TaN/Ta barrier layer (which prevents Cu from diffusing into the dielectric and a Cu seed layer (Fig. 2.6d). The vias and trenches are then filled with Cu by electroplating (Fig. 2.6e). The excess metal over the field regions is removed by chemical mechanical polishing (CMP). The final step is deposition of an SiCN capping layer, that protects the Cu from oxidation (Fig. 2.6f). These steps are repeated for each metal level. After the last metal layer is fabricated, thick dielectric passivation layers are deposited and vias are opened to the bond pads. Note that from the design perspective, there are a number of important differences between Al interconnects and Cu interconnects. Because Cu is patterned by polishing, there are more restrictions on pattern density compared to Al technology, and dummy metal shapes are required to minimize differences in pattern density across a chip [24]. In addition, the Cu must be capped with hermetic barrier layers (SiN or SiCN) to protect it from oxidation during processing or during device operation. These materials have much higher dielectric constants than that of the interlevel dielectric; for SiN, $k \sim 7$, and for SiCN, $k$ ranges from $\sim 4$ to 5, depending on the processing [25]. Hence, the effective dielectric constant is typically 10% higher than that of the interlevel dielectric.

## 3.2 Low-k dielectrics

Initially, $SiO_2$ was used as the interlevel dielectric surrounding the Cu wires (Fig. 2.7 and Table 2.1).

For process integration, $SiO_2$ has many good properties [26–28]. It is thermally and chemically stable, and therefore does not degrade during processing. It is mechanically rigid (i.e., it has a high elastic modulus) and is relatively impermeable to moisture (at least at the operating temperature of integrated circuits), which simplifies packaging. In addition, high quality films can be deposited by plasma enhanced chemical vapor deposition (PECVD). Of course, the disadvantage of using $SiO_2$ is that the dielectric constant is higher than desired.

In many ways, F-doped $SiO_2$ (fluorosilicate glass, FSG) is the ideal replacement for $SiO_2$. With fluorine doping, Si—F bonds replace Si—OH and Si—O bonds. The Si—F bonds have lower polarizability than either Si—OH or Si—O, resulting in a lower dielectric constant ($\sim 3.5$) [29]. However, the mechanical, chemical, and thermal properties of FSG are similar to $SiO_2$. Therefore, processing and packaging of die with FSG dielectrics is relatively easy (at least compared to other low-$k$ dielectrics). As a result, FSG replaced $SiO_2$ as the interlevel dielectric at the 130 nm technology node.

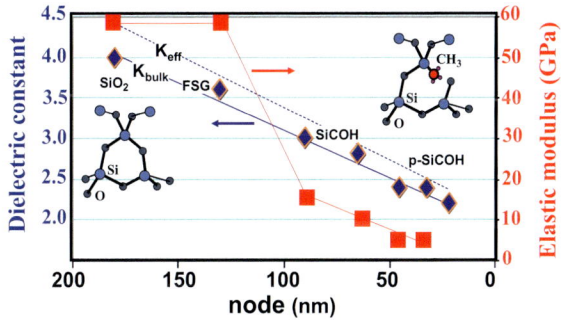

**Figure 2.7** Trend for low-$k$ dielectric scaling [21].

Table 2.1 Technology trend for low-k dielectrics [26]. * = Author's estimates; ** = Data from Chen [26a].

| Node (nm) | Dielectric | k | Modulus | Thermal cond. | Porosity | Reference |
|---|---|---|---|---|---|---|
| 180 | SiO$_2$ | 4 | 60 GPa | 1.0 W/m-K | 0% | Gambino |
| 130 | FSG | 3.6 | 60 GPa | 1.0 W/m-K** | 0% | Gambino |
| 90 | SiCOH | 3.1 | 15 GPa | 0.59 W/m-K** | 0% | Grill |
| 65 | SiCOH | 2.7 | 8.0 GPa | 0.55 W/m-K* | 7% | Priyadarshini |
| 45 | p-SiCOH | 2.45 | 6.6 GPa | 0.48 W/m-K* | 19% | Priyadarshini |
| 32-14 | p-SiCOH | 2.55 | 6.6 GPa | 0.45 W/m-K* | 14% | Priyadarshini |
| 45-7 | p-SiCOH | 2.4 | 7.0 GPa | 0.47 W/m-K* | 20% | Nguyen |
| 14 | Air gap | 2.1 | 3.5 GPa* | 0.3 W/m-K* | 50%* | Fischer |

But further reductions in dielectric constant are required as the device dimensions are reduced. The dielectric constant of $SiO_2$ can be further reduced by using carbon doping instead of fluorine [18,19,26a]. Bridging Si—O bonds are replaced by nonbridging Si—$CH_3$ bonds (Fig. 2.7), resulting in a lower density, and hence a lower dielectric constant. In addition, the Si—C bonds have lower polarizability than Si—O bonds. The C-doped $SiO_2$ is often called SiCOH, which corresponds to the chemical components in the film. The dielectric constant of nonporous SiCOH is typically 2.7—3.0. However, even lower dielectric constants (2.2 or less) are possible by adding pores to the SiCOH [30,30a,31]. Because of the improved performance associated with the lower dielectric constant, nonporous SiCOH is used at the 90 and 65 nm technology nodes, and porous SiCOH is used at the 45 nm node and below. However, the integration of Cu interconnects in SiCOH dielectrics requires many process changes, both during wafer processing and during packaging, especially for porous SiCOH.

SiCOH films used in manufacturing are deposited by plasma enhanced chemical vapor deposition (PECVD) [18,19]. It is essential to have small (∼1 nm diameter) isolated pores for a number of reasons [18,19,32—34]. Isolated pores are desirable to prevent water and other contaminants from diffusing into the dielectric step during wet clean steps prior to metalization and during CMP. If pore connectivity is too high (Fig. 2.8), water may be absorbed in the dielectric, resulting in higher dielectric constant and/or degraded reliability for dielectric breakdown [33,34].

Because of all the problems associated with porous low-*k* dielectrics, there is renewed interest in using air gap technology [35—38]. There are two basic air gap approaches: the localized airgap method (Fig. 2.9a—c) and the global air gap method (Fig. 2.9d—f).

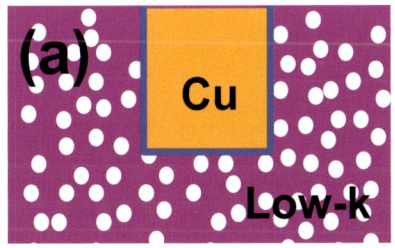

**Figure 2.8** Schematic of porous low-*k* material with (a) closed pore and (b) open pore structure.

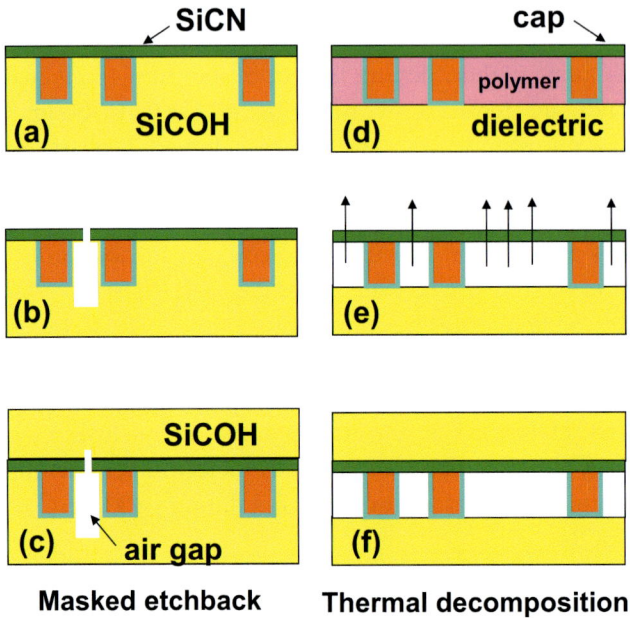

**Figure 2.9** Schematic of localized airgap (a–c) and global airgap (d–f) processes [21].

The localized air gap method is the preferred approach, though an extra mask is required for each level where air gaps are used [37]. With the localized method, air gaps are only formed in critical regions of the circuit. Hence, the mechanical integrity and thermal conductivity are maintained in most regions of the die, allowing the use of conventional wafer processes and packaging processes. The localized air gap approach was first demonstrated on a 65 nm microprocessor, with effective $k$ values as low as 2.0. Recently, air gaps have been introduced into 14 nm technology, providing a 17% capacitance reduction compared to a nonair gap structure [39,39a]. The disadvantage of the air gap process compared to porous low-$k$ materials is that there is extra cost associated with lithography and etching of the air gaps. However, there are also extra costs associated with processing low-$k$ materials, so air gap technology is an attractive option for the 14 nm node and below.

## 3.3 Dielectric patterning

Dual damascene patterning is generally used for Cu interconnect technology due to lower cost compared to single damascene processes. The dual damascene process with a resist mask can use either a trench-first or via-first sequence The trench-first process has a relatively simple etch, but the via lithography is difficult, because a large depth of focus is required to print vias in trenches [20,22,40–43]. In contrast, for the via-first process, the via lithography is relatively simple, but the trench etch is difficult [40,43]. In the advanced technology nodes (14 nm and below), there is an additional complication because double, triple, or quad patterning is required for each of the tightest pitch interconnect layers [44–47].

The via-first process is difficult because the SiN or SiCN etch stop layer at the bottom of the via must be preserved during the via etch and the trench etch, to avoid exposing the underlying Cu to oxidizing resist strip chemistries [43]. An organic anti-reflective coating (ARC) is generally used during trench lithography (Fig. 2.3b) [48]. The organic ARC layer planarizes the topography from the vias, protects the etch stop layer from erosion during the trench etch, and can help prevent resist poisoning, by slowing down amine diffusion in the resist [48–50]. The amount and uniformity of the ARC fill and the trench etch process must both be optimized to ensure good etch profiles are achieved in the vicinity of vias [40,43]. If the ARC is filled too high and if the trench etch has high selectivity to the ARC, then residues or "fences" can remain around the perimeter of the via [51–53] (see below). However, if the ARC fills insufficiently, then the underlying etch stop may be removed during the trench etch.

Via and trench patterns for damascene processing are etched into the dielectrics using reactive ion etching (RIE) in fluorocarbon chemistries such as $C_4F_8$ or $CF_4$ [22,54–57]. High etch selectivity is required for via etching; the via is etched in SiCOH, stopping on SiCN (Fig. 2.3a). Etch selectivity is achieved by forming a polymer film on the etch stop layer. The C in the fluorocarbon chemistry promotes polymer formation whereas the F promotes etching of Si-containing materials (i.e., by forming volatile Si–F species). The polymer forms more easily on SiCN than SiCOH, because oxygen in the SiCOH reacts with the polymer to form volatile CO and $CO_2$. The thick polymer formation on the SiCN slows down the etch rate, resulting in a much lower etch rate for SiCN compared to SiCOH.

An number of problems can occur during dual damascene patterning of $SiO_2$ and SiCOH dielectrics, including fence formation (Fig. 2.10b), chamfering of the top corner of the trenches (Fig. 2.10c), microtrenching (Fig. 2.10d), and RIE lag (Fig. 2.10e).

Fences can cause problems with yield (opens) and reliability (electromigration fails) because the volume of Cu surrounding the via is reduced. A

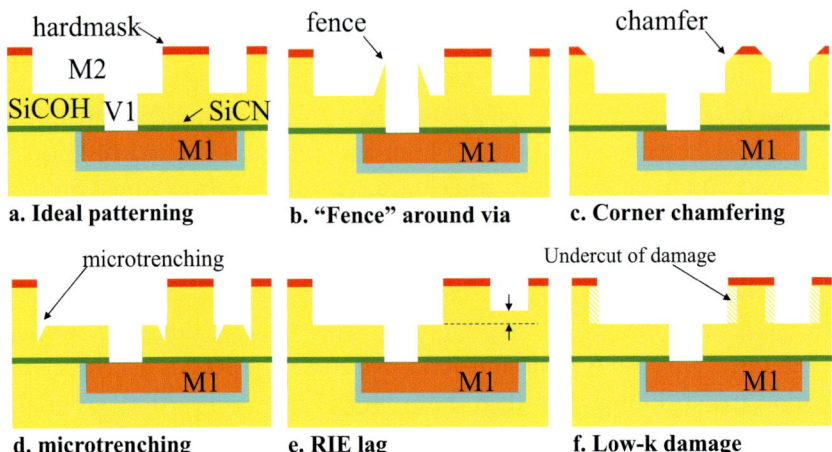

**Figure 2.10** Schematic of problems associated with dual damascene patterning; (a) ideal patterning, (b) "fence" around via, (c) corner chamfering, (d) microtrenching, (f) RIE lag, and (g) undercut or damage of low-*k* dielectric.

fence can form around the via in a via-first process in two different ways. One mechanism is by resist poisoning. During the trench lithography process, amines in the dielectric stack can diffuse from the vias into the deep UV resist during bakes, and neutralize the photo-acid catalyst [49,50]. As a result, the resist is incompletely developed and the trench etch is masked adjacent to the vias. Fences can also form when there is no resist poisoning, if the ARC in the via is not recessed sufficiently during the trench etch. If the ARC protrudes above the bottom of the trench, then the trench etch will be masked resulting in a fence surrounding the via. Corner chamfering and line edge roughness are caused by excessive resist erosion during the dielectric etch. Resist erosion can be minimized by using a more polymerizing chemistry and by reducing ion bombardment [58].

Microtrenching can cause poor liner coverage, and hence can result in reliability problems (i.e., void nucleation or Cu diffusion into the dielectric). Microtrenching is caused by ion reflection from the sidewalls of features [59]. Microtrenching can be minimized by making the sidewall more vertical (i.e., fewer ion reflections) or by operating the etch in a neutral-limited regime (where the etch rate is determined by the neutral flux to the surface, rather than by the ion flux to the surface).

Aspect ratio dependent etching (ARDE) or RIE lag results in a reduced etch rate as feature size decreases, for via holes and trenches [60,61]. RIE lag can cause a systematic variation in wire sheet resistance for different line widths (i.e., higher sheet resistance for narrow lines compared to wide lines). RIE lag can be caused by a number of factors, including reduced flux of reactive ions or neutrals in narrow features, or due to increased formation of nonvolatile reaction products (polymer) in narrow features. A number of approaches can be used to minimize RIE lag during dielectric etching, including using a less polymerizing chemistry (to minimize polymer formation in narrow features) and using a lower pressure (to increase ion bombardment in narrow features) [61].

One of most difficult challenges of patterning the SiCOH dielectric is to minimize damage from the reactive ion etch and resist strip processes. The resist strip processes are especially damaging, because ions and radicals in the resist strip process remove methyl groups from the surface of the SiCOH (Fig. 2.11) [62]. The surface becomes hydrophilic, resulting in water absorption and an increase in the dielectric constant of the material. A number of processes must be optimized to minimize damage from resist strip, including the C content and bonding in the SiCOH [63–65], the resist strip chemistry [66,67] and use of silylation to repair damage [62].

In oxygen-containing plasmas, damage to the sidewalls of trenches and vias occurs due to oxygen ions and radicals that diffuse into the structure and react with Si—$CH_3$ bonds [64]. SiCOH films with higher C concentration, and in particular, films with higher order hydrocarbons in the side chains, are less susceptible to damage, because the oxygen species react with the hydrocarbon chains rather than the Si—$CH_3$ bonds Small pore size is also important, to minimize diffusion of oxygen species into the film.

Conventional resist strips in oxygen plasmas can damage the low-k material, even if the C concentration and pore size have been optimized.

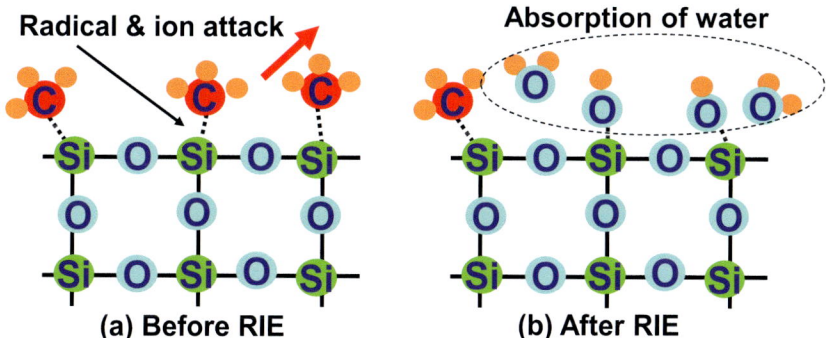

**Figure 2.11** Resist strip damages SiCOH dielectric by removing methyl groups from the surface [62].

Hence, the process integration and resist strip conditions must be chosen carefully to minimize etch damage [66–70]. There are two basic integration approaches for dual damascene patterning at the 32 nm node and below; the metal hardmask method [66,70,71] and the multilayer resist method [49,66] (Figs. 2.12 and 2.13). These complicated methods are needed for patterning small features because the resist thickness must be reduced as feature size decreases to ensure an adequate process window for lithography [72]. In the metal hardmask approach, the resist is stripped prior to the trench etch and via etch into the SiCOH, so there is minimal resist strip damage [71]. However, there are a number of problems with the metal hardmask approach [66,70]. Polymer can form on the sidewalls of the trenches during the trench etch; this polymer must be removed without damaging the low-$k$ material to ensure high yield. Metal residues can form on the etched surfaces and block etching of the low-$k$ material. Finally, stress in the metal layer must be minimized to avoid pattern deformation after the etch. The multilayer resist approach avoids the metal residue and metal stress problems associated with the metal hardmask approach [49,66]. However, the low-$k$ material is fully exposed to the resist strips. Hence, resist strips with low damage must be used with the multilayer resist approach.

There are two different approaches for resist strip for low-$k$ materials; downstream $H_2$ chemistry [68,73–75] and CO- or $CO_2$-based reactive ion etching (RIE) [67,76,77]. Direct exposure of the low-$k$ materials to $O_2$, $N_2$, or $H_2$ plasmas (i.e., with ion bombardment) causes significant damage [73], with more damage for porous materials compared to nonporous materials. With downstream plasma exposure (i.e., no ion bombardment), the

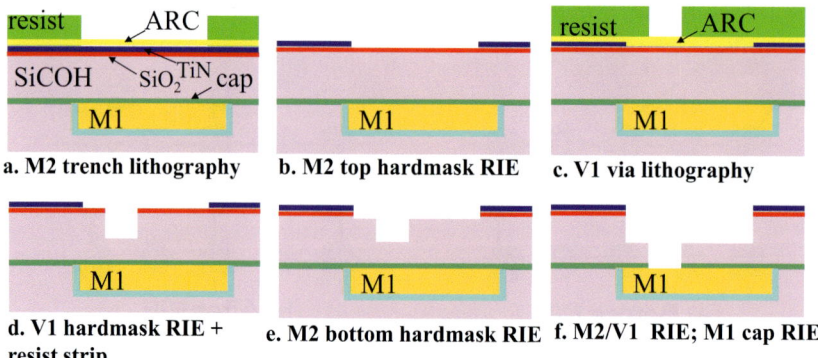

**Figure 2.12** Dual damascene patterning with a metal hardmask [71].

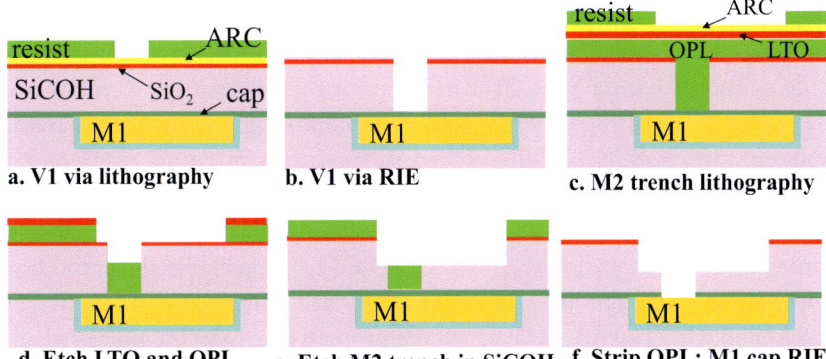

**Figure 2.13** Dual damascene patterning with multilayer resist [49].

damage to the low-$k$ material is significant for $O_2$ plasmas, but greatly reduced for $N_2$ plasmas, and there is no measurable damage for $H_2$ plasmas. Acceptable resist etch rates (>100 nm/min) can be achieved by using a high temperature (260°C) down-stream $H_2$ plasma. It has been reported that residues are left after a downstream $H_2$ plasma strip, which must be removed with a wet clean [75]. The downstream $H_2$ plasma only reacts with H in the low-$k$ film, in a replacement reaction, without altering the stoichiometry of the film [74]. Hence, there is no change in the film thickness or dielectric constant.

The CO or $CO_2$-based resist strip approaches are generally run in RIE tools [67,76,77]. Argon is often added to the strip chemistry and a bias is applied to the wafer to enable a high removal rate of resist [76]. During patterning, the top surface of the low-$k$ dielectric is typically capped with a hard dielectric such as $SiO_2$ (i.e., in the multilayer resist approach, Fig. 2.13). So the regions at risk for damage are the sidewalls of vias and trenches and the bottoms of trenches. The sidewalls are exposed to very little ion bombardment and primarily react with neutral species in the plasma. The low damage associated with $CO_2$-based resist strips is at least partly due to the lower amount of atomic oxygen present in the plasma compared to $O_2$ resist strips [76]. Another possible reason for low damage with CO- and CO-based strips is formation of a C-rich passivation layer on pores and sidewalls of the low-$k$ material [67].

Even if plasma damage is minimized, the removal of methyl groups from the surface of the SiCOH is likely to occur. With the loss of methyl groups, the surface becomes hydrophilic and absorbs water [62,78,79]. The absorbed water can cause problems with reliability, such as stress-induced voiding [33]. Hence, it may be necessary to restore the hydrophobic surface of the patterned SiCOH material prior to metalization. A number of silylation methods have been reported, consisting of high temperature exposre (150−350°C) of the etched surfaces to a silylating agent such as hexamethyldisilazane (HMDS), trimethysilyl-dimethylamine (TMSDMA), or tetramethylcyclotetrasiloxane (TMCATS) [78]. Recently, a "plasma protection" method has been demonstrated to protect the dielectric from plasma etch damage [80]. The pores of the fully cured dielectric are backfilled with a sacrificial layer, which protects the dielectric during plasma processes. The filler material is then removed from the pores after patterning, thereby restoring the low dielectric constant.

An additional complication for fine pitch interconnect fabrication (beyond the 22 nm node) is the need for advanced patterning processes [81−84]. There are a number of options to choose from including Single Expose (SE), Double Patterning with Litho-Etch-Litho-Etch (2× LE) (Fig. 2.14), Triple patterning (3× LE), Self-Aligned Double Patterning (SADP), Self-Aligned Quadruple Patterning (SAQP). Each of these options can be implemented either with 193 nm immersion lithography (i193) or with Extreme Ultra-Violet Lithography (EUV) (Fig. 2.15).

The choice of lithography option depends on capability (for a given metal pitch) and cost (Fig. 2.16).

## 3.4 Metallization

The process sequence for copper metalization is more complicated than for Al metalization, because electroplating is used to fill the high aspect ratio vias and trenches (Fig. 2.6e). In addition, the Cu must be surrounded by a

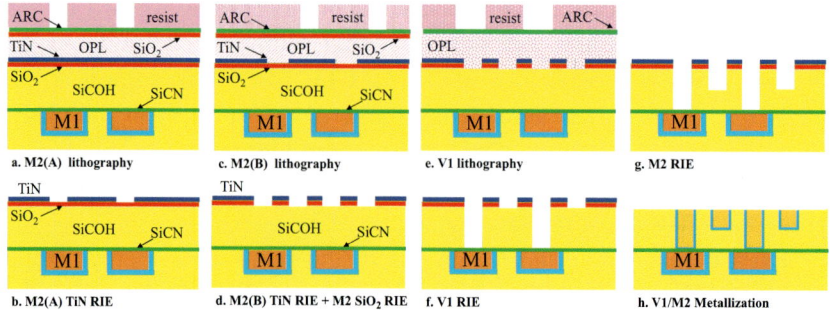

**Figure 2.14** Trench-first dual damascene patterning for Litho-etch-Litho-etch process for M2, along with self-aligned V1 patterning (after [81,83]).

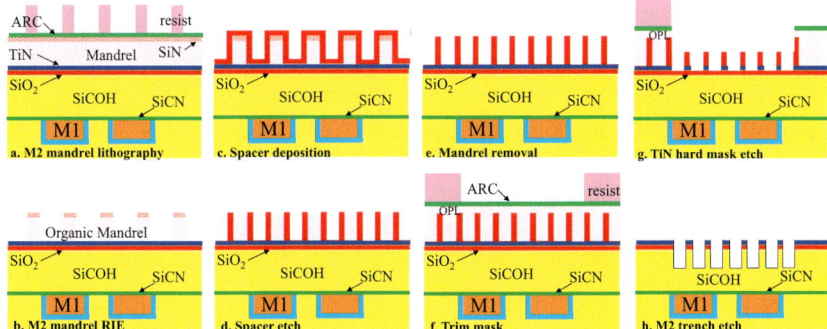

**Figure 2.15** Trench-first dual damascene for self-aligned-double-patterning for M2. Self-aligned via patterning (not shown) occurs between steps g and h (after [81,84]).

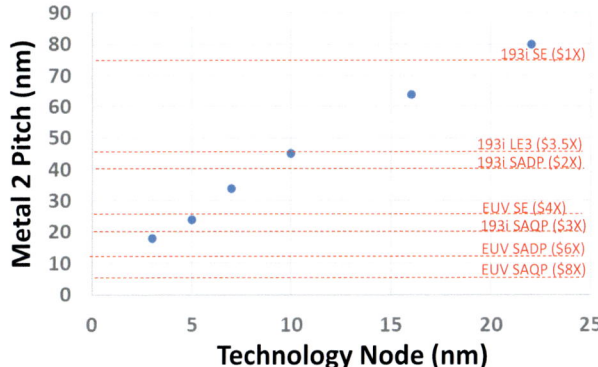

**Figure 2.16** Minimum metal two pitch versus technology node showing the smallest possible pitch for various patterning options. The cost numbers are normalized to the cost for 193 nm immersion single exposure (193i SE) [82,84].

good diffusion barrier (Fig. 2.6d) to prevent Cu diffusion into the dielectric [12,20,22]. However, by using electroplating, both vias and wires can be formed with the same metalization step, resulting in lower cost. In addition, very small, high aspect ratio features can be filled without voids by Cu plating, resulting in high reliability. Hence, Cu interconnect technology is used for memory as well as logic circuits for advanced technology nodes [13,85].

The final step in the dual damascene patterning process is removal of the SiN or SiCN etch stop layer at the bottom of the via, using a fluorocarbon RIE chemistry, such as $CF_4$ or $CHF_3$. The RIE overetch results in contamination of Cu at the bottom of the via with C, F, and oxygen

[86,87]. Prior to metalization, the contamination on the Cu at the bottom of the via must be removed. This is typically achieved by using a wet clean, such as dilute HF, followed by an in-situ Ar sputter clean or $H_2$ plasma clean in the metalization tool [86–91]. The wet etch removes most of the C and F from the RIE process, as well as Cu oxides. However, the Cu reoxidizes during the water rinse and exposure to air. Hence, an in-situ clean is required in the metalization tool to remove Cu oxides from the bottom of the via. Initially, an Ar sputter clean was used to remove the Cu oxides at the bottom of the via [86,88,91]. The energy and time of the Ar sputter clean must be carefully controlled. Excessive Ar sputtering can lead to chamfering at the tops of vias and trenches (which can lead to increased leakage current between neighboring wires) and to resputtering of Cu onto the sidewalls of the via [88,91]. One way to minimize the detrimental effects of the Ar sputter clean is to use a "barrier-first" process. In the barrier-first process, TaN is deposited, then the sputter clean is used to etch through the TaN and the contamination at the bottom of the via, then Ta is deposited. The presence of the TaN reduces the chamfering at the top corner of vias and trenches, and prevents resputtered Cu from contaminating the dielectric [88].

Reactive sputter cleaning using a $He/H_2$ plasma is an alternative to Ar sputter cleaning, that minimizes problems with chamfering and resputtering of Cu [89,90]. Cu oxides such as CuO and $Cu_2O$ can be reduced to metallic Cu at temperatures higher than 150°C [92]. The reactive sputter also results in less damage to low-$k$ materials. For porous SiCOH materials, a remote plasma source is used to minimize damage to the dielectric [89].

The metal deposition in dual damascene structures consists of barrier layer and Cu seed layer deposition by sputtering, followed by Cu electroplating. For metal pitch greater than ~50 nm, the barrier layer is typically a TaN/Ta bilayer [93]. Both Ta and TaN are good diffusion barriers for Cu. TaN provides good adhesion to the dielectric while Ta provides a surface with good wet ability of the Cu seed layer. A smooth, continuous Cu seed layer is required for void-free Cu plating [94,95]. In order to form a smooth, continuous seed layer, it is essential to have good wetting of the Cu seed on the barrier and to have a low deposition temperature (100°C or less) [94]. Recently, Cu reflow has been demonstrated for filling narrow features [96–99] (Fig. 2.17). Cu reflow occurs in narrow lines (<50 nm line width) providing an improved process window for Cu plating. Ru is preferred (instead of Ta) for the wetting layer for the Cu seed, enabling better reflow of Cu into narrow features. However, some

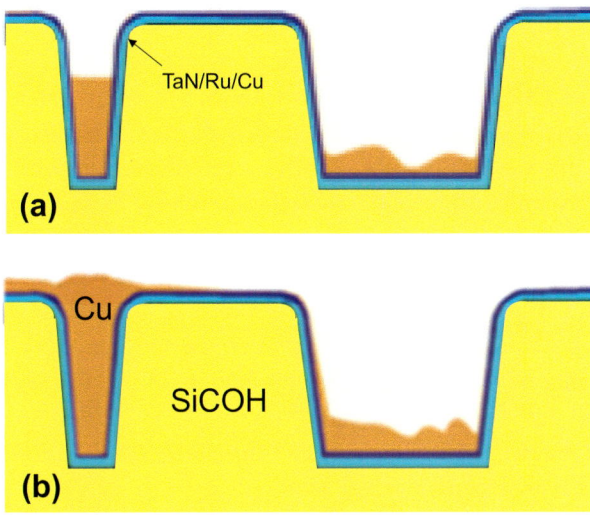

**Figure 2.17** Schematic of Cu reflow. (a) Initial Cu deposition and reflow, (b) additional Cu deposition and reflow time [96].

optimization of the Ru layer is need to achieve long electromigration lifetime [96].

Sputter deposition is the preferred method for depositing barrier and seed layers, because it can produce high purity films (i.e., such as Ta, which are essential for good wetting of Cu) at relatively low cost [Forster]. The biggest challenge for barrier and Cu seed layer deposition is ensuring adequate conformality in high aspect ratio vias and trenches. Good step coverage of sputter deposited films is possible by using ionized physical vapor deposition (ionized PVD) [100–102]. A two-step process is used to provide good sidewall coverage. The first step uses magnetron sputtering, where the sputtered metal is ionized and directionally deposited onto the substrate. The directional deposition results in a thicker film at the bottoms of trenches than on the sidewalls. The second step uses an Ar plasma to resputter some of the material from the bottom of the features onto the sidewalls. By using ionized PVD, good barrier and seed layer coverage have been demonstrated for 35 nm wide trenches with ~5:1 aspect ratio [85,103].

One approach for achieving a thin liner is to use CVD Mn [104,105]. Mn reacts with the dielectric and forms a $MnSi_xO_y$ silicate, which is a barrier to Cu diffusion, so TaN is not required, resulting in a lower resistance for the Cu interconnect.

Some alternatives to Ta-based barrier layers are Ti, Ru and Co. Ti barrier layers can be used instead of Ta to reduce cost [103,106]. For large Cu structures (>1 μm width), such as TSVs, Cu pillars, or redistribution layers, a Ti barrier can be in direct contact with the Cu seed layer [107,108]. For small Cu structures (<1 μm width), a multilayer film of Ti/TiN/Ti is required; TiN prevents excessive reaction between Ti and Cu, which can increase the resistivity of the wire. However, Cu wetting on TiN is poor, so a thin Ti layer is required on top of the TiN for good wetting of Cu [106].

Ru and Co are of interest as a replacement for Ta at the advanced technology nodes (below 14 nm node) because they have lower resistivity than Ta (which allows thinning or elimination of the Cu seed layer) and because Cu has better wettability on Ru or Co compared to Ta [96,109–119] (Fig. 2.18). Ru and Co are not good diffusion barriers for Cu. So a bilayer structure is still required, such as TaN/Ru or TaN/Co [111,112]. For line widths less than ~30 nm, there it is possible to have

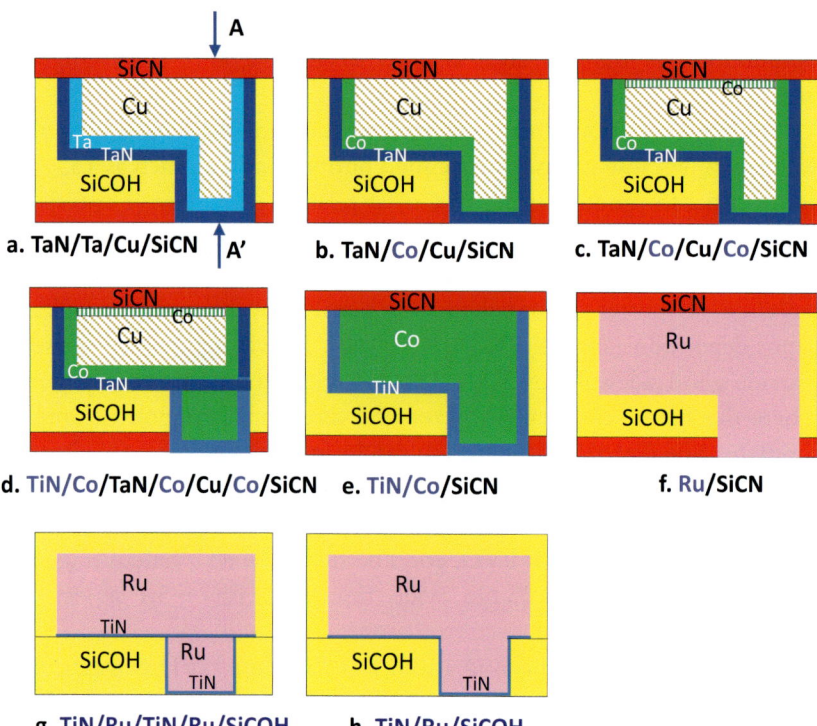

**Figure 2.18** Metalization options for interconnect line widths less than ~30 nm. Note that (a–d) use Cu CMP, (e) uses Co CMP, and (f–h) use Ru RIE for metal patterning.

lower resistance with Co or Ru interconnects rather than Cu. The exact cross-over point depends on the barrier thickness used. With Co and Ru, it may be possible to use a thinner barrier (or even no barrier) compared to Cu, resulting in a lower resistance compared to Cu at small enough dimensions (Fig. 2.19). Another option with Ru is to use RIE for metal patterning (Fig. 2.18), rather than CMP. This eliminates the need for an adhesion layer on the sidewall of the metal and also provides easier integration with airgaps [120–122].

The barrier layer deposition is more difficult for porous low-$k$ dielectrics. Deposition of thin, continuous metal barrier layers (such as TaN/Ta) is more difficult as pore size increases; incomplete barrier coverage can result in Cu diffusion into the dielectric [32]. The target for barrier layer thickness at the 22 nm node is ~3 nm. Hence, even for well-designed porous low-$k$ materials (i.e., with isolated pores less than 2 nm in diameter) the metal thickness is approaching the pore size. To ensure reliability, it may be necessary to seal the pores prior to metalization, using plasma treatments or conformal dielectric deposition [32,123,124]. A number of dielectrics have been examined as pore sealing materials including SiCH, SiOC, SiO$_2$ [123], and divinyl-siloxane benzocyclobutene polymer (p-BCB) [124]. The drawback to using an additional pore sealing layer is that the RC delay will be increased [125]. An additional problem is that moisture trapped in the porous dielectric can oxidize the TaN liner (Fig. 2.20a–c) [126–128]. Copper adheres poorly to oxidized Ta.

Hence the oxidized TaN/Ta barrier can cause poor yield and degrade reliability. A number of approaches have been used to minimize problems associated with barrier oxidation. One method is to increase the nitrogen

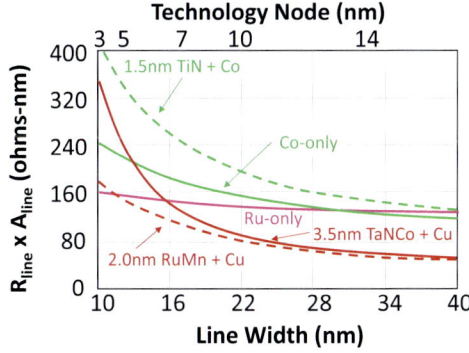

**Figure 2.19** Normalized resistance (resistance per unit length × cross-section area on interconnect) versus line width for different metalization options [117].

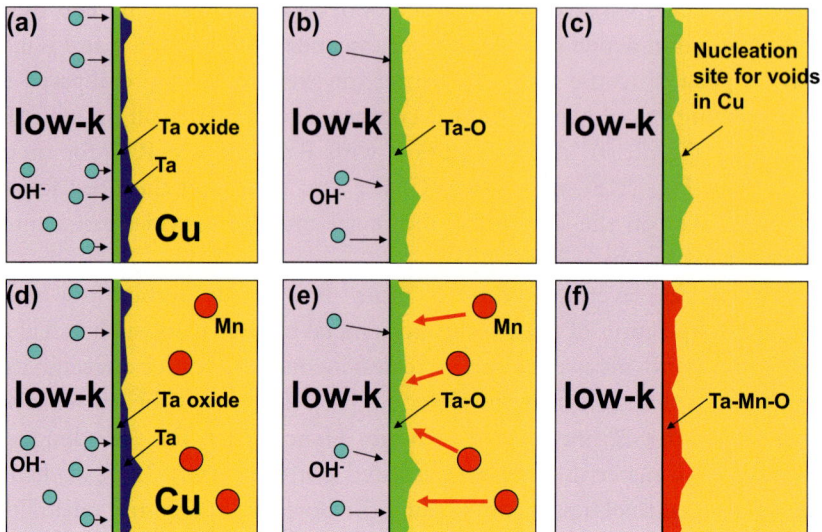

**Figure 2.20** Schematic of barrier oxidation (a–c) and Cu–Mn seed layer barrier restoration (d–f) [94].

content of the TaN; the oxidation is greatly reduced when TaN stoichiometry is changed from 4:1 to 2:1 (i.e., higher nitrogen content) [128]. Another approach is to use a Mn-doped alloy seed layer to restore the barrier (Fig. 2.20d–f) [127]. This process begins with sputter deposition of a Ta-based barrier, like the conventional process. The CuMn alloy seed layer is deposited, followed by plating and annealing. If the thin Ta barrier is oxidized, the Mn will segregate at the interface and form a Ta–Mn–O phase, thereby enhancing the barrier (Fig. 2.20f).

Electroplating of Cu provides void-free fill in high aspect ratio features, with low resistivity and high reliability [129,130]. Electroplating is performed by immersing the wafers into a solution containing cupric ions, sulfuric acid, and trace organic additives [129]. Electrical contact is made to the seed layer and current is passed that drives the following reaction at the surface of the wafer:

$$Cu^{2+} + 2e^- \rightarrow Cu\ (solid) \qquad (2.4)$$

The additives consist of suppressors, which reduce the plating rate at the tops of features, and accelerators, which enhance the plating rate at the bottom of features, and levelers, which reduce the plating rate at the upper corners of vias and trenches (Fig. 2.21) [129a]. The correct combination of

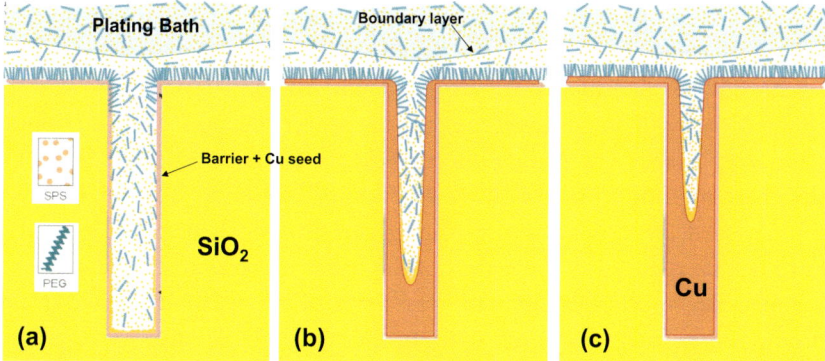

**Figure 2.21** Schematic of Cu plating "superfilling" mechanism. (a) During initial immersion in plating bath, accelerators (SPS) diffuse into narrow features faster than suppressors (PEG); (b, c) As plating proceeds, accelerator "piles up" at bottom of feature, resulting in an even higher plating rate at the bottom of the feature compared the field regions; i.e., bottom-up plating (after Kiegler [129a]).

these additives results in "bottom up", void-free filling of vias and trenches, which is commonly called "superfilling". Accelerators, such as dimercaptopropane sulfonic acid (SPS), contain sulfide and thiol like functional groups, which strongly absorb on Cu surfaces. The presence of SPS on the Cu surface may act as a charge transfer site for the reduction of $Cu^{2+}$ to $Cu^+$, and thereby enhances Cu deposition [130]. The SPS has a high solubility in the plating bath, so rather than being incorporated in the growing film, it continues to accelerate the reaction at the bottom of vias and trenches. Suppressors are polymers such as polyethylene glycol (PEG) that slow down the plating reaction. Possible mechanisms for the slower plating rate in the presence of suppressors are blocking of growth sites on the surface of the Cu and slower diffusion of Cu ions to the surface.

The concentrations of accelerators, suppressors, and levelers are in the ppm range, and must be automatically controlled in manufacturing to ensure void-free plating [131,132]. The composition of the plating bath changes over time, because the additives are consumed during the process either due to "drag-out" as wafers are withdrawn from the bath or due to breakdown of the additives in the bath. Many different techniques can be used to monitor the bath composition, but the most common method is cyclic stripping voltammetry (CVS). CVS is an electrochemical measurement method, based on cyclic plating and stripping of metal from an electrode, resulting in measurement of total charge associated with the plating and stripping process. With proper calibration, CVS can be used for quantitative measurement of the organic additive concentrations.

As device dimensions are scaled down, the Cu seed layer must be thinned, to avoid pinching off the tops of the trenches and vias. The thinner seed layer makes Cu plating more difficult for two reasons. Plated Cu thickness uniformity across the wafer is more difficult to achieve with a thinner seed layer, because the plating current (and hence the deposition rate) will be reduced in the center of the wafer if the seed layer resistance is comparable to the plating bath resistance. One method to avoid this problem is to increase the resistivity of the plating bath, either by lowering the acid concentration in the bath or by placing resistive membranes in the plating bath adjacent to the wafer [130]. Another method is to add a second cathode around the perimeter of the wafer, to draw current away from the very edge of the wafer [129]. The other problem with thin seed layers is that it is difficult to ensure continuity of the seed layer. If there are pinholes in the seed layer, plating will be delayed in these regions and voids may be trapped in the structure. A number of approaches are used to improve Cu plating on thin seed layers. One approach is to adjust the acid concentration in the plating bath [133,134] (Table 2.2). High acid content results in a faster onset of superfill behavior, however low acid content improves across wafer uniformity due to higher bath resistivity. Another approach is to apply a plating current as soon as the wafer is immersed in the plating bath, to avoid dissolving the seed layer in the plating bath [129].

Although Cu plating processes are commonly used for IC fabrication, some modifications are required for filling TSVs, including prewetting (to ensure no air is trapped in vias), pretreatment of additives on the surface of the seed layer, and a combination of conformal and bottom-up plating. The

**Table 2.2** Copper plating bath chemistries for different technology nodes [134].

| Parameter | High acid bath | Low acid bath | Medium acid bath |
|---|---|---|---|
| Technology node | 180 nm | 90 nm | 90 nm |
| $H_2SO_4$ | 175 g/L | 10 g/L | 80 g/L |
| Copper ion | 17 g/L | 40 g/L | 50 g/L |
| Cl | 50 ppm | 50 ppm | 50 ppm |
| pH | 0.64 | 1.41 | 0.85 |
| Conductivity | 510 mS/cm | 65 mS/cm | ~270 mS/cm |
| Accel/suppress/leveler | 2/8/1.5 mL/L | 6/2/2.5 mL/L | 9/2/2.5 mL/L |

initial deposition uses conformal plating, which helps to reduce the total deposition time (because plating occurs along the sidewalls). The final deposition steps use bottom-up plating, which provides void-free fill and a small overburden [135].

As the dimensions of Cu interconnects are reduced, the resistivity increases due to surface scattering, grain boundary scattering, and an increasing fraction of refractory metal liner in the trench (Fig. 2.22) [13,136].

What can be done to avoid problems associated with the increasing resistivity of Cu as wire dimensions are reduced? In principle, increasing the grain size of Cu in narrow lines would be very beneficial. However, in practice, it is difficult to achieve large grain size in narrow lines, because grain growth of Cu in trenches is inhibited at small dimensions [136]. A more promising approach is to reduce the thickness of the refractory metal liner (Fig. 2.22). Thinner liners are possible by using improved sputtering methods [100–102], atomic layer deposition (ALD) instead of sputter deposition [137], or by using self-forming barrier layers such as Mn silicate [138]. Some additional options are to use Co vias (with no metal barrier) to reduce the via resistance [139] and to use Cu RIE (instead of Cu damascene) to allow better grain growth and thinner metal barriers [140]. There are also design solutions to this problem. The interconnects with the largest increase in resistivity are at the lowest levels, where the length is typically short, so the resistance increase is less critical [141]. For layers where the high resistivity of the Cu is critical, changes in the design may be necessary, such as increasing the wiring pitch or adding a metal layer.

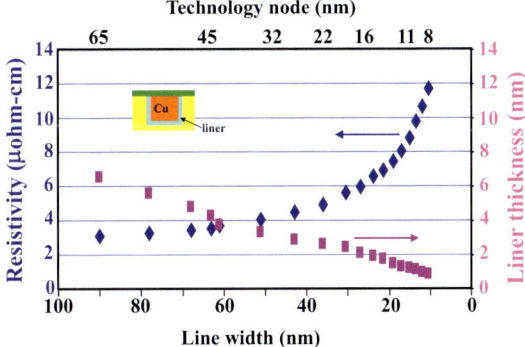

**Figure 2.22** Trend for effective line resistivity and for metal barrier thickness in trench.

## 3.5 Chemical mechanical polishing (CMP)

CMP is used to pattern the Cu and barrier layer after metalization of the dual damascene structure. The wafers are placed face-down on a rotating pad on which the slurry is dispensed [142]. Copper CMP typically requires at least two steps [22,143]. The first step is Cu removal, stopping on the barrier layer, and the second step is the barrier removal, stopping on the dielectric. Overpolishing is required to ensure that all metal is removed from the field regions in all parts of the wafer. During the overpolish, there will be thinning of the Cu in regions with high Cu pattern density. This thinning results in variations in wire resistance. To minimize the variation in wire resistance caused by differences in local pattern density, design rules are required which restrict the local Cu pattern density [24,144,145] (Fig. 2.23). In addition, low down force process are required to minimize Cu and low-$k$ dielectric erosion during the overpolish step [146].

The basic mechanism for metal CMP is (1) creation of a soft, passivation metal oxide and (2) removal of the passivation layer when it comes in contact with abrasive in the slurry (Fig. 2.24) [147–149].

There are a wide variety of Cu CMP slurries reported in the literature (Table 2.3) [150–153]. The basic components are (1) abrasive, (2) pH control, (3) oxidizer, (4) corrosion inhibitor, (5) complexing agents, and (6) surfactants. Colloidal silica is commonly used as an abrasive because of which it can be prepared with well controlled particle size at relatively low

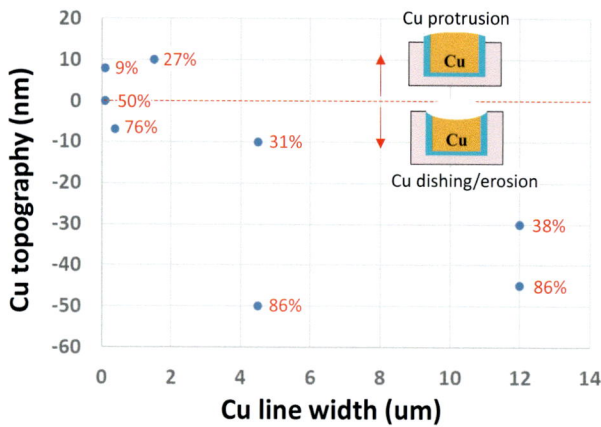

**Figure 2.23** Copper topography versus Cu line width (and Cu pattern density). Note that the barrier CMP process has high selectivity to Cu, so Cu can protrude above the dielectric at low pattern densities and small line widths [144].

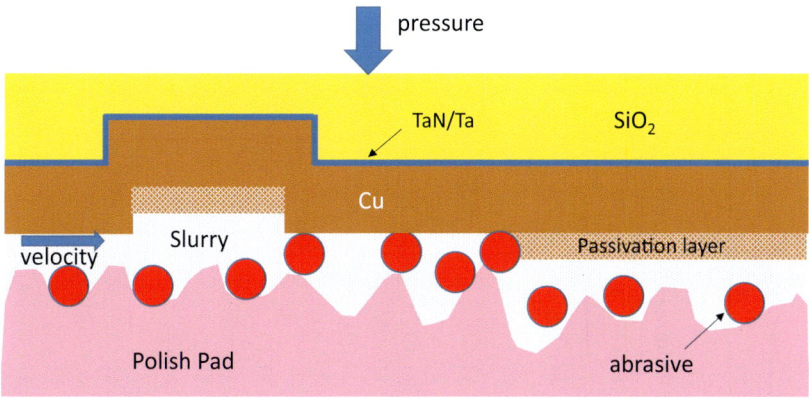

**Figure 2.24** Schematic of Cu CMP mechanism. Cu removal occurs by formation of passivation layer followed by mechanical removal by abrasive in slurry (after Lee [148]).

cost [152]. Copper slurries generally use alkaline pH, because Cu etching and corrosion is thermodynamically favorable in acidic solutions. $H_2O_2$ is often used as the oxidizer, and when combined with a corrosion inhibitor, such as BTA, forms a passivation layer on the copper. Complexing agents such as glycine, are needed to keep the dissolved Cu in solution, and prevent it from precipitating onto other regions of the wafer [147,154,155]. Surfactants are common additives in CMP slurries to prevent agglomeration of particles, and thereby reduce the risk of scratches [156].

There are a number of problems with Cu CMP in a porous low-$k$ structure, including Cu dishing and insulator erosion, cracking and adhesion loss in the dielectric stack, and scratching or contamination of the low-k material by components or the slurry or reaction by-products [157–161]. The problems with dishing/erosion and cracking/adhesion loss can be minimized by reducing the downforce during CMP and improving the adhesion between layers in the stack [159]. There are two basic integration schemes for Cu CMP with porous low-k structures; the permanent polish stop method (Fig. 2.25a–c) [157,159,162] and the direct CMP method (Fig. 2.25d–f) [162]. In the permanent polish stop approach, a relatively dense material, such as $SiO_2$ [159] or nonporous SiCOH [157,162], is used on top of the porous low-k material. The advantage of this approach is that the porous low-k material is protected from CMP-related scratches and contamination. The disadvantage of the permanent polish stop is that the effective dielectric constant of the stack increases, so most manufacturers use the direct CMP method. Hence, there is much research on minimizing

Table 2.3 Slurry compositions for Cu CMP. ADS: ammonium dodecyl sulfate, FA/O, FA/OII:R(NH$_2$) (OH) product name @hebei university, BTA: benzotriazole, TT-LYK:2,2′-[[(Methyl-1H-benzotriazol-1-yl)methyl]iminoïbisethanol, TAZ: 1,2,4-triazole.

| Type | Abrasive | Corrosion Inhibitor | Additive | Complexing agent | Oxidizer | pH | Reference |
|---|---|---|---|---|---|---|---|
| One-step CMP | Zirconia, 20 nm, 1 wt% | None | K$_4$Fe(CN)$_6$ | Arginine | H$_2$O$_2$ 1.5 wt% | 10 | [150] |
| Two-step (Cu CMP) | Colloidal silica, 80 –90 nm, 2 wt% | ADS | Alcohol polyoxyethylene ether | FA/O II | H$_2$O$_2$ | 10.5 | [150] |
| Two-step (Cu CMP) | Colloidal silica, 60 nm, 1 wt% | BTA | None | Glycine | H$_2$O$_2$ 2.0 wt% | 8.5 | [150] |
| Two-step (Cu CMP) | Colloidal silica, 80 nm | TT-LYK | None | Glycine | H$_2$O$_2$ 10 mM | 10 | [150] |
| Two-step (Cu CMP) | Colloidal silica, 95 –104 nm, 2 wt% | None | Alcohol polyoxyethylene ether | FA/O | H$_2$O$_2$ 2.0 wt% | 8.8–9.2 | [150] |
| Two-step (Cu CMP) | Colloidal silica, 15 –90 nm, 0.5 wt% | BTA 0.018 wt% | Hexadecyltrimethyl ammonium bisulfate | Glycine 1.0 wt% | H$_2$O$_2$ 1.0 vol% | 7.6–8.4 | [151] |
| Two-step (Ru CMP) | Colloidal silica, 85 nm, 5 wt% | TAZ | None | Guanidine Carbonate | H$_2$O$_2$ | 9 | [150] |
| Two-step (Co CMP) | Colloidal silica, 70 –90 nm, 0.5 wt% | TAZ | None | FA/O | H$_2$O$_2$ 0.5 vol% | 10 | [150] |
| Two-step (Co/TaN CMP) | Colloidal silica, 60 nm, 5 wt% | TT-LYK | None | Potassium Tartrate | H$_2$O$_2$ 1.0 wt% | 9 | [150] |
| Two-step (Co CMP) | Fumed silica, 2 wt% | BTA | None | None | NaOCl 0.5 wt% | 9 | [150] |

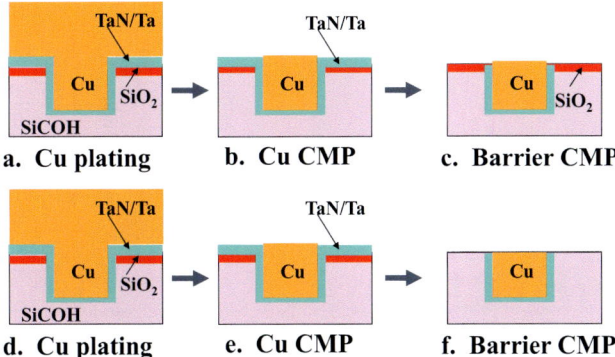

**Figure 2.25** Schematic of CMP options; (a–c) polish stop method and (d–f) direct CMP method [162].

damage and contamination when the polish stops directly on the porous low-k dielectric [158,162–164].

For direct CMP on the low-k dielectric, the first requirement is a low removal rate for the porous low-k material during the polish [165]. Organic compounds such as surfactants are used to lower the polish rate of the low-k material with respect to the metal layer [162–164,166]. The surfactants selectivity segregate to the surface of the SiCOH and thereby reduce the polish rate with respect to the metals [166]. However, it is often observed that the presence of surfactants in the slurry increases the dielectric constant of the porous SiCOH [162,164]. Additional $-CH_2$ and C–H bonds are observed in the bulk of the porous SiCOH after exposure to surfactants in the slurry, which are responsible for the increase in dielectric constant [164]. Significant diffusion of both linear and branched surfactants into porous low-k materials has been observed at room temperature, consistent with this model [163]. There are a number of approaches to minimize the change in dielectric constant. One approach is to optimize the slurry to prevent residues from forming in the pores [162,164]. Another approach is to do a post-CMP anneal at 350°C to restore the dielectric constant [162,164]. A third approach is to use a bilayer porous SiCOH film, with the near surface region (that is exposed to CMP) having a lower porosity than the bulk of the film [34].

There are a variety of defects associated with Cu CMP (Fig. 2.26) [154,167]. For larger dimension and $SiO_2$ or FSG dielectric, the main challenges of Cu CMP are scratches, particles, and dishing. Additional challenges with low-k dielectrics are delamination (requiring lower down

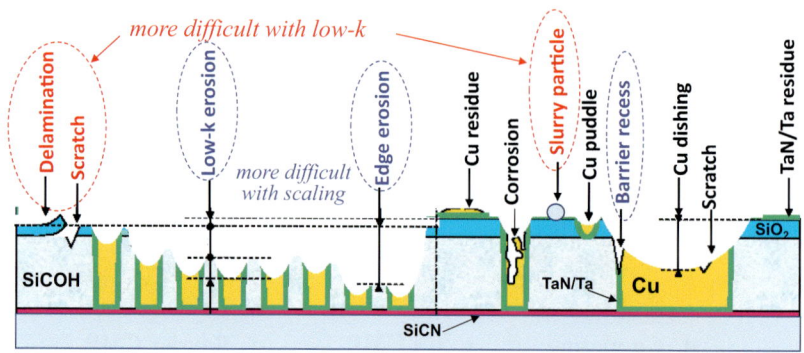

**Figure 2.26** Defects associated with Cu CMP and low-k dielectrics (based on [167]).

force during CMP) and water marks (requiring IPA treatment after DI H₂O rinsing). Additional challenges with Cu CMP at small device dimension are control of Cu dishing and minimizing barrier recess (especially with Co-based adhesion layers) [147,154].

## 4. Reliability

One of the main reasons for switching from Al to Cu interconnects was the improved reliability of Cu (for electromigration in particular) [12]. For the first generations of Cu technology, the improvement in electromigration was more than adequate for the needs of circuit designers. However, as device dimensions shrink, the electromigration lifetime of conventional Cu is no longer adequate. In addition, there are new reliability problems associated with the small dimensions, such as dielectric breakdown (Fig. 2.27). Low-k dielectrics cause additional problems for device

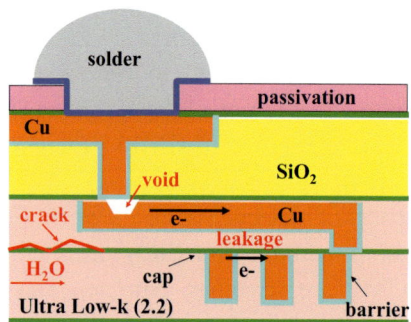

**Figure 2.27** Possible failure mechanisms for Cu interconnects in low-k dielectrics [21].

reliability, especially in terms of package reliability, due to the low mechanical strength and brittleness of these materials. In this section, we will describe reliability challenges for Cu interconnects in low-k dielectrics.

## 4.1 Electromigration

Electromigration is the migration of metal atoms in a conductor due to an electrical current (Fig. 2.27) [168]. The electrons moving toward the anode impart momentum to the atoms in the lattice, so that the atoms preferentially migrate toward the anode. For Cu interconnects, the TaN/Ta barrier layers at the bottom of the via act as blocking boundaries. Hence, during an electromigration stress, metal atoms will be depleted at the upstream side of the wire, and eventually voids will form. If the voids grow large enough (i.e., so that the void spans the wire or the via), the resistance will greatly increase and the circuit will fail (Fig. 2.28). At the downstream end of the wire, metal will accumulate, resulting in a hydrostatic stress. This stress produces a back flux of atoms, that is opposite in direction to the flux from electromigration [169], and is called the Blech effect. For short wires (i.e., below a critical threshold of current density times length), the back flux of atoms prevents killer voids from forming and the wires are immortal [170]. Therefore, it is possible to avoid electromigration problems in local interconnects by limiting the length of the wires. However, if the stress is high enough and the dielectrics are weak, metal extrusions may form (i.e., if the critical stress for metal extrusion is greater than the critical stress for void nucleation) [171], which can cause leakage between neighboring wires.

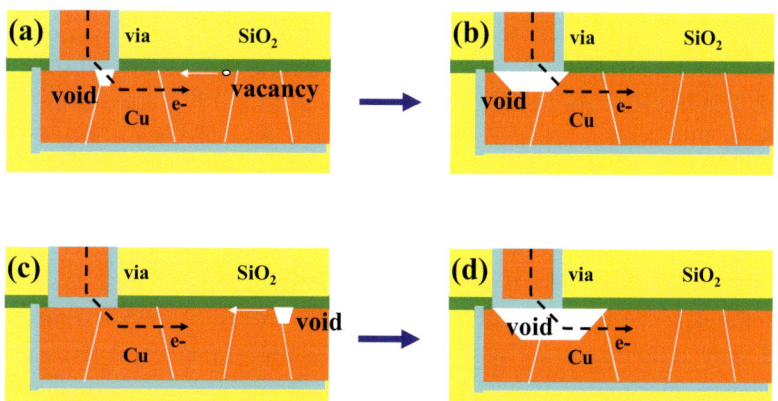

**Figure 2.28** Schematic of void formation in Cu during an electromigration stress for kinetics limited by (a) and (b) void nucleation and (c) and (d) void growth and migration [26].

The electromigration lifetime test uses high current densities and high temperatures to accelerate the fails. The test structures typically consist of simple via chains, where the resistance is monitored as a function of the stress time for a sample of test devices [172]. The length of the Cu wire to be tested must be sufficiently long (typically 200 µm or more) so that the Blech effect is minimized [171]. The failures are recorded as a function of time (where a fail is defined as a fixed change in resistance or a fixed percentage change in resistance), and the data (number of fails vs. time) is plotted using a log-normal distribution. The median time to fail, $t_{50}$, is extrapolated back to the use conditions using Black's equation [173,174]:

$$t_{50} = c j^{-n} \exp (E/kT) \qquad (2.5)$$

where $j$ is the current density, $E$ is the activation energy for diffusion, $k$ is Boltzman's constant, $T$ is temperature, and $c$ is a constant. The current exponent, $n$, provides information on the kinetics of electromigration; $n = 1$ corresponds to kinetics limited by void growth, whereas $n = 2$ corresponds to kinetics limited by void nucleation [175]. Intermediate values of $n$ (i.e., $n$ between 1 and 2) indicate that both nucleation and growth are occurring during the electromigration stress. The kinetics depend on the initial site for void nucleation (Fig. 2.28). For a dual damascene structure, with electron flow in the downstream direction, if the void initially nucleates directly under the via (Fig. 2.28a), then failure will occur soon after nucleation, and hence the kinetics are nucleation limited. In contrast, if the void initially forms in the wire far from the via (Fig. 2.28c), then considerable void growth (and diffusion to the via) will be required before failure occurs, and the kinetics are limited by void growth [175]. In-situ observations of electromigration in Cu dual damascene structures are consistent with kinetics that are limited by void growth [176,177], corresponding to $n \sim 1$. The voids nucleate in the wire (away from the via), then migrate toward the via at the cathode end of the test structure.

The electromigration lifetime for Cu is much greater than that for Al, by $>100\times$ [12], due to the lower diffusivity of Cu compared to Al. This allows circuit designers to use higher current densities in circuits, and thereby achieve higher switching speeds. However, as device dimensions and wire dimensions are reduced, it is desirable to increase the electromigration lifetime of Cu [72]. As device dimensions decrease, the drive current in the devices increases and the switching speed increases. At the same time, the dimensions of the minimum size wires (i.e., used for local wiring) decreases. Hence, ideally, a higher current density is required in the

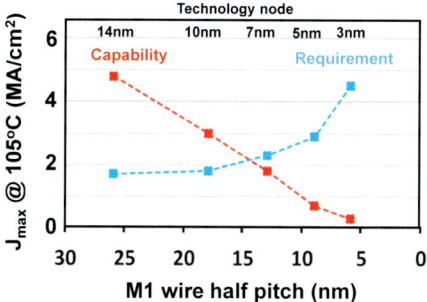

**Figure 2.29** Technology requirements for electromigration based on ITRS roadmap [72]. The maximum current density that the process is capable of decreases with technology node (because of decreasing wire dimensions), whereas the required maximum current density increases (because of high drive current of the devices). Hence, a short time is required to form a "killer" void (Fig. 2.30) [123,178]. In addition, for narrow Cu wires (<0.2 μm), the grain size decreases with line width [98]. Because of the smaller grain size, grain boundary diffusion can be significant during an electromigration stress, resulting in a lower electromigration lifetime [179—181].

wires (Fig. 2.29). However, the electromigration lifetime decreases as wire dimensions decrease (Fig. 2.30) for a number of reasons. First, the void size required to cause a fail decreases as the via size and wire size decreases [182].

Low-k dielectrics also contribute to lower electromigration lifetime. Low-k materials have lower thermal conductivity than $SiO_2$ (Table 2.1). Hence, there will be more joule heating for a given current density [183], resulting in a higher temperature for the wire, and therefore a faster rate of

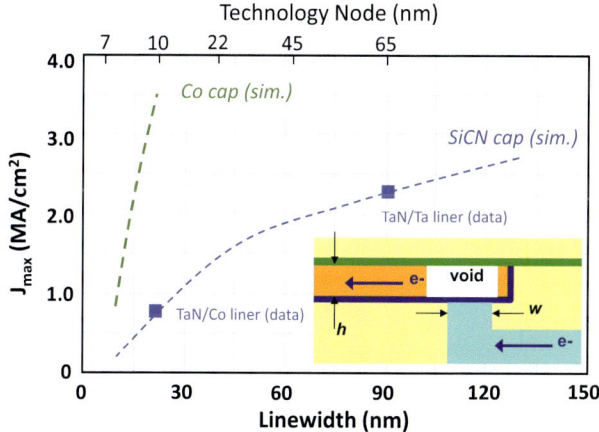

**Figure 2.30** Effect of wire dimensions on electromigration lifetime [178,182].

electromigration. The low-k dielectrics also have a lower modulus then $SiO_2$. Because of the lower modulus, the Blech effect will be reduced, and the critical length for line immortality will be reduced [184]. The barrier layers often have weak adhesion to low-k materials; the weak adhesion can result in extrusion fails during an electromigration stress [185].

There are a number of ways to improve the electromigration lifetime of Cu. The electromigration lifetime for Cu interconnects is determined by mass transport at the interface between Cu and the capping layer [182], and can be improved by increasing the adhesion between these layers [186]. Hence, the capping layer process is critical to achieving a long electromigration lifetime. A typical dielectric capping process (Fig. 2.31a–d) consists of a plasma clean to remove Cu oxides, a brief $SiH_4$ exposure to form a thin Cu silicide layer for improved adhesion, and finally, the dielectric deposition (either SiN or SiCN) [187,188].

A further reduction in interface diffusion can be achieved by using a metal capping layer (Fig. 2.31e–g), rather than a dielectric capping layer [182,186,187,189]. Improvements in electromigration lifetime of over $300\times$ have been reported with a Co-based capping layers [189–192]. The activation energy for diffusion for bamboo lines (i.e., no grain boundary diffusion) increases from $\sim 1.0$ eV for an SiCN cap to $\sim 2.0$ eV for a

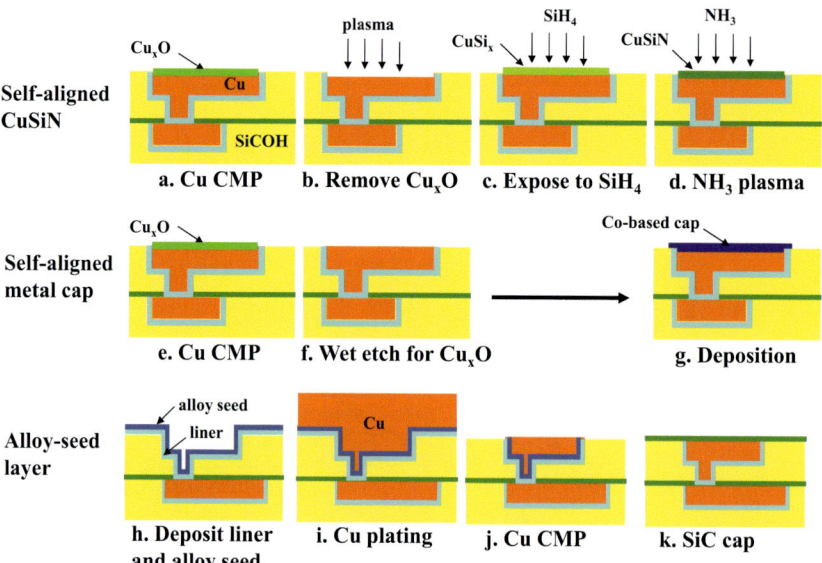

Figure 2.31 Methods to improve electromigration lifetime of Cu interconnects [26].

**Table 2.4** Action energies for stress-induced voiding (SIV), electromigration (EM), and diffusion in copper [26].

| Test | Activation Energy (eV) | ILD | Cap | Reference |
| --- | --- | --- | --- | --- |
| SIV | 0.74 | FSG | SiN | Ogawa |
| SIV | 0.75 | None | SiN | Gan |
| EM | 0.9 | SiLK | SiCN | [189] |
| EM | 1.9 | SiLK | CoWP | [189] |
| EM | 0.87 | SiCOH | SiCN | [193] |
| Cu bulk | 2.2 | | | [182] |
| Cu g.b. | 0.8–1.0 | | | [182] |
| Cu/SiN | 0.8–1.1 | | | [182] |

CoWP cap [189,193] (Table 2.4), suggesting that the diffusion mechanism changes (for example, from interface diffusion to bulk diffusion). The interface cleanliness and interface bonding between the Cu and the capping layer are critical to achieving a long electromigration lifetime. Table 2.4 also shows activation energies for stress-induced voiding [194,195] where void growth is also controlled by diffusion at the interface between the Cu and the capping layer.

Another method to improve the electromigration lifetime is to dope the Cu with impurities, such as Al [196,197], Ag [198], or Mn [199–201] [293], [202]. The dopants are typically introduced into the Cu seed layer (Fig. 2.32h–k). During subsequent anneals, the impurities segregate at grain boundaries and interfaces, including the critical interface between the Cu and the capping layer [196,199]. The presence of the impurities at the interfaces reduces Cu diffusion, resulting in an enhancement of electromigration lifetimes of over 10× [197], with higher doping concentrations resulting in higher electromigration lifetimes. The main problem with this

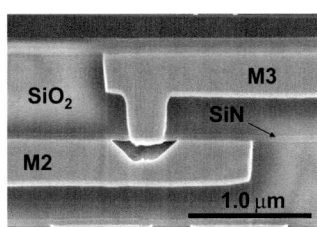

**Figure 2.32** SEM cross-section of failing via chain after stress migration test at 225°C for 1000 h [203].

approach is that the impurities increase the resistivity of Cu. The impurity concentration in the seed layer is relatively low; 0.5−2.0 atm% for Al [196], 1.0 wt% for Ag [197], and 4.0 atm% for Mn [199]. However, increases in resistivity of 3%−10% are observed for most of these impurities [196,198]. Manganese has some advantages compared to the other dopants; low solubility in Cu and high affinity for oxygen. Hence, for Cu−Mn seed layers, by optimizing the post-metal anneal the increase in resistivity can be minimized [199].

## 4.2 Stress-induced voiding

Voids can form in passivated Cu interconnects during annealing at moderate temperatures ($\sim$200−250°C), due to tensile stress in the metal. There are two mechanisms for the stress in the metal; thermal stress, due to thermal expansion mismatch between the metal and the insulator, and growth stress, due to grain growth in the metal [195,204,205]. Voids will form if the tensile stress is above the critical stress. If the voids grow large enough (i.e., spanning the width of a line or a via), the resistance will increase and the circuit will fail (Fig. 2.32). The lifetime for stress-induced voids is monitored using simple via chain structures, where the resistance is measured as a function of time at the stress temperature [203] (Fig. 2.33).

Thermal stress in Cu wires is associated with high temperature processes (typically cap and interlevel dielectric deposition) after Cu CMP [204,205].

After Cu CMP, the thermal stress in the wires is relatively low, because the metal processing occurs at low temperatures (typically <300°C) and the top surface of the Cu is unconstrained. However, the deposition of the capping layers (SiN or SiCN) and interlevel dielectrics occurs at a relatively high temperature (>300°C). Copper in the trenches expands due to heating at the deposition temperature, then contracts during cooling to

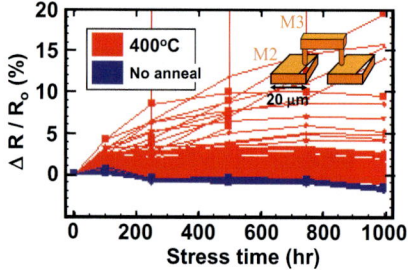

**Figure 2.33** Resistance versus time data during stress migration test for M3−V2−M2 via chain, for samples with or without post-metal anneal [203].

room temperature. (The expansion during heating is especially dramatic for Cu in TSVs, where the "Cu pumping" can cause mechanical damage to overlying structures [206,207].) But the Cu is constrained by the dielectric capping layer, so it cannot shrink to the original dimensions [208] (Fig. 2.34). As a result, there is a tensile stress in the Cu wires after deposition of the capping layer. Assuming there is no deformation of the dielectric during cooling, the stress in the z-direction, $\sigma_z$, in the Cu wire after capping layer deposition is given by the following:

$$\sigma_z = E_m \, \Delta\alpha \, \Delta T \tag{2.6}$$

where $E_m$ is the bulk modulus of the metal (120 GPa for Cu), $\Delta\alpha$ is the difference in thermal expansion mismatch between the metal and the surrounding dielectric layer ($\alpha_{Cu} = 17$ ppm/°K, $\alpha_{SiO2} = 3$ ppm/°K), and $\Delta T$ is the difference in temperature between the dielectric deposition temperature and the temperature of interest (i.e., for example, the operating temperature of the device) [205]. Hence, the stress in Cu will be higher as the temperature of the capping layer deposition increases.

Another contribution to tensile stress in Cu is confined grain growth [195]. The copper plating process occurs at low temperatures (<100°C), so that as-deposited Cu has a small grain size. Subsequent dielectric depositions occur at much higher temperatures (up to 400°C). Hence, grain growth will occur in the Cu. Grain growth eliminates excess free space in the Cu, resulting in excess vacancies and shrinkage of the metal lines. If grain growth occurs before cap layer deposition, the vacancies can diffuse to the surface and are annihilated. However, if grain growth occurs after cap deposition, the excess vacancies will be trapped in the Cu. The shrinkage of the metal lines results in an additional component of tensile stress in the metal. To minimize void formation associated with confined grain growth, it is common practice in the industry to anneal the Cu after plating [195,209].

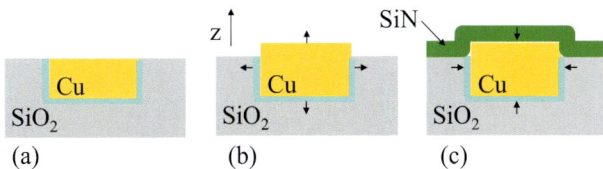

**Figure 2.34** Tensile stress in Cu due to cap deposition. (a) After Cu CMP (25°C); (b) during SiN deposition (~400°C); (c) after cooling back to room temperature [208].

The tensile stress in the metal can be relieved by the formation of voids in the wire. If the voids grow large enough (i.e., causing a large enough resistance shift), the circuit will fail. In order for a void to form, the strain energy release (associated with void formation) must exceed the change in interface energy (i.e., assuming the void forms at an interface) [210–212]. There exists a critical stress for void nucleation, above which a void will be thermodynamically stable, and growth is limited only be kinetics. The critical stress depends on the adhesion between Cu and the various barrier layers, and on the modulus of the dielectric. The critical stress is lower for Cu in a SiCOH dielectric ($E = 6$ GPa) compared to Cu in F-doped $SiO_2$ ($E = 71$ GPa) [212,213]. Hence, stress-induced voiding is more likely for Cu in SiCOH than for Cu in F-doped $SiO_2$ (even though the tensile stress is lower for Cu in SiCOH [212]. The nucleation barrier for void formation is also reduced if there are pre-existing defects in the structure, such as undercut from the via process [209] or poor fill during the plating process [214].

For stress-induced voids that originate from thermal expansion mismatch (i.e., during cap layer deposition), the maximum rate of void growth in Cu occurs below the "stress-free" temperature [195]. The stress-free temperature is related to the deposition temperature of the cap layer and subsequent processes (such as interlevel dielectric deposition). At temperatures close to the stress-free temperature, the tensile stress, $\sigma$, in the metal is low, so the void growth rate is low. At low temperatures ($<150\,°C$), the Cu diffusivity, $D$, is low, so the void growth rate is low (Fig. 2.35). Significant void growth only occurs at intermediate temperatures ($\sim 200-250\,°C$), where there is both significant tensile stress in the metal and significant diffusivity of Cu. The rate of void growth, $R$, is given by the following [195]:

$$R = C\,(T_o - T)^N \exp\,(-Q/kT) \qquad (2.7)$$

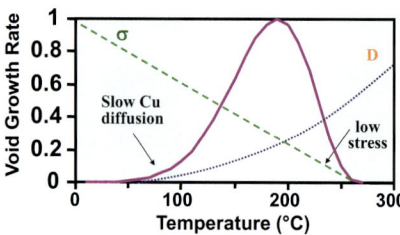

**Figure 2.35** Void growth rate as a function of temperature [195].

where $T_o$ is the stress-free temperature, $N$ is the creep exponent, $Q$ is the activation energy, $k$ is the Boltzmann constant, and $C$ is a proportionality constant. Measured values of the activation energy are typically ∼0.75 eV. The activation energies for bulk Cu diffusion, grain boundary diffusion, and interface diffusion are 2.2 eV, 0.8–1.0 eV, and 0.8–1.1 eV, respectively [182]. This suggests that the dominant diffusion mechanisms for stress-induced voiding are grain boundary and/or interfacial diffusion [194,195].

Stress-induced voids in Cu are typically observed at grain boundaries [215], under vias [195], or inside vias [216]. Voids form preferentially at grain boundaries and interfaces (i.e., often at the intersection of a grain boundary and the capping layer), because these are fast diffusion paths compared to bulk diffusion [211]. Void formation under vias is favorable due to the stress gradient in the underlying Cu and due to the presence of the via-metal interface. Modeling results show that for a via landing on a metal line, there is a high tensile stress in the metal at the edge of the via [195,212]. A void will nucleate if the tensile stress exceeds the critical stress. If there is weak adhesion between the barrier metal and the underlying Cu at the bottom of the via, the critical stress will be reduced, making void nucleation even more favorable. Once a void nucleates, the stress field surrounding the void becomes less tensile, and the resulting stress gradient favors vacancy diffusion toward the void, resulting in further growth [217]. Voids form in vias because the tensile stress in vias and narrow lines is lower than the tensile stress in wide lines. Vacancies will diffuse to the regions of lower tensile stress in the via. If there is a defect in the via (due to poor seed layer coverage, for example), then void nucleation will be further enhanced in the vias.

The line and via dimensions have a large effect on the failure rate due to stress-induced voids. The failure rate for stress-induced voids in Cu increases with increasing line width (Fig. 2.36), opposite to what is observed with Al [195,205]. There are two reasons for this line-width dependence. For Cu lines, the hydrostatic stress increases with line width (due to higher growth stress) [204], so there is a larger driving force for void formation for wide lines compared to narrow lines. In addition, for wide lines, the kinetics of void formation are enhanced because a greater number of vacancies are available within a diffusion length of the via [195]. In order to cause a fail, the void must be large enough to span the bottom of the via. The "active diffusion volume" is larger for wide lines than for narrow lines, so it takes less time to form a "killer void". The fail rate due to

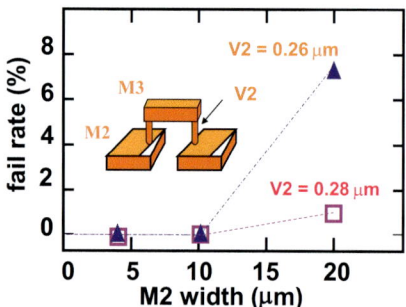

**Figure 2.36** Fail rate for stress-induced voids versus M2 line width after stressing at 225°C for 1000 h [208].

stress-induced voids increases as via size decreases [218] (Fig. 2.36), for similar reasons as for electromigration; The void size required to cause a fail decreases as the via size. Because stress-induced voids are mainly a problem for wide lines (rather than narrow lines), a simple design solution is to use redundant vias [219,220].

To minimize the fail rate from stress-induced voids, it is important to have good metal barrier coverage on the bottoms and sidewalls of trenches and vias, to prevent Cu from diffusing into the dielectric. Having adequate barrier coverage is more difficult as dimensions shrink, due to the requirement to thin down the barrier layer (from 9 nm at the 65 nm node to 3 nm at the 22 nm node) [125], to ensure low wire resistance. An additional challenge is ensuring adequate metal coverage on the sidewalls of porous low-k materials. Pore sealing prior to metalization is required to ensure good metal coverage [221]. Because the barrier layers are deposited by physical vapor deposition, good control of the via and trench profiles is also critical in order to achieve adequate metal coverage [222].

The copper fill is also important to prevent failure due to stress-induced voids. If there are pre-existing voids because of incomplete Cu fill, the devices will fail after very short stress times. Achieving void-free Cu fill is more challenging as device dimensions scale down, because the Cu seed layer thickness must be reduced (to prevent pinch-off at the tops of trenches and vias) and because the resulting aspect ratio for Cu plating is increased. Improvements to the PVD tools are required as device dimensions scale down, to ensure good uniformity of the Cu seed layer across the wafers [103]. In addition, the Cu plating process must be optimized, for example, by optimizing the additives in the plating bath [223].

Annealing also affects the formation of stress-induced voids. An anneal is required after Cu plating to maximize grain size prior to cap layer deposition (i.e., to prevent confined grain growth). But after cap layer deposition, the maximum anneal temperature must be limited. High temperature anneals after cap deposition can lead to high rates of stress-induced void formation (Fig. 2.33), due to either confined grain growth or due to increased stress in the Cu [208,215].

Because the kinetics of void formation are controlled by interface and grain-boundary diffusion [224], the rate of void growth can be reduced by using metal capping layers [225] or by alloying the Cu [196,197], similar to methods used to improve electromigration lifetime (Fig. 2.31). As with electromigration, it is important to form a strongly adhering interface between the Cu and the capping layer, in order to minimize vacancy diffusion along this fast-diffusion path.

## 4.3 Time-dependent dielectric breakdown (TDDB)

During a prolonged stress at high electric fields, damage can occur in dielectrics, eventually resulting in a conducting path and electrical breakdown [168,226,227]. Historically, this was mainly a problem for gate dielectrics, because the spacing between metal wires was relatively large, so the electric field across the BEOL dielectric was low. However, as device dimensions are reduced, the lateral electric field across the BEOL dielectric increases (Fig. 2.37).

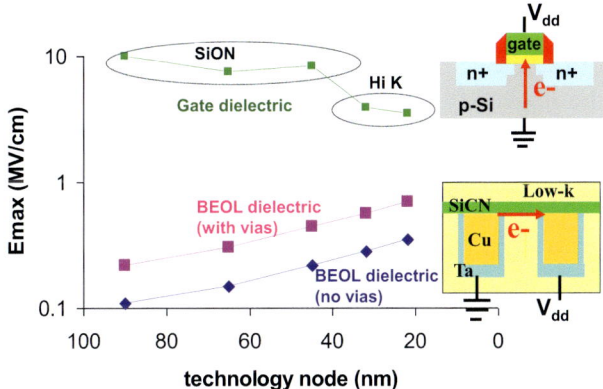

**Figure 2.37** Maximum electrode field across gate dielectric and between minimum pitch interconnects as a function of technology node based on ITRS roadmap [26].

Although the maximum electric field across the BEOL dielectric is still considerably lower than that across the gate dielectric, the breakdown strength of the BEOL dielectric is considerably less. There are a number of reasons for the low breakdown strength of BEOL dielectrics, including a high density of defect sites in the as-deposited dielectric (especially for low-k materials) [228,229], damage or contamination of the dielectric from processes such as CMP [228,230], Cu diffusion into the dielectric through the barrier layers [231], and patterning problems such as line-edge-roughness or via misalignment [232–234]. As a result, dielectric reliability becomes more challenging as device dimensions shrink and as the dielectric constant of the BEOL insulator is reduced.

Dielectric reliability is assessed using a time-dependent breakdown test (TDDB). The test structure consists of a comb-comb or comb-serpent layout at the appropriate metal level (typically M1, because this has the smallest pitch) [232,235,236]. One electrode (serpent, for example) is grounded and a constant positive voltage is applied to the other electrode (comb, for example). The TDDB measurements are typically conducted over a range of voltages (i.e., on the order of 10× higher than the use conditions), then the voltage acceleration is used to extrapolate back to the use conditions. The TDDB is usually tested at the device operating temperature [237]. Very high stress fields are required ($\sim 5$ MV/cm) in order to complete the test in a reasonable amount of time. The current is monitored as a function of time, with a sharp increase in current corresponding to breakdown. The initial decrease in current is due to trapping of charge, followed by stress-induced leakage, and finally breakdown [235,238].

Dielectric breakdown is a "weakest-link" reliability problem [168]. The fail will occur at a flaw in the dielectric that grows with time. Hence, the reliability of the dielectric scales with area; the lifetime of the device decreases as the insulator area increases. Dielectric data is generally analyzed using a Weibull distribution [239–241], because the Weibull distribution (unlike the log-normal distribution) scales with area [239]. For a cumulative failure probability as a function of time, $F(t)$, for the Weibull distribution is given by Ref. [168]:

$$F(t) = 1 - exp\,\text{-}(t/t_{63})^\beta \qquad (2.8)$$

where $t_{63}$ is the characteristic failure time at which 63% of the population has failed and $\beta$ is the Weibull shape parameter [168] (i.e., the slope of the Weibull plot). Assuming the defects are randomly distributed, the extrapolation from the test structure area to the chip area can be made using Poisson area scaling [239,240], given by:

$$1-F = \exp(-D/A) \tag{2.9}$$

where $F$ is the fraction of failed devices, $D$ is the defect density, and $A$ is the device area. The probability of failure for the chip can then be determined using the following:

$$1-F_C = (1-F_T)^{Ac/At} \tag{2.10}$$

where $F_C$ and $F_T$ are the probabilities of failure and $Ac$ and $At$ are the areas, of the chip and test structure, respectively.

For comb-serp test structures at small metal spacing (<100 nm), it is often observed that Weibull plots of TDDB data are nonlinear [226,236,242] (i.e., the slope, $\beta$, varies with fraction failed). The nonlinearity in slope makes it difficult to extrapolate to the use conditions. The nonlinearity is often due to variations in the line-to-line spacing (which are more significant at small dimensions). In order to determine to TDDB lifetime, the populations with different spacing must be analyzed separately [242]. By doing this, linear Weibull slopes are obtained.

To determine the fail rate for TDDB at use conditions, the high field stress data must be extrapolated to the lower fields at the use conditions. There are two models commonly used for this extrapolation, the "E-model" [235,242] and the "ÖE-model" [237,238].

The E-model is a field-driven model. It is assumed that the interaction of the dipole moment of the bond (for example, the Si—Si bond at an oxygen vacancy in $SiO_2$) with the electric field weakens the bond, so that it is more easily broken either by thermal energy or by hole capture. The natural logarithm of time-to-breakdown is directly proportional to the electric field [235,243]. The time-to-breakdown is written as:

$$t_{bd} = A \exp(\Delta H_o / kT) \exp(-\gamma / E) \tag{2.11}$$

where $A$ is a materials-dependent constant, $\Delta H_o$ is the zero-field activation energy for bond breakage in the dielectric, $k$ is the Boltzmann constant, and $T$ is the temperature in degrees Kelvin. The field-acceleration parameter, $\gamma$, is given by

$$\gamma = p_{eff} / kT = -(d \ln(t_{bd}) / dE) \tag{2.12}$$

where $p_{eff}$ is the effective dipole moment for the molecule in the material. Note that the by using the linear relation between ln $t_{bd}$ and $E$, it is possible to determine the lifetime at use conditions.

The ÖE-model is a current driven model. It is assumed that damage in the dielectric is proportional to the total amount of charge carriers injected into the material, and that breakdown occurs at a critical charge level (i.e., the "charge-to-breakdown"). The amount of charge injected in the dielectric is proportional to the leakage current. Assuming a constant leakage current, the charge-to-breakdown, $Q_{bd}$, is related to the time-to-breakdown, $t_{bd}$, as follows:

$$Q_{bd} \alpha J\, t_{bd} \quad (2.13)$$

where $J$ is the current density injected into the dielectric during the stress. (i.e., charge-to-breakdown is proportional to current density).

For line-to-line leakage in Cu interconnect structures (SiCOH dielectric with SiCN cap), the leakage is typically due to Schottky emission at low fields (<1.4 MV/cm) and due to Poole–Frenkel emission at high fields (>1.4 MV/cm) [243,244] (Fig. 2.38). The leakage current densities for Schottky emission, $J_{SE}$, and Frenkel–Poole emission, $J_{FP}$, are given as follows [246];

$$J_{SE} \sim A^* T^2 \exp\{-(q/kT)[\phi_B - (qE/4\pi\varepsilon_i)^{1/2}]\} \quad (2.14)$$

$$J_{FP} \sim E \exp\{-(q/kT)[\phi_B - (qE/4\pi\varepsilon_i)^{1/2}]\} \quad (2.15)$$

where $A^*$ is the effective Richardson constant, $\phi_B$ is the barrier height at the metal-insulator interface, and $\varepsilon_i$ is the dielectric constant of the insulator. Schottky emission is due to electrons that are thermally excited over the potential barrier at the metal-insulator interface (Fig. 2.38a). Frenkel–Poole emission is due to field-enhanced thermal excitation of trapped electrons into the conduction band of the insulator (Fig. 2.38b).

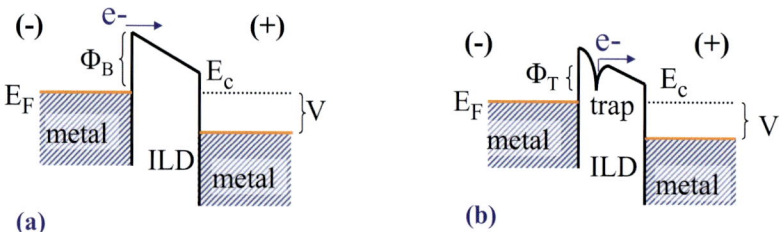

**Figure 2.38** Schematic of (a) Schottky emission and (b) Poole-Frenkel emission [245].

Assuming leakage is due to Schottky emission, the time-to-breakdown is obtained by inserting Eq. (2.8) into Eq. (2.7), and rearranging [237,245];

$$t_{bd} \propto Q_{bd}/J_{SE} \propto exp\ \{q/kT[\phi_B - (qE/4\pi\varepsilon_i)^{1/2}]\} \quad (2.16)$$

From Eq. (2.10), it can be seen that the natural logarithm of time-to-breakdown is directly proportional to the square root of the electric field. Note that Eq. (2.10) also applies if the leakage is due to Frenkel–Poole emission, because the exponential terms are the same for both Schottky emission and Frenkel–Poole emission. The field-acceleration parameter, for the ÖE-model, is given by Ref. [238];

$$\gamma = (q/\pi\varepsilon_i)^{1/2}\ /\ kT = -\ (d\ \ln\ (t_{bd})\ /\ d\ddot{O}E\ ) \quad (2.17)$$

There are different models for how the injected charge creates damage in the dielectric. Some models [237,247] assume that the injected charge results in ionization of Cu in the interconnect, resulting in drift into the dielectric, and creation of traps (Fig. 2.39). An alternate model [247] assumes that damage occurs due to acceleration of electrons in the electric field.

The two models predict much different lifetimes at the use conditions, with the E-model being the more conservative in terms of lifetime prediction. Hence, it is important to determine which model is valid for the lifetime extrapolation. Unfortunately, it is difficult to distinguish between these models at the high field conditions typically used for TDDB tests. It is only possible to distinguish between the models at low fields, which requires very long test times (on the order of years). There are a number of recent reports on TDDB at low field conditions; one supports the E-model [248] and two support the ÖE-model [237,247]. Hence, there continues to

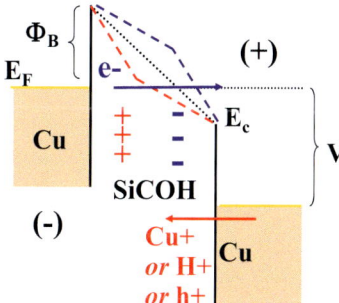

**Figure 2.39** Schematic of Cu drift into dielectric during TDDB stress.

be disagreement on which of these models is best for extrapolation of TDDB lifetime for Cu interconnects in low-k dielectrics [249].

Dielectric breakdown between neighboring Cu wires generally occurs at the interface between the capping layer and the dielectric [241,247]. The electric field is highest at this location because the Cu wires are generally tapered (wider at the top than at the bottom), so the space is smallest at the top of wires. In addition, the interface is expected to have a higher trap density than the bulk dielectrics, due to bond mismatch between the different materials or due to contaminants from the Cu CMP process [160,231]. It is expected that the interface between the capping layer and the dielectric is the preferred leakage path due to the combination of the high electric field and the high defect density.

The TDDB lifetime is very sensitive to the materials and processes used to form the interconnect layers. The TDDB lifetime typically decreases as the dielectric constant of the material decreases [228]. In particular, porous materials have lower TDDB lifetime than nonporous materials. Possible reasons for the lower TDDB lifetime are weaker bonds, higher trap densities, or lower barrier heights at the metal−insulator interface. Despite the lower TDDB lifetime of porous materials compared to nonporous materials, the intrinsic reliability is still adequate for integrated circuits. The main problem with achieving adequate TDDB lifetime is minimizing early fails, associated with nonoptimized processing. The basic methodology for ensuring adequate TDDB reliability is to minimize electric field enhancement and minimize Cu diffusion into the dielectrics and capping layers. Field enhancement can be minimized by reducing line-edge roughness [236], and having good dimensional control of trenches and vias [240]. Copper diffusion into the dielectric can be minimized by using adequate metal barrier layers [233,241], minimizing residues after post-CMP cleaning [64], and minimizing air exposure prior to capping of the copper [231,241].

## 4.4 Package reliability

There are two basic packaging methods to connect the die to the packaging substrate, wire bonding and flip chip solder (Fig. 2.40). The wirebond process is generally cheaper, and is used when the number of input/output (I/O) devices is low. The flip chip solder process [250] is more expensive (Fig. 2.41), but has a number of advantages compared to wirebond packages, including lower lead inductance, higher I/O density, and smaller form factor (which is desirable for portable devices).

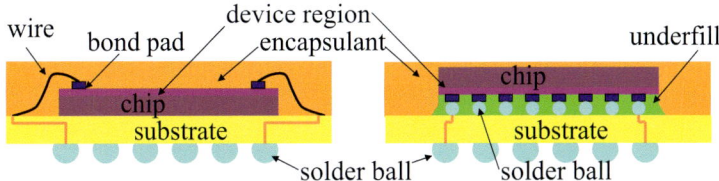

**Figure 2.40** The two basic package types are (a) wirebond and (b) flip chip solder attach.

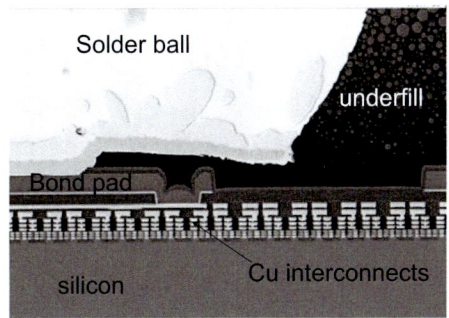

**Figure 2.41** SEM micrograph of solder ball and underfill attached to devices with Cu/low-k BEOL [251].

Packaging processes such as dicing, wirebonding, and flip-chip die attach can damage the low-k dielectrics due to mechanical stress [251–259]. Standard tests for assessing the reliability of packaged parts include high temperature storage (HTS), temperature-humidity-bias (THB), high temperature operating life (HTOL), and thermal cycle (T/C) (Table 2.5). These tests are designed to accelerate fails associated with mechanical damage in the die, the wirebond, the flip-chip solder bump, or in the encapsulant.

**Table 2.5** Tests for package reliability.

| Test | Conditions |
| --- | --- |
| High temperature storage | 125°C, no bias, 1000 h |
| Temperature-humidity-bias (THB) | 85°C, 85% RH*, $V_{dd}$ + 20%, 1000 h |
| High temperature operating life (HTOL) | 85°C, $V_{dd}$ + 20%, 1000 h |
| Thermal cycle | −55°C to +125°C, 1000 cycles |

Flip-chip attach is a good example of the mechanical stress problems that can occur during packaging. The substrate typically has a much higher coefficient of thermal expansion (CTE) than the silicon die. The die is joined to the substrate at the solder reflow temperature ($\sim 180°C$ for Pb-based eutectic solder, $\sim 220°C$ for Pb-free solder). During cooling back to room temperature, there is stress in the solder and in the chip (especially at the corners of the die) due to the thermal expansion mismatch between the die and the substrate [260] (Fig. 2.42). Cracks can form in the die or in the solder during chip joining or during subsequent thermal cycling, resulting in device failure. Historically, underfills have been used to reduce the stress in the solder bumps (Fig. 2.43) [257]. Underfills are epoxy-based materials that are typically dispensed between the die and the substrate after chip joining. With $SiO_2$ dielectrics, a high modulus underfill (>9 GPa) can be used, which minimizes the stress on the solder. However, if a high modulus underfill is used, the stress on the die is increased, and can crack low-k dielectrics. Hence, a lower modulus underfill must be used to avoid cracking the die. Note that if the modulus of the underfill is too low (5 GPa), then fails will occur in the solder. Hence, underfills with intermediate values of modulus must be used, to minimize stress in the die as well as in the solder bump [257].

The stress on the die in a flip-chip package is even higher when Pb-free are solders are used. Pb-free solders have a higher melting point and higher elastic modulus than Pb-based solders (Table 2.4). Hence, additional modifications are required to assure reliable device operation, including optimizing the solder composition, solder reflow conditions, and pad layout [261].

The modulus of SiCOH dielectrics is much lower than that of $SiO_2$ (Fig. 2.7) and the films are brittle. Hence, a number of design and process

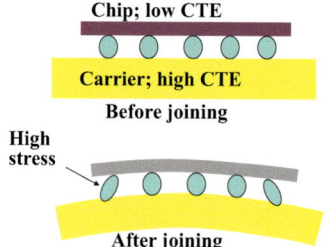

**Figure 2.42** Solder balls at the perimeter of the die experience high stress during cooling from the solder reflow, due to the CTE mismatch between the silicon and the substrate.

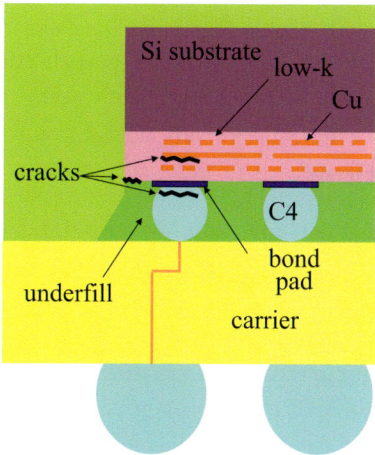

**Figure 2.43** Schematic of cracks that can form in the die or in the solder due to stress from the die attach process [26]. *T. Pan et al., IMAPS workshop, Dec. 2003 (www.kns.com). www.amkor.com.*

changes must be made to allow reliable packaging of these die. Design solutions include improved layout of the crack stop and edge seal [253,256], and the bond pads [259]. An example is the addition of dummy vias underneath bond pads to mechanically reinforce the dielectric stack. Packaging process changes include optimizing the dicing process (two-step dicing or laser dicing) [255,258,262] (Fig. 2.44), the underfill (lower modulus) [257], and the molding compound (lower coefficient of thermal expansion, CTE) [252]. In addition, $SiO_2$ is used as the interlevel dielectric (rather than a low-k material) for the last one or two metal levels, to provide increased mechanical strength underneath the bond pads [254,263].

The low density of SiCOH and porous SiCOH can cause problems during device operation as well. The diffusivity of $H_2O$ is very fast in both of these materials [263,264], and is a potential reliability problem during

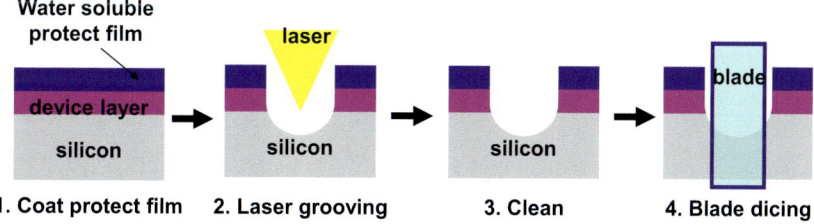

**Figure 2.44** Schematic of two-step dicing, using laser dicing to remove the low-k dielectric followed by conventional dicing to dice through the silicon [263].

device operation. To ensure reliability, each Cu layer is capped with hermetic barrier layers (such as SiN or SiCN) and an edge seal is used around the perimeter of the chip to block $H_2O$ diffusion.

## 5. Future directions

Interconnect scaling is already running into fundamental limits for resistivity and electromigration lifetime. So what can be done to enable continuing improvements circuit performance and density? There are three basic options; (1) incremental improvements in the existing interconnect technology, (2) 3D integration, and (3) new interconnect materials.

This chapter has focused on improvements in existing Cu interconnect technology (option 1). For the foreseeable future (i.e., down to the 2 nm node), the industry will continue using copper interconnect technology, with improvements in barrier layers, capping layers, and design to address problems associated with increasing resistivity and lower electromigration lifetime of sub-30 nm wide Cu interconnects. A recent publication shows Cu interconnects at all layers for the 4 nm node (Fig. 2.45) [192].

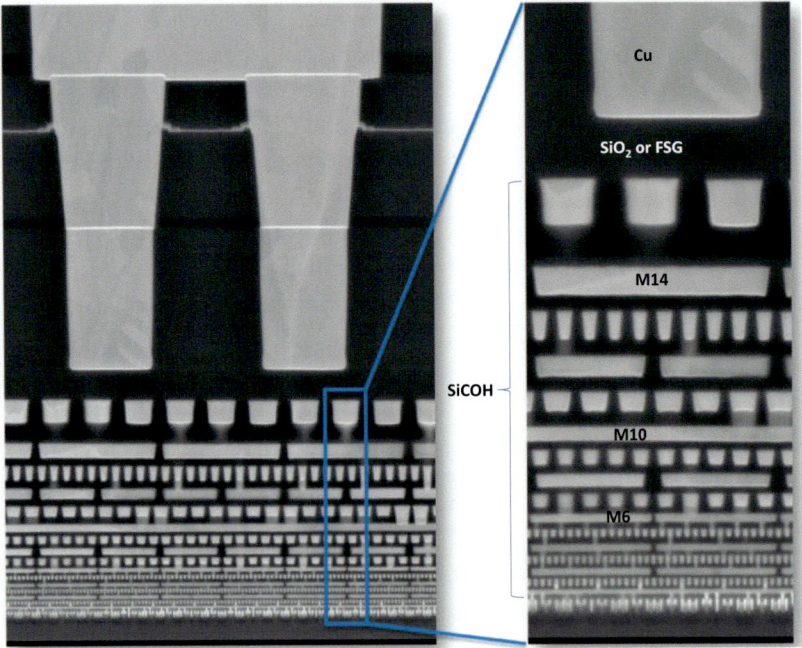

**Figure 2.45** BEOL structure at the 4 nm node. Note that Cu interconnects are used at all metal levels [192].

3D integration (option 2) has many potential benefits, including higher circuit densities and the heterogeneous integration. An additional benefit is reduced interconnect length, with short vertical connections replacing long horizontal connections. The shorter interconnect lengths provide higher speed communications, higher numbers of interconnections and lower power level communication links between circuits. Already, there are a number of products that use 3D integration such as Field Programmable Gate Arrays (FPGAs) on Si interposers [265], stacked CMOS image sensors [266,267] and stacked DRAM [268,269]. Copper technology plays an important role in 3D stacked die (Fig. 2.46), where vertical connections are typically made either with Cu-filled TSVs [17] or with Cu–Cu hybrid bonding [270–272].

A simpler implementation of 3D integration compared to die stacking is the use of buried (power rail) interconnects or backside interconnects (Fig. 2.47) [273–282]. Buried interconnects can be formed during the STI process (Fig. 2.48). The main challenge with buried interconnects is that the metal is exposed to high temperature ($\sim 1100°C$) anneals and oxidations, requiring the use of refractory metals such as W or Ru, and the use a capping layer that prevents oxidation of the metal. The benefit of the buried interconnects compared to conventional frontside interconnects is

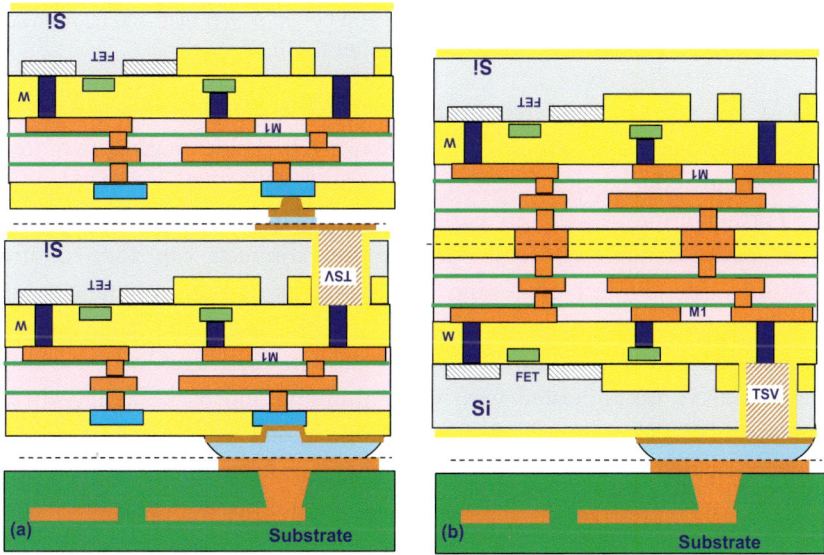

**Figure 2.46** Schematic of stacked die using (a) Cu filled TSVs and (b) Cu–Cu hybrid bonding [270].

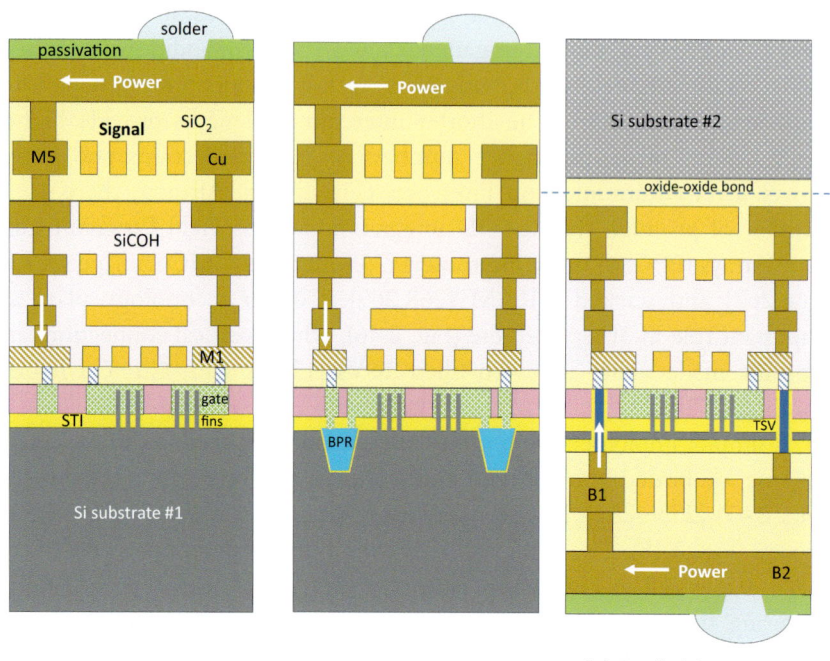

(a) Frontside Power    (b) Frontside Power + buried power rail.    (c) Backside Power

**Figure 2.47** Interconnect options for on-chip power delivery; (a) conventional frontside interconnects; (b) frontside interconnects and buried power rails; and (c) frontside and backside interconnects, with nano-TSVs (after [273]).

that a low resistance interconnect is available in close proximity to the devices for power distribution (without using frontside routing channels), resulting in reduced area ($\sim 20\%$) for the logic cells [281].

Further improvements in performance and cell area can be achieved with backside interconnects. Backside interconnects can formed after frontside interconnect processing (Fig. 2.49) by using permanent wafer bonding, wafer thinning (Si thickness $\sim 350$ nm), and "nano" TSVs (diameter $\sim 100$ nm) [280]. The main process challenges with backside interconnects are associated with wafer thinning and backside alignment. Additional challenges are higher cost and increased self-heating of the devices compared to conventional interconnects or buried interconnects (i.e., bulk Si is a relatively good thermal conductor and provides heat dissipation for devices formed on bulk Si). The advantages of backside interconnects compared to frontside interconnects are improved performance and reduced area ($\sim 30\%$) for the logic cells [281].

Process integration for on-chip interconnects

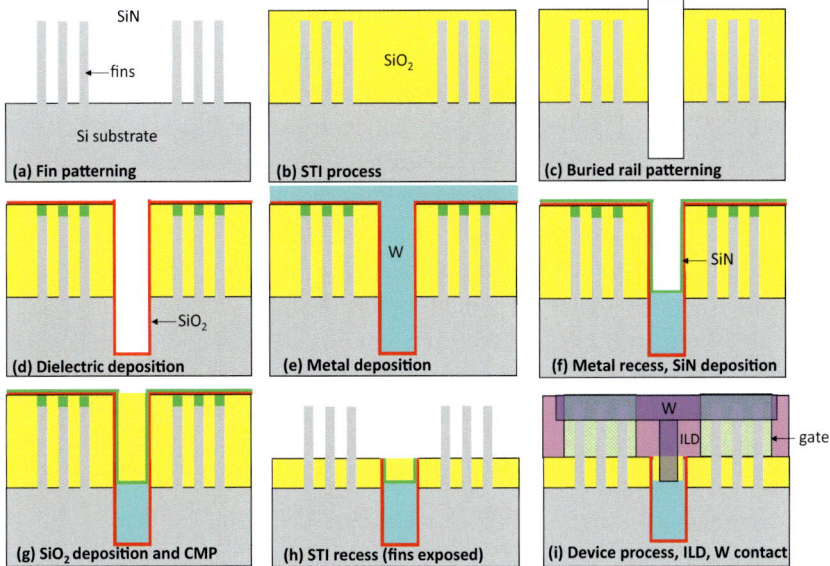

**Figure 2.48** Fabrication steps for buried power rail interconnects formed during the STI process. Note that a refractory metal, such as W, Ru, or Mo, must be used for the buried power rail in step (e) because of subsequent high temperature processes. In addition, an SiN cap layer is used on top of the buried power rail metal in step (f) to protect the metal from high temperature oxidations. The W shown in step (i) represents a connection from the buried interconnect to the source/drain contact (after [278]).

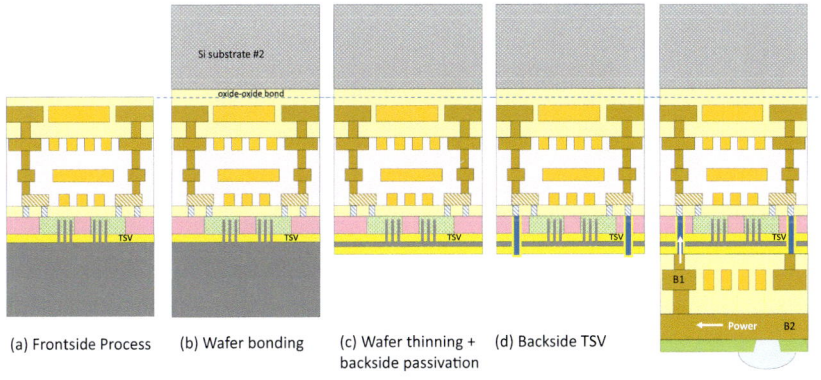

**Figure 2.49** Process flow for backside interconnects. Note that Cu can be used for backside interconnects B1 and B2 (after [280]).

New interconnect materials (Option3) include metals, graphene, and carbon nanotubes. Metals such as Ru, Co, W, and NiSi have a higher melting point than Cu and therefore a much greater electromigration lifetime [8,282a]. The resistivities of these metals are all much higher than that of Cu, but no barrier layer is required. If it is not possible to thin the barrier layer, then at small dimensions (line width <10 nm), these higher melting point metals could have a comparable interconnect resistance to that of Cu interconnects. A possible scenario for future devices is that the new interconnect materials would be used only for local wiring (line width <10 nm) and Cu interconnects would still be used for intermediate and global wiring (line width >10 nm).

The two most promising metals for new interconnect materials are Co and Ru. Co has the advantage in that it is a cheaper material than Ru. In addition, Co is already used for contacts and one or two interconnect levels by some manufacturers [119]. Ru has the advantage in that it can be patterned by RIE, and high resistivity barrier/adhesion layers are not required on the sidewalls (Fig. 2.50), resulting in a larger percentage of low resistivity material in the interconnect. An additional advantage is that air gaps are easy to implement with metal patterning by RIE.

Carbon nanotubes (CNTs) and graphene have a number of desirable properties for interconnects including improved reliability for electromigration and reduced resistance (Table 2.6) [283–287]. However, the ideal properties shown in Table 2.4 are difficult to achieve in real interconnect structures. Some of the issues with carbon nanotubes are (1) ensuring the CNTs are metallic rather than semiconducting, (2) achieving

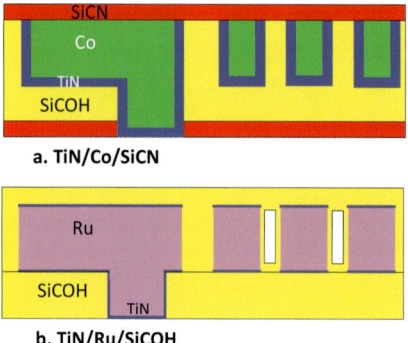

**Figure 2.50** (a) Cobalt interconnect compared to (b) Ruthenium interconnect. Ruthenium interconnects can be patterned by RIE, eliminating the need for high resistivity barrier/adhesion layers on the sidewalls.

**Table 2.6** Properties of carbon interconnect materials compared to Cu and W [283]. The mean free path for electron transport through the material gives a measure of the conductivity (longer mean free path → lower resistance).

| Property | W | Cu | MWCNT | Graphene |
|---|---|---|---|---|
| Maximum current density (A/cm$^2$) | $10^8$ | $>10^7$ | $>10^9$ | $>10^8$ |
| Melting point (°K) | 3695 | 1357 | 3800 (graphite) | |
| Mean free path (nm) at room temperature | 33 | 40 | $>10^4$ | $10^3$ |

high packaging density, and (3) growth of long, horizontal CNT bundles [285]. Hence, the work on CNTs for interconnects has focused on vias. For horizontal interconnects, graphene is more controllable than CNT, because of it has a two-dimensional structure and it can be more easily patterned. However, there are still many issues with using graphene for interconnects, including (1) growth or transfer of large area graphene layers onto a 300 mm device wafer, (2) line edge roughness (which can greatly reduce electron mobility), (3) need for multiple layers of graphene to reduce total resistance, and (4) high contact resistance of graphene [285,287,288]. It is unlikely that carbon-based interconnects will be ready for production in the foreseeable future (i.e., 3 nm node will still use Cu interconnects).

Many processes must be optimized to fabricate Cu interconnects with high yield and reliability. There are many challenges with implementation of Cu interconnects at the 2 nm node and beyond, including increased resistivity, integration with porous low-k materials, and reliability. During development, it is important to consider the design options in parallel with the materials options to provide the best overall solution for the product, i.e., Design Technology Co-Optimization (DTCO) [274,289–292]. Interconnect technology will continue to be an active area of research and development for advanced technology nodes.

## References

[1] S.E. Thompson, S. Parthasarathy, Moore's law: the future of Si microelectronics, Mater. Today 9 (2006) 20–25.
[2] National Science and Engineering Council, www.nfg.gov/crssprgm/nano/reports/omb_nifty50.jsp.
[3] R.D. Isaac, The future of CMOS Technology, IBM J. Res. Dev. 44 (2000) 369–378.
[4] M. Bohr, MOS Transistors: Scaling and Performance Trends, Semiconductor International, 1995, pp. 75–80.
[5] R.H. Dennard, F.H. Gaensslen, H.-N. Yu, V.L. Rideout, E. Bassous, A.R. LeBlanc, Design of ion-implanted MOSFET's with very small physical dimensions, IEEE J. Sol. State Circuits SC-9 (1974) 256–268.
[6] G. Schindler, W. Steinhogl, G. Steinlesberger, M. Traving, M. Engelhardt, Scaling of parasitics and delay times in backend-of-line, Microelectron. Eng. 70 (2003) 7–12.
[7] R. Ho, K.W. Mai, M.A. Horowitz, The future of wires, Proc. IEEE 89 (2001) 490–504.
[8] Z. Tokei, End of Cu roadmap and beyond Cu, in: IEEE Int. Tech. Conf. Proc, 2016.
[9] R.H. Havemann, J.A. Hutchby, High-performance interconnects: an integration overview, Proc. IEEE 89 (2001) 586–601.
[10] V. Huang, D. Shim, H. Simka, A. Naeemi, From interconnect materials and processes to chip level performance: modeling and design for conventional and exploratory concepts, in: IEEE International Electron Devices Meeting (IEDM), 2020, pp. 32.6.1–32.6.4.
[11] H.B. Bakoglu, Circuits, Interconnections, and Packaging for VLSI, Addison-Wesley, New York, 1990.
[12] D. Edelstein, J. Heidenreich, R. Goldblatt, W. Cote, C. Uzoh, N. Lustig, P. Roper, T. McDevitt, W. Motsiff, A. Simon, J. Dukovic, R. Wachnik, H. Rathore, R. Schulz, L. Su, S. Luce, J. Slattery, Full copper wiring in a sub-0.25μm CMOS ULSI technology, in: IEEE International Electron Device Meeting Proc, 1997, pp. 773–776.
[13] H.B. Lee, J.W. Hong, G.J. Seong, J.M. Lee, H. Park, J.M. Baek, K.I. Choi, B.L. Park, J.Y. Bae, G.H. Choi, S.T. Kim, U.I. Chung, J.T. Moon, J.H. Oh, J.H. Son, J.H. Jung, S. Hah, S.Y. Lee, A highly reliable Cu interconnect technology for memory devices, in: IEEE International Interconnect Technology Conf. Proc, 2007, pp. 64–66.
[14] A.K. Stamper, A.K. Chinthakindi, D.D. Coolbaugh, K. Downes, E.E. Eshun, M. Ertuk, R.A. Groves, P. Lindgren, Z.-X. He, V. Ramachandran, Advanced analog metal and passives integration, in: Proc. Advanced Metallization Conf. 2004, MRS, 2005, pp. 37–43.
[15] J. Gambino, J. Adkisson, T. Hoague, M. Jaffe, R. Leidy, R.J. Rassel, J. Kyan, D. McGrath, D. Sackett, Optimization of Cu interconnect layers for CMOS image sensor technology, in: Proc. Advanced Metallization Conf. 2005, MRS, 2006, pp. 151–157.
[16] T.R. Efland, C.-Y. Tsai, S. Pendharkar, Lateral thinking about power devices (LDMOS), in: IEEE Int. Electron Device Meeting. Proc, 1998, pp. 679–682.
[16a] T. Pinili, R. Manolo, A. Denoyo, B. Yabut, D. Moore, B. Cowell, J. Jenson, K. Truong, J. Gambino, R. Watkins, W. Qin, G. Brizar, J. De Clerq, Copper Wire Bond Optimization for Power Devices, 2018 IEEE International Symposium on the Physical and Failure Analysis of Integrated Circuits (IPFA), Singapore, 2018, pp. 1–5.
[17] J.P. Gambino, S.A. Adderly, J.U. Knickerbocker, An overview of through-silicon-via technology and manufacturing challenges, Microelectron. Eng. 135 (2015) 73–106.

[17a] Z. Tőkei, V. Vega, G. Murdoch, M. O'Toole, K. Croes, R. Baert, M. Van der Veen, C. Adelmann, J.P. Soulié, J. Boemmels, C. Wilson, S.H. Park, K. Sankaran, G. Pourtois, J. Sweerts, S. Paolillo, S. Decoster, M. Mao, F. Lazzarino, J. Versluijs, V. Blanco, M. Ercken, E. Kesters, Q.-T. Le, F. Holsteyns, N. Heylen, L. Teugels, K. Devriendt, H. Struyf, P. Morin, N. Jourdan, S. Van Elshocht, I. Ciofi, A. Gupta, H. Zahedmanesh, K. Vanstreels, M.H. Na, Inflection points in interconnect research and trends for 2nm and beyond in order to solve the RC bottleneck, in: IEEE International Electron Devices Meeting (IEDM), 2020, pp. 32.2.1–32.2.4.
[18] A. Grill, Low and ultralow dielectric constant films prepared by plasma-enhanced chemical vapor deposition, in: M. Baklonov, M. Green, K. Maex (Eds.), Dielectric Films for Advanced Microelectronics, John Wiley & Sons, 2007, pp. 1–32.
[19] A. Grill, Porous pSiCOH ultralow-k dielectrics for chip interconnects prepared by PECVD, Ann. Rev. Mater. Sci. 39 (2009) 49–69.
[20] G.A. Dixit, R.H. Havemann, Overview of interconnect – copper and low-k integration, in: R. Doering, Y. Nishi (Eds.), Handbook of Semiconductor Manufacturing Technology, second ed., CRC Press, N.Y., 2008. Chap. 2.
[21] J. Gambino, F. Chen, J. He, Copper interconnect technology for the 32nm node and beyond, in: IEEE Custom Integrated Circuits Conf. Proc, 2009, pp. 141–148.
[22] M. Quirk, J. Serda, in: Semiconductor Manufacturing Technology, Prentice-Hall, Upper Saddle, River, NJ, 2001.
[23] J.D. Plummer, M.D. Deal, P.B. Griffin, Silicon VLSI Technology, Prentice Hall, Upper Saddle Ridge, NJ, 2000.
[24] H.S. Landis, J.-T. Sucharitaves, Changing density requirements for semiconductor manufacturing, in: A.J. McKerrow, Y. Sacham-Diamand, S. Shingubara, Y. Shimogaki (Eds.), Advanced Metallization Conf. Proc. 2007, MRS, 2008, pp. 535–542.
[25] L.M. Matz, T. Tsui, E.R. Engbrecht, K. Taylor, G. Haase, S. Ajmera, R. Kuan, A. Griffin, R. Kraft, A.J. McKerrow, Structural characterization of silicon carbide dielectric barrier materials, in: S.H. Brongersma, T.C. Taylor, M. Tsujimura, K. Masu (Eds.), Advanced Metallization Conf. Proc. 2005, MRS, Warrendale, PA, 2006, pp. 437–443.
[26] J. Gambino, T.C. Lee, F. Chen, T.D. Sullivan, Reliability challenges for advanced copper interconnects: electromigration and time-dependent dielectric breakdown (TDDB), in: Proc. IEEE Int. Symp. on the Physical & Failure Analysis of Integrated Circuits, 2009, pp. 677–684.
[26a] F. Chen, J. Gill, D. Harmon, T. Sullivan, B. Li, A. Strong, H. Rathore, D. Edelstein, C.-C. Yang, A. Cowley, L. Clevenger, Measurements of effective thermal conductivity for advanced interconnect structures with various composite low-K dielectrics, IRPS Proc (2004) 68–73.
[27] J. Gambino, A. Stamper, T. McDevitt, V. McGahay, S. Luce, T. Pricer, B. Porth, C. Senowitz, R. Kontra, M. Gibson, H. Wildman, A. Piper, C. Benson, T. Standaert, P. Biolsi, E. Cooney, E. Webster, R. Wistrom, A. Winslow, E. White, Integration of copper with low-k dielectrics for 0.13 μm technology, in: Proc. IEEE Int. Symp. on the Physical & Failure Analysis of Integrated Circuits, 2002, pp. 111–117.
[28] M.J. Shapiro, S.V. Nguyen, T. Matsuda, D. Dobuzinsky, CVD of fluorosilicate glass for ULSI applications, Thin Solid Films 270 (1995) 503–507.
[29] S.W. Lim, Y. Shimogaki, Y. Nakano, K. Tada, H. Komiyama, Changes in the orientational polarization and structure of silicon dioxide film by fluorine addition, J. Electrochem. Soc. 146 (1999) 4196–4202.
[30] A. Grill, S.M. Gates, T.E. Ryan, S.V. Nguyen, D. Priyadarshini, Progress in the development and understanding of advanced low k and ultralow k dielectrics for very

large-scale integrated interconnects—state of the art, Appl. Phys. Rev. 1 (2014) 011306.
[30a] E.T. Ryan, D. Priyadarshini, S.M. Gates, H. Shobha, J. Chen, K. Virwani, A. Madan, E. Adams, E. Huang, E. Liniger, D. Collins, M. Stolfi, K.S. Yim, A. Demos, A. Grill, Optimizing ULK film properties to enable BEOL integration with TDDB reliability, in: IEEE Int. Tech. Conf. Proc, 2015, pp. 349–352.
[31] D. Priyadarshini, S.V. Nguyen, H. Shobha, E. Liniger, J.H.-C. Chen, H. Huang, S.A. Cohen, A. Grill, Advanced single precursor based pSiCOH k = 2.4 for ULSI interconnects, J. Vac. Sci. Technol. B35 (2017) 021201.
[32] S. Chikaki, K. Kinoshita, T. Nakayama, K. Kohmura, H. Tanaka, M. Hirakawa, E. Soda, Y. Seino, N. Hata, T. Kikkawa, S. Saito, 32 nm node ultralow-k (k=2.1)/Cu damascene multilevel interconnect using high-porosity (50%) high-modulus (9 GPa) self-assembled porous silica, in: IEEE Int. Electron Device Meeting Proc, 2007, pp. 969–972.
[33] F. Ito, T. Takeuchi, H. Yamamoto, T. Ohdaira, R. Suzuki, Y. Hayashi, Pore-connectivity dependence of moisture absorption into porous low-k films by positron-annihilation lifetime spectroscopy, in: A.J. McKerrow, Y. Sacham-Diamand, S. Shingubara, Y. Shimogaki (Eds.), Proc. Advanced Metallization Conf. 2007, MRS, Pittsburgh, PA, 2008, pp. 465–470.
[34] T. Seo, Y. Oka, K. Seo, K. Goto, H. Chibahara, H. Korogi, S. Suzuki, M. Hamada, N. Suzumura, K. Tsukamoto, A. Ueki, T. Furuhashi, D. Kodama, S. Kido, J. Izumitani, K. Tomita, E. Kobori, A. Ikeda, Y. Kawano, T. Ueda, Direct CMP process with advanced ELK for 45 nm half pitch interconnects, in: IEEE Int. Interconnect Technology Conf. Proc, 2010 paper 5.5.
[35] R. Daamen, P.H.L. Bancken, D.E. Badaroglu, J. Michelon, V.H. Nguyen, G.J.A.M. Verheijden, A. Humbert, J. Waeterloos, A. Yang, J.K. Cheng, L. Chen, T. Martens, R.J.O.M. Hoofman, Multi-level air gap integration for 32/22 nm nodes using a spin-on thermal degradable polymer and SiOC CVD hard mask, in: IEEE Int. Interconnect Technology Conf. Proc, 2007, pp. 61–63.
[36] J.P. Gueneau de Mussy, C. Bruynsereade, Z. Tokei, G.P. Beyer, K. Maex, Novel selective sidewall airgap process, in: IEEE Int. Interconnect Technology Conf. Proc, 2005, pp. 150–152.
[37] S. Nitta, S. Ponoth, G. Breyta, M. Colburn, L. Clevenger, D. Horak, M. Bhushan, J.E.S. Cohen, J. Colt, P. Flaitz, E. Fluhr, N. Fuller, A.E. Huang, C.K. Hu, K. Kumar, H. Landis, B. Li, W.-K. Li, E. Liniger, A. Lisi, X. Liu, J.R. Lloyd, I. Melville, J. Muncy, T. Nogami, V. Ramachandran, D.L. Rath, T. Standaert, J.-T. Sucharitaves, D. Tumbull, E. Crabbe, B. McCredie, M. Lane, S. Purushothaman, D. Edelstein, A multilevel copper/low-k/airgap BEOL technology, in: A.J. McKerrow, Y. Sacham-Diamand, S. Shingubara, Y. Shimogaki (Eds.), Advanced Metallization Conf. Proc. 2007, MRS, Pittsburgh, PA, 2008, pp. 329–336.
[38] J. Noguchi, K. Sato, N. Konishi, S. Uno, T. Oshima, K. Ishikawa, H. Ashihara, T. Saito, M. Kubo, T. Tamaru, Y. Yamada, H. Aoki, T. Fujiwara, Process and reliability of air-gap Cu interconnect using 90-nm node technology, IEEE Trans. Elec. Dev. 52 (2005) 352–359.
[39] K. Fischer, M. Agostinelli, C. Allen, D. Bahr, M. Bost, P. Charvat, V. Chikarmane, Q. Fu, C. Ganpule, M. Haran, M. Heckscher, H. Hiramatsu, E. Hwang, P. Jain, I. Jin, R. Kasim, S. Kosaraju, K.S. Lee, H. Liu, R. McFadden, S. Nigam, R. Patel, C. Pelto, P. Plekhanov, M. Prince, C. Puls, S. Rajamani, D. Rao, P. Reese, A. Rosenbaum, S. Sivakumar, B. Song, M. Uncuer, S. Williams, M. Yang, P. Yashar, S. Natarajan, Low-k interconnect stack with multi-layer air gap and tri-metal-insulator-metal capacitors for 14nm high volume manufacturing, in: IEEE Int. Tech. Conf. Proc, 2015, pp. 5–8.

[39a] K. Fischer, H.K. Chang, D. Ingerly, I. Jin, H. Kilambi, J. Longun, R. Patel, C. Pelto, C. Petersburg, P. Plekhanov, C. Puls, L. Rockford, I. Tsameret1, M. Uncuer, P. Yashar, Performance enhancement for 14nm high volume manufacturing microprocessor and system on a chip processes, in: IEEE Int. Tech. Conf. Proc, 2015, pp. 5—7.

[40] S. Deshpande, X. Shao, J. Lamb, N. Brakensiek, J. Johnson, X. Wu, G. Xu, B. Simmons, Advancements in organic anti-reflective coatings for dual damascene processes, in: N.T. Sullivan (Ed.), Metrology, Inspection, and Process Control for Microlithography XIV, 3998, SPIE Proc., 2000, pp. 797—805.

[41] J. Gambino, T. Stamper, H. Trombley, S. Luce, F. Allen, C. Weinstein, B. Reuter, M. Dunbar, V. Samek, P. McLaughlin, T. Kane, Dual damascene process for fat wires in copper/FSG technology, in: A.J. McKerrow, J. Leu, O. Kraft, T. Kikkawa (Eds.), Materials, Technology and Reliability for Advanced Interconnects and Low-K Dielectrics — 2003, 766, MRS Proc., 2003, pp. 71—76.

[42] J. Kriz, C. Angelkort, M. Czekalla, S. Huth, D. Meinhold, A. Pohl, S. Schulte, A. Thamm, S. Wallace, Overview of dual damascene integration schemes in Cu BEOL integration, Microelectron. Eng. 85 (2008) 2128—2132.

[43] M. Maenhoudt, D. Van Goidsenhoven, I. Pollentier, K. Ronse, M. Lepage, H. Struyf, M. Van Hove, Lithography aspects of dual damascene interconnect technology, in: C.A. Mack, T. Stevenson (Eds.), Lithography for Semiconductor Manufacturing II, 4404, SPIE Proc., 2001, pp. 1—13.

[44] S.-T. Chen, H. Tomizawa, K. Tsumura, M. Tagami, H. Shobha, M. Sankarapandian, O. Van der Straten, J. Kelly, D. Canaperi, T. Levin, S. Cohen, Y. Yin, D. Horak, M. Ishikawa, Y. Mignot, C.-S. Koay, S. Burns, S. Halle, H. Kato, G. Landie, Y. Xu, A. Scaduto, E. Mclellan, J.C. Arnold, M. Colburn, T. Usui, T. Spooner, 64 nm pitch Cu dual-damascene interconnects using pitch split double exposure patterning scheme, in: IEEE Int. Tech. Conf. Proc, 2011, pp. 1—3.

[45] J.S. Chawla, R. Chebiam, R. Akolkar, G. Allen, C.T. Carver, J.S. Clarke, F. Gstrein, M. Harmes, T. Indukuri, C. Jezewski, B. Krist, H. Lang, A. Myers, R. Schenker, K.J. Singh, R. Turkot, H.J. Yoo, Demonstration of a 12 nm-half-pitch copper ultralow-k interconnect process, in: IEEE Int. Tech. Conf. Proc, 2013, pp. 1—3.

[46] L. Liebmann, J. Zeng, X. Zhu, L. Yuan, G. Bouche, J. Kye, Overcoming scaling barriers through design technology cooptimization, VLSI Technology Symp (2016) 978—979.

[47] M. He, C. Ordonio, C.H. Low, P. Welti, G. Lobb, A. Clancy, J. Shu, A. Hamouda, J. Stephens, K. Shah, A. Chandrasekhar, M.C. Silvestre, P. Periasamy, A.S. Km Mahalingam, S. Pal, C. Child, 10nm local interconnect challenge with iso-dense loading and improvement with ALD spacer process, in: IEEE Int. Tech. Conf. Proc, 2016, pp. 15—17.

[48] S. Takei, T. Shinjo, Y. Sakaida, Study of high etch rate bottom antireflective coating and gap fill materials using dextrin derivatives in ArF lithography, Jpn. J. Appl. Phys. 46 (2007) 7279—7284.

[49] W. Cote, D. Edelstein, C. Bunke, P. Biolsi, W. Wille, H. Baks, R. Conti, T. Dalton, W.-K. Li, Y.-H. Lin, D. Restaino, T. Van Kleeck, S. Vogt, T. Houghton, S. Moskowitz, T. Ivers, Non-poisoning dual damascene patterning scheme for low-k and ultra low-k BEOL, in: S.W. Russell, M.E. Mills, A. Osaki, T. Yoda (Eds.), Proc. Advanced Metallization Conf. 2006, MRS, Pittsburgh, PA, 2007, pp. 289—294.

[50] S. Takei, Resist poisoning studies of gap fill materials for patterning metal trenches in via-first dual damascene process, Jpn. J. Appl. Phys. 47 (2008) 8766—8770.

[51] W. Jin, H.H. Sawin, Profile evolution simulation of oxide fencing during via-first dual damascene etching processes, J. Electrochem. Soc. 150 (2003) G711—G717.

[52] D.L. Kiel, B.A. Helmer, S. Lassig, Review of trench and via plasma etch issues for copper dual damascene in undoped and fluorine-doped silicate glass oxide, J. Vac. Sci. Technol. B 21 (2003) 1969–1985.
[53] R.F. Schnabel, D. Dobuzinsky, J. Gambino, K.P. Muller, F. Wang, D.C. Perng, H. Palm, Dry etch challenges of 0.25 μm dual damascene structures, Microelectron. Eng. 37/38 (1997) 59–65.
[54] X. Hua, X. Wang, D. Fuentevilla, G.S. Oehrlein, F.G. Celli, K.H.R. Kirmse, Study of $C_4F_8/N_2$ and $C_4F_8/Ar/N_2$ plasmas for highly selective organosilicate glass etching over $Si_3N_4$ and SiC, J. Vac. Sci. Technol. A 21 (2003) 1708–1716.
[55] L. Ling, X. Hua, X. Li, G.S. Oehrlein, F.G. Celli, K.H.R. Kirmse, P. Jiang, H.M. Anderson, Study of $C_4F_8/CO$ and $C_4F_8/Ar/CO$ plasmas for highly selective etching of organosilicate glass over $Si_3N_4$ and SiC, J. Vac. Sci. Technol. A 22 (2004) 236–244.
[56] M. Schaepkens, T.E.F.M. Standaert, N.R. Rueger, P.G.M. Sebel, G.S. Oehrlein, J.M. Cook, Study of the $SiO_2$-to-$Si_3N_4$ etch selectivity mechanism in inductively coupled fluorocarbon plasmas and a comparison with the $SiO_2$-to-Si mechanism, J. Vac. Sci. Technol. A 17 (1999) 26–37.
[57] T.E.F.M. Standaert, C. Hedlund, E.A. Joseph, G.S. Oehrlein, T.J. Dalton, Role of fluorocarbon film formation in etching of silicon, silicon dioxide, silicon nitride, and amorphous hydrogenated silicon carbide, J. Vac. Sci. Technol. A 22 (2004) 53–60.
[58] N. Negishi, H. Takesue, M. Sumiya, T. Yoshida, Y. Momonoi, M. Izawa, Deposition control for reduction of 193 nm photoresist degradation in dielectric etching, J. Vac. Sci. Technol. B 23 (2005) 217–223.
[59] D. Keil, B.A. Helmer, G. Mueller, E. Wagganer, Oxide dual damascene trench etch profile control, J. Electrochem. Soc. 148 (2001) G383–G388.
[60] R.A. Gottscho, C.W. Jurgensen, D.J. Vitkavage, Microscopic uniformity in plasma etching, J. Vac. Sci. Technol. B 10 (1992) 2133–2147.
[61] O. Joubert, G.S. Oehrlein, Y. Zhang, Fluorocarbon high density plasma. V. Influence of aspect ratio on the etch rate of silicon dioxide in an electron cyclotron resonance plasma", J. Vac. Sci. Technol. A 12 (1994) 658–664.
[62] A. Kojima, N. Nakamura, N. Matsunaga, H. Hayashi, K. Kubota, R. Asako, K. Maekawa, H. Shibata, T. Yoda, T. Ohiwa, Silylation gas restoration subsequent to all-in-one RIE process without air exposure for porous low-k SiOC/copper dual-damascene interconnects, in: S.W. Russell, M.E. Mills, A. Osaki, T. Yoda (Eds.), Proc. Advanced Metallization Conf. 2006, MRS, Pittsburgh, PA, 2007, pp. 301–305.
[63] S.M. Gates, A. Grill, C. Dimitrakopoulos, V. Patel, S.T. Chen, T. Spooner, E.T. Ryan, S.A. Cohen, E. Simonyi, E. Liniger, Y. Ostrovski, R. Bhatia, Integration compatible porous SiCOH dielectrics from 45 to 22 nm, in: M. Naik, R. Shaviv, T. Yoda, K. Ueno (Eds.), Proc. Advanced Metallization Conf. 2008, MRS, Pittsburgh, PA, 2009, pp. 531–536.
[64] Y. Hayashi, H. Ohtake, J. Kawahara, M. Tada, S. Saito, N. Inoue, F. Ito, M. Tagami, M. Ueki, N. Furutake, T. Takeuchi, H. Yamamoto, M. Abe, Comprehensive chemistry designs in porous SiOCH film stacks and plasma etching gases for damageless Cu interconnects in advanced ULSI devices, IEEE Trans. Semicond. Manuf. 21 (2008) 469–480.
[65] N. Inoue, N. Furutake, F. Ito, H. Yamamoto, T. Takeuchi, Y. Hayashi, Impact of barrier metal sputtering on physical and chemical damages in low-k SiOCH films with various hydrocarbon content, Jpn. J. Appl. Phys. 47 (2008) 2468–2472.
[66] T. Chevolleau, N. Posseme, T. David, R. Bouyssou, J. Ducote, F. Bailly, M. Darnon, M. El Kodadi, M. Besacier, C. Licitra, M. Guillermet, A. Ostrovsky, C. Vervore, O. Joubert, Etching process scalability and challenges for ULK materials, in: IEEE Int. Interconnect Technology Conf. Proc, 2010 paper 5.1.

[67] H. Shi, H. Huang, J. Im, P.S. Ho, Y. Zhou, J.T. Pender, M. Armacost, D. Kyser, Minimization of plasma ashing damage to OSG low-k dielectrics, in: IEEE Int. Interconnect Technology Conf. Proc, 2010 paper 8.12.
[68] M.R. Baklanov, A. Urbanowicz, G. Mannaert, S. Vanhaelemeersch, Low dielectric constant materials; challenges of plasma damage, in: Proc. 8th Int. Conf. Solid-State Integrated Circuits Technology, 2006, pp. 291–294.
[69] O.V. Braginsky, A.S. Kovalev, D.V. Lopaev, Y.A. Mankelevich, E.M. Malykhin, O.V. Proshina, T.V. Rakhimova, A.T. Rakhimova, A.N. Vasilieva, D.G. Voloshin, S.M. Zyryanov, M.R. Baklanov, Interaction of O and H Atoms with low-k SiCOH films pretreated in He plasma, in: M. Gall, A. Grill, F. Iacopi, J. Koike, T. Usui (Eds.), Materials, Processes and Reliability for Advanced Interconnects for Micro- and Nanoelectronics – 2009, 1156, MRS, Pittsburgh, PA, 2009 paper D01-06.
[70] V. Travaly, J. Van Aelst, V. Truffert, P. Verdonck, T. Dupont, E. Camerotto, O. Richard, H. Bender, C. Kroes, D. De Roest, G. Vereecke, M. Claes, Q.T. Le, E. Kesters, M. Van Cauwenberghe, J. Beynet, S. Kaneko, H. Struyf, M. Baklanov, K. Matsushita, N. Kobayashi, H. Sprey, G. Beyer, Key factors to sustain the extension of a MHM-based integration scheme to medium and high porosity PECVD low-k materials, in: IEEE Int. Interconnect Technology Conf. Proc, 2008, pp. 52–54.
[71] O. Hinsinger, R. Fox, E. Sabouret, C. Goldberg, C. Verove, W. Besling, P. Brun, E. Josse, C. Monget, O. Belmont, J. Van Hassel, B.G. Sharma, J.P. Jacquemin, P. Vannier, A. Humbert, D. Bunel, R. Gonella, E. Mastromatteo, D. Reber, A. Farcy, J. Mueller, P. Christie, V.H. Nguyen, C. Cregut, T. Berger, Demonstration of an extendable and industrial 300mm BEOL integration for the 65-nm technology node, in: IEEE Int. Electron Devices Meeting Proc, 2004, pp. 317–320.
[72] International Technology Roadmap for Semiconductors, Interconnect, 2011. http://www.itrs.net/.
[73] X. Hua, M. Kuo, G.S. Oehrlein, P. Lazzeri, E. Iacob, M. Anderle, C.K. Inoki, T.S. Kuan, P. Jiang, W. Wu, Damage of ultralow k materials during photoresist mask stripping process, J. Vac. Sci. Technol. B. 24 (2006) 1238–1247.
[74] P. Lazzeri, G.S. Oehrlein, G.J. Stueber, R. McGowan, E. Busch, S. Pederzoli, C. Jeynes, M. Bersani, M. Anderle, Interactions of photoresist stripping plasmas with nanoporous organo-silicate ultra low dielectric constant dielectrics, Thin Solid Films 516 (2008) 3697–3703.
[75] O. Louveau, C. Bourlot, A. Marfoure, I. Kalinovski, J. Su, G. Hills, D. Louis, Dry ashing process evaluation for ULK films, Microelectron. Eng 73–74 (2004) 351–356.
[76] M.-S. Kuo, A.R. Pal, G.S. Oehrlein, P. Lazzeri, M. Anderle, Mechanistic study of ultralow $k$-compatible carbon dioxide *in situ* photoresist ashing processes. I. Process performance and influence on ULK material modification, J. Vac. Sci. Technol. B. 28 (2010) 952–960.
[77] J. Lee, W.-J. Park, D.-H. Kim, J. Choi, K. Shin, I. Chung, Low-k film damage-resistant CO chemistry-based ash process for low-k/Cu interconnection in flash memory devices, Thin Solid Films 517 (2009) 3847–3849.
[78] K. Kinoshita, S. Chikaki, E. Soda, K. Tomioka, H. Tanaka, K. Kohmura, T. Nakayama, T. Kikkawa, S. Saito, A. Kojima, Process induced damages and recovery by silylation for low-k/Cu interconnects with highly-porous self-assembled silica film, in: A.J. McKerrow, Y. Sacham-Diamand, S. Shingubara, Y. Shimogaki (Eds.), Proc. Advanced Metallization Conf. 2007, MRS, Pittsburgh, PA, 2008, pp. 513–520.
[79] S.V. Nitta, S. Purushothaman, N. Chakrapani, O. Rodriguez, N. Kymko, E.T. Ryan, G. Bonilla, S. Cohen, S. Molis, K. McCullough, Use of diffunctional silylation agents for enhanced repair of post plasma damaged porous low k dielectrics, in: S.H. Brongersma,

T.C. Taylor, M. Tsujimura, K. Masu (Eds.), Proc. Advanced Metallization Conf. 2005, MRS, Pittsburgh, PA, 2006, pp. 325—331.
[80] H. Huang, K. Lionti, W. Volksen, T. Spooner, H. Shobha, J. Lee, J.H.-C. Chen, T. Magbitang, B. Peethala, E.G. Liniger, C.K. Hu, E. Huang, D.F. Canaperi, T.E. Standaert, D.C. Edelstein, A. Grill, G. Dubois, G. Bonilla, Post porosity plasma protection integration at 48 nm pitch, in: IEEE Int. Tech. Conf. Proc, 2016, pp. 153—155.
[81] S.-T. Chen, N.A. Lanzillo, S. Van Nguyen, T. Nogami, A.H. Simon, Interconnect processing: integration, dielectrics, metals, in: M. Rudan (Ed.), Springer Handbook of Semiconductor Devices, Springer Nature Switzerland AG, 2023. Chap. 5.
[82] D. Volger, The roadmap to 5nm, Semicond. Eng. (2015). https://semiengineering.com/the-roadmap-to-5nm/.
[83] P. Brun, F. Bailly, M. Guillermet, E. Aparico, N. Posseme, Plasma etch challenges at 14nm and beyond technology nodes in the BEOL, in: 2015 IEEE International Interconnect Technology Conference and 2015 IEEE Materials for Advanced Metallization Conference (IITC/MAM), Grenoble, France, 2015, pp. 21—24.
[84] S. Thibaut, A. Raley, N. Mohanty, S. Kal, E. Liu, A. Ko, D. O'Meara, K. Tapily, P. Biolsi, Self-aligned quadruple patterning using spacer on spacer integration optimization for N5, in: Proc. SPIE 10149, Advanced Etch Technology for Nanopatterning VI, 101490I, April 4, 2017, https://doi.org/10.1117/12.2258173.
[85] R.H. Havemann, G.A. Antonelli, G.K. Arendt, M. Danek, A.J. McKerrow, R.S. Weinberg, Copper BEOL solutions for advanced memory, Solid State Technol. (2009) 10—13, 31.
[86] J. Gambino, E. Cooney, S. Barkyoumb, J. Robbins, A. Rutkowski, A. Piper, M. Moon, C. Benson, E. Walton, C. Johnson, B. Laughlin, M. Gibson, J. Coffin, H. Wildman, Precleans for copper vias in and FSG process, in: A.J. McKerrow, Y. Shacham-Diamond, S. Zaima, T. Ohba (Eds.), Proc. Advanced Metallization Conf. 2001, MRS, Pittsburgh, PA, 2002, pp. 49—55.
[87] K. Ueno, V.M. Donnelly, T. Kikkawa, Cleaning of $CHF_3$ plasma-etched $SiO_2$/SiN/Cu via structures with dilute hydrofluoric acid solutions, J. Electrochem. Soc. 144 (1997) 2565—2572.
[88] G.B. Alers, R.T. Rozbicki, G.J. Harm, S.K. Kailasam, G.W. Ray, M. Danek, Barrier-first integration for improved reliability in copper dual damascene interconnects, in: IEEE Int. Interconnect Technology Conf. Proc, 2003, pp. 27—29.
[89] X. Fu, J. Forster, J. Yu, P. Gopalraja, A. Bhatnagar, S. Ahn, A. Demos, P. Ho, Advanced preclean for integration of PECVD SiCOH (k=2.5) dielectrics with copper metallization beyond 45nm technology, in: IEEE Int. Interconnect Technology Conf. Proc, 2006, pp. 51—53.
[90] R.P. Mandal, D. Cheung, W.-F. Yau, B. Cohen, S. Rengarajan, E. Chou, Comparison of $\kappa<3$ silicon oxide-based dielectric pre-copper metallization preclean processes using black diamond, in: IEEE/SEMI Advanced Semiconductor Manufacturing Conf. Proc, 1999, pp. 299—303.
[91] Z. Tokei, F. Lanckmans, G. Van den bosch, M. Van Hove, K. Maex, H. Bender, S. Hens, J. Van Landuyt, Reliability of copper dual damascene influenced by preclean, in: IEEE International Symp. on the Physical and Failure Analysis on Int. Circuits, 2002, pp. 118—123.
[92] M.R. Baklanov, D.G. Shamiryan, Z. Tokei, G.P. Beyer, T. Conard, S. Vanhaelemeersch, K. Maex, Characterization of Cu surface cleaning by hydrogen plasma, J. Vac. Sci. Technol. B 19 (2001) 1201—1211.
[93] D. Edelstein, C. Uzoh, C. Cabral Jr., P. DeHaven, P. Buchwalter, A. Simon, E. Cooney, S. Malhotra, D. Klaus, H. Rathore, B. Agarwala, D. Nguyen, An optimal liner for copper damascene interconnects, in: A.J. McKerrow, Y. Shacham-Diamond,

S. Zaima, T. Ohba (Eds.), Proc. Advanced Metallization Conf. 2001, 2002, pp. 541—547.

[94] E.C. Cooney, D.C. Strippe, J.W. Korejwa, A.H. Simon, C. Uzoh, Effects of collimator aspect ratio and deposition temperature on copper sputtered seed layers, J. Vac. Sci. Technol. A 17 (1999) 1898—1903.

[95] J. Reid, S. Mayer, E. Broadbent, E. Klawuhn, K. Ashtiani, Factors influencing damascene feature fill using copper PVD and electroplating, Solid State Technol. 43 (2000) 86—94.

[96] R.-H. Kim, B.H. Kim, T. Matsuda, J.N. Kim, J.M. Baek, J.J. Lee, J.O. Cha, J.H. Hwang, S.Y. Yoo, K.-M. Chung, K.H. Park, J.K. Choi, E.B. Lee, S.D. Nam, Y.W. Cho, H.J. Choi, J.S. Kim, S.Y. Jung, D.H. Lee, I.S. Kim, D.W. Park, H.B. Lee, S.H. Ahn, S.H. Park, M.-C. Kim, B.U. Yoon, S.S. Paak, N.-I. Lee, J.-H. Ku, J.S. Yoon, H.-K. Kang, E.S. Jung, Highly reliable Cu interconnect strategy for 10nm node logic technology and beyond, in: IEEE Int. Electron Device Meeting, 2014, pp. 768—771.

[97] K. Motoyama, O. van der Straten, H. Tomizawa, J. Maniscalco, S.T. Chen, Novel Cu reflow seed process for Cu/Low-k 64nm pitch dual damascene interconnects and beyond, in: IEEE International Interconnect Technology Conference (IITC), 2012, pp. 1—3.

[98] Z. Wu, F. Chen, G. Shen, Y. Hu, S. Pethe, J.J. Lee, J. Tseng, W. Suen, R. Vinnakota, K. Kashefizadeh, M. Naik, Pathfinding of Ru-liner/Cu-reflow interconnect reliability solution, in: IEEE International Interconnect Technology Conference (IITC), 2018, pp. 51—53.

[99] C.-C. Yang, F.R. McFeely, B. Li, R. Rosenberg, D. Edelstein, Low-temperature reflow anneals of Cu on Ru, IEEE Electron. Device Lett. 32 (2011) 806—808.

[100] J. Forster, P. Gopalraja, T.J. Gung, A. Sundarrajan, X. Fu, N. Hammond, J. Fu, U. Kelkar, A. Bhatnagar, A PVD based barrier technology for the 45 nm node, Microelec. Eng. 82 (2005) 594—599.

[101] S.M. Rossnagel, Physical vapor deposition, in: R. Doering, Y. Nishi (Eds.), Handbook of Semiconductor Manufacturing Technology, second ed., CRC Press, N.Y., 2008. Chap. 15.

[102] A.H. Simon, T. Bolom, T.J. Tang, B. Baker, C. Peters, B. Rhoads, P.L. Flaitz, S. Sankaran, S. Grunow, Extendability study of a PVD Cu seed process with $Ar^+$ Rf-plasma enhanced coverage for 45nm interconnects, Mater. Res. Soc. Proc. 1079 (2008) paper N03-04.

[103] N. Kumar, K. Moraes, M. Narasimhan, P. Gopalraja, Advanced Metallization Needs Copper, Semiconductor International, 2008, pp. 26—33.

[104] Y.K. Siew, N. Jourdan, Y. Barbarin, J. Machillot, S. Demuynck, K. Croes, J. Tseng, H. Ai, J. Tang, M. Naik, P. Wang, M. Narasimhan, M. Abraham, A. Cockburn, J. Bömmels, Z. Tőkei, CVD Mn-based self-formed barrier for advanced interconnect technology, in: IEEE Int. Tech. Conf. Proc, 2013, pp. 1—3.

[105] Y.K. Siew, N. Jourdan, I. Ciofi, K. Croes, C. Wilson, B. Tang, S. Demuynck, Z. Wu, H. Ai, D. Cellier, A. Cockburn, J. Bommels, Z. Tokei, Cu wire resistance improvement using Mn-based self-formed barriers, in: IEEE Int. Tech. Conf. Proc, 2014, pp. 311—313.

[106] W. Wu, H.J. Wu, G. Dixit, R. Shaviv, M. Gao, T. Mountsier, G. Harm, A. Dulkin, N. Fuchigami, S.K. Kailasam, E. Klawuhn, R.H. Havemann, Ti-Based barrier for Cu interconnect applications, in: Proc. IEEE Int. Interconnect Technology Conf, 2008, pp. 202—204.

[107] D. Henry, F. Jacquet, M. Neyret, X. Baillin, T. Enot, V. Lapras, C. Brunet-Manquat, J. Charbonnier, B. Aventurier, N. Sillon, Through silicon vias technology for CMOS image sensors packaging, in: IEEE Elec. Comp. Tech. Conf, 2008, pp. 556—562.

[108] F. Battegay, M. Fourel, Barrier material selection for TSV last, flipchip & 3D – UBM & RDL integrations, in: IEEE Elec. Comp. Tech. Conf, 2015, pp. 1183–1192.
[109] H.Y. Huang, C.H. Hsieh, S.M. Jeng, H.J. Tao, M. Cao, Y.J. Mii, A new enhancement layer to improve copper performance, in: IEEE Interconnect Technology Conf. Proc, 2010 paper 4.2.
[110] J. Rullan, T. Ishizaka, F. Cerio, S. Mizuno, Y. Mizusawa, T. Ponnuswamy, J. Reid, A. McKerrow, C.-C. Yang, Low resistance wiring and 2×nm void free fill with CVD Ruthenium liner and DirectSeed copper, in: IEEE Interconnect Technology Conf. Proc, 2010 paper 8.5.
[111] M. Tagami, N. Furutake, S. Saito, Y. Hayashi, Highly-reliable low-resistance Cu interconnects with PVD-Ru/Ti barrier metal toward automotive LSIs, in: IEEE Interconnect Technology Conf. Proc, 2008, pp. 205–207.
[112] C.-C. Yang, S. Cohen, T. Shaw, P.-C. Wang, T. Nogami, D. Edelstein, Characterization of ultrathin-Cu/Ru(Ta)/TaN liner stack for copper interconnects, IEEE Elec. Dev. Lett. 31 (2010) 722–724.
[113] T. Standaert, G. Beique, H.-C. Chen, S.-T. Chen, B. Hamieh, J. Lee, P. McLaughlin, J. McMahon, Y. Mignot, F. Mont, K. Motoyama, S. Nguyen, R. Patlolla, B. Peethala, D. Priyadarshini, M. Rizzolo, N. Saulnier, H. Shobha, S. Siddiqui1, T. Spooner, H. Tang, O. van der Straten, E. Verduijn, Y. Xu, X. Zhang1, J. Arnold, D. Canaperi, M. Colburn, D. Edelstein, V. Paruchuri, G. Bonilla, BEOL process integration for the 7 nm technology node, in: IEEE Int. Tech. Conf. Proc, 2016, pp. 2–4.
[114] T. Nogami, M. He, X. Zhang, K. Tanwar, R. Patlolla, J. Kelly, D. Rath, M. Krishnan, X. Lin, O. Straten, H. Shobha, J. Li, A. Madan, P. Flaitz, C. Parks, C.-K. Hu, C. Penny, A. Simon, T. Bolom, J. Maniscalco, D. Canaperi, T. Spooner, D. Edelstein, CVD-Co/Cu(Mn) integration and reliability for 10 nm node, in: IEEE Int. Tech. Conf. Proc, 2013, pp. 1–3.
[115] M.H. van der Veen, O. Varela Pedreira, N. Jourdan, S. Park, H. Struyf, Z. Tőkei, C. Leal Cerantes, F. Chen, X. Xie, Z. Wu, A. Jansen, J. Machillot, A. Cockburn, Low resistance Cu vias for 24nm pitch and beyond, in: 2022 IEEE International Interconnect Technology Conference (IITC), San Jose, CA, USA, 2022, pp. 129–131, https://doi.org/10.1109/IITC52079.2022.988.
[116] T. Nogami, R. Patlolla, J. Kelly, B. Briggs, H. Huang, J. Demarest, J. Li, R. Hengstebeck, X. Zhang, G. Lian, B. Peethala, P. Bhosale, J. Maniscalco, H. Shobha, S. Nguyen, P. McLaughlin, T. Standaert, D. Canaperi, D. Edelstein, V. Paruchuri, Cobalt/copper composite interconnects for line resistance reduction in both fine and wide lines, in: 2017 IEEE International Interconnect Technology Conference (IITC), Hsinchu, Taiwan, 2017, pp. 1–3, https://doi.org/10.1109/IITC-AMC.2017.7968961.
[117] I. Ciofi, P.J. Roussel, R. Baert, A. Contino, A. Gupta, K. Croes, C.J. Wilson, D. Mocuta, Z. Tokei, RC benefits of advanced metallization options, IEEE Trans. Electron. Dev. 66 (2019) 2339–2345.
[118] H.W. Kim, Recent trends in copper metallization, Electronics 11 (2022) 2914.
[119] P. Singer, Intel 4 Process Drops Cobalt Interconnect, Goes with Tried and Tested Copper with Cobalt Liner/Cap, Semiconductor Digest, 2022.
[120] G. Murdoch, Z. Tokei, S. Paolillo, O.V. Pedreira, K. Vanstreels, C.J. Wilson, Semidamascene interconnects for 2nm node and beyond, in: 2020 IEEE International Interconnect Technology Conference (IITC), San Jose, CA, USA, 2020, pp. 4–6, https://doi.org/10.1109/IITC47697.2020.9515597.
[121] R. Baert, I. Ciofi, S. Patli, O. Zografos, S. Sarkar, B. Chehab, D. Jang, A. Spessot, J. Ryckaert, Z. Tokei, Interconnect design-technology co-optimization for sub-3nm

technology node, in: IEEE International Interconnect Technology Conference (IITC), 2020, pp. 28−30.

[122] C. Penny, K. Motoyama, S. Ghosh, T. Bae, N. Lanzillo, S. Sieg, J. Lee, C. Park, L. Zou, H. Lee, D. Metzler, J. Lee, S. Cho, M. Shoudy, S. Nguyen, A. Simon, K. Park, L. Clevenger, B. Anderson, C. Child, T. Yamashita, J. Arnold, T. Wu, T. Spooner, K. Choi, K.-I. Seo, D. Guo, Subtractive Ru interconnect enabled by novel patterning solution for EUV double patterning and top via with embedded airgap integration for post Cu interconnect scaling, International Electron Devices Meeting (IEDM) (2022) 12.1.1−12.1.4.

[123] A. Furuya, K. Yoneda, E. Soda, T. Yoshie, H. Okamura, M. Shimada, N. Ohtsuka, S. Ogawa, Ultrathin pore-seal film by plasma enhanced chemical vapor deposition SiCH from tetramethylsilane, J. Vac. Sci. Technol. B23 (2005) 2522−2525.

[124] M. Tada, T. Tamura, F. Ito, H. Ohtake, M. Narihiro, M. Tagami, M. Ueki, K. Hijioka, M. Abe, N. Inoue, T. Takeuchi, S. Saito, T. Onodera, N. Furutake, K. Arai, M. Sekine, M. Suzuki, Y. Hayashi, Robust porous SiOCH/Cu interconnects with ultrathin sidewall protection liners, IEEE Trans. Elec. Dev. 53 (2006) 1169−1179.

[125] M. Gallitre, L.G. Gosset, A. Farcy, B. Blampey, R. Gras, C. Bermond, B. Flechet, J. Torres, Performance prediction of prospective air gap architectures for the 22 nm node, in: IEEE Int. Interconnect Technology Conf. Proc, 2007, pp. 132−134.

[126] M. Hamada, K. Ohmori, K. Mori, E. Kobori, N. Suzumura, R. Etou, K. Maekawa, M. Fujisawa, H. Miyatake, A. Ikeda, Highly reliable 45-nm-half-pitch Cu interconnects incorporating a Ti/TaN multilayer barrier, in: IEEE Int. Interconnect Technology Conf. Proc, 2010, p. 13.4.

[127] A. Haneda, T. Tabira, H. Sakai, H. Kudo, M. Sunayama, Ohtsuka, A. Tsukune, N. Shimizu, Self-restored barrier using Cu-Mn alloy, in: A.J. McKerrow, Y. Sacham-Diamand, S. Shingubara, Y. Shimogaki (Eds.), Proc. Advanced Metallization Conf. 2007, MRS, Pittsburgh, PA, 2008, pp. 59−65.

[128] A.H. Simon, F. Baumann, T. Bolom, J.G. Park, C. Child, B. Kim, P. DeHaven, R. Davis, O. Ogunsola, M. Angal, Effect of TaN stoichiometry on barrier oxidation and defect density in 32nm Cu/Ultra-Low K interconnects, in: J.W. Bartha, C.L. Borst, S. DeNardis, H. Kim, A. Naeemi, A. Nelson, S.S. Papa Rao, H.W. Ro, D. Toma (Eds.), Advanced Interconnects and Chemical Mechanical Planarization for Micro- and Nanoelectronics, 1249, MRS., Pittsburgh, PA, 2010, pp. F01−F02.

[129] J. Reid, A. McKerrow, S. Varadarajan, G. Kozlowski, Copper electroplating approaches for 16nm technology, Solid State Technol. 53 (2010).

[129a] A. Keigler, Z. Liu, J. Chiu, J. Drexler, Sematech 3D Equipment Challenges: 300mm Copper Plating, 2008.

[130] J. Reid, Damascene copper electroplating, in: R. Doering, Y. Nishi (Eds.), Handbook of Semiconductor Manufacturing Technology, second ed., CRC Press, N.Y., 2008. Chap. 16.

[131] R. Carpio, A. Jaworski, Review—Management of copper damascene plating, J. Electrochem. Soc. 166 (2019) D3072−D3096.

[132] T. Ritzdorf, Monitoring and control, in: M. Schlesinger, M. Paunovic (Eds.), Modern Electroplating, fifth ed., John Wiley & Sons, Inc., 2010, pp. 527−554.

[133] S. Dasilva, T. Mourier, P.H. Haumesser, M. Cordeau, K. Haxaire, G. Passemard, E. Chainet, Gap fill enhancement with medium acid electrolyte for the 45nm node and below, in: S.H. Brongersma, T.C. Taylor, M. Tsujimura, K. Masu (Eds.), Proc. Advanced Metallization Conf. 2005, MRS, Pittsburgh, PA, 2006, pp. 513−517.

[134] C. Witt, J. Srinivasana, X. Lina, R. Carpio, Effect of electrolyte acidity on copper plating process performance, ECS Trans. 2 (2007) 107−115.

[135] A. Keigler, Z. Liu, J. Chiu, Optimized TSV Filling Processes Reduce Costs, Semiconductor International, 2009.
[136] W. Steinhogl, G. Schindler, G. Steinlesberger, M. Traving, M. Engelhardt, Comprehensive study of the resistivity of copper wires with lateral dimensions of 100 nm and smaller, J. Appl. Phys. 97 (2005), 0237061 − 0237067.
[137] K. Namba, T. Ishigami, M. Enomoto, S. Kondo, H. Shinriki, D. Jeong, A. Shimizu, N. Saitoh, W.-M. Li, S. Yamamoto, T. Kawasaki, T. Nakada, N. Kobayashi, PEALD of Ru layer on WNC ALD barrier for Cu/porous low-k integration, in: S. W Russell, M.E. Mills, A. Osaki, T. Yoda (Eds.), Proc. Advanced Metallization Conf. 2006, MRS, Pittsburgh, PA, 2006, pp. 269−274.
[138] T. Usui, H. Nasu, J. Koike, M. Wada, S. Takahashi, N. Shimizu, T. Nishikawa, M. Yoshimura, H. Shibata, Low resistive and highly reliable Cu dual-damascene interconnect technology using self-formed $MnSi_xO_y$ barrier layer, in: IEEE Int. Interconnect Technology Conf. Proc, 2005, pp. 188−190.
[139] M.H. van der Veen, K. Vandersmissen, D. Dictus, S. Demuynck, R. Liu, X. Bin, P. Nalla, A. Lesniewska, L. Hall, K. Croes, L. Zhao, J. Bömmels, A. Kolics, Z. Tökei, Cobalt bottom-up contact and via prefill enabling advanced logic and DRAM technologies, in: IEEE Int. Tech. Conf. Proc, 2015, pp. 25−27.
[140] L. Wen, F. Yamashita, B. Tang, K. Croes, S. Tahara, K. Shimoda, T. Maeshiro, E. Nishimura, F. Lazzarino, I. Ciofi, J. Bömmels, Z. Tökei, Direct etched Cu characterization for advanced interconnects, in: IEEE Int. Tech. Conf. Proc, 2015, pp. 173−175.
[141] R. Sarvari, A. Naeemi, R. Venkatesan, J.D. Meindl, Impact of size effects on the resistivity of copper wires and consequently the design and performance of metal interconnect networks, in: IEEE Int. Interconnect Technology Conf. Proc, 2005, pp. 197−199.
[142] M. Tsujimura, Chemical mechanical polishing (CMP) removal rate uniformity and role of carrier parameters, in: Advances in Chemical Mechanical Planarization (CMP), Elsevier, 2016. Chap. 16.
[143] Y. Kamigata, Y. Kurata, K. Masuda, J. Amanokura, M. Yoshida, M. Hanazono, Why abrasive free Cu slurry is promising? Mater. Res. Soc. Proc. 671 (2001) paper M1.3.
[144] M. Mellier, T. Berger, R. Duru, M. Zaleski, M.C. Luche, M. Rivoire, C. Goldberg, G. Wyborn, K.-L. Chang, Y. Wang, V. Ripoche, S. Tsai, M. Thothadri, W.-Y. Hsu, L. Chen, Full copper electrochemical mechanical planarization (Ecmp) as a technology enabler for the 45 and 32nm nodes, in: IEEE Int. Interconnect Technology Conf. (IITC), 2007, pp. 70−72.
[145] J.-Y. Lai, N. Saka, J.-H. Chun, Evolution of copper-oxide damascene structures in chemical mechanical polishing. II. Copper dishing and oxide erosion, J. Electrochem. Soc. 149 (2002) G41−G50.
[146] T. Kanki, T. Shirasu, S. Takesako, M. Sakamoto, A.A. Asneil, N. Idani, T. Kimura, T. Nakamura, M. Miyajima, On the elements of high Throughput Cu-CMP slurries compatible with low step heights, in: Proc. IEEE Int. Interconnect Technology Conf, 2008, pp. 79−81.
[147] M. Krishnan, M.F. Lofaro, Copper chemical mechanical planarization (Cu CMP) challenges in 22 nm back-end-of-line (BEOL) and beyond, in: Advances in Chemical Mechanical Planarization (CMP), Elsevier, 2016. Chap. 2.
[148] D. Lee, H. Lee, H. Jeong, Slurry components in metal chemical mechanical planarization (CMP) process: a review, Int. J. Precis. Eng. Manuf. 17 (2016) 1751−1762.
[149] H. Lee, H. Kim, H. Jeong, Approaches to sustainability in chemical mechanical polishing (CMP): a review, Int. J. Precis. Eng. Manuf. Green Technol. 9 (2022) 349−367.

[150] S.-S. Yun, Y.-H. Son, G.- P Jeong, J.-H. Lee, J.-H. Jeong, J.-Y. Bae, S.-I. Kim, J.-H. Park, J.-G. Park, Dishing-free chemical mechanical planarization for copper films, Colloids Surf. A Physicochem. Eng. Asp. 616 (2021) 12614.

[151] S. Armini, C.M. Whelan, M. Moinpour, K. Maex, Copper CMP with composite polymer core—silica shell abrasives: a defectivity study, J. Electrochem. Soc. 156 (2009) H18—H26.

[152] J. Seo, U. Paik, Preparation and characterization of slurry for chemical mechanical planarization (CMP), in: Advances in Chemical Mechanical Planarization (CMP), Elsevier, 2016. Chap. 11.

[153] S. Tamilmani, W. Huang, S. Raghavan, R. Small, Potential-pH diagrams of interest to chemical mechanical planarization of copper, J. Electrochem. Soc. 149 (2002) G638—G642.

[154] W.-T. Tseng, Approaches to defect characterization, mitigation, and reduction, in: Advances in Chemical Mechanical Planarization (CMP), Elsevier, 2016. Chap. 17.

[155] C. Gabrielli, L. Beitone, C. Mace, E. Ostermann, H. Perrot, Electrochemistry on microcircuits. II: Copper dendrites in oxalic acid, Microelectron. Eng. 85 (2008) 1686—1698.

[156] K. Pate, P. Safier, Chemical metrology methods for CMP quality, in: Advances in Chemical Mechanical Planarization (CMP), Elsevier, 2016. Chap. 12.

[157] L.L. Chapelon, H. Chaabouni, G. Imbert, P. Brun, M. Mellier, K. Hamioud, M. Vilmay, A. Farcy, J. Torres, Dense SiOC cap for damage-less ultra low k integration with direct CMP in C45 architecture and beyond, Microelectron. Eng 85 (2008) 2098—2101.

[158] N. Heylen, E. Camerotto, H. Volders, Y. Travaly, G. Vereecke, G.P. Beyer, Z. Tokei, CMP process optimization for improved compatibility with advanced metal liners, in: IEEE Int. Interconnect Technology Conf. Proc, 2010, pp. 17—19.

[159] S. Kondo, B.U. Yoon, S. Tokitoh, K. Misawa, S. Sone, H.J. Shin, N. Ohashi, N. Kobayashi, Low-pressure CMP for 300-mm ultra low-k (k=1.6-1.8)/Cu integration, in: IEEE Int. Electron Devices Meeting Proc, 2004, pp. 151—154.

[160] D. Oshida, T. Takewaki, M. Iguchi, T. Taiji, T. Morita, Y. Tsuchiya, S. Yokogawa, H. Kunishima, H. Aizama, N. Okada, Quantitative analysis of correlation between insulator surface copper contamination and TDDB lifetime based on actual measurement, in: IEEE Int. Interconnect Technology Conf. Proc, 2008, pp. 222—224.

[161] M. Ueki, T. Onodera, A. Ishikawa, S. Hoshino, Y. Hayashi, Defectless Monolithic low-k/Cu interconnects produced by chemically controlled chemical mechanical polishing process with in situ end-point-detection technique, Jpn. J. Appl. Phys. 49 (2010) paper 04C029.

[162] S. Gall, C. Euvard, S. Chhun, S. Maitrejean, M. Assous, P.-H. Haumesser, M. Rivoire, Investigation of ULK (k=2.5) damage by direct CMP process for C45 technology node, in: A.J. McKerrow, Y. Sacham-Diamand, S. Shingubara, Y. Shimogaki (Eds.), Proc. Advanced Metallization Conf. 2007, MRS, Pittsburgh, PA, 2008, pp. 115—120.

[163] T.-S. Kim, T. Konno, T. Yamanaka, R.H. Dauskardt, Quantitative roadmap for optimizing CMP of ultra-low-k dielectrics, in: IEEE Int. Interconnect Technology Conf. Proc, 2008, pp. 171—173.

[164] M. Kodera, T. Takahashi, G. Mimamihaba, Evaluation of dielectric constant through direct chemical mechanical planarization of porous low-$k$ film, Jpn. J. Appl. Phys. 49 (2010) paper 04DB07.

[165] J. Nalaskowski, S.S. Papa Rao, Ultra low-k materials and chemical mechanical planarization (CMP), in: Advances in Chemical Mechanical Planarization (CMP), Elsevier, 2016. Chap. 4.

[166] J. Bian, Surfactants in controlling removal rates and selectivity in barrier slurry for Cu CMP, in: G. Zwicker, C. Borst, L. Economikos, A. Philipossian (Eds.), Advances and Challenges in Chemical Mechanical, 991, MRS., Pittsburgh, PA, 2009 paper C09-03.
[167] S. Rader, CMP Users Group, October 2002.
[168] M. Ohring, Reliability and Failure Analysis of Electronic Materials and Devices, Academic Press, NY, 1998.
[169] I.A. Blech, C. Herring, Stress generation by electromigration, Appl. Phys. Lett. 29 (1976) 131–133.
[170] C. Christiansen, B. Li, J. Gill, Blech effect and lifetime projection for Cu/low-k interconnects, in: IEEE Int. Interconnect Technology Conf. Proc, 2008, pp. 114–116.
[171] F.L. Wei, C.L. Gan, T.L. Tan, C.S. Hau-Riege, A.P. Marathe, J.J. Vlassak, C.V. Thompson, Electromigration-induced extrusion failures in Cu/low-k interconnects, J. Appl. Phys. 104 (2008) paper 023529.
[172] E.T. Ogawa, K.-D. Lee, V.A. Blaschke, P.S. Ho, Electromigration reliability issues in dual-damascene Cu interconnections, IEEE Trans. Rel. 51 (2002) 403–419.
[173] J.R. Black, Electromigration failure Modes in Aluminum metallization for. Semiconductor devices, Proc. IEEE 57 (1969) 1587–1594.
[174] J.C. Blair, P.B. Ghate, C.T. Haywood, Concerning electromigration in thin films, Proc. IEEE 59 (1971) 1023–1024.
[175] J.R. Lloyd, Black's law revisited - nucleation and growth in electromigration failure, Microelec. Rel. 47 (2007) 1468–1472.
[176] Z.-S. Choi, R. Monig, C.V. Thompson, Effects of microstructure on the formation, shape, and motion of voids during electromigration in passivated copper interconnects, J. Mater. Res. 23 (2008) 383–391.
[177] A.V. Vairagar, S.G. Mhaisalkar, K.N. Tu, A.M. Gusak, M.A. Meyer, E. Zschech, In situ observation of electromigration-induced void migration in dual-damascene Cu interconnect structures, Appl. Phys. Lett. 85 (2004) 2502–2504.
[178] H. Zahedmanesh, O. Varela Pedreira, C. Wilson, Z. Tőkei, K. Croes, Copper Electromigration; Prediction of Scaling Limits, IITC/MAM, 2019.
[179] C.-K. Hu, L. Gignac, B. Baker, E. Liniger, R. Yu, Impact of Cu microstructure on electromigration reliability, in: IEEE Int. Interconnect Technology Conf. Proc, 2007, pp. 93–95.
[180] A.S. Oates, Strategies to ensure electromigration reliability of Cu/low-k interconnects at 10 nm, ECS J. Solid State Sci. Technol. 4 (2015) N3168–N3176.
[181] B. Li, C. Christiansen, D. Badami, C.-C. Yang, Electromigration challenges for advanced on-chip Cu interconnects, Microelectron. Reliab. 54 (2014) 712–724.
[182] C.-K. Hu, L. Gignac, R. Rosenberg, Electromigration of Cu/low dielectric constant interconnects, Microelec. Rel. 46 (2006) 213–231.
[183] K. Mosig, V. Blaschke, Electromigration reliability of Cu/spin-on porous ultra low-k interconnects, in: A.J. McKerrow, Y. Shacham-Diamond, S. Zaima, T. Ohba (Eds.), AMC Proc, MRS, Pittsburgh, PA, 2001, pp. 427–432, 2002.
[184] S.P. Hau-Riege, C. Thompson, The effects of the mechanical properties of the confinement material on electromigration in metallic interconnects, J. Mater. Res. 15 (2000) 1797–1802.
[185] K.-D. Lee, X. Lu, E.T. Ogawa, H. Matsuhashi, P.S. Ho, Electromigration study of Cu/low K dual-damascene interconnects, in: IRPS Proc, 2002, pp. 322–326.
[186] M.W. Lane, E.G. Liniger, J.R. Lloyd, Relationship between interfacial adhesion and electromigration in Cu metallization, J. Appl. Phys. 93 (2003) 1417–1421.
[187] L.G. Gosset, S. Chhun, J. Guillan, R. Gras, J. Flake, R. Daamen, J. Michelon, P.-H. Haumesser, S. Olivier, T. Decorps, J. Torres, Self aligned barrier approach: overview on process, module integration, and interconnect performance

improvement challenges, in: IEEE Int. Interconnect Technology Conf. Proc, 2006, pp. 84−86.

[188] A.K. Stamper, H. Baks, E. Cooney, L. Gignac, J. Gill, C.-K. Hu, T. Kane, E. Liniger, Y.-Y. Wang, J. Wynne, Damascene copper integration impact on electromigration and stress migration, in: S.H. Brongersma, T.C. Taylor, M. Tsujimura, K. Masu (Eds.), Proc. Advanced Metallization Conf, MRS, Pittsburgh, PA, 2005, pp. 727−733, 2006.

[189] C.K. Hu, L. Gignac, R. Rosenberg, E. Liniger, J. Rubino, C. Sambucetti, A. Stamper, A. Domenicucci, X. Chen, Reduced Cu interface diffusion by CoWP surface coating, Microelec. Rel. 70 (2003) 406−411.

[190] H.K. Jung, H.-B. Lee, M. Tsukasa, E. Jung, J.-H. Yun, J.M. Lee, G.-H. Choi, S. Choi, C. Chung, Formation of Highly Reliable Cu/Low-K Interconnects by Using CVD Co Barrier in Dual Damascene Structures, IRPS, 2011, pp. 307−311.

[191] A.H. Simon, T. Bolom, C. Niu, F.H. Baumann, C.-K. Hu, C. Parks, J. Nag, H. Kim, J.Y. Lee, C.-C. Yang, S. Nguyen, H.K. Shobha, T. Nogami, S. Guggilla, J. Ren, D. Sabens, J.F. Buchon, Electromigration comparison of selective CVD cobalt capping with PVD Ta(N) and CVD cobalt liners on 22nm-groundrule dual-damascene Cu interconnects, IRPS (2013) 3F.4.1−3F.4.6.

[192] B. Sell, S. An, J. Armstrong, D. Bahr, B. Bains, R. Bambery, K. Bang, D. Basu, S. Bendapudi, D. Bergstrom, R. Bhandavat, S. Bhowmick, M. Buehler, D. Caselli, S. Cekli, V.R.S.K. Chaganti, Y.J. Chang, K. Chikkadi, T. Chu, T. Crimmins, G. Darby, C. Ege, P. Elfick, T. Elko-Hansen, S. Fang, C. Gaddam, M. Ghoneim, H. Gomez, S. Govindaraju, Z. Guo, W. Hafez, M. Haran, M. Hattendorf, S. Hu, A. Jain, S. Jaloviar, M. Jang, J. Kameswaran, V. Kapinus, A. Kennedy, S. Klopcic, D. Krishnan, J. Leib, Y.-T. Lin, N. Lindert, G. Liu, O. Loh, Y. Luo, S. Mani, M. Mleczko, S. Mocherla, P. Packan, M. Paik, A. Paliwal, R. Pandey, K. Patankar, L. Pipes, P. Plekhanov, C. Prasad, M. Prince, G. Ramalingam, R. Ramaswamy, J. Riley, J.R. Sanchez Perez, J. Sandford, A. Sathe, F. Shah, H. Shim, S. Subramanian, S. Tandon, M. Tanniru, D. Thakurta, T. Troeger, X. Wang, C. Ward, A. Welsh, S. Wickramaratne, J. Wnuk, S.Q. Xu, P. Yashar, J. Yaung, K. Yoon, N. Young, Intel 4 CMOS technology featuring advanced FinFET transistors optimized for high density and high-performance computing, in: Symp. On VLSI Technology & Circuits Digest of Technical Papers, 2022, pp. 282−283.

[193] B. Li, C. Christiansen, J. Gill, T. Sullivan, E. Yashchin, R. Filippi, Threshold electromigration failure time and its statistics for Cu interconnects, J. Appl. Phys. 100 (2006) 114516.

[194] D. Gan, B. Li, P.S. Ho, Stress-induced void formation in passivated Cu films, in: Materials, Technology and Reliability of Advanced Interconnects − 2005, MRS Proc, 863, 2005, pp. 259−264.

[195] E.T. Ogawa, J.W. McPherson, J.A. Rosal, K.J. Dickerson, T.-C. Chiu, L.Y. Tsung, M.K. Jain, T.D. Bonifield, J.C. Ondrusek, Stress-induced voiding under vias connected to wide Cu metal leads, in: IEEE Int. Rel. Phys. Symp. Proc, 2002, pp. 312−321.

[196] K. Maekawa, K. Mori, K. Kobayashi, N. Kumar, S. Chu, S. Chen, G. Lai, D. Diehl, M. Yoneda, Improvement in reliability of Cu dual-damascene interconnects using Cu-Al alloy seed, in: D. Erb, P. Ramm, K. Masu, A. Osaki (Eds.), AMC Proc, MRS, Warrendale, PA, 2004, pp. 221−226.

[197] S. Yokogawa, H. Tsuchiya, Effects of Al doping on the electromigration performance of damascene Cu interconnects, J. Appl. Phys. 101 (2007) 013513.

[198] A. Isobayashi, Y. Enomoto, H. Yamada, S. Takahashi, S. Kadomura, Thermally robust Cu interconnects with Cu-Ag alloy for sub 45nm node, in: IEEE Int. Electron Device Meeting Proc, 2004, pp. 953−956.

[199] J. Koike, M. Haneda, J. Iijima, M. Wada, Cu alloy metallization for self-forming barrier process, in: IEEE Int. Interconnect Technology Conf. Proc, 2006, pp. 161–163.
[200] Y. Ohoka, Y. Ohba, A. Isobayashi, T. Hayashi, N. Komai, S. Arakawa, R. Kanamura, S. Kadomura, Integration of high performance and low cost Cu/ultra low-k SiOC(k=2.0) interconnects with self-formed barrier technology for 32 nm-node and beyond, in: IEEE Int. Interconnect Technology Conf. Proc, 2007, pp. 67–69.
[201] T. Usui, K. Tsumura, H. Nasu, Y. Hayashi, G. Minamihaba, H. Toyoda, H. Sawada, S. Ito, H. Miyajima, K. Watanabe, H. Shimada, A. Kojima, Y. Uozima, H. Shibata, High performance ultra low-k (k=2.0/$k_{eff}$=2.4)/Cu dual-damascene interconnect technology with self-formed $MnSi_xO_y$ barrier layer for 32 nm-node, in: IEEE Int. Interconnect Technology Conf. Proc, 2006, pp. 216–218.
[202] M. Hauschildt, C. Hennesthal, G. Talut, O. Aubel, M. Gall, K.B. Yeap, E. Zschech, Electromigration early failure void nucleation and growth phenomena in Cu and Cu(Mn) interconnects, IEEE Int. Rel. Phys. Symp. (2013) 2C1.1–2C1.6.
[203] J. Gambino, T.C. Lee, D. Meatyard, S. Mongeon, B. Li, F. Chen, The effect of post-metallization annealing on the reliability of copper interconnects, in: Int. Semicond. Tech. Conf. Proc, 2008, pp. 13–19.
[204] J.-M. Paik, I.-M. Park, Y.-C. Joo, K.-C. Park, Linewidth dependence of grain structure and stress in damascene Cu lines, J. Appl. Phys. 99 (2006) 024509.
[205] T.D. Sullivan, Stress-induced voiding in microelectronic metallization: void growth models and refinements, Ann. Rev. Mater. Sci. 26 (1996) 333–364.
[206] J. De Messemaeker, O.V. Pedreira, B. Vandevelde, H. Philipsen, I. De Wolf, E. Beyne, K. Croes, Impact of post-plating anneal and through-silicon via dimensions on Cu pumping, in: IEEE Electronic Components and Technology Conference (ECTC), 2013, pp. 586–591.
[207] J. An, K.-J. Moon, S. Lee, D.-S. Lee, K. Yun, B.-L. Park, H.-J. Lee, J. Sue, Y.-L. Park, G. Choi, H.-K. Kang, C. Chung, Annealing process and structural considerations in controlling extrusion-type defects Cu TSV, in: IEEE International Interconnect Technology Conference (IITC), 2012, pp. 1–3.
[208] J.P. Gambino, T.C. Lee, F. Chen, T.D. Sullivan, Reliability of copper interconnects: stress-induced voids, Electrochem. Soc. Trans. 18 (2009) 205–211.
[209] A.H. Fischer, A. von Glasow, S. Penka, F. Ungar, Process optimization – the key to obtain highly reliable Cu interconnects, in: IEEE Int. Interconnect Technology Conf. Proc, 2003, pp. 253–255.
[210] P.A. Flinn, S. Lee, J. Doan, T.N. Marieb, J.C. Bravman, M. Madden, Void phenomena in passivated metal lines: recent observations and interpretation, in: H. Okabayashi, S. Shingubara, P.S. Ho (Eds.), Stress Induced Phenomena in Metallization, 1998, pp. 250–261.
[211] R.J. Gleixner, B.M. Clemens, W.D. Nix, Void nucleation in passivated interconnect lines: effects of site geometries, interfaces, and interface flaws, J. Mater. Res. 12 (1997) 2081–2090.
[212] C.J. Zhai, H.W. Yao, P.R. Besser, A. Marathe, R.C. Blish II, D. Erb, C. Hau-Riege, Sidharth, K.O. Taylor, Stress modelling of Cu/low-k BEOL – appliations to stress migration, in: Int. Rel. Phys. Symp. Proc, 2004, pp. 234–239.
[213] C.S. Hau-Riege, S.P. Hau-Riege, A.P. Marathe, The effect of interlevel dielectric on the critical tensile stress to void nucleation for the reliability of Cu interconnects, J. Appl. Phys. 96 (2004) 5792–5796.
[214] K. Arita, N. Ito, N. Hosoi, H. Miyamoto, Development of a two-step electroplating process with a long-term stability for applying to Cu metallization of 0.1 μm generation Logic ULSIs, in: IEEE Semicond. Manu. Symp, 2001, pp. 155–158.

[215] T.M. Shaw, L. Gignac, X.-H. Liu, R.R. Rosenberg, E. Levine, P. McLaughlin, P.-C. Wang, S. Greco, G. Biery, Stress voiding in wide copper lines, in: S.P. Baker, M.A. Korhonen, E. Arzt, P.S. Ho (Eds.), Stress-Induced Phenomena in Metallization, 2002, pp. 177–183.

[216] K.Y.Y. Doong, R.C.J. Wang, S.C. Lin, L.J. Hung, S.Y. Lee, C.C. Chiu, D. Su, K. Wu, K.L. Young, Y.K. Peng, Stress-induced voiding and its geometry dependency characterization, in: IEEE Int. Rel. Phys. Symp. Proc, 2003, pp. 156–160.

[217] S. Orain, A. Fuchsmann, V. Fiori, X. Federspiel, Reliability issues in Cu/low-k structures regarding the initiation of stress-voiding or crack failure, in: Proc. EuroSime, 2006, pp. 1–6.

[218] T. Oshima, K. Hinode, H. Yamaguchi, H. Aoki, K. Torii, T. Saito, K. Ishikawa, J. Noguchi, M. Fukui, T. Nakamura, S. Uno, K. Tsugane, J. Murata, K. Kikushima, H. Sekisaka, E. Murakami, K. Okuyama, T. Iwasaki, Suppression of stress-induced voiding in copper interconnects, in: IEEE Int. Electron Device Meeting Proc, 2002, pp. 757–760.

[219] K. McCullen, Redundant via insertion in restricted topology layouts, in: Proc. of the 8th International Symposium on Quality Electronic Design, 2007, pp. 821–828.

[220] K. Yoshida, T. Fujimaki, K. Miyamoto, T. Honma, H. Kaneko, H. Nakazawa, M. Morita, Stress-induced voiding phenomena for an actual CMOS LSI interconnects, in: IEEE Int. Electron Device Meeting Proc, 2002, pp. 753–756.

[221] S. Arakawa, I. Mizuno, Y. Ohoka, K. Nagahata, K. Tabuchi, R. Kanamura, S. Kadomura, Breakthrough integration of 32nm-node Cu/ultra low-k SiOC (k=2.0) interconnects by using advanced pore-sealing and low-k hard mask technologies, in: IEEE Int. Interconnect Technology Conf. Proc, 2006, pp. 210–212.

[222] F. Chen, B. Li, T. Lee, C. Christiansen, J. Gill, M. Angyal, M. Shinosky, C. Burke, W. Hasting, R. Austin, T. Sullivan, D. Badami, J. Aitken, Technology reliability qualification of a 65nm CMOS Cu/low-k BEOL interconnect, in: IEEE Int. Symp. On the Physical & Failure Analysis of Integrated Circuits, 2006, pp. 97–105.

[223] Y.-C. Huang, X. Lin, B. Zheng, C.S. Ngai, V. Paneccasio, J. Behnke, C. Witt, J. Dukovic, A. Rosenfeld, High performance copper plating process for 65nm and 45nm technology nodes, in: S.H. Brongersma, T.C. Taylor, M. Tsujimura, K. Masu (Eds.), Advanced Metallization Conf. Proc. 2005, MRS, Warrendale, PA, 2006, pp. 507–511.

[224] D. Gan, P.S. Ho, Y. Pang, R. Huang, J. Leu, J. Maiz, T. Scherban, Effect of passivation on stress relaxation in electroplated copper films, J. Mater. Res. 21 (2006) 1512–1518.

[225] T. Ishigami, T. Kurokawa, Y. Kakuhara, B. Withers, J. Jacobs, A. Kolics, I. Ivanov, M. Sekine, K. Ueno, High reliability Cu interconnection utilizing a low contamination CoWP capping layer, in: IEEE Int. Interconnect Technology Conf. Proc, 2004, pp. 75–77.

[226] M. Kimura, Oxide breakdown mechanism and quantum physical chemistry for time-dependent dielectric breakdown, in: IEEE Int. Rel Phys. Symp. Proc, 1997, pp. 190–200.

[227] E.T. Ogawa, J. Kim, G.S. Haase, H.C. Mogul, J.W. McPherson, Leakage, breakdown, and TDDB characteristics of porous low-k silica-based interconnect dielectrics, in: IEEE Int. Rel. Phys. Symp. Proc, 2003, pp. 166–172.

[228] J. Noguchi, N. Ohashi, T. Jimbo, H. Yamaguchi, K. Takeda, K. Hinode, Effect of $NH_3$-plasma treatment and CMP modification on TDDB improvement in Cu metallization, IEEE Trans. Elec. Dev. 48 (2001) 1340–1345.

[229] S.-C. Lee, A.S. Oates, Reliability limitations to the scaling of porous low-K dielectrics, IEEE Int. Rel. Phys. Symp. (IRPS) (2011) 155–159.

[230] J. Noguchi, N. Miura, M. Kubo, T. Tamaru, H. Yamaguchi, N. Hamada, K. Makabe, R. Tsuneda, K. Takeda, Cu-ion-migration phenomena and its influence on TDDB lifetime in Cu metallization, in: IEEE Int. Rel. Phys. Symp. Proc, 2003, pp. 287–292.
[231] Z. Tokei, V. Sutcliffe, S. Demuynck, F. Iacopi, P. Roussel, G.P. Beyer, R.J.O.M. Hoofman, K. Maex, Impact of the barrier/dielectric interface quality on reliability of Cu porous-low-k interconnects, in: IEEE Int. Rel. Phys. Symp. Proc, 2004, pp. 326–332.
[232] F. Chen, J.R. Lloyd, K. Chanda, R. Achanta, O. Bravo, A. Strong, P.S. McLaughlin, M. Shinosky, S. Sankaran, E. Gabraselesie, A.K. Stamper, Z.X. He, Line edge roughness and spacing effect on low-k TDDB characteristics, in: IEEE Int. Rel. Phys. Symp. Proc, 2008, pp. 132–137.
[233] K. Ueno, A. Kameyama, A. Matsumoto, M. Iguchi, T. Takewaki, D. Oshida, H. Toyoshima, N. Kawahara, S. Asada, M. Suzuki, N. Oda, Time-dependent dielectric breakdown characterization of 90- and 65-nm-node Cu/SiOC interconnects with via plugs, Jpn. J. Appl. Phys. 46 (2007) 1444–1451.
[234] S.-C. Lee, A.S. Oates, A new methodology for copper/low-K dielectric reliability prediction, IEEE Int. Rel. Phys. Symp. (IRPS) (2014) 3A.3.1–3A.3.7.
[235] F. Chen, O. Bravo, K. Chanda, P. McLaughlin, T. Sullivan, J. Gill, J. Lloyd, R. Kontra, J. Aitken, A comprehensive study of low-k SiCOH TDDB phenomena and its reliability lifetime model development, in: IEEE Int. Rel. Phys. Symp. Proc, 2006, pp. 46–53.
[236] G.S. Haase, E.T. Ogawa, J.W. McPherson, Reliability analysis method for low-k interconnect dielectrics breakdown in integrated circuits, J. Appl. Phys. 98 (2005) 034503.
[237] K.-Y. Yiang, H.W. Yao, A. Marathe, TDDB kinetics and their relationship with the E- and ÖE-models, in: IEEE Int. Interconnect Technology Conf. Proc, 2008, pp. 168–170.
[238] F. Chen, K. Chanda, J. Gill, M. Angyal, J. Demarest, T. Sullivan, R. Kontra, M. Shinosky, J. Li, L. Economikos, M. Hoinkis, S. Lane, D. McHerron, M. Inohara, S. Boettcher, D. Dunn, M. Fukasawa, B.C. Zhang, K. Ida, T. Ema, G. Lembach, K. Kumar, Y. Lin, H. Maynard, K. Urata, T. Bolom, K. Inoue, J. Smith, Y. Ishikawa, M. Naujok, P. Ong, A. Sakamoto, D. Hunt, J. Aitken, Investigation of CVD SiCOH low-k time-dependent dielectric breakdown at 65nm node technology, IEEE Int. Rel. Phys. Symp. Proc. (2005) 501–507.
[239] F. Chen, P. McLaughlin, J. Gambino, E. Wu, J. Demarest, D. Meatyard, M. Shinosky, The effect of metal area and line spacing on TDDB characteristics of 45nm low-k SiCOH dielectrics, IEEE Int. Rel. Phys. Symp. Proc. (2007) 382–389.
[240] G.S. Haase, J.W. McPherson, Modeling of interconnect dielectric lifetime under stress conditions and new extrapolation methodologies for time-dependent dielectric breakdown, in: IEEE Int. Rel. Phys. Symp. Proc, 2007, pp. 390–398.
[241] W.R. Hunter, The analysis of oxide reliability data, Int. Rel. Workshop Final Report (1998) 114–134.
[242] J.W. McPherson, H.C. Mogul, Underlying physics of the thermochemical E model in describing low-field time-dependent dielectric breakdown in $SiO_2$ thin films, J. Appl. Phys. 84 (1998) 1513–1523.
[243] K.Y. Yiang, Q. Guo, W.J. Woo, A. Krishnamoorthy, Study of leakage mechanisms of the copper/black diamond(TM) damascene process, Thin Solid Films 462–463 (2004) 330–333.
[244] Y. Li, Z. Tokei, T. Mandrekar, B. Mebarki, G. Groeseneken, K. Maex, Barrier integrity effect on leakage mechanism and dielectric reliability of copper/OSG

interconnects, in: Materials, Technology and Reliability of Advanced Interconnects — 2005, MRS Proc, 863, 2005, pp. 265—270.
[245] N. Suzumura, S. Yamamoto, D. Kodama, K. Makabe, J. Komori, E. Murakami, S. Maegawa, K. Kubota, A new TDDB degradation model based on Cu ion drift in Cu interconnect dielectrics, in: IEEE Int. Rel. Phys. Symp. Proc, 2006, pp. 484—489.
[246] S.M. Sze, Physics of Semiconductor Devices, John Wiley & Sons, Inc., NY, 1981.
[247] J.R. Lloyd, E. Liniger, T.M. Shaw, Simple model for time-dependent dielectric breakdown in inter- and intralevel low-k dielectrics, J. Appl. Phys. 98 (2005) 084109.
[248] J. Kim, E.T. Ogawa, J.W. McPherson, Time dependent dielectric breakdown characteristics of low-k dielectric (SiOC) over a wide range of test areas and electric fields, IEEE Int. Rel. Phys. Symp. Proc. (2007) 399—404.
[249] T.K.S. Wong, Time dependent dielectric breakdown in copper low-k interconnects: mechanisms and reliability models, Materials 5 (2012) 1602—1625.
[250] C. Muzzy, D. Danovitch, H. Gagnon, R. Hannon, E. Kinser, P.V. McLaughlin, G. Mongeau, J.-G. Quintal, J. Sylvestre, E. Turcotte, J. Wright, Chip package interaction evaluation for a high performance 65nm and 45nm CMOS technology in a stacked die package with C4 and wirebond interconnections, in: Electronic Components and Tech. Conf. Proc, 2008, pp. 1472—1475.
[251] W. Landers, D. Edelstein, L. Clevenger, S. Das, C.-C. Yang, T. Aoki, F. Beaulieu, J. Casey, A. Cowley, M. Cullinan, T. Daubenspeck, C. Davis, J. Demarest, E. Duchesne, L. Guerin, D. Hawkin, T. Ivers, M. Lane, X. Liu, T. Lombardi, C. McCarthy, C. Muzzy, J. Nadeau-Filteau, D. Questad, W. Sauter, T. Shaw, J. Wright, Chip-to-package interaction for a 90 nm Cu/PECVD low-k technology, in: IEEE Int. Tech. Conf. Proc, 2004, pp. 108—110.
[252] C. Goldberg, S. Downey, V. Fiori, R. Fox, K. Hess, O. Hinsinger, A. Humbert, J.-P. Jacquemin, S. Lee, J.-B. Lhuillier, S. Orain, S. Pozder, L. Proenca, F. Querica, E. Sabouret, T.A. Tran, T. Uehling, Integration of a mechanically reliable 65-nm node technology for low-k and ULK interconnects with various substrates and package types, in: IEEE Int. Tech. Conf. Proc, 2005, pp. 3—5.
[253] T.C. Huang, C.T. Peng, C.H. Yao, C.H. Huang, S.Y. Li, M.S. Liang, Y.C. Wang, W.K. Wan, K.C. Lin, C.C. Hsia, M.-S. Liang, Evaluation and numerical simulation of optimal structural designs for reliable packaging of ultra low k process technology, in: IEEE Int. Tech. Conf. Proc, 2006, pp. 92—94.
[254] J.W. Jang, C.Y. Liu, P.G. Kim, K.N. Tu, A.K. Mal, D.R. Frear, Interfacial morphology and shear deformation of flip chip solder joints, J. Mater. Res. 15 (2000) 1679—1687.
[255] J. Li, H. Hwang, E.-C. Ahn, Q. Chen, P. Kim, T. Lee, M. Chung, T. Chung, Laser dicing and subsequent die strength enhancement technologies for ultra-thin wafer, in: Proc. IEEE Elec. Comp. Tech. Conf, 2007, pp. 761—766.
[256] M. Saran, R. Cox, C. Martin, G. Ryan, T. Kudoh, M. Kanasugi, J. Hortaleza, M. Ibnabdeljalil, M. Murtuza, D. Capistrano, R. Roderos, R. Macaraeg, Elimination of bond-pad damage through structural reinforcement of intermetal dielectrics, in: IEEE Int. Rel. Phys. Symp. Proc, 1998, pp. 225—231.
[257] M. Tagami, H. Ohtake, M. Abe, F. Ito, T. Takeuchi, K. Ohto, T. Usami, M. Suzuki, T. Suzuki, N. Sashida, Y. Hayashi, Comprehensive process design for low-cost chip packaging with circuit-under-pad (CUP) structure in porous-SiCOH film, in: IEEE Int. Tech. Conf. Proc, 2005, pp. 12—14.
[258] P.-H. Tsao, C. Huang, M.-J. Lii, B. Su, N.-S. Tsai, Underfill characteristics for low-k dielectric/Cu interconnect IC flip-chip package reliability, in: Proc. IEEE Elec. Comp. Tech. Conf, 2004, pp. 767—769.

[259] W. ZhiJie, S. Wang, J.H. Wang, S. Lee, Y. SuYing, R. Han, Y.Q. Su, 300mm low k wafer dicing saw study, in: Proc. IEEE Conf. Electronic Packaging Tech, 2005, pp. 262–268.
[260] R.A. Susko, T.H. Daubenspeck, T.A. Wassick, T.D. Sullivan, W. Sauter, J. Cincotta, Solder bump electromigration and CPI challenges in low-k devices, Electrochem. Soc. Trans. 16 (2009) 51–60.
[261] S.M. Sullivan, Current evolution of wafer thinning and dicing, Electrochem. Soc. Trans. 18 (2009) 745–750.
[262] T. Furusawa, K. Goto, J. Izumitani, M. Matsuura, M. Fujisawa, N. Kawanabe, T. Hirose, E. Hayashi, S. Baba, Y. Asano, T. Ichiki, Y. Takata, in: IEEE Int. Interconnect Technology Conf. Proc, 2010, p. 9.2.
[263] T.M. Shaw, D. Jimerson, D. Haders, C.E. Murray, A. Grill, D.C. Edelstein, D. Chidambarrao, Moisture and oxygen uptake in low-k/copper interconnect structures, in: G.W. Ray, T. Smy, T. Ohta, M. Tsujimura (Eds.), Advanced Metallization Conf. Proc. 2003, MRS, Warrendale, PA, 2004, pp. 77–84.
[264] L.M. Matz, T. Tsui, E. R. Engbrecht, K. Taylor, G. Haase, S. Ajmera, R. Kuan, A. Griffin, R. Kraft, A.J. McKerrow, in: S.H. Brongersma, T.C. Taylor, M. Tsujimura, K. Masu (Eds.), Proc. Advanced Metallization Conf. 2005, MRS, Warrendale, PA, 2006, pp. 437–443.
[265] B. Banijamali, S. Ramalingam, K. Nagarajan, R. Chaware, Advanced reliability study of TSV interposers and interconnects for the 28nm technology FPGA, in: IEEE Elec. Comp. Tech. Conf. (ECTC), 2011, pp. 285–290.
[266] S. Sukegawa, T. Umebayashi, T. Nakajima, H. Kawanobe, K. Koseki, I. Hirota, T. Haruta, M. Kasai, K. Fukumoto, T. Wakano, K. Inoue, H. Takahashi, T. Nagano, Y. Nitta, T. Hirayama, N. Fukushima, A 1/4-inch 8Mpixel back-Illuminated stacked CMOS image sensor, in: IEEE Sol. St. Circuits Conf. (ISSCC), 2013, pp. 484–486.
[267] K. Shiraishi, Y. Shinozuka, T. Yamashita, K. Sugiura, N. Watanabe, R. Okamoto, T. Ashitani, M. Furuta, T. Itakura, 1.2e- Temporal Noise 3D-stacked CMOS image sensor with Comparator-based multiple-Sampling PGA, in: IEEE Sol. St. Circuits Conf. (ISSCC), 2016, pp. 122–124.
[268] U. Kang, H.-J. Chung, S. Heo, S.-H. Ahn, H. Lee, S.-H. Cha, J. Ahn, D.M. Kwon, J.H. Kim, J.-W. Lee, H.-S. Joo, W.-S. Kim, H.-K. Kim, E.-M. Lee, S.-R. Kim, K.-H. Ma, D.-H. Jang, N.-S. Kim, M.-S. Choi, S.-J. Oh, J.-B. Lee, T.-K. Jung, J.-H. Yoo, C. Kim, 8Gb 3D DDR3 DRAM using through-silicon-via technology, in: IEEE Sol. St. Circuits Conf. (ISSCC), 2009, pp. 130–132.
[269] H.-Y. Son, T. Oh, J.-W. Hong, B.-D. Lee, J.-H. Shin, S.-H. Kim, N.-S. Kim, in: IEEE Elec. Comp. Tech. Conf. (ECTC), 2016, pp. 356–360.
[270] J.P. Gambino, R. Winzenread, K. Thomas, R. Muller, H. Truong, D. Defibaugh, D. Price, K. Goshima, T. Hirano, Y. Watanabe, M. Breen, N. Oldham, Reliability of hybrid bond interconnects, in: IEEE International Interconnect Technology Conference (IITC), 2017.
[271] P. Enquist, G. Fountain, C. Petteway, A. Hollingsworth, H. Grady, Low cost of ownership scalable copper direct bond interconnect 3D IC technology for three dimensional integrated circuit applications, in: IEEE 3D Systems Integration Conf. (3DIC), 2009.
[272] S. Lhostis, A. Farcy, E. Deloffre, F. Lorut, S. Mermoz, Y. Henrion, L. Berthier, F. Bailly, D. Scevola, F. Guyader, F. Gigon, C. Besset, S. Pellissier, L. Gay, N. Hotellier, M. Arnoux, A.-L. Le Berrigo, S. Moreau, V. Balan, F. Fournel, A. Jouve, S. Chéramy, B. Rebhan, G.A. Maier, L. Chitu, Reliable 300 mm wafer level hybrid bonding for 3D stacked CMOS image sensors, in: IEEE Electronic Components and Technology Conference (ECTC), 2016, pp. 869–876.

[273] S. Ramamurthy, New Ways to Wire and Integrate Chips, 2022. https://ir.appliedmaterials.com/.
[274] J. Smith, Design technology co-optimization approaches for integration and migration to CFET and 3D logic, in: Proc. Surf. Preparation Cleaning Conf. (SPCC). New Delhi, India: Linx, 2019. Available: https://www.linx-consulting.com/spcc-2019-technical-program/.
[275] D. Prasad, S.S. Teja Nibhanupudi, S. Das, O. Zografos, B. Chehab, S. Sarkar, R. Baert, A. Robinson, A. Gupta, A. Spessot, P. Debacker, D. Verkest, J. Kulkarni, B. Cline, S. Sinha, Buried Power Rails and Back-Side Power Grids: Arm® CPU Power Delivery Network Design beyond 5nm, IEDM, 2019, pp. 446−449.
[276] E. Beyne, J. Ryckaert, Integrated circuit chip with power delivery network on the backside of the chip, US Patent Number (2020), 10,636,739.
[277] B. Cline, D.P.E. Beyne, O. Zografos, Next-gen chips will be powered from below, IEEE Spectr. (2021).
[278] A. Gupta, O. Varela Pedreira, G. Arutchelvan, H. Zahedmanesh, K. Devriendt, H. Mertens, Z. Tao, R. Ritzenthaler, S. Wang, D. Radisic, K. Kenis, L. Teugels, F. Sebai, C. Lorant, N. Jourdan, B.T. Chan, S. Subramanian, F. Schleicher, T. Hopf, A. Premkumar Peter, N. Rassoul, H. Debruyn, I. Demonie, Y.K. Siew, T. Chiarella, B. Briggs, X. Zhou, E. Rosseel, A. De Keersgieter, E. Capogreco, E. Dentoni Litta, G. Boccardi, S. Baudot, G. Mannaert, N. Bontemps, A. Sepulveda, S. Mertens, M.-S. Kim, E. Dupuy, K. Vandersmissen, S. Paolillo, D. Yakimets, B. Chehab, P. Favia, C. Drijbooms, J. Cousserier, M. Jaysankar, F. Lazzarino, P. Morin, E. Altamirano, J. Mitard, C.J. Wilson, F. Holsteyns, J. Boemmels, S. Demuynck, Z. Tokei, N. Horiguchi, Buried power rail integration with FinFETs for ultimate CMOS scaling, IEEE Trans. Electron Devices 67 (2020) 5349−5354.
[279] A. Jourdain, F. Schleicher, J. De Vos, M. Stucchi, E. Chery, A. Miller, G. Beyer, G. Van der Plas, E. Walsby, K. Roberts, H. Ashraf, D. Thomas, E. Beyne, Extreme wafer thinning and nano-TSV processing for 3D heterogeneous integration, in: IEEE Electronic Components and Technology Conference (ECTC), 2020, pp. 42−48.
[280] A. Jourdain, M. Stucchi, G. Van der Plas, G. Beyer, E. Beyne, Buried power rails and nano-scale TSV: technology boosters for backside power delivery network and 3D heterogeneous integration, in: IEEE Electronic Components and Technology Conference (ECTC), 2022, pp. 1531−1538.
[281] J. Ryckaert, A. Gupta, A. Jourdain, B. Chava, G. Van der Plas, D. Verkest, E. Beyne, Extending the roadmap beyond 3nm through system scaling boosters: a case study on buried power rail and backside power delivery, in: Electron Devices Technology and Manufacturing Conference (EDTM), 2019, pp. 50−52.
[282] A. Veloso, A. Jourdain, D. Radisic, R. Chen, G. Arutchelvan, B. O'Sullivan, H. Arimura, M. Stucchi, A. De Keersgieter, M. Hosseini, T. Hopf, K. D'have, S. Wang, E. Dupuy, G. Mannaert, K. Vandersmissen, S. Iacovo, P. Marien, S. Choudhury, F. Schleicher, F. Sebaai, Y. Oniki, X. Zhou, A. Gupta, T. Schram, B. Briggs, C. Lorant, E. Rosseel, A. Hikavyy, R. Loo, J. Geypen, D. Batuk, G.T. Martinez, J.P. Soulie, K. Devriendt, B.T. Chan, S. Demuynck, G. Hiblot, G. Van der Plas, J. Ryckaert, G. Beyer, E. Dentoni Litta, E. Beyne, N. Horiguchi, Scaled FinFETs connected by using both wafer sides for routing via buried power rails, IEEE Trans. Electron Devices 69 (2022).
[282a] K.L. Lin, S.A. Bojarski, C.T. Carver, M. Chandhok, J.S. Chawla, J.S. Clarke, M. Harmes, B. Krist, H. Lang, M. Mayeh, S. Naska, J.J. Plombon, S.H. Sung, H.J. Yoo, Nickel silicide for interconnects, in: IEEE International Interconnect Technology Conference (IITC), 2015, pp. 169−172.
[283] H. Li, C. Xu, N. Srivasta, K. Banerjee, Carbon nanomaterials: the ideal interconnect technology for next-generation ICs, IEEE Des. Test Comput. (2010) 20−31.

[284] K.-J. Lee, H. Park, J. Kong, A.P. Chandrakasan, Demonstration of a subthreshold FPGA using monolithically integrated graphene interconnects, IEEE Trans. Elec. Dev. 60 (2013) 383–390.

[285] H. Li, C. Xu, N. Srivasta, K. Banerjee, Carbon nanomaterials for next-generation interconnects and passives: physics, status, and prospects, IEEE Trans. Elec. Dev. 56 (2009) 1799–1821.

[286] C. Xu, H. Li, K. Banerjee, Modeling, analysis, and design of graphene nano-ribbon interconnects, IEEE Trans. Elec. Dev. 56 (2009) 1567–1578.

[287] S. Rakheja, V. Kumar, A. Naeemi, Evaluation of the potential performance of graphene nanoribbons as on-chip interconnects, Proc. IEEE 101 (2013) 1740–1765.

[288] I. Asselberghs, M. Politou, B. Soree, S. Sayan, D. Lin, P. Pashaei, C. Huyghebaert, P. Raghavan, I. Radu, Z. Tokei, Graphene wires as alternative interconnects, in: IEEE International Interconnect Technology Conference (IITC), 2015, pp. 317–319.

[289] E.M. Bazizi, A. Pal, J. Kim, L. Jiang, V. Reddy, B. Alexander, B. Ayyagari-Sangamalli, Materials to systems co-optimization platform for rapid technology development targeting future generation CMOS nodes, IEEE Trans. Electron. Dev. 68 (2021) 5358–5363.

[290] V. Moroz, X.-W. Lin, P. Asenov, D. Sherlekar, M. Choi, B. Cheng, S. Parikh, P.-W. Chan, J.J. Lee, Can we ever get to a 100nm tall library? Power rail design for 1nm technology node, in: VLSI Symposium on Circuits and Technology, 2020 paper JFS3.2.

[291] H. Park, K. Chang, J. Jeong, J. Ahn, K.-S. Chung, T. Kim, Challenges on DTCO methodology towards deep submicron interconnect technology, in: International SoC Design Conference (ISOCC), 2021, pp. 215–218.

[292] Y. Sun, M. Wang, X. Li, S. Hu, Z. Liu, Y. Liu, X. Li, Y. Shi, Improved MEOL and BEOL parasitic-aware design technology co-optimization for 3 nm gate-all-around nanosheet transistor, IEEE Trans. Electron. Dev. 69 (2022) 462–468.

[293] C. Christiansen, B. Li, M. Angyal, T. Kane, V. McGahay, Y.Y. Wang, S. Yao, Electromigration-resistance enhancement with CoWP or CuMn for advanced Cu interconnects, IEEE Int. Rel. Phys. Symp. (2011) 312–316.

# CHAPTER 3

# Sputter processing

**Andrew H. Simon**
IBM Research, Albany, NY, United States

## 1. Introduction

Sputtering is one of the most widely used thin-film fabrication techniques, used in such diverse industries as semiconductor processing, surface finishing, and jewelry making. The most widespread industrial application is in the deposition of metals, but it is also used for insulating materials. In its most basic form, sputtering is a process in which ionized atoms are accelerated into a surface in order to eject atoms from that surface. The ejected atoms can then be condensed onto a sample to nucleate a thin film of the ejected material. This process is called sputter deposition. The same type of physical process can also be performed to remove unwanted material from a sample, in which case the ejected atoms can be collected on the chamber shielding. The latter process is called sputter etching.

Sputter deposition has many advantages over other methods for depositing metals, such as evaporation, plating, or chemical vapor deposition (CVD). Historically, sputter deposition and evaporation have been broadly grouped under the term "physical vapor deposition" (PVD). In this chapter, we will use the term PVD as a synonym for sputter deposition since sputter deposition has largely supplanted evaporation in the semiconductor industry. Sputter deposition is distinguished from evaporation in that it produces high-energy flux which has high surface mobility and can thus condense into smooth, dense, conformal, and continuous films more easily than can evaporated films. The sputtering rates of metals of technological interest are all within an order of magnitude from the lowest to the highest [1]. Thus, unlike evaporation or CVD, sputtering preserves the stoichiometry of the target source since the physical bombardment mechanism of particle ejection results in a consistent stoichiometry on the sample surface.

In this fifth edition of the Handbook of Thin-Film Deposition, we have retained all the fundamental background material from the third and fourth editions which pertains to fundamentals of sputtering science and process tooling as it is used in the semiconductor industry. Section 3.6 has been updated once again to cover recent semiconductor applications down to technology nodes of 5 nm and smaller and will bring the reader up to date on trends in sputter processing and associated thin-film processes (CVD, ALD) which are now widely used in combination.

The treatment here is not intended as an encyclopedic listing of all sputtering science and tooling. The interested reader is referred to the full-length monographs by Mahan [1] and Mattox [2] and review articles by Rossnagel [3–5] for more comprehensive, in-depth treatments of many of the topics covered here.

## 2. Energy and kinematics of sputtered atoms

In order to understand sputtering processes, a review of the physical mechanisms of sputtering is helpful. Fig. 3.1 shows a schematic illustration of typical sputtering phenomena. A bombardment ion is accelerated into a substrate as a result of acceleration through an accelerating potential in the

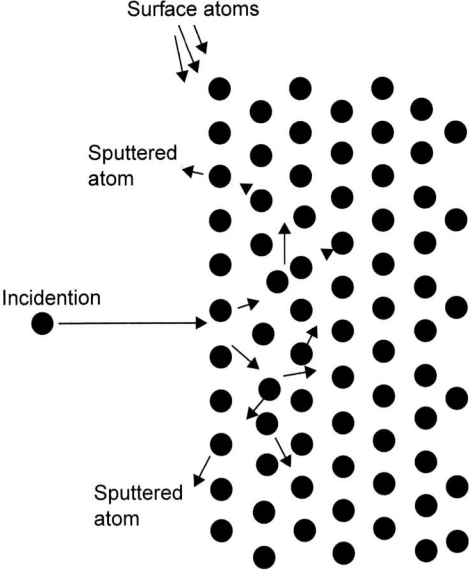

**Figure 3.1** Schematic of a physical sputtering process [5].

sputter chamber. For large-scale industrial applications in integrated-circuit fabrication, the species used for bombardment is typically $Ar^+$, due to its chemical inertness and low cost. Argon also has the advantage that its atomic mass is similar to that of many of the metals which are widely used in integrated circuit (IC) manufacturing, such as Ti, Al, and Cu. Applications involving noninert gases (reactive sputtering) and self-sputtering of metals will be discussed in later sections. In sputtering processes of interest, the accelerating energy of the bombardment ion is typically provided by the potential drop between the ionized plasma gas and the bombarded surface, which is referred to as the sputter target in typical deposition applications.

Once the bombardment ion collides with the target surface, atoms from the target can be ejected to condense on a substrate to form a thin film. A key metric to characterize the sputtering event is the sputter yield, Y, which is a measure of the number of atoms ejected from the target for each bombardment ion.

$$Y = \frac{\text{Number of sputtered atoms ejected}}{\text{Number of sputtering atoms incident}}$$

## 3. Energy dependence of sputtering

Sputter yields will generally show a characteristic dependence on the energy of the bombarding ion (Fig. 3.2), which can be broken down into several regimes:

1. Low-energy (subthreshold) sputtering: at ion energies below the surface binding energy of the cathode material, typically <50 eV, sputter yields are orders of magnitude less than unity, in the range of $10^2-10^6$, since the bombarding ions can only eject the most loosely bound surface atoms or adsorbed molecular species.
2. Knock-on sputtering: ion energies in the range of ~10 eV to 1 keV are of prime interest for commercial and industrial applications of sputtering. Once the energy of the sputtering ions is greater than the surface binding energy of the cathode material, it is energetically possible to dislodge surface and near-surface atoms from their equilibrium sites. These dislodged atoms then in turn set in motion recoil collisions which eventually result in the ejection of atoms from the cathode surface. The key hallmark of this energy regime is the roughly linear dependence of the sputter yield on the ion bombardment energy and the ion current.

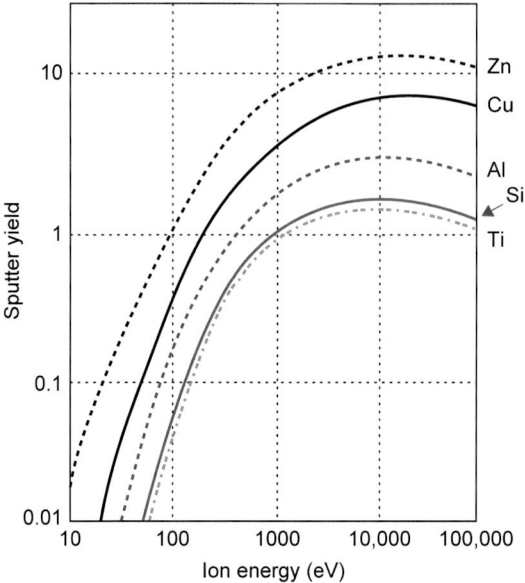

**Figure 3.2** Energy dependence of sputter yield for different metals [5,6].

Sputter yields in this regime are generally in the range of 0.1–3.0 for most materials of technological interest. This region is thus referred to as the linear-cascade regime by some authors [1,7,8].
3. Above an ion threshold energy of ~1 keV, collision-cascade (nonlinear cascade) sputtering behavior is observed, in which the incident ions have enough energy to dislodge multiple cathode atoms. Sputter yields in this regime will be in the range of ~5–50 and higher. Due to the high energies required and the high ejected energies of the sputtered atoms, this regime is usually not of industrial interest. Incident ion energies above 50 keV result in deep-ion implantation into the cathode and a reduction in net sputter yield.

We will briefly cover some analytical expressions which capture the kinematics of sputtering events. For a more in-depth discussion of these topics, the reader is referred to Mahan [1], Chapter VII. The semi-empirical expression published by Bohdansky et al. [6,9] approximates the typical shape of sputter yield curves as a function of energy:

$$Y = (6.4 \times 10^{-3}) m_r \gamma^{5/3} E^{0.25} \left(1 - \frac{E_{th}}{E}\right)^{3.5}$$

where $Y$ is the sputter yield, $E$ is initial energy of the incident atom in electron volts, $m_r$ is the recoil mass and $m_p$ the projectile mass in atomic mass units, and $\gamma$ is the energy transfer mass factor defined by:

$$\gamma = \frac{4 m_r m_p}{\left(m_r + m_p\right)^2}$$

$E_{th}$ is the threshold energy for sputtering, defined by Bohdansky's expression:

$$E_{th} = \frac{U_{sb}}{\gamma(1-\gamma)} \quad \text{for} \quad \frac{m_p}{m_r} < 0.3$$

and

$$E_{th} = 8 U_{sb} \left(\frac{m_p}{m_r}\right)^{2/5} \quad \text{for} \quad \frac{m_p}{m_r} > 0.3$$

where $U_{sb}$ is the surface binding energy of the target atoms and is effectively the heat of sublimation per particle (Ref. [1], Chapter VII).

## 3.1 Cosine sputtering law

In cases of normal incidence of the projectile atoms onto the target surface, the angular distribution of the sputtered species emitted from the target surface can typically be approximated by a cosine distribution:

$$j_\Omega(\theta) = Y\phi\left(\frac{\cos\theta}{\pi}\right)$$

where $j_\Omega(\theta)$ is the emission flux angular distribution as a function of the angle $\theta$ (measured from the vertical) into the differential solid angle $d\Omega(\theta) = \sin\theta \, d\theta \, d\vartheta$, with $\vartheta$ being the azimuthal angle, $Y$ the sputter yield emitted from the surface, and $\varphi$ is the local ion flux incident onto the surface. This result can be derived analytically if the recoil velocities of the sputtered atoms are assumed to be isotropic (Ref. [1], Chapter VII). Deviations from the ideal cosine distribution are observed at low sputter-ion energies (undercosine, or "flatter" distribution) and high ion energies (over cosine), with a more strongly forward-peaked distribution (Fig. 3.3).

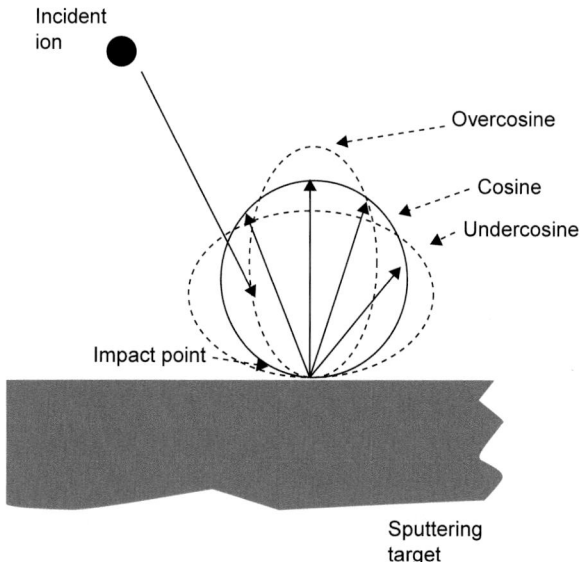

**Figure 3.3** Illustration of cosine-law angular distribution [5].

## 4. Plasmas and sputtering systems

In order to understand the principles of sputter plasmas and how they affect tooling design, we will provide a brief review of some basic types of sputter apparatus.

### 4.1 DC diode plasmas

The simplest type of sputtering apparatus, the DC diode, consists of two plates, a vacuum chamber, and a power supply (see Fig. 3.4). A sputtering gas, typically argon, is introduced at a pressure in the millitorr range, and voltage is applied across the plates. Above a threshold breakdown voltage (depending on sputter gas, pressure, and cathode material), a plasma discharge forms in which positively charged ionized gas atoms are drawn to the negatively charged cathode. The lighter mass of the electrons relative to the gas ions gives the electrons a much higher velocity in the plasma. The result is that a thin sheath layer forms next to the cathode. This sheath layer is depleted of electrons, and most of the potential drop between anode and cathode occurs in the sheath, also referred to as the "cathode fall."

The heavier ions are accelerated through this potential drop into the cathode, causing newly ejected (secondary) electrons to be emitted from

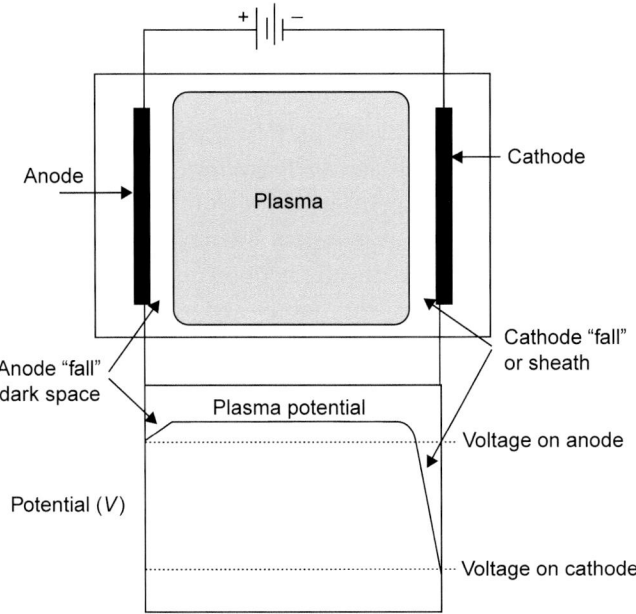

**Figure 3.4** Schematic illustration of a diode plasma with accompanying plot of the potential distribution along the centerline of the apparatus [4].

the cathode surface. The secondary electrons are then accelerated away from the cathode, across the plasma sheath, giving them sufficient energy to ionize more sputtering gas atoms through two mechanisms: (1) by direct collisions with neutral atoms in the plasma, and (2) by elevating the Maxwellian velocity distribution (temperature) of electrons already present in the plasma through electron–electron scattering. In the case of the latter mechanism, ionization of neutral atoms by the high-energy population (in the electron-volt range) of the Maxwellian electron distribution predominates over direct collisional ionization by secondary electrons, owing to the comparatively low proportion of secondary electrons and the decrease of electron-ionization cross-sections at higher electron energies [4]. The secondary-electron yield (typically ∼5%–10% for materials of interest in semiconductor manufacturing [3]) must be roughly the inverse of the net number of ions created for a steady-state plasma to exist. When the ionization rate due to secondary-electron collisions surmounts this threshold, the plasma is self-sustaining, and a constant current can flow through the discharge. The mechanism of sputter-gas ionization by means of secondary electrons is a key part of creating a stable plasma condition.

In diode plasmas, secondary electrons which do not ionize the sputter gas after being ejected from the cathode can travel the full length of the apparatus, where they are lost to the anode or the sidewalls and can no longer contribute to ionization. Since the ionization cross-section peaks for electron energies of ~100 eV and then declines at higher energies [6,10], the process cannot be scaled up by applying more power. For these reasons, diode plasmas are no longer of industrial interest.

It should be noted that the sheath phenomenon seen at the cathode also occurs to a smaller extent at the anode and sidewalls of the discharge chamber. The high mobility of the electrons relative to ions results in a modest positive potential ("anode fall" or "dark space" in the case of the anode) between the electrically neutral, conductive interior of the plasma and any conducting surface, where electrons are lost to the plasma. This sheath effect is present regardless of any externally applied potential and arises due to the local space-charge distribution caused by electron depletion near the conducting surface. This phenomenon is indicated schematically by the potential diagram associated with the diode discharge in Fig. 3.4.

## 4.2 RF plasmas

Owing to the limited ability of the DC diode apparatus to achieve high levels of gas ionization and sputtering of the cathode, an evolution of the DC diode apparatus is to replace the DC power supply depicted in Fig. 3.4 with an alternating-current radio frequency (RF) source and associated impedance-matching hardware. The typical RF supplied to the electrodes is 13.56 MHz or some multiple. The alternating RF power couples to the electron motion in the plasma, resulting in longer residence times in the plasma, higher collisional ionization, and higher plasma densities. Adjustable impedance circuitry is used to tune the output impedance of the power supply and impedance-matching network to the plasma's impedance for maximum energy transfer. In addition to providing higher plasma densities for metals sputtering, RF plasmas enable the sputtering of insulating materials such as silicon dioxide and alumina because the alternating polarity of the cathode (target) prevents charge build-up on the cathode surface. Similarly, the alternating potential of the anode (sample) allows for sputter-cleaning and planarization via resputtering of the deposited film via ion bombardment from the sputter gas. This type of process is referred to as bias sputtering and has been adapted for use in modern DC discharges as well and will be discussed later.

## 4.3 Magnetron sputtering

A significant advance in the efficiency of sputter tooling is the magnetron source developed in the 1970s. The magnetron uses strong magnetic fields, typically from permanent magnets, to keep secondary electrons spatially confined in the vicinity of the target surface. By confining the secondaries near the target surface, their residence time in the plasma is greatly lengthened, resulting in greater ionization of the sputter-gas atoms, a denser plasma, and higher plasma currents and deposition rates.

In a magnetron sputter source, the high electric field arising from the cathode fall potential accelerates secondary electrons in a direction normal to the target surface (Fig. 3.5a). The magnetic field configuration is typically engineered so that the field lines are parallel to the target surface, resulting in an $E \times B$ drift force which acts on the secondary electrons. The electrons are thus confined to move in cycloidal drift orbits parallel to the target surface, resulting in additional collisional ionization of the sputter-gas atoms and higher overall plasma currents (Fig. 3.5b). This magnetic confinement of the secondary electrons has analogs in cyclotron motion and the Hall Effect. Early published measurements by Rossnagel and Kaufman [11] on 150-mm sources indicated a secondary-electron current approximately $5\times$ that of the discharge current, indicating multiple orbits of the sputter source can occur. Magnetrons typically operate at pressures ranging from <1 mTorr, for directional and self-sustained sputtering, to >10 mTorr, in which the velocities of the sputtered atoms will be randomized and thermalized due to collisions with the gas atoms.

In current industrial practice, the permanent magnet is mounted behind the target. If the magnet position were to be kept fixed, this would lead to highly nonuniform local erosion of the target in areas where the fields are strongest, and correspondingly nonuniform deposition on the sample

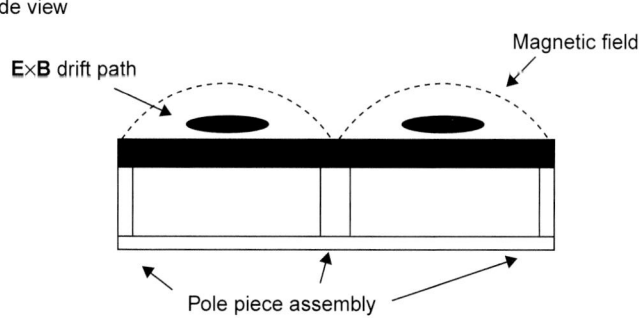

**Figure 3.5(a)** Magnetic field configuration of a planar magnetron (side view) [5].

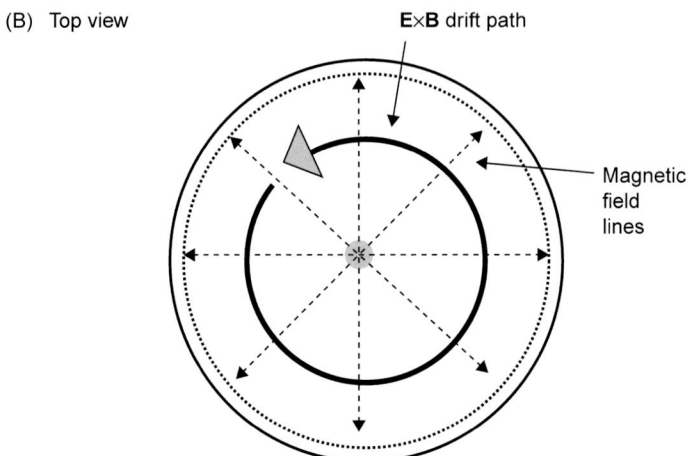

**Figure 3.5(b)** Magnetic field configuration of a planar magnetron (top view), showing the orbital ExB drift path of secondary electrons in the plasma [5].

surface. The high variance of the local magnetic field strength and field configuration of any given permanent magnet configuration make it impractical to achieve both high field strength and uniformity across the entire target surface using a static magnet arrangement. As a result, virtually all sputter systems for industrial use have permanent magnets mounted on motor-driven mechanisms so that the magnetic fields can be swept over the surface of the target in a repeating, orbital motion.

### 4.3.1 Magnetron designs

The simplest and most common magnetron configuration is comprised of a circular, planar target with motorized magnets mounted behind the target [1–5] (Fig. 3.5a and b). Common magnet configurations in industrial use are a cardioid-shaped magnet rotating in a circular orbit. More complex, proprietary, orbital patterns using smaller and stronger magnet configurations are now offered by sputter-equipment manufacturers for some applications and are an area of continual development.

Modern semiconductor processing typically requires powers of >10 kW for 300-mm wafer systems. A key advantage of magnetron designs is that large amounts of cooling water can be flowed through the back side, separating the cooling water loop from the vacuum system and heat sinking the target.

Various other shaped magnetron designs have been used in industrial or research settings. The common feature in all designs is that the $E \times B$ drift effect keeps the secondary electrons in confined closed paths such that they can ionize several sputter-gas atoms, i.e., they are designed so that the magnetic fields are perpendicular to the strong electric fields present at the cathode fall. Proprietary designs have been produced commercially for the semiconductor industry which have targets with rectangular (racetrack), conical, and cylindrical (hollow cathode) magnetron shapes [4,5,12–14].

## 5. Reactive sputter deposition

Sputter deposition of metallic and insulating compounds is of considerable technological interest. Sputter-deposited compounds commonly used in semiconductor applications include TiN, TaN, $Al_2O_3$, and $SiO_2$.

Sputtering a compound target presents several difficulties. If the sputtered material is an insulator or resistive metal, the only method to deposit the material is RF sputtering, with concomitant problems of thermal build-up, cracking, and bonding of insulating materials. Where feasible, an alternate and preferred method for most semiconductor applications is to sputter a pure metallic target (typically Ti, Ta, Al) and to react the sputtered metal with the appropriate compound gas, typically nitrogen or oxygen, at the sample surface. This method, known as reactive sputtering, also has the advantage of providing more control over the stoichiometry of the deposited film by adjusting the gas flow. A major drawback of reactive sputtering is that the target surface can react with the compound gas, creating an insulating surface film in situ. This behavior needs to be managed carefully in the design and operation of reactive sputter processes.

### 5.1 Current–voltage hysteresis in reactive sputtering systems

A typical reactive sputtering chamber is similar to the DC magnetron discussed previously in Section 3.4. The salient difference is the capability to flow a reactive gas such as nitrogen or oxygen. Titanium nitride (TiN) provides an example with wide application in the semiconductor industry.

The main operating principles in reactive sputtering are illustrated schematically by the graph in Fig. 3.6, which plots the deposition rate, target voltage, and chamber pressure versus the reactive gas flow (nitrogen, in the case of TiN). At all times during the reactive sputter process, it is understood that a constant flow of argon gas, resulting in an argon partial pressure of several millitorr, will be present to create a stable plasma to achieve sputtering of Ti metal. The target voltage in this baseline, argon-only process condition, corresponds to the lower branch (increasing from zero reactive gas flow up to point "A") on the graphs in Fig. 3.6.

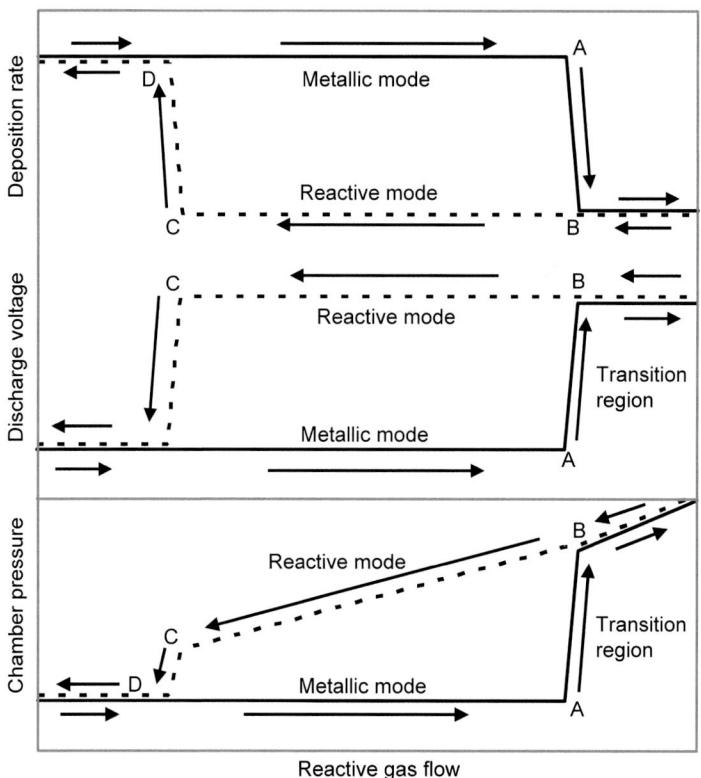

**Figure 3.6** Hysteresis behavior in reactive gas-flow versus chamber pressure, discharge voltage, and deposition rate in reactive sputtering (schematic-not to scale).

There is little change in the deposition behavior at low flows of nitrogen, as the nitrogen atoms are gettered by the chamber shielding and are incorporated by the deposited film. The target voltage will remain little changed up until the nitrogen flow reaches the transition point "A" at which point the target voltage and chamber pressure show a very abrupt rise, corresponding to point "B." This rise in the target voltage is accompanied by a substantial, several-fold drop in the deposition rate, indicated by the corresponding points "A" and "B" on the deposition-rate chart. The critical change that has occurred is that the target surface has become nitridized and is no longer purely metallic.

Further increases in nitrogen flow result in an essentially flat response on the lower branch of the deposition-rate graph (from point "B" rightwards). If the nitrogen flow is lowered, the plasma behavior does not immediately revert to the metallic behavior observed between points "A" and "B." The target voltage will continue to stay elevated, and the deposition rate will remain at the reduced level for reactive gas flows well below the threshold flow that nitrided the target (point "B"). Only when the target surface has been sputtered clean of the nitrided film does the target voltage fall back to its metallic-regime value, with the deposition rate rising to its original metallic-state condition (point "C").

This type of curve is called a hysteresis curve, due to the history dependence of the output variables (target voltage, deposition rate, pressure) on the input variables (nitrogen gas flow in this case). Similar shapes are familiar from discussions of magnetization phenomena.

The abrupt behavior of the system in the transition region (knee) between points "A" and "B" requires particular care in the design and operation of these processes. If the desired film properties cannot be obtained operating in one of the stable plasma regimes (metallic or reactive), other means of maintaining control and reproducibility of the films may be needed, such as increased pumping speed or feedback-controlled schemes for reactive gas flow [1,2,14]. In routine semiconductor industrial use, virtually all reactive sputter systems require periodic maintenance of the target ("pasting"), in which extended depositions without the reactive gas flow are run on dummy wafers or shutters in order to ensure reproducible cathode conditions.

Not all materials will follow the example illustrated above by TiN. A case in point is TaN, which is used as a barrier layer for copper interconnects: since TaN and Ta have similar sputter yields, minimal hysteresis is observed with the Ta—TaN system [15,16].

## 6. Sputter-tool design and applications for semiconductor technology

We will review some design elements of sputter tooling which have found common use for semiconductor manufacturing applications.

### 6.1 Batch/planetary systems

In batch-processing systems, the samples being deposited are mounted on disk- or dome-shaped (planetary) sample holders, which are capable of holding multiple samples, and can be rotated past the target. This configuration is no longer commonly used in the semiconductor industry. There are multiple reasons why it is not feasible with current industry requirements: larger-diameter wafers make the systems dimensionally unfeasible, the need to vacuum cluster non-PVD processes (e.g., cleans, CVD, or atomic layer deposition (ALD) layers) with PVD, the dimensional demands of high-directionality sputtering, and the economic consequences of product loss if a tool fault occurs during batch processing. Planetary systems are still used for other industrial applications where the sample dimensions and processing sequences make it practical and economical.

## 6.2 Single-wafer systems

The great majority of present-day integrated-circuit sputter processing is done with clustered vacuum tools, in which silicon wafers are processed individually through sequential processing steps in separate, dedicated sputter chambers. The separate processing chambers are mounted on a main transfer chamber with an ultrahigh vacuum, in which a mechanical handler moves each wafer from one processing chamber to the next without breaking vacuum.

Clustered vacuum tools have several advantages: (1) They permit better vacuum isolation of the most sensitive process steps. (2) Process faults or aborts affect only one wafer. (3) They allow clustering of PVD processes with non-PVD processes, such as CVD, ALD, degas, sputter etch, chemical cleans, and plasma cleans. (4) Specialized source designs with more demanding dimensional requirements (to achieve ionized sputtering, collimation, and long-throw sputtering) are more feasible in single-wafer chambers.

The issue of vacuum isolation forces critical constraints on base pressures. For metals, in particular, there is a strong sensitivity to interface oxidation and the associated surface cleaning steps. If the base pressure is in the range of $10^{-6}$ Torr, it takes about one second for a surface to be covered with a monolayer of oxygen. In modern PVD sources, deposition rates in the range of $\sim 5-50$ Å/s are common, which would result in an oxygen concentration of $\sim 1\%$ or higher in the deposited film if the $10^{-6}$ Torr base pressure was the best that could be achieved. As a result, base pressures of $10^{-8}$ to $10^{-9}$ Torr are needed for processes that include deposition of multiple layers or surface cleans without vacuum break. Stainless steel construction, heated chamber shielding, and extended automated bake-out sequences after chamber servicing are the norm.

The high base-pressure requirements mean that cryopumping is used for all chambers in which inert or nontoxic species like argon or nitrogen are used. Gas flows in hundreds of standard cubic centimeters per minute (sccm) are common in modern sputter-processing chambers, which result in frequent regeneration schedules for cryopumps.

For hazardous gases or processes with exceptionally high flows, cryopumps typically cannot be used, and turbopumping is usually substituted. Modern processing sequences frequently require the clustering of sputtered metals processes with processes which use hazardous gases, such as CVD, ALD, and chemical or reactive gas cleans. Precautions must be taken in pump down and valve sequencing to make sure no transfer of hazardous gases into the cryopumped chambers can occur.

### 6.2.1 Clustered sputter-tool layout

A typical sputter-tool layout consists of the following elements (see Fig. 3.7):

1. *Mainframe*: has central handler which moves wafers individually between cassettes/loadlocks and process chambers.
2. *Loadlocks*: modern 300-mm tooling now pumps down individual wafers rather than entire cassettes.
3. *Degas*: once introduced into to the vacuum system, wafers are typically subjected to a high temperature bake in an inert ambient, typically argon, in order to desorb aqueous and volatile species from the wafer.

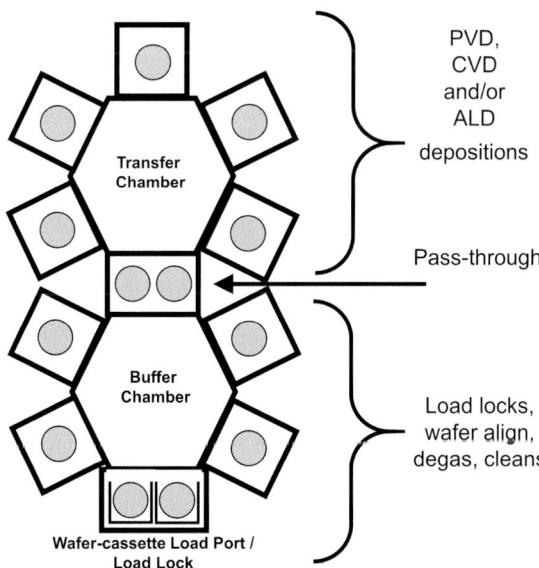

**Figure 3.7** Schematic of a typical configuration for a single-wafer sputter process tool which can process vacuum-clustered PVD, CVD and/or ALD process flows. Central robots in the transfer and buffer chambers move wafers from position to position. Processes that are more sensitive to base vacuum, such as pure metals, will usually be positioned on the transfer chamber, away from the load lock.

This step should be the hottest step in the deposition sequence, while keeping in mind the overall thermal dose relative to other steps in the process flow. Residual gas analyzer (RGA) monitoring of the degas chamber can be useful in estimating the necessary time and temperature, based on the outgassing species in the RGA spectrum.
4. *Buffer and transfer chambers*: these chambers perform similar functions in that they contain handlers which move the wafers from chamber to chamber. The most sensitive processes to interfacial oxidation should be put on the back to provide the most isolation from oxidizing species coming from the wafer loadlocks.
5. *Cleans*: prior to metal deposition, native oxide, hydrocarbon, or fluorocarbon removal is critical to obtaining high-quality interfaces. Cleaning chambers can consist of physical cleans, such as argon sputter etches, and chemical cleans which are typically developed for specific applications.

Sputter etching is the simplest way of removing surface layers from samples. Instead of depositing a layer on the sample by sputtering from a target and onto a sample, the sample becomes the target. It is essentially the RF sputter source operated in reverse, with the sample taking the role of the cathode and the chamber shielding taking the role of the anode. In sputter-etch chambers used for semiconductor fabrication, the $Ar^+$ sputter species are typically generated by a radio frequency ionization coil. The ions in the $Ar^+$ plasma then bombard the wafer surface under the influence of AC bias applied to the wafer chuck, which is typically in the kHz–MHz range. Sputter etching has the disadvantage that even modest accumulations of sputter-etched material on chamber shielding are prone to delamination, resulting in particulate contamination falling on the product wafer.

Non-sputter processes are now frequently clustered together with sputter deposition on clustered vacuum mainframes in order to perform chemical cleaning in the increasing number of applications in which sputter etching would be too damaging to surrounding structures. Examples which have found widespread adoption in the semiconductor industry include chambers designed to perform hydrogen fluoride etching in situ to replace batch HF wet cleans for pre-silicide depositions [17,18] and reactive hydrogen plasmas to reduce native oxides on metal contact surfaces, which are now widely used in copper interconnects [19,20].

## 6.3 Directional sputter deposition

The most challenging sputter processes currently used in the semiconductor industry must fill high aspect-ratio (∼2:1 up to ∼10:1 or more) features and are exemplified by dual-damascene processes used in interconnects and aggressive contact and silicide schemes. Specific material sets that have wide application are Ti/TiN liners for PVD aluminum interconnects or CVD W plugs, Ta(N)/Cu liners for Cu BEOL interconnects, and Ni-refractory metal alloys for silicides. In the cases of tungsten and copper fill, the process used for filling the structures (CVD W or electroplating of Cu) is not a PVD process, but the high conformality and step coverage of modern PVD technologies are needed to deposit the thin and conformal liner layers which often have minimum thicknesses of $\sim 20-30$ Å.

The demanding feature dimensions pose ongoing problems for the extendibility of sputter processes and tooling. In conventional magnetron plasmas, the cosine theta distribution of sputtered material results in a relatively isotropic distribution of metal-atom flux at the wafer surface. The result is that attempting to cover high aspect-ratio features using a conventional, planar PVD source leads to problems of too much coverage overhang on the top corners of features (Fig. 3.8) and poor coverage at the lower sidewalls of features. Thus, a prime focus of sputter-tool development is the need to make the sputter-deposition process more directional, in order to cover the bottoms and lower sidewalls of features.

### 6.3.1 Collimation

The simplest way to increase directionality is to move the target further away from the wafer. Geometrically, the angle subtended by the wafer will define the degree of directional selection. There are several limitations to longer source-wafer ("long-throw") spacings:

1. There is considerable asymmetry in the deposition at the wafer edge, due to the larger flux of metal atoms from the center of the target

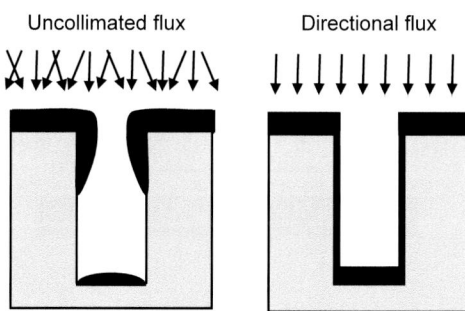

**Figure 3.8** Schematic illustration of sputtered thin-film coverage in a high aspect-ratio feature using uncollimated deposition flux versus a directional deposition flux.

relative to the flux of atoms from the edge. This issue cannot be resolved without making the target and sputter source impractically large. As a result, inboard/outboard deposition asymmetries of ~2–5× on the lower sidewalls at wafer edge have been reported in the literature [21,22].

2. Moreover, even if the target could be scaled up indefinitely, the limitations of traditional magnetron operation (generally not sustainable for most metals below ~0.1 mTorr) mean that scattering of the metal atoms by the argon sputter gas makes it pointless to extend the target-wafer spacing much beyond ~20 cm, since the straight line-of-sight deposition trajectory will no longer hold [3]. Thus, for modern 300-mm systems, long-throw geometries are generally of limited usefulness.

An alternate method of directional selection is to use a physical collimation [23], in which the angular distribution of the sputtered species is selected by physically blocking off normal atoms from reaching the substrate. The directional selection is done by interposing a physical collimator in between the target and substrate, so that normal operating pressures and target–substrate distances can be maintained (Fig. 3.9a).

Physical collimators are essentially tubes of a predetermined aspect ratio (more typically, honeycomb structures fabricated from metal sheet), which are used for the directional flux selection. Typical collimators used in commercial sputtering systems might have heights of the order of 1–2 cm

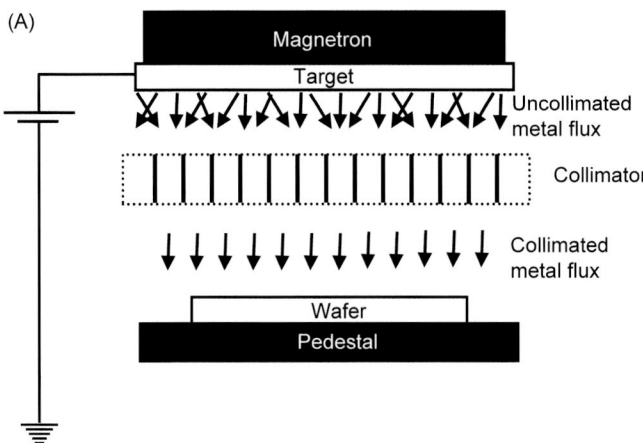

**Figure 3.9(a)** Schematic illustration of a collimated sputter-deposition PVD source [3,21].

and comparable cell dimensions. Thus a 2:1 aspect-ratio collimator might be fabricated from sheet metal 2 cm high such that the lateral cell dimensions are 1 cm.

As with long-throw sputtering, much of the off-axis metal flux from the target is essentially discarded and is deposited on the collimator itself. As the deposition accumulates on the collimator, the collimator blocks an increasing amount of flux from the target, resulting in a gradual drop in the deposition over the life of the collimator. This downward drift in deposition rate must be corrected for with increases in deposition times in order to maintain constant deposition thickness on the wafer. The impingement of the sputtered metal flux on the collimator can also result in considerable heat transfer, necessitating water cooling of the collimator fixturing in some applications in order to prevent heat build-up during operation. The effect of collimator aspect ratio on angular selection of the flux is illustrated by Rossnagel [5] (Fig. 3.9b).

### 6.3.1.1 Ionized and self-ionized sputtering

Collimation has obvious drawbacks in that the interposition of the collimator between wafer and target can result in particles and wastage of target material through deposition on the collimator. In addition, the collimator's aspect ratio can be altered between beginning and end of kit life as deposited material accumulates on it.

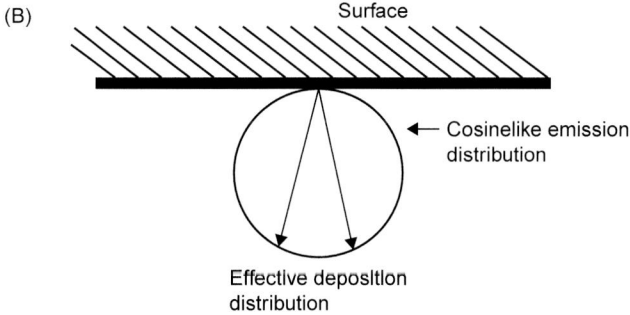

**Figure 3.9(b)** Collimator angular flux versus aspect ratio [3].

A more efficient solution to the problem of achieving directional deposition in high aspect-ratio features is to ionize the metal atoms on their way to the sample and use the plasma potential, possibly along with external bias, to give the metal ions a strong velocity component normal to the sample surface. As long as the acceleration potential is significantly larger than the thermal energy of the metal ions, the metal will be deposited into the feature at near-normal incidence, resulting in conformal coverage of patterned structures. This type of deposition is referred to as ionized PVD, and a typical configuration for ionized PVD [24–27] is shown in Fig. 3.10.

The ionized PVD apparatus shown in Fig. 3.10 consists of a DC magnetron source which is used to generate the flux to metal atoms in the conventional way, as described previously. In order to ionize the metal atoms on their way to the sample, a second plasma is generated in the space between the target and the sample. A high plasma density ($n >> 10^{11} cm^{-3}$) enables collisional ionization of the metal-atom flux by means of electron–metal-atom collisions. The second plasma uses the same sputter gas (typically argon) as the magnetron plasma at the top of the source. In the example shown, the second plasma is generated via inductive coupling through RF coils typically driven at 13.56 MHz, which encircle the space in between the target and the wafer. In this configuration, plasma operation

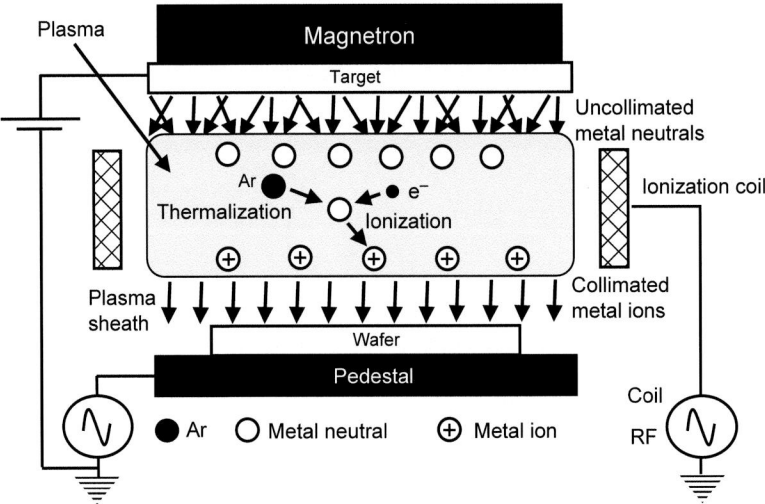

**Figure 3.10** Ionized sputter source for metals. An ionization coil encircles the region between the source and substrate to create a dense plasma. Sputtered atoms are thermalized by collisions with the argon sputter gas, ionized by electron impact and collimated by the sheath potential and wafer bias [3,4,22].

at pressures of well above 10 mTorr can be optimal, in contrast to the 1—10 mTorr operating regime typical of non-ionized sources. Metal-atom ionization efficiencies can be quite high, owing to the relatively high ionization energy of argon (15.7 eV) relative to the typical metal-atom species of interest (<10 eV).

Similar to what occurs at the target surface, a net positive plasma potential (typically tens of electron volts) will occur between the body of the ionization plasma and the sample itself, due to the lower mass and higher mobility of electrons in the plasma. The directionality of ionized PVD is due to the accelerating potential the metal ions experience between the plasma potential and the plasma sheath that exists at the sample. In most commercially available systems used for semiconductor manufacturing, the accelerating effect of the plasma potential is typically augmented by an external AC bias applied to the sample pedestal in order to increase the directionality of the metal ions as they are deposited on the wafer.

The ionized PVD configuration discussed above is not unique, and commercially available ionized PVD systems are available which achieve high ionization densities by other means, such as electron—cyclotron resonance or through unique configurations of the DC magnetron. An example of the latter is described in Refs. [12,13], in which a proprietary hollow-cathode (bucket-shaped) magnetron is encircled by a series of electromagnet coils extending from the top of the target to the space in between the target and the wafer. A combination of stacked electromagnets and cylindrical target shape results in a magnetic field configuration in which the field lines are parallel to the sidewalls of the target and form a dense ($n > 10^{12}$ cm$^{-3}$) plasma in which collisional ionization of the metal species is achieved solely from DC power sources, although an AC bias power supply to the pedestal is still used to adjust the directionality of the flux at the wafer.

One benefit of the high secondary-electron densities present in ionized PVD chambers is that conditions similar to those which produce dense metallic plasmas can also be exploited to produce argon plasmas in the same chamber. In combination with the biasable wafer chuck, the argon plasma process can be used as a sputter-etch process to remove or resputter material on the wafer. Depending on the PVD source design and the process conditions desired, the argon sputter-etch condition can typically be done at a DC magnetron power that will be as much as an order of magnitude lower than the ionized PVD deposition condition, but with an RF pedestal bias of several hundred watts or more for a 300-mm wafer. In chamber designs with inductively coupled RF coils, the RF coils will be used to ionize the argon plasma, but similar effects can be achieved with other source designs under appropriate process conditions [28–30].

With both deposition and etch conditions being achievable in the same PVD chamber, it is possible to construct multistep process sequences in which a metallic layer is first deposited using ionized deposition conditions and then etched or resputtered using an $Ar^+$ plasma. This type of deposition-etch sequence has been exploited extensively in copper-interconnect applications [28–34].

### 6.3.1.2 Self-sustained sputtering

An alternate way of achieving ionization of the metal-atom species is to operate the sputter source in a regime in which the metal atoms are self-sputtering, i.e., the plasma discharge is capable of steady-state operation without a sputter gas like argon [35–39].

The condition for a self-sustaining self-sputtering was formulated by Hosokawa et al. [35]:

$$\alpha \beta Y_S \geq 1$$

where $\alpha$ is the ionization probability of the sputtered atoms, $\beta$ is the probability of the ion returning to the cathode (target), and $Y_S$ is the yield for self-sputtering. This type of sputter deposition is unusual, in that it requires a high sputter yield and low ionization energy (i.e., a high ionization probability) for the discharge to be self-sustaining.

Among the metals of interest to the semiconductor industry, copper, which has a self-sputtering yield of 2.3 in the regime of interest, is most readily adapted to self-sustained sputtering. Peak plasma densities of $\sim 10^{17}-1^8$ $cm^{-3}$ are achieved by employing high magnetic field strengths in the active region of the magnetron, with plasma current densities of the order of $\sim 100$ $mA/cm^2$ or more in the active region [37,38].

In commercial sputtering systems, self-sustained sputtering is typically initiated in the same way as for a conventional magnetron discharge, with argon flow and the application of DC power to the magnetron being used to ignite the plasma. Once plasma ignition is achieved, the argon flow is stopped, and the DC magnetron power is ramped up so that the self-sustained regime ($\sim 50-100$ $W/cm^2$ in the case of Cu) is achieved. These high local power densities require high cooling efficiency in the sputter source and high thermal conductivity in the target material. Self-sustained sputtering is now widely used for advanced copper sputtering sources in the semiconductor industry.

In closing, it should be emphasized that the most advanced sputter sources in industrial use rely on proprietary designs which often incorporate several of the process developments mentioned here (e.g., ionized sputtering, bias sputtering, collimation, inductive plasmas, electromagnetic coils, and shaped magnetrons) into one design. In the semiconductor industry, commercially available sputter-tooling designs will typically be optimized for a deposition of specific metal to be used in a specific application and often for the needs of a specific technology node. Due to the ever-stricter demands imposed by the continual scaling of semiconductor technology to smaller ground rules, it is not unusual for a sputter source design to have a service life of only one or two technology nodes before it becomes obsolete or requires significant upgrades.

## 6.4 Current applications: Nanometer-scale engineering using PVD

Sputter deposition has several well-established applications in the semiconductor industry. The best-known are now found primarily in the middle-of-line (MOL) contact and back-end-of line (BEOL) interconnect levels. In the MOL contact application, ionized PVD is widely used to deposit Ti adhesion layers for the tungsten MOL local interconnects. In the BEOL copper interconnect levels, ionized PVD TaN and/or PVD Ta layers have long been employed as barrier and adhesion layers, with ionized PVD copper used for the seedlayer application. More recent applications have included PVD TiN as a metal hardmask layer, which is widely used for advanced interconnect patterning at groundrules below 45—32 nm. In contrast, modern front-end-of-line (FEOL) fabrication techniques, such as fin field-effect transistor (FinFET) replacement metal gate, have largely moved to CVD and ALD deposition techniques to fill the minimum-dimension features.

As with all sectors of the semiconductor equipment industry, the design of PVD deposition sources for advanced groundrule technologies is a focus of continual, ongoing development. Sputter-source development is now very application-specific and proprietary, but incremental advances in directional deposition are occurring and are built upon the general principles and techniques described previously in Section 3.6.3.

A prototypical example of sputter deposition that is widely used in the semiconductor industry is as a barrier/liner and seedlayer for damascene interconnects [40–43]. A schematic illustration is shown Fig. 3.11a–e. In Fig. 3.11a, we see an unfilled damascene structure as it appears immediately after being etched into a dielectric layer, with a vertical via structure (for interlevel electrical connection) and a horizontal trench structure (for intralevel electrical connection.) The structure will then be processed through a multi-chamber single-wafer deposition tool like the one shown in Fig. 3.7 which enables multiple process steps without breaking vacuum. Following a degas step (to volatilize adsorbed water, etc.) and a preclean (to remove surface oxide on exposed metal), a barrier layer of TaN, or more optimally a TaN/Ta bilayer [15,16], is deposited using PVD. This deposition will normally be ionized for optimal sidewall and bottom coverage (Fig. 3.11b). The wafer then moves under vacuum to the PVD Cu seedlayer deposition (Fig. 3.11c). In modern technology nodes, the PVD Cu seedlayer process will typically be self-ionized, as described in Section 3.6.3. Once the barrier/liner/seed deposition is complete, the wafer is removed from the vacuum system and is moved to Cu electroplating, where the Cu-seeded damascene features are filled with electroplated Cu and some

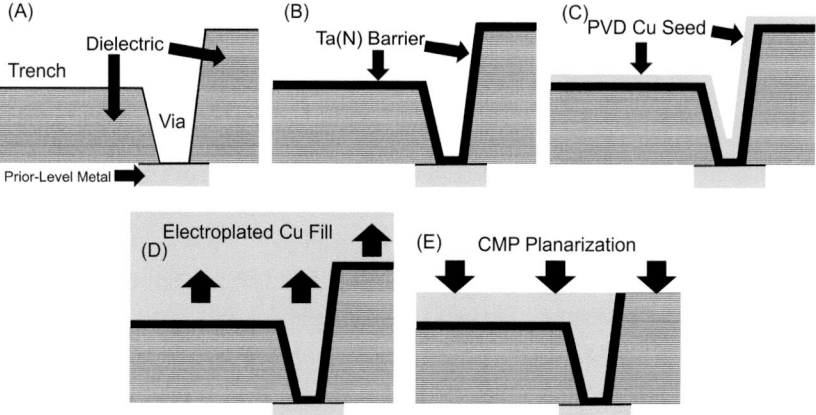

**Figure 3.11** Example of the process flow for a dual-damascene Cu interconnect process using sequential PVD TaN/Ta and PVD Cu layers (Schematic-not to scale). (a) Shows the unfilled, etched damascene structure as it is introduced into the clustered vacuum sputter tool. The via structure will provide electrical continuity to the previous wiring layer. (b) Depicts the structure after the deposition of the TaN single-layer or TaN/Ta bilayer barrier, typically with ionized PVD. (c) The structure is deposited with a Cu seedlayer, which will typically be done with self-ionized PVD at modern ground rules. (d) The feature is filled with electroplated Cu, which is plated to a thickness sufficient to create an overburden of Cu on top. (e) The overburden is polished off and the structure is polished down to the specified height using CMP planarization.

additional overburden is plated on top (Fig. 3.11d). Finally, the entire structure is subjected to chemical-mechanical polishing (CMP) (Fig. 3.11e), in which the plated Cu overburden polished off and the Cu-filled structures are polished down to their specified height.

In the third edition of this handbook, we cited the specific example of a multistep ionized sputter deposition/Ar+ sputter-etch process sequence in the TaN/Ta liner chamber to illustrate how modern volume-manufacturing sputter tooling and processes can be used to solve performance, yield, and reliability challenges in Cu interconnects [28–34]. This use of sputter deposition-etch sequencing was used successfully at larger groundrules but has largely fallen out of use at BEOL groundrules of 32 nm and below, due to problems with feature chamfering and the difficulty of mitigating damage to low-K dielectrics when Ar+ sputter etching is employed. New sputter-processing methods and sequences have been developed that are better suited to the needs of more recent technology nodes. In the following sections, we will discuss selected applications that highlight State-of the art applications of PVD, combined with ALD, CVD, and other process technologies, that either have recently entered wide use in the semiconductor industry or offer the potential to do so in future technology nodes at 3 nm and beyond.

### 6.4.1 PVD copper-alloy seedlayers for self-capping layers for Cu interconnects

A key challenge in scaling Cu interconnects down to current nodes smaller than ~65–45 nm groundrule has been the suppression of electromigration failure caused by Cu diffusion along the top-surface of wires, where there has traditionally been only a dielectric cap layer in contact with the copper. Solutions to this challenge have focused on achieving improved adhesion by fabricating a top-surface metallic interface between the dielectric capping layer and the copper conductor metal.

One solution that has gained wide usage is to modify the composition of the PVD copper seedlayer. Traditional PVD copper seedlayers have been deposited with ultrapure copper (99.995% or higher) sputter targets. By inserting a chemically reactive minority-alloy component, into the PVD copper sputter target, the diffusivity of the minority component during subsequent heat cycling can be exploited to create an adhesion layer between the copper conductor metal and the dielectric cap. The choice of which element to alloy with the Cu in the seedlayer is dictated by the solubility and diffusivity of the element in Cu and by favorable energetics for oxide formation relative to $SiO_2$ [44–52]. Two elements that meet these criteria that have been used in the industry are aluminum and manganese, typically at concentrations of $\leq 1.0\%$ in copper. "Self-capping" adhesion layers of this type are composed of the minority-alloy metal chemically bonded to the dielectric cap layer, such as manganese oxide in the case of a manganese-doped copper target. This self-capping behavior has multiple advantages in that it does not require any additional processing steps (the copper seedlayer deposition is necessary in any case) or tooling (the alloy target is simply installed in the existing PVD Cu chamber in place of ultrapure Cu). The remainder of the process flow is essentially unchanged.

The process flow is illustrated schematically in Fig. 3.12. Like the previous example, a dual-damascene via and trench structure is shown in a low-k dielectric. The PVD Ta(N) barrier/liner and PVD copper-alloy seedlayer are shown already in place and have been deposited in a conventional clustered-vacuum process sequence (Fig. 3.12a). The dopant (minority alloy component) atoms are indicated schematically in the illustration. After the wafer is removed from the vacuum environment the seeded structures are filled with electroplated copper, with additional

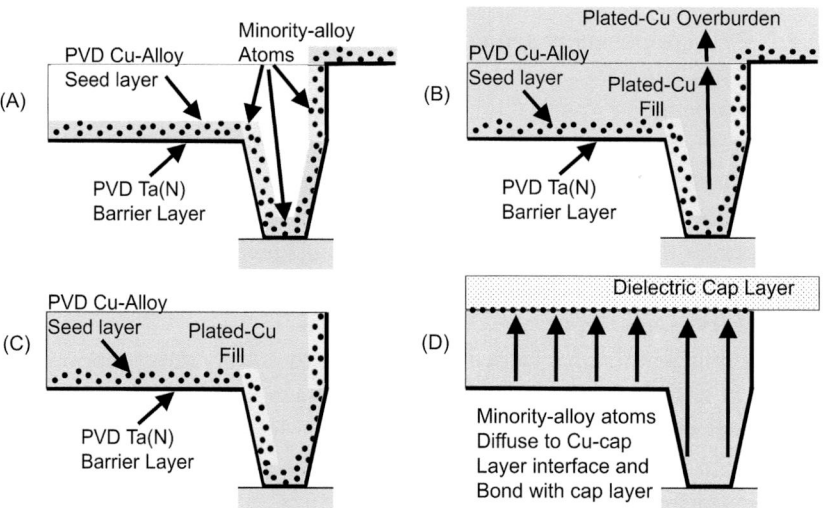

**Figure 3.12** Schematic (not to scale) illustration of self-capping phenomenon with PVD Cu-alloy seedlayers. (a) Following PVD Ta(N) barrier and Cu-alloy seed deposition, the feature is filled with electroplating (b) and planarized (c). After CVD dielectric cap-layer deposition, the minority alloy-component in the seedlayer (typically Mn or al) segregates to the Cu-capping layer interface (d).

copper ("overburden") beyond the minimum needed for feature-fill being plated in order to help stabilize the grain structure (Fig. 3.12b). Following a low-temperature anneal to recrystallize the plated copper, the entire wafer is subjected to chemical-mechanical polishing (CMP) to polish off the overburden and planarize the entire interconnect structure to achieve the specified trench height (Fig. 3.12c). At this point, the minority-alloy species are still contained within the volume of the PVD alloy seedlayer as originally deposited. Following planarization, the structure is deposited with a CVD capping layer, typically $SiC_xN_yH_z$ or a related compound, at a temperature in excess of 200°C. This heat treatment drives the minority-alloy component to the top surface, where it reacts with oxygen in the capping layer to form an adhesive bond (Fig. 3.12d). The gettering of the dopant out of the copper build by the self-capping reaction enhances electromigration while reducing interconnect resistance to levels approaching (within ~5%) those seen with pure Cu.

Fig. 3.13 shows transmission electron microscopy (TEM)/energy dispersive X-ray analysis (EDX) images of trench structures fabricated using PVD CuAl and CuMn seedlayers, illustrating the top-surface segregation of Al and Mn at the copper- $SiC_xN_yH_z$ interface. Electromigration lifetime enhancements ranging from 3× up to nearly 100× have been reported in the literature using PVD alloy seedlayers [50].

As copper interconnect linewidths shrink below 40 nm, the gapfill capabilities of the traditional PVD TaN/PVD Ta/PVD Cu barrier/liner/seed system have been challenged by the increasing difficulty of plating defect-free copper inside features seeded with thin PVD Cu on top of PVD Ta. A solution which has found wide applicability in the industry is the

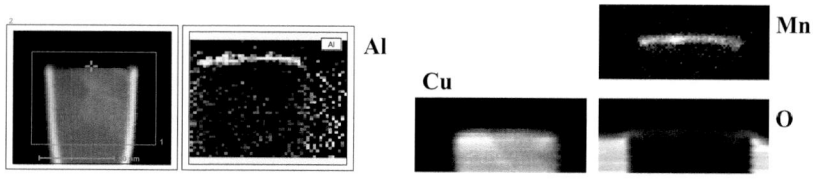

**Al Segregation at Cu Top Surface    Mn Segregation at Cu Top Surface**

**Figure 3.13** TEM/EDX images of copper interconnect structures built using PVD CuAl (left) and PVD CuMn (right) alloy seedlayers. The self-capping behavior of the minority-alloy components are illustrated by the Al and Mn elemental maps. *(From Nogami, et al., High reliability 32 nm Cu/ULK BEOL based on PVD CuMn seed, and its extendibility, Proceedings of the International Electron Devices Meeting (IEDM), 2010, p. 33–35.1, reprinted with permission.)*

replacement of the PVD Ta liner layer with a conformal CVD cobalt liner [53,54]. The PVD TaN/CVD Co/PVD Cu barrier/liner/seed combination has been shown to give improved copper gapfill due to the improved adhesion and wettability of copper on cobalt as opposed to tantalum.

The adoption of CVD cobalt in substitution of PVD Ta as a liner layer for copper has the drawback that the self-capping capability of seedlayer dopants like Mn or Al can be partially neutralized in the presence of the CVD cobalt layer [54–56]. This is believed to be due to the high oxygen content of the carbonyl compounds used as CVD cobalt precursors. As a result, when a CVD Cobalt liner is used, a selective cobalt cap must also be used for top-surface adhesion in lieu of the alloy seedlayer. A schematic of the process flow is shown in Fig. 3.14a–d, with TEM/EDX images of the cobalt liner and cap in Fig. 3.15 [55]. This combination of PVD TaN barrier/CVD Cobalt liner/PVD Copper seed with electroplated copper fill and selective CVD Cobalt capping layer has been shown to enhance electromigration performance by up to 100x at 22 nm groundrules [55] and continues to be extended down to the 3 nm technology node and beyond.

At technology nodes of 7 nm and smaller, the scaling of copper interconnects faces challenges in maintaining the performance advantages they have long demonstrated over other possible interconnect metals.

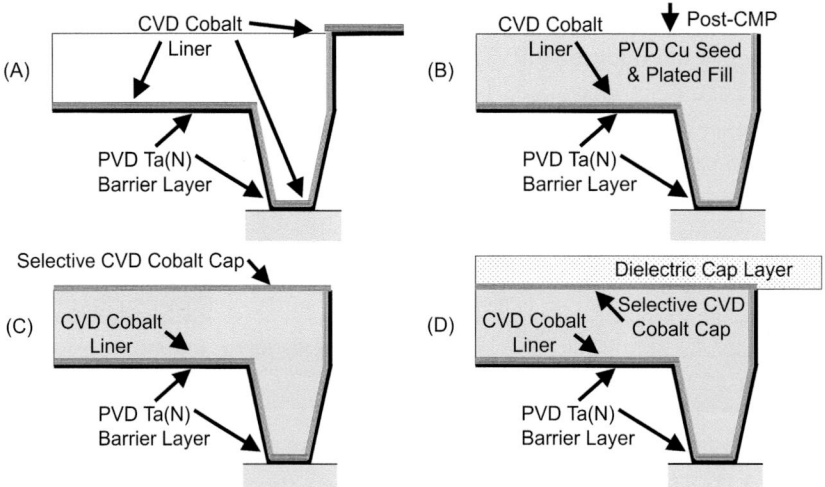

**Figure 3.14** Copper interconnect fabrication sequence using a hybrid PVD/CVD barrier/liner (Schematic-not to scale). (a) PVD TaN and CVD Co liner deposition (b) following PVD Cu seed, plating and CMP, the top-surface Cu is exposed. (c) A CVD Co layer selective to Cu is deposited, which enhances the adhesion between the copper and the dielectric diffusion barrier cap (d).

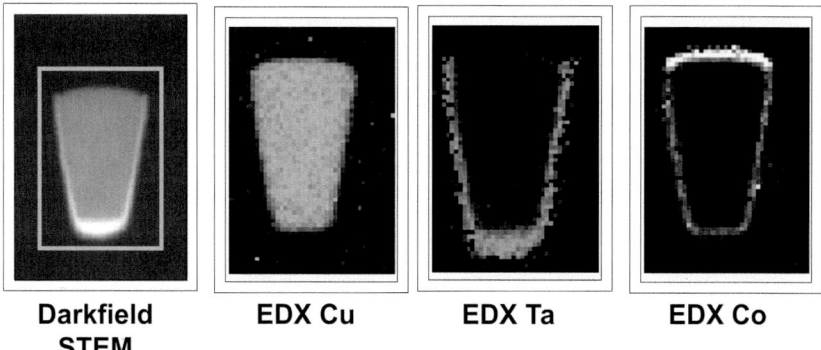

**Figure 3.15** STEM dark field and EDX elemental maps of a Cu interconnect fabricated with PVD TaN, CVD cobalt liner PVD Cu seed and selective CVD cobalt cap. The Cu, Ta and Co maps illustrate the CVD Co coverage on all sides of the Cu conductor. *(From Simon, et al., Electromigration comparison of selective CVD cobalt capping with PVD Ta(N) and CVD cobalt liners on 22nm-groundrule dual-damascene Cu interconnects, Proceedings of the International Reliability Physics Symposium (IRPS), IEEE, 2013, 3F.4.1, reprinted with permission.)*

Modern commercially available PVD Ta(N) sources are capable of depositing ionized layers of ~10−15 Å reproducibly, but this capability may be only of theoretical interest if the minimum thicknesses needed for a hermetic diffusion barrier are in the range of ~20−30 Å. With future technology groundrules projecting linewidths (half-pitch) ~20 nm or less, interconnect trenches could have barrier/liner layers that occupy 25% or more of the cross-sectional area, regardless of whether they are deposited by PVD, CVD, or ALD. In addition, advanced-groundrule vias at these dimensions and the trend toward unidirectional BEOL patterning will place further focus on via resistance, and particularly on resistive barrier/liner layers, as performance limiters for BEOL interconnects [57,58].

The continued focus on finding replacements for separate PVD barrier layers in the interest of lowering interconnect resistance will likely drive future solutions in which previously discussed process technologies (CVD wetting layers, alloy seed layers) are combined together in order to meet performance and reliability challenges associated with interconnect scaling [46−48,59−61].

### 6.4.2 PVD copper fill of advanced-groundrule interconnects using reflow

As copper interconnects scale to wiring dimensions smaller than 15−20 nm, a key scaling challenge faces the traditional process flow for filling damascene structures. The minimum thickness for effective barrier (TaN and/or Ta), and liner (CVD Co or Ru, etc.) for a process flow like the one depicted in Fig. 3.11 is typically of order 1—2 nm on the feature sidewalls for each layer. Similarly, the minimum thickness for a copper seedlayer capable of being plated is of order 2—3 nm sidewall coverage. The combined barrier, liner, and seed layers will have an aggregate thickness of at least 4 nm on each side of an unfilled damascene structure for a total of 8 nm when the opposite side of the structure is considered. Thus, before the electroplating process can take place, a structure that is $\leq$ 15—20 nm wide will have roughly half of its volume occupied by the barrier, liner, and seed layers, with accompanying difficulties in achieving mass flow of plated copper into the feature. This example is an idealized best-case in that it assumes that sidewall coverage of the barrier, liner, and seedlayers is uniform over the full height of the feature, which will not be true for even the best ionized PVD processes. Additionally, minimum-thickness PVD Cu seedlayers face difficulties upon air exposure due their high sheet resistance, which limits plating current, and the ready formation of native copper oxide, which is soluble in sulfuric-acid based copper plating baths.

A key solution to these problems is to fill the smallest ($\leq$15—20 nm) features entirely with ionized PVD Cu by exploiting controlled heating, surface mobility, surface tension, and capillary action to flow the thin PVD Cu layer into the smallest features [62—85]. A schematic of this type of process is shown in Fig. 3.16a—d. In Fig. 3.16a, we see the feature after a conventional deposition of the deposition of barrier, liner and ionized PVD Cu. (The barrier and liner layers are illustrated as one layer for clarity.) In current semiconductor technologies, the barrier will typically be ALD TaN and/or PVD Ta(N), the liner will be CVD Co, and the Cu will usually be deposited on a cooled or room-temperature chuck. As depicted in Fig. 3.16a, the ionized PVD Cu will have a characteristic bottom-heavy directional coverage in the damascene feature, but judicious choices of deposition thickness and pedestal bias can enable a continuous copper layer from the feature bottom up to the top surface, as long as the feature does not have an undercut or re-entrant (concave) profile. In Fig. 3.16b, we see the copper coverage change during the heating the wafer to ~250—300°C, typically by lamps in the chamber. Capillary action in the smallest features, combined with surface tension in the continuous copper layer on the sidewalls and top, causes a wicking action of the copper down into the feature. The result is partial or complete copper feature fill without the use of electroplating. After cooling the wafer down to the original deposition

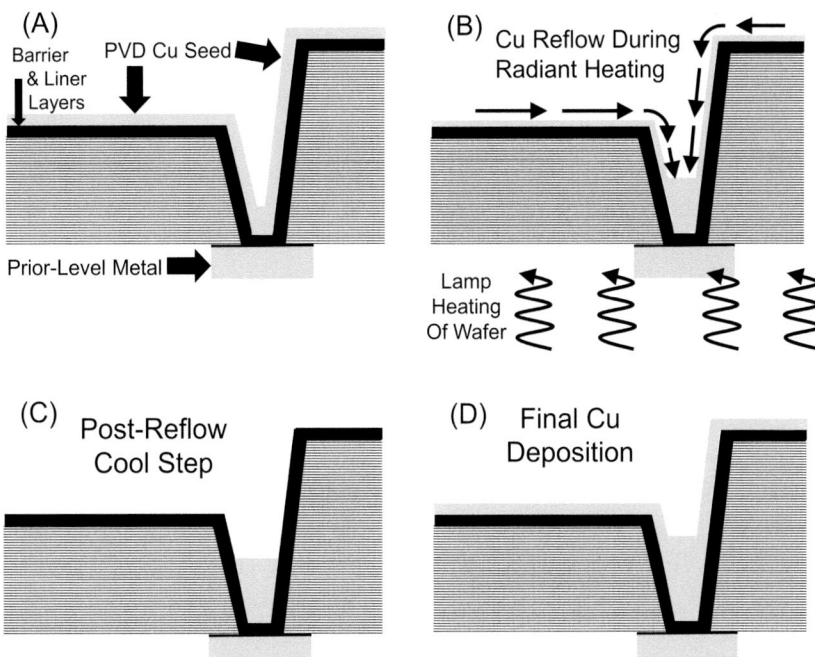

**Figure 3.16** Example of the process flow for a dual-damascene Cu interconnect process using ionized PVD Cu combined with heated reflow to fill small features with dimensions ~15–20A and smaller (Schematic-not to scale). (a) We see a damascene structure which has been deposited with PVD, ALD, or CVD liner and barrier layers followed by a PVD Cu seed layer. (b) The wafer is subjected to heating, typically using in-situ lamps. The heating increases Cu mobility on the surface, causing a wicking of the Cu down into the smallest features as a result of capillary action. (c) The wafer is cooled down to room temperature, and the deposition and reflow cycle can be repeated, if needed, or a final Cu deposition can be done (d). The wafer will subsequently be plated and polished using CMP.

temperature (Fig. 3.16c), this cycle can be repeated multiple times if needed for more complete feature fill in the smallest structures. After a final cooling cycle, a final copper layer is deposited (Fig. 3.16d) so that the top surface can support a plating process that will fill larger features and create an overburden for CMP.

Feature fill using PVD Cu reflow is now industry standard for interconnect features ≤15–20 nm wide. Various published studies at wiring dimensions down to ~18–20 nm dimensions [62–82] indicate improved feature fill, yields, and reliability (electromigration) when compared to PVD seed with electroplated fill. One aspect of reflow PVD Cu feature fill that is superior to plated copper is the self-annealing and grain growth of the copper fill metal in minimum-dimension features due to the elevated temperature of the PVD reflow process. Kikuchi backscattering studies at 7 nm ground rules of the Cu grain size [81,82] show a 2x larger grain size in minimum-feature structures with a concomitant improvement of 10x in the earliest electromigration failure times, and 2x in the median failure times.

While PVD copper reflow is a key method of extending copper damascene technology, it puts constraints on process and feature engineering are, if anything, even more stringent than in the case for thin PVD copper seed with plated fill. As is the case with thin seed and electroplating, it is of utmost importance that all the surfaces in the feature be non-reentrant, which is to say that they have no concave surfaces, undercuts, or overhangs. These profile requirements are needed to ensure that there is no shadowing of the line-of-sight ionized PVD barrier, liner, and copper depositions. The continuous, unbroken copper layer is critical to maintain copper surface tension, which enables the capillary action to wick the copper down into the smallest features. In addition, the deposition thicknesses and bias settings need to be carefully engineered so that the thickest copper is in the damascene feature bottoms, and copper dewetting or agglomeration does not occur on the top surfaces.

### 6.4.3 Via-resistance reduction using selective ALD barrier sequences

A major performance issue in the scaling copper interconnects is the resistance of the PVD and/or ALD TaN barrier layer at the via bottom. Previous studies have shown this layer to be the single largest vertical series resistance in the via [58]. In the third edition of this handbook, we reviewed the use of in-situ Ar+ sputter-etch processes as a means of reducing the via resistance by thinning the barrier layer and enlarging the via contact area [28–34]. Unfortunately, Ar+ sputter etching is of diminishing utility at technology nodes smaller than 45 nm, due to the high damage that occurs to low-K dielectrics and the unacceptable feature enlargement that occurs due to the beveling of unfilled damascene structure corners with Ar+ sputter removal.

At current groundrules smaller than the 7 nm technology node, a new means of reducing the damascene via resistance is achievable by using a self-assembled monolayer (SAM) to suppress nucleation of the ALD TaN barrier layer specifically on the exposed metal contact, which reduces the via resistance without compromising overall barrier performance [83–85].

A schematic diagram of this process flow is shown in Fig. 3.17a—e. In Fig. 3.17a, we see in the unfilled damascene feature incoming into the vacuum-clustered deposition tool. The structure is degassed, and the exposed Cu contact from the previous wiring level is cleaned of native oxides using a reactive hydrogen plasma. In Fig. 3.17b, the structure is exposed to a gas-phase self-assembled monolayer (SAM) treatment which nucleates an organic layer preferentially on the exposed Cu while leaving the dielectric surfaces untouched. In Fig. 3.17c, the ALD TaN layer is deposited. The ALD TaN covers the dielectric surfaces as normal, but adsorption on the exposed Cu metal contact is blocked by the SAM layer. Following the ALD TaN step, the SAM layer is removed by plasma treatment, leaving behind the full-thickness ALD TaN layer on the dielectric sidewalls, but leaving the exposed Cu contact essentially unchanged (Fig. 3.17d). The following steps in the vacuum process flow will typically include PVD Ta layer deposition or densification treatment, CVD Co wetting layer, and Cu seedlayer and reflow fill of the smallest features. Plating and CMP will follow once the wafer is removed from the vacuum cluster (Fig. 3.17e). This method of enhancing ALD barrier layer selectivity is often referred to as "reverse selective barrier" (RSB).

Published results on 44 nm pitch [83] Cu interconnects show a 50% reduction in via resistance with the RSB barrier sequence compared to control splits with no SAM layer prior to ALD TaN deposition. Reliability

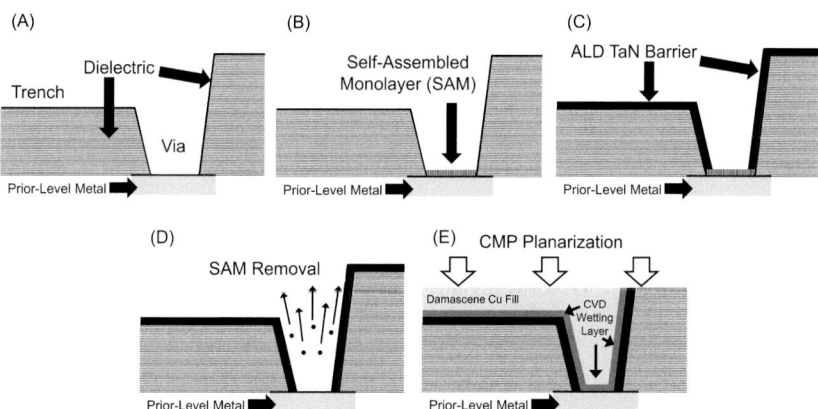

**Figure 3.17** (a)—(e) steps in the use of self-assembled monolayers (SAM) to suppress ALD TaN nucleation on the exposed Cu contact in a dual-damascene Cu structure (Schematic-not to scale). The final structure (e) has ALD TaN barrier on the sidewalls and trenches, but not on the via contact area, where it would cause increased resistance (see text).

performance is unaffected, with no degradation in time-dependent dielectric breakdown (TDDB) or electromigration when RSB splits are compared to non-RSB controls. This type of selectivity enhancement to ALD barrier layers is a critical part of extending Cu damascene interconnects and is in use at advanced technology nodes below 7 nm groundrule.

# 7. Contamination and metrology

Quality control monitoring of sputter-deposited films is key to successful operation of any semiconductor fabrication facility. While there are a number of measurement and characterization techniques which have been used in research settings, only a relatively small number of them are used for in-line monitoring in volume production or development facilities. A brief summary of methods which are currently in wide use are given below.

## 7.1 Metrology of sputtered films
### 7.1.1 Resistance/four-point probe measurement

The most established metrology technique for thin metallic films is the four-point probe resistance measurement, in which four in-line probe tips are used to measure sheet resistance. For a rectangular block of conducting material of resistivity $\rho$, length $L$, thickness $t$, and width $w$, the resistance $R$ will be given in formula $R = \rho L/(tw)$.

For the special case where the sample's width is equal to the length, $w = L$, the resistance expression simplifies to $R = \rho/t$. The resistance in this case is referred to as the sheet resistance per unit square and is quoted in terms of $\Omega$/square. We thus see that if the bulk resistivity of the material is known with certainty, the thickness can be determined using the four-point probe sheet resistance measurement. Alternatively, if the thickness is measured using some separate technique, the four-point probe measurement can be used to determine the bulk resistivity of the film.

Typically, the two outer probes are operated in current source mode, with the two inner probes measuring the voltage drop across the current path in the sample (Fig. 3.18). This arrangement eliminates any confounding effects due to contact resistance. Assuming that the dimensions of the sample are much greater than the probe-tip spacing, the geometrical correction factor to convert the current and voltage measurements is $R = 4.532\ V/I$, where $V$ is the voltage between the inner-probe tips, and $I$ is the current forced through the outer probe tips. Probe-tip spacings of $\sim 0.5-2.0$ mm are typical for semiconductor applications.

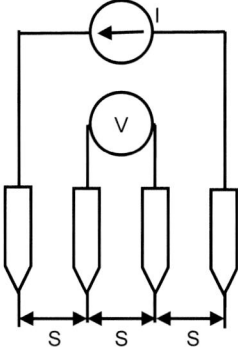

**Figure 3.18** Probe configuration for a four-point probe resistance measurement. For thin metal films, the outer probes are typically operated in current-source mode, with the inner two probes used for voltage measurement. For a thin metal film, the sheet resistance be measured from the voltage and current values (see text).

Commercially available four-point probe tools for the semiconductor industry will typically have a user-selectable probing pattern that samples the wafer center, equally azimuthally spaced points at the wafer edge (at a user-specified maximum radius $r$), and concentric rings of points at some fractional radii in between (most often $r/3$ and $2r/3$). This concentric-ring sampling enables standardized measurements of sputtered-film uniformity which are widely accepted in the industry for process benchmarking.

Modern commercial instruments can be programmed to adjust the probe current automatically so as to give a suitably large inner-probe voltage reading that results in minimal error, enabling measurements ranging from $\sim 1$ to $\sim 1$ m$\Omega$/square. One limitation for four-point probe measurements is that the probe tips can punch through films of $\sim 50$ Å or thinner, thus leading to spurious readings indicative of the substrate or prior layer rather than the film itself. Similarly, it should be noted that if one is measuring a multilayer film stack, the measured film of interest should be the lowest resistance film in the stack.

### 7.1.2 Non-destructive thickness measurements

In modern semiconductor applications, the deposited metal films are often thin enough that surface-scattering and grain-morphology effects make the effective resistivity of thin PVD films significantly larger than bulk values found in references. In addition, the physical contact required for four-point probing frequently makes it an unsuitable measurement technique for product wafers. Here we review several alternative thickness measurement techniques which are noncontact and independent of resistivity effects, and which have been adapted for commercial semiconductor applications.

### 7.1.2.1 X-ray fluorescence

X-ray fluorescence (XRF) is a well-established technique for materials analysis that has been adapted for in-line semiconductor industry use [86]. The sample under analysis is illuminated by X-rays or gamma rays, which results in the excitation of core-level electrons to excited states. The radiative decay of these electrons from the excited states back to their respective ground states results in the emission of fluorescent or "secondary" X-rays that are characteristic of the energy levels of each atomic species and thus serve as a spectroscopic fingerprint for each element present in the sample. Product-wafer spot sizes can be as small as ∼100 nm or less.

Since the XRF signal intensity for each atomic species correlates directly to the number of atoms present, the XRF signal can be used as a direct measurement of the thickness of metals and alloys. The primary strength of XRF as a measurement technique lies in its ability to assess thin-film thicknesses and alloy concentrations independent of any numerical modeling techniques. The high-frequency transparency of metals means that samples of several microns thickness can be measured. At the opposite extreme, XRF can, in theory, be used to measure arbitrarily thin layers of <10 nm thickness. The main challenge in measuring very thin layers is the long acquisition time needed for the XRF detector to acquire a statistically significant number of fluorescence counts and the care needed to deconvolve any spectral overlap coming from substrates, underlayers, etc.

There are some limitations to XRF. For rigorous quantitative results, the XRF signal should be calibrated against known thickness standards. XRF also is less useful for measurement of elements with low atomic numbers, typically $Z < 11$, due to weak fluorescence from these species. The X-ray transparency of the films in question means that XRF generally cannot be used for depth profiling: it can measure the thicknesses of stacked films but typically cannot tell which one is on top of the other. Finally, care needs to be taken in selecting which spectral lines to sample since strong spectral signals from substrates or underlayers can potentially overlap with the thin-film signals being measured, leading to error in the estimation of the signal strength of the latter.

### 7.1.2.2 X-ray reflectance

X-ray reflectance (XRR) is an adaptation of the well-known phenomenon of fringes which occur due the constructive and destructive interference between top- and bottom-surface reflections in thin films. In this respect, XRR is analogous to the visible-to ultraviolet-wavelength spectrophotometry of dielectric films which is widely used in the industry.

In XRR, the sample being measured is illuminated with X-rays, and the reflected signal intensity is measured at different incident angles. The resulting interference fringes enable film thicknesses, interface roughnesses, and densities of thin-film layers on the sample to be inferred from mathematical modeling of the optical path lengths in the sample stack. The dependence of XRR on mathematical modeling makes it necessary to have a starting estimate of the film-stack parameters and composition. The modeling of the refractive index at X-ray wavelengths is sensitive to the electron density in the material, and materials with similar electron densities will not be easily distinguishable from each other with this technique.

Depending on the material, single layers less than $\sim 30-100$ Å might not be measurable due to the lack of interference fringes. End users of commercially available XRR equipment will typically be using proprietary, commercial software packages for the layer analysis and need to be aware of any limitations of the modeling software for their application.

### 7.1.2.3 Time-resolved picosecond ultrasound

Another technique for thin-film thickness measurement, which has found widespread commercial application, is time-resolved ultrasonic pulses [87–90]. In this technique, the film under measurement is illuminated with laser pulses of femtosecond duration. The optical pulses cause local heating in the film sample which results in the optical energy being converted into acoustic pulses which reflect off the bottom of the film sample, and which can also be partially transmitted to the layers below. A second "probe" laser detects the reflected pulse when it returns to the top surface of the measured film. Once the speed of sound is calibrated in the sample film, the technique serves as a rapid, in-line film thickness measurement. More complex, multiple-layer stack measurements are also possible if the film-stack characteristics are known and properly modeled.

Limitations of picosecond ultrasound are somewhat similar to those for XRR: the inferred measurements are extracted from algorithms which numerically model the reflection phenomena and will require proper calibration and interpretation in initial setup. Ultra-thin layers might be too thin to cleanly resolve the reflected pulses.

Regardless of which non-contact method is used to measure thin-film thickness, the method should be calibrated against reference samples using an independent technique such as TEM.

## 7.2 Contamination control and prevention in sputtering systems

Particle monitoring, whether done on product wafers or blankets, is the standard means by which the cleanliness level of a sputter-deposition chamber is monitored. Sputter-system contamination can be intrinsic (due to sputtered material delaminating off of chamber shielding) or extrinsic (particles or contaminants brought into the system from other process sectors, e.g., photoresist). Methods for minimizing either type of contamination are often specific to proprietary equipment designs and cleaning techniques, as well as the specific material being sputtered. Nevertheless, some general guidelines are well established.

### 7.2.1 Tooling and shielding considerations

A primary consideration in the design of magnetron sputter sources is that there are no regions on the sputter target that are redeposited with target material during deposition [91]. A magnetron design that fully erodes the entire target surface is said to have full-face erosion. If a magnetron design does not provide for full-face erosion, sustained operation will lead to a build-up of redeposited target material on regions of the target which are not eroded. The redeposited material will be loosely adhering, leading to target flaking and erratic deposition rates when the redeposited material becomes sufficiently thick. Complex magnetron shapes or designs might require special burn-ins ("pasting," i.e., extended depositions onto shutters or dummy wafers) at regular intervals to clear redeposited material off the full target surface.

Proper shielding design and surface treatment are crucial to optimal particle performance. Regardless of the material being sputtered, the shielding should be designed in a way that minimizes sharp- or small-radius corners. Sharp-radius corners create high local concentrations of sputtered-film stress, which has the effect of initiating cracking and causing delamination of the deposited film off of the shielding. By designing corners in the shielding to have as large a radius of curvature as possible, high local concentrations of film stress are avoided.

Surface treatment can be critical to particle performance as well. For lower-stress materials, grit blasting of stainless steel or aluminum shielding can be sufficient to ensure good sputtered-film adhesion throughout kit life. For high-stress refractory metals such as Ti(N), Ta(N), and TiW, it is typically necessary to coat the shielding with a coating of a more adhesive metal, typically aluminum, to prevent delamination of the sputtered films (Fig. 3.19).

While adhesion might be enhanced, a dense, sputtered layer of aluminum would not prevent the stress build-up of the high-stress sputtered material from the target. However, application of the adhesion layer using plasma flame spraying can introduce substantial porosity and a high degree of surface roughness into the aluminum, which provides significant stress-relief within the aluminum layer and serves to further prolong the service life of the shields (Fig. 3.19). Combining aluminum flame spraying with optimized corner shapes can increase shielding kit life by ~10%—30% or more [92—94].

### 7.2.2 Extrinsic contamination control: RGA monitoring

Contamination from other processes sectors can have an adverse effect on sputter-tool cleanliness. Most notable is organic contamination from

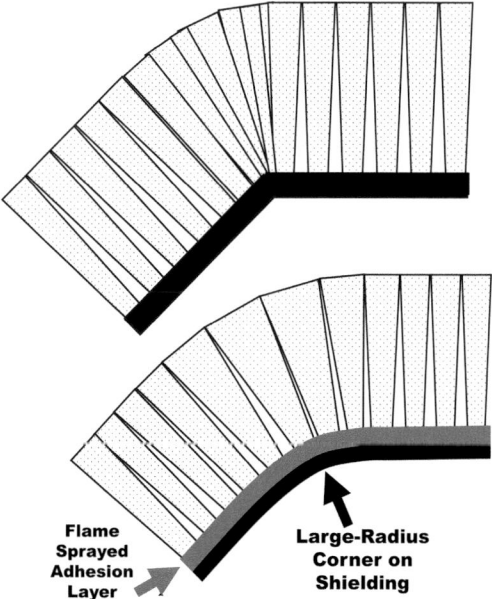

**Figure 3.19** Diagram illustrating practices for reducing delamination of accumulated metal deposited on sputter-tool shielding. In the lower diagram, the enlarged radius of curvature and the flame-sprayed adhesion layer result in improved particulate performance [92].

incomplete photoresist stripping, which presents either an immediate (high-level) or pernicious (low-level) threat to the ultrahigh vacuum environment in sputter tooling. Considerable work has been devoted to developing automated contamination detection of incoming product wafers using RGAs.

Analysis of the outgassing species from a large population of product wafers enables identification of the molecular species associated with specific sources of contamination such as photoresist and outgassing dielectrics. The resulting profile of molecular weights corresponding to each contamination source can then be stored in a system database and compared against the outgassing profiles of incoming wafers. By sampling the outgassing species from each incoming wafer during the degas step, contaminated wafers can be flagged and stopped through a real-time feedback loop so that the ultra-high vacuum (UHV) deposition chambers are not contaminated. The specific system configuration for data acquisition and feedback varies according to the systems infrastructure in each facility. Two examples are described in the literature by Xu et al. [95] and Rampf and McCafferty [96].

## 8. Future directions

Sputter deposition is still the preferred deposition technique for thin film applications where purity, density, and stoichiometry are critical to the film's functionality, especially for metals. Evolutionary advances and refinements to sputter deposition tooling and processes continue to proceed in rough cadence with semiconductor industry technology nodes.

Some of the trends discussed in this and previous editions of this chapter are now widespread in the semiconductor industry. Many process sequences in high-volume manufacturing (HVM) now require the vacuum clustering of PVD depositions with other process types, such as ALD, CVD, and plasma or chemical clean steps. Copper damascene interconnects at advanced groundrules of 16 nm and below typically use ionized PVD Ta or TaN combined with some combination of ALD TaN barrier or CVD Co wetting layers, as discussed previously in this chapter. As was discussed in Section 3.6, one significant advance in PVD deposition is the use of PVD Cu copper combined with heated reflow steps to completely fill the smallest wiring features (<20 nm), which circumvents the difficulties of scaling thin PVD Cu seedlayers and electro-plating at those dimensions. Self-capping Cu-alloy seedlayers have already served multiple roles as capping and seedlayers for several technology generations.

In the middle-of-line (MOL) contact area, scaling of CVD or ALD TiN layers to serve as fluorine barriers for WF6 deposition of tungsten is reaching scaling limits, and substitute liner candidates include ALD or CVD W-based nucleation layers or ionized PVD W to replace the dissimilar Ti or TiN liner layers.

Certain overarching trends will dictate how and which PVD techniques will be used at technology groundrules below the 2 nm node. For BEOL copper interconnect applications, the effect of sidewall scattering causes the effective resistivity of the copper in minimum-dimension wires to increase much more rapidly once the linewidths are reduced below the bulk electron mean-free path (MFP) of 39 nm. For linewidths below the bulk MFP, the effective copper resistivity increases rapidly with decreasing linewidth, as sidewall scattering increasingly dominates electron mobility in constrained dimensions. Simulations and experiments showing a greater-than-twofold increase in the effective copper resistivity as the linewidth shrinks from >100 nm down to 20 nm have been reported [97−105].

At via and wiring dimensions below 10 nm, the effective resistivity of copper becomes so high that alternate candidates [95−117], such as cobalt, tungsten, molybdenum, and the platinum-group metals (i.e., Ru, Ir, Rh) become potentially attractive candidates as the conductor metal. The platinum-group metals in particular have favorable scaling properties at linewidths <10 nm and have minimal need for diffusion barriers due to their chemical inertness. They also have high melting points, which suggest good electromigration performance.

In addition to the damascene flows discussed previously, subtractive process flows are possible with many of these "post-copper" metal candidates [112−117], in that unlike copper, they are readily capable of being patterned with reactive ion etch (RIE) chemistries. One example of such an alternative process flow is shown in Fig. 3.20. In comparison to damascene process flows, in which the dielectric layer is first deposited and etched with wiring features which are then filled with metal, a subtractive metal process flow does the opposite. The blanket PVD metal is deposited and annealed first, and advanced patterning and RIE is used to etch away the unwanted metal so that dielectrics or airgap structures can be deposited between the wires. The deposition and annealing of the metal layer prior to patterning enables grain growth larger than the linewidths of the final structure, which is difficult in damascene process flows. By using separate patterning steps, via and wire structures can be fabricated from the same initial metal layer, which enables a via-wire interface without any high resistance barrier or

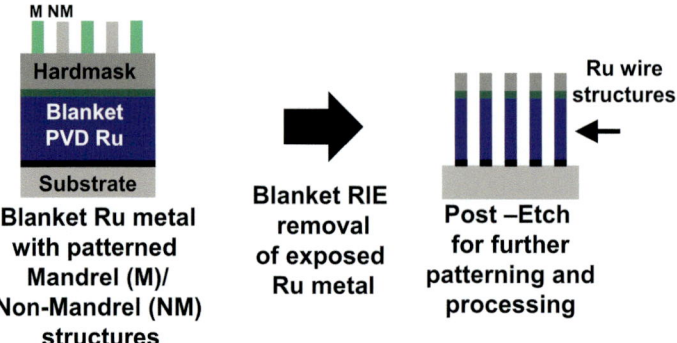

**Figure 3.20** Example of a process flow for forming interconnects using a PVD metal layer with mandrel (N)/non-mandrel (NM) patterning and subtractive RIE to remove the unwanted metal and form interconnects.

adhesion layers which would increase the via resistance [117]. Various versions of these process flows exemplify new types of PVD-based metal deposition sequences which could replace established damascene process flows in the future.

## References

[1] J.E. Mahan, Physical Vapor Deposition of Thin Films, Wiley-Interscience, New York, 2000 (For detailed discussions of sputtering kinematics, refer to chapter VII).
[2] D.M. Mattox, Handbook of Physical Vapor Deposition (PVD) Processing, second ed., William Andrew, Norwich, 2010.
[3] S. M Rossnagel, Sputter deposition for semiconductor manufacturing, IBM J. Res. Dev. 43 (1999) 163.
[4] S.M. Rossnagel, Thin film deposition with physical vapor deposition and related technologies, J. Vac. Sci. Technol. A 21 (5) (2003) S74.
[5] S.M. Rossnagel, Sputtering and sputter deposition, in: K. Seshan (Ed.), Handbook of Thin Film Deposition (Materials and Processing Technology), second ed., William Andrew, Norwich, 2002, pp. 319−348 (Chapter 8).
[6] J. Bohdansky, A universal relation for the sputtering yield of monatomic solids at normal ion incidence, Nucl. Instrum. Methods B2 (1984) 587.
[7] N. Matsunami, et al., Energy dependence of the ion-induced sputtering yields of monatomic solids, At. Data Nucl. Data Tables 31 (1984) 1.
[8] R. Kelly, The mechanisms of sputtering part I. Prompt and slow collisional sputtering, Radiat. Eff. 80 (1984) 273.
[9] J. Bohdansky, J. Roth, H.L. Bay, An analytical formula and important parameters for low-energy ion sputtering, J. Appl. Phys. 51 (5) (1980) 2861.
[10] D. Chapman, Glow Discharge Processes: Sputtering and Plasma Etching, Wiley-Interscience, New York, 1980.
[11] S.M. Rossnagel, H.R. Kaufman, Induced drift currents in circular planar magnetrons, J. Vac. Sci. Technol. A 5 (1987) 88.

[12] E. Klawuhn, G.C. D'Couto, K.A. Ashtiani, P. Rymer, M.A. Biberger, K.B. Levy, Ionized physical-vapor deposition using a hollow-cathode magnetron source for advanced metallization, J. Vac. Sci. Technol. A 18 (4) (2000) 1546.
[13] K.A. Ashtiani, E. Klawuhn, D. Hayden, M. Ow, K.B. Levy, M. Danek, A new hollow-cathode magnetron source for 0.10μm copper applications, in: Proceedings of the International Interconnect Technology Conference (IITC), IEEE, 2000, p. 37.
[14] J. Musil, J. Vicek, P. Baroch, Magnetron discharges for thin films plasma processing, in: Y. Pauleau (Ed.), Materials Surface Processing by Directed Energy Techniques (European Materials Research Society Series), Elsevier, Oxford, 2006, pp. 67–110 (Chapter 3).
[15] D. Edelstein, et al., A high performance liner for copper damascene interconnects, in: Proceedings of the IEEE International Interconnect Technology Conference (IITC), 2001, p. 9.
[16] D. Edelstein, et al., An optimal liner for copper damascene interconnects, in: Proceedings of the Advanced Metallization Conference (AMC), 2001, p. 541.
[17] J. Lei, S.E. Phan, X. Lu, C.T. Kao, K. Lavu, Advantage of Siconi(TM) preclean over wet clean for pre silicide applications beyond 65nm node, in: IEEE International Symposium on Semiconductor Manufacturing (ISSM), 2006, p. 393.
[18] R. Yang, N. Su, P. Bonfanti, J. Nie, J. Ning, T. Li, Advanced in situ pre-Ni silicide (Siconi) cleaning at 65nm to resolve defects in $NiSi_x$ modules, J. Vac. Sci. Technol. B 28 (1) (2010) 56.
[19] X. Fu, et al., Advanced preclean for integration of PECVD SiOCH (k=2.5) dielectrics with copper metallization beyond 45nm technology, in: IEEE International Interconnect Technology Symposium (IITC), 2006, p. 51.
[20] C. Lin, Reactive barrier/seed preclean process for damascene process, US Patent 7,273,808 B1.
[21] J.N. Broughton, C. Backhouse, M. Brett, S. Dew, G. Este, Long throw sputter deposition of Ti at low pressure, in: Proceedings of the 12th International VLSI Multilevel Interconnection Conference, 1995, p. 201.
[22] A.A. Mayo, S. Hamaguchi, J.H. Joo, S.M. Rossnagel, Across-wafer nonuniformity of long throw sputter deposition, J. Vac. Sci. Technol. B 15 (1997) 1788.
[23] S.M. Rossnagel, D. Mikalsen, H. Kinoshita, J.J. Cuomo, Collimated magnetron sputter deposition, J. Vac. Sci. Technol. A 9 (1991) 261.
[24] J. Hopwood, Ionized physical vapor deposition of integrated circuit interconnects, Phys. Plasmas 5 (5) (1998) 1624.
[25] D. Mao, K. Tao, J. Hopwood, Ionized physical vapor deposition of titanium nitride: plasma and film characterization, J. Vac. Sci. Technol. A 20 (2) (2002) 379.
[26] J. Hopwood (Ed.), Ionized Physical Vapor Deposition, Academic Press, Waltham, 2000.
[27] S.M. Rossnagel, J. Hopwood, Metal ion deposition from ionized magnetron sputtering discharge, J. Vac. Sci. Technol. B 12 (1994) 449.
[28] A.H. Simon, C.E. Uzoh, Open-bottomed via liner structure and method for fabricating same, US Patents 5,933,753 and 6,768,203.
[29] R.M. Geffken, S.E. Luce, Method of forming a self-aligned copper diffusion barrier in vias, US Patent 5,985,762.
[30] S.G. Malhotra, A.H. Simon. Method for depositing a metal layer on a semiconductor interconnect structure. US Patent 6,949,461.
[31] G.B. Alers, et al., Electromigration improvement with PDL TiN(Si) barrier in copper dual damascene structures, in: Proceedings of the IEEE 41st International Reliability Physics Symposium (IRPS), 2003, p. 151.
[32] C.-C. Yang, et al., Simultaneous native oxide removal and metal neutral deposition method, US Patent 6,784,105.

[33] N. Kumar, et al., Improvement in parametric and reliability performance of 90nm dual-damascene interconnects using Ar+ punch-Thru PVD Ta(N) barrier process, in: Proceedings of the Advanced Metallization Conference (AMC), 2004, p. 247.
[34] C.-C. Yang, T. Shaw, A. Simon, D. Edelstein, Effects of contact area on mechanical strength, electrical resistance, and electromigration reliability of Cu/Low-k interconnects, Electrochem. Solid-State Lett 13 (6) (2010) H197.
[35] N. Hosokawa, T. Tsukada, H. Kitahara, Effect of discharge current and sustained self-sputtering, in: Proceedings of the Eighth International Vacuum Congress, Cannes, France, vol. 1, 1980, p. 11.
[36] W.M. Posadowski, Z.J. Radzimski, Sustained self-sputtering using a direct current magnetron source, Vac. Sci. Technol. A 11 (1993) 2980.
[37] Z.J. Radzimski, O.E. Hankins, J.J. Cuomo, W.P. Posadowski, S. Shingubara, Optical emission spectroscopy of high density metal plasma formed during magnetron sputtering, J. Vac. Sci. Technol. B 15 (1997) 202.
[38] Z.J. Radzimski, W.M. Posadowski, S.M. Rossnagel, S. Shingubara, Directional copper deposition using dc magnetron self-sputtering, J. Vac. Sci. Technol. B 16 (1998) 1102.
[39] M. La Hemann, J. Bohlmark, A. Ehiarson, J.T. Gudmundsson, Ionized physical vapor deposition (IPVD): a review of technology and applications, Thin Solid Films 513 (1−2) (2006) 1−24.
[40] D. Edelstein, et al., Full copper wiring in a sub-0.25μm CMOS ULSI technology, in: Technical Digest—International Electron Devices Meeting, IEDM, 1997, p. 773.
[41] J. Heidenreich, et al., Copper dual damascene wiring for sub-0.25 μm CMOS technology, in: Proceedings of the IEEE, International Interconnect Technology Conference (IITC), 1998, p. 151.
[42] E.G. Colgan, P.M. Fryer, Structure and method of making Alpha-Ta in thin films, US Patent 5,281,485.
[43] A. Simon, et al., Temperature control and Ta/Cu interface quality in advanced Cu BEOL, in: Proceedings of the Advanced Metallization Conference (AMC), 2005, p. 429.
[44] D.C. Edelstein, J.M.E. Harper, C.-K. Hu, A.H. Simon, C.E. Uzoh, Copper interconnection structure incorporating a metal seed layer, US Patent 6,181,012.
[45] D.C. Edelstein, J.M.E. Harper, C.-K. Hu, A.H. Simon, C.E. Uzoh, Copper interconnection structure incorporating a metal seed layer, US Patent 6, 399,496.
[46] J. Koike, M. Wada, Self-forming diffusion barrier layer in Cu−Mn alloy metallization, Appl. Phys. Lett. 87 (2005) 41911.
[47] T. Usui, et al., Highly reliable copper dual-damascene interconnects with self-formed MnSi$_x$O$_y$ barrier Layer, in: IEEE Transactions on Electron Devices v.53, 2006, p. 2492.
[48] H. Kudo, et al., Copper wiring encapsulation with ultra-thin barriers to enhance wiring and dielectric reliabilities for 32-nm nodes and beyond, in: International Electron Devices Meeting (IEDM) Technical Digest, 2007, p. 513.
[49] A.H. Simon, et al., Mn-dopant segregation as an indicator of barrier integrity in 32nm. Groundrule Cu/Ultra-Low K interconnects, in: Proceedings of the Advanced Metallization Conference (AMC), 2011.
[50] T. Nogami, et al., High reliability 32 nm Cu/ULK BEOL based on PVD CuMn seed, and its extendibility, in: Proceedings of the International Electron Devices Meeting (IEDM), 2010, pp. 33−35.1.
[51] D.C. Edelstein, et al., Copper interconnect structure and its formation, U.S. Patent 8,969,197B2.
[52] T. Nogami, et al., Electromigration extendibility of Cu (Mn) alloy-seed interconnects, and understanding the fundamentals, in: Proceedings of the International Electron Devices Meeting (IEDM), 2012 p.33.37.1.

[53] J.E. Mueller, et al., Nucleation and wetting of PVD Cu seed on ultra-thin Ta, Co, Ru, and Ta-Ru liners, SRC Report P019881 (2007). https://www.src.org/library/publication/p019881/.

[54] T. Nogami, et al., CVD Co and its application to Cu damascene interconnections, in: Proceedings of the International Interconnect Technology Conference (IITC), IEEE, 2010.

[55] A.H. Simon, et al., Electromigration comparison of selective CVD cobalt capping with PVD Ta(N) and CVD cobalt liners on 22nm-groundrule dual-damascene Cu interconnects, in: Proceedings of the International Reliability Physics Symposium (IRPS), IEEE, 2013, 3F.4.1.

[56] F.H. Baumann, T. Bolom, C.-K. Hu, K. Motoyama, C. Niu and A.H. Simon, Copper interconnect with CVD liner and metallic cap, US Patent 9,111,938.

[57] J. Nag, et al., ALD TaN barrier for enhanced performance with low contact resistance for 14nm technology node Cu interconnects, ECS Trans. 69 (7) (Oct. 2015) 161–169.

[58] A.H. Simon, et al., Via-resistance and TaN/Ta liner properties in advanced-groundrule interconnects, in: Proceedings of the Advanced Metallization Conference (AMC), 2015.

[59] Y.K. Siew, et al., CVD Mn-based self-formed barrier for advanced interconnect technology, in: Proceedings of the International Interconnect Technology Conference (IITC), IEEE, 2013, pp. S02–S03.

[60] T. Nogami, et al., Performance of ultrathin alternative diffusion barrier metals for next-generation BEOL technologies, and their effects on reliability, in: Proceedings of the International Interconnect Technology Conference (IITC), IEEE, 2014, p. 223.

[61] T. Nogami, et al., Through-Cobalt Self Forming Barrier (tCoSFB) for Cu/ULK BEOL: a novel concept for advanced technology nodes, in: Proceedings of the International Electron Devices Meeting (IEDM), vol. 8.1.1, 2015.

[62] R. Brain, Capillary-Driven Reflow of Thin Cu Films with Submicron, High Aspect Ratio Features, Ph.D. Thesis, California Institute of Technology, CA, 1996.

[63] H. Kim, et al., Cu wettability and diffusion barrier property of Ru thin film for Cu metallization, J. Electrochem. Soc. 152 (8) (2005) G594–G600.

[64] K. Shima, H. Shimizu, T. Momose, Y. Shimogaki, Study on the adhesion strength of CVD-Cu films with ALD-Co(W) Underlayers made using amidinato precursors, ECS J. Solid-State Sci. Technol. 4 (2) (2015) 20–29, https://doi.org/10.1149/2.005402ssl.

[65] K. Shima, H. Shimizu, T. Momose, Y. Shimogaki, Comparative study on Cu-CVD nucleation using β-diketonato and Amidinato precursors for Sub-10-nm-thick continuous film growth, ECS J. Solid-State Sci. Technol. 4 (8) (2015) P305–P313.

[66] S.G. Malhotra, et al., Integration of direct plating of Cu onto a CVD Ru liner, in: Proceedings of the Advanced Metallization Conference (AMC), 2004, p. 525.

[67] K. Suzuki, et al., Thin CVD Ru film performance as Cu diffusion barrier and for direct plating, in: Proceedings of the Advanced Metallization Conference (AMC), 2005, p. 469.

[68] C.C. Yang, et al., Physical, electrical, and reliability characterization of Ru for Cu interconnects, in: Proceedings of the International Interconnect Technology Conference (IITC), IEEE, 2006, p. 187.

[69] J. Rullan, et al., Low resistance wiring and 2Xnm void free fill with CVD Ruthenium liner and DirectSeed(TM) copper, in: Proceedings of the International Interconnect Technology Conference (IITC), IEEE, 2010, p. 8.5.

[70] O. Yokoyama, et al., Copper wiring forming method with Ru liner and Cu alloy fill, U.S Patent 9406557B2.

[71] C.C. Yang, Y. Loquet, B. Li, P. Flaitz, D. Edelstein, Reflow of copper on Ruthenium, in: Proceedings of the Advanced Metallization Conference (AMC), 2011.

[72] C.C. Yang, et al., Ultrathin-Cu/Ru(Ta)/TaN liner stack for copper interconnects, IEEE Electron. Device Lett. 31 (2010) 722.
[73] T. Ishizaka, et al., Cu dry-fill on CVD Ru liner for advanced gap-fill and lower resistance, in: Interconnect Technology Conference and 2011 Materials for Advanced Metallization (IITC/MAM), IEEE, 2011, p. 8.5.
[74] T. Matsuda, et al., Superior Cu fill with highly reliable Cu/ULK integration for 10nm node and beyond, in: Proceedings of the International Electron Devices Meeting (IEDM), vol. 29.2, 2013.
[75] R.-H. Kim, et al., Highly reliable Cu interconnect strategy for 10nm node logic technology and beyond, in: Proceedings of the International Electron Devices Meeting (IEDM) vol. 32.2.1, 2014.
[76] Y. Kikuchi, et al., Electrical properties and TDDB performance of Cu interconnects using ALD Ta(Al)N barrier and Ru liner for 7nm node and beyond, in: IEEE International Interconnect Technology Conference/Advanced Metallization Conference (IITC/AMC), 2016, p. 99.
[77] Z. Wu, et al., Pathfinding of Ru-liner/Cu-reflow interconnect reliability solution, in: 2018 IEEE International Interconnect Technology Conference (IITC), Santa Clara, CA, USA, 2018, pp. 51–53, https://doi.org/10.1109/IITC.2018.8430464.
[78] K. Motoyama, O. van der Straten, H. Tomizawa, J. Maniscalco, S.T. Chen, Novel Cu reflow seed process for Cu/low-k 64nm pitch dual damascene interconnects and beyond, in: 2012 IEEE International Interconnect Technology Conference, San Jose, CA, 2012, pp. 1–3, https://doi.org/10.1109/IITC.2012.6251656.
[79] K. Motoyama, O. van der Straten, J. Maniscalco, M. He, PVD Cu reflow seed process optimization for defect reduction in nanoscale Cu/low-k dual damascene interconnects, J. Electrochem. Soc. 160 (12) (2013) D3211–D3215.
[80] P.S. Bhosale, J. Maniscalco, N. Lanzillo, T. Nogami, D. Canaperi, K. Motoyama, H. Huang, P. McLaughlin, R. Shaviv, M. Stolfi, R. Vinnakota, G. How, S. Pethe, B. Sheu, X. Xie, L. Chen, Modified ALD TaN barrier with Ru liner and dynamic Cu reflow for 36nm pitch interconnect integration, in: 2018 IEEE International Interconnect Technology Conference (IITC), Santa Clara, CA, 2018, pp. 43–45, https://doi.org/10.1109/IITC.2018.8430474.
[81] K. Motoyama, O. van der Straten, J. Maniscalco, K. Cheng, S. DeVries, C.-K. Hu, H. Huang, K. Park, Y. Kim, S. Hosadurga, N. Lanzillo, A. Simon, L. Jiang, B. Peethala, T. Standaert, T. Wu, T. Spooner, K. Choi, EM enhancement of Cu interconnects with Ru liner for 7 nm node and beyond, in: 2019 IEEE International Interconnect Technology Conference/Materials for Advanced Metallization (IITC/MAM), 2019.
[82] K. Motoyama, O. van der Straten, J. Maniscalco, K. Cheng, S. DeVries, H. Huang, T. Shen, N. Lanzillo, S. Hosadurga, K. Park, T. Bae, H. Seo, T. Wu, T. Spooner, K. Choi, Co-doped Ru liners for highly reliable cu interconnects with selective Co cap, in: 2020 IEEE International Interconnect Technology Conference (IITC), Santa Clara, CA, 2020 (Virtual), 2020.
[83] S. You, et al., Selective barrier for Cu interconnect extension in 3nm node and beyond, in: 2021 IEEE International Interconnect Technology Conference (IITC), Kyoto, Japan, 2021, pp. 1–3, https://doi.org/10.1109/IITC51362.2021.9537559.
[84] P. Bhosale, et al., Composite interconnects for high-performance computing beyond the 7nm node, in: 2020 IEEE Symposium on VLSI Technology, Honolulu, HI, USA, 2020, pp. 1–2, https://doi.org/10.1109/VLSITechnology18217.2020.9265021.
[85] P. Bhosale, N. Lanzillo, K. Motoyama, T. Nogami, A. Simon, H. Huang, K. Chen, Y. Mignot, D. Edelstein, N. Loubet, B. Haran, S. Parikh, R. Tao, M. Gage, S. Reidy, L. Chen, M. Stolfi, J. Lee, F. Chen, T. Ha, A. Yeoh, S. Natarajan, Dual damascene

BEOL extendibility with Cu reflow/selective TaN and Co/Cu composite, in: 2021 Symposium on VLSI Technology, 2021, pp. 1–2.
[86] For an in-depth treatment, see, in: B. Beckhoff, N. Langhoff, B. Kanngiefer, R. Wedell, H. Wolff (Eds.), Handbook of Practical X-Ray Fluorescence Analysis, Springer, Berlin, Heidelberg, 2006.
[87] C.J. Morath, G.J. Collins, R.G. Wolf, R.J. Stoner, Ultrasonic multilayer metal film metrology, Solid State Technol. 40 (6) (1997) 85.
[88] R.J. Stoner, et al., Noncontact ultrasonic ULSI process metrology using ultrafast lasers, Proc. SPIE 3269 (1998) 104.
[89] M. Colgan, C. Morath, G. Tas, M. Grief, An ultrasonic laser sonar technique for copper damascene CMP metrology, Solid State Technol. 44 (2) (2001) 67.
[90] S.L. Manikonda, et al., Methodology to estimate TSV film thickness using a novel inline 'adaptive pattern registration' method, in: SEMI Advanced Semiconductor Manufacturing Conf. (ASMC), 2016, p. 129.
[91] J. Fu, J. van Gogh, Sputter target for eliminating redeposition on the target sidewall, US Patent 6,059,945.
[92] F.O. Armstrong, B.B. Jeffreys, Methods and systems for shielding in sputtering chambers, US Patent 5,482,612.
[93] J. Sasserath, R. Yenchik, Superior particle control for PVD TiW processes through improved chamber shield design, in: Microcontamination Conference, Proc. of the SPIE, 2334, 1994, p. 35.
[94] R.W. Rosenberg, Increasing PVD Tool Uptime and Particle Control with Twin-Wire- Arc Spray Coatings. www.micromagazine.com/archive/01/03/rosenberg.html.
[95] Y. Xu, J. Byrne, H. Clark, J. Parker, Successful application of residual gas analysis in IBM's 300-mm wafer fabrication facility, Semicond. Int. 27 (9) (2004) 46.
[96] G. Rampf, R. McCafferty, Devising an APC strategy for metal sputtering using residual gas analyzers, http://www.micromagazine.com/archive/02/07/rampf.html.
[97] W. Steinhögl, et al., Size-dependent resistivity of metallic wires in the mesoscopic range, Phys. Rev. B 66 (2002) 075414.
[98] M. Wada, et al., A study on resistivity increase of copper interconnects with the dimension comparable to electron mean free path utilizing Monte Carlo simulations, in: Proceedings of the Advanced Metallization Conference (AMC), 2009, pp. 1–10.
[99] L. Carbonell, et al., Metallization of sub-30 nm interconnects: comparison of different liner/seed combinations, in: Proceedings of the International Interconnect Technology Conference (IITC), IEEE, 2009, p. 200.
[100] C. Cabral Jr., et al., Metallization opportunities and challenges for future back-end-of-the-line technology, in: Proceedings of the Advanced Metallization Conference (AMC), 2010.
[101] International Technology Roadmap for Semiconductors, 2015. http://www.itrs2.net/.
[102] J.H.-C. Chen, et al., Interconnect performance and scaling strategy at the 5 nm Node, in: IEEE International Interconnect Technology Conference/Advanced Metallization Conference (IITC/AMC), vol. 3.2, 2016.
[103] Z. Tőkei, End of Cu roadmap and beyond Cu, in: IEEE International Interconnect Technology Conference/Advanced Metallization Conference (IITC/AMC), 2016, p. 1.
[104] J. Chawla, et al., Resistance and electromigration performance of 6 nm wires, in: IEEE International Interconnect Technology Conference/Advanced Metallization Conference (IITC/AMC) vol. 9.3, 2016.
[105] D. Gall, Electron mean free path in elemental metals, J. Appl. Phys. 119 (2016) 085101.

[106] C. Adelmann, et al., Alternative metals for advanced interconnects, in: IEEE International Interconnect Technology Conference/Advanced Metallization Conference (IITC/AMC) vol. 4.1, 2014.
[107] M.H. van der Veen, et al., Barrier/liner stacks for scaling the Cu interconnect metallization, in: IEEE International Interconnect Technology Conference and IEEE Materials for Advanced Metallization Conference (IITC/MAM), 2015, p. 25.
[108] J.J. Kelly, et al., Experimental study of nanoscale Co damascene BEOL interconnect structures, in: IEEE International Interconnect Technology Conference/Advanced Metallization Conference (IITC/AMC) vol. 5.6, 2016.
[109] V. Kamenini, et al., Tungsten and cobalt metallization: a material study for MOL local interconnects, in: IEEE International Interconnect Technology Conference/Advanced Metallization Conference (IITC/AMC) vol. 15.2, 2016.
[110] C.J. Jezewski, et al., Cobalt based interconnects and methods of fabrication thereof, US Patent 9,514,983 B2.
[111] R. Shaviv, et al., Methods for forming cobalt interconnects, US Patent Application US20160309596 A1.
[112] X. Zhang, et al., Ruthenium interconnect resistivity and reliability at 48 nm pitch, in: IEEE International Interconnect Technology Conference/Advanced Metallization Conference (IITC/AMC) vol. 5.3, 2016.
[113] L.G. Wen, et al., Ruthenium metallization for advanced interconnects, in: IEEE International Interconnect Technology Conference/Advanced Metallization Conference (IITC/AMC) vol. 5.4, 2016.
[114] D. Wan, et al., Subtractive etch of Ruthenium for sub-5nm interconnect, in: 2018 IEEE International Interconnect Technology Conference (IITC), Santa Clara, CA, USA, 2018, pp. 10−12, https://doi.org/10.1109/IITC.2018.8454841.
[115] G. Murdoch, Z. Tokei, S. Paolillo, O.V. Pedreira, K. Vanstreels, C.J. Wilson, Semidamascene interconnects for 2nm node and beyond, in: 2020 IEEE International Interconnect Technology Conference (IITC), San Jose, CA, USA, 2020, pp. 4−6, https://doi.org/10.1109/IITC47697.2020.9515597.
[116] G. Murdoch, M. O'Toole, G. Marti, A. Pokhrel, D. Tsvetanova, S. Decoster, S. Kundu, Y. Oniki, A. Thiam, Q.T. Le, O. Varela Pedreira, A. Lesniewska, G. Martinez-Alanis, S. Park, Z. Tokei, First demonstration of two metal level semi-damascene interconnects with fully self-aligned vias at 18MP, in: 2022 IEEE Symposium on VLSI Technology and Circuits (VLSI Technology and Circuits), 2022, pp. 1−2.
[117] C. Penny, et al., Subtractive Ru interconnect enabled by novel patterning solution for EUV double patterning and TopVia with embedded airgap integration for post Cu interconnect scaling, in: 2022 International Electron Devices Meeting (IEDM), San Francisco, CA, USA, 2022, pp. 12.1.1−12.1.4, https://doi.org/10.1109/IEDM45625.2022.10019479.

# CHAPTER 4

# Epitaxial growth processes for high performance advanced CMOS devices

**Shogo Mochizuki**
IBM Research, Albany, NY, United States

## 1. Introduction

### 1.1 MOSFET device scaling

In recent decades, progress in advanced information technology has been supported by semiconductor large scale integrated circuit (LSI) technology. LSIs are incorporated into all kinds of electronic devices, and their performance depends on the LSI's performance. A metal-oxide-semiconductor field-effect transistor (MOSFET), which is one of the basic elements constituting an LSI, is a switching element for calculating digital signals. Improvements in the performance of advanced Si semiconductor devices have been achieved by reducing the dimensions of LSI unit elements, a technique called "scaling" proposed by Dennard et al. in 1974 [1]. Fig. 4.1 shows the concept of scaling when the electric field is constant (constant electric field scaling). Dimensional changes are indicated by the dimensionless scaling factor $\kappa$ ($>1$). The principle of constant electric field scaling is to increase the substrate doping concentration and decrease the device dimensions and applied voltage with the same scaling factor $\kappa$, so that the electric field applied to the device is unaffected. In this way, ideal scaling can be achieved while avoiding short channel effects caused by potential changes from the drain bias. According to this theory, circuit delay time is reduced by a factor of $1/\kappa$, power consumption is reduced by a factor of $1/\kappa^2$, and circuit density is improved by a factor of $\kappa^2$ due to reduction in lateral dimensions such as gate length/width. Also, vertical dimensions such as gate oxide thickness and source/drain junction depth are reduced by a factor of $1/\kappa$. Furthermore, the impurity concentration in the channel

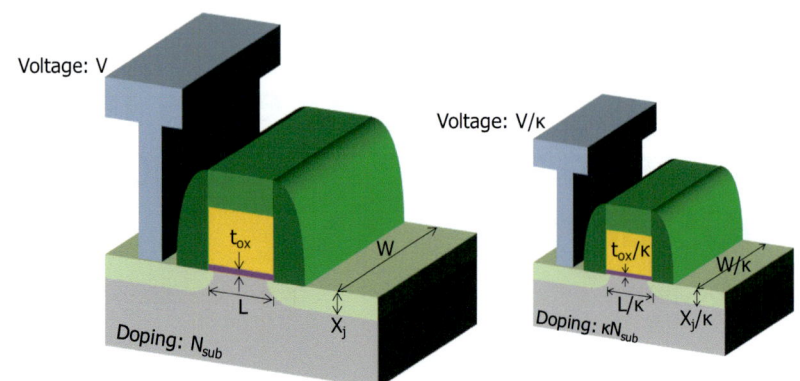

**Figure 4.1** Change in MOS device dimensions with factor κ (>1) under constant electric field scaling.

increases by a factor of κ, and the applied voltage decreases by a factor of 1/κ. As a result, high speed, low power consumption, and high integration, which are the most important parameters of integrated circuits, can be achieved simultaneously. Table 4.1 summarizes the effects of constant electric field scaling on device parameters. One of the most important parameters for device performance is the drain current ($I_d$), shown below.

$$I_d = \mu Q_{inv} \frac{W}{L} V = Q_{inv} W \mu E, \qquad (4.1)$$

where $\mu$ is the carrier mobility, $Q_{inv}$ is the inversion layer charge density per unit area, $W$ is the gate width, $L$ is the gate length, $V$ is the voltage applied

**Table 4.1** Relationship between MOS device dimensions and device characteristics with constant electric field scaling [1].

| Device or circuit parameter | Scaling factor |
| --- | --- |
| Gate Length $L$, gate width $W$ | $1/\kappa$ |
| Gate oxide thickness $t_{ox}$, junction depth $X_j$ | |
| Supply voltage ($V$) | |
| Drive current ($I_d$) | |
| Gate capacitance ($C_g$) | |
| Circuit delay time ($t \sim C_g V/I_d$) | |
| Doping concentration ($N_{sub}$) | $\kappa$ |
| Power consumption ($P \sim VI_d$) | $1/\kappa^2$ |
| Circuit density (1/A) | $\kappa^2$ |
| Power density (P/A) | 1 |

to the channel, and $E$ is the lateral electric field in the channel. As a result of scaling, $I_d$ is reduced by a factor of $1/\kappa$. However, since both the drain current and the applied voltage are scaled by the same factor, the on-state channel resistance of the scaled device does not change. Capacitance (C) is proportional to area and inversely proportional to vertical dimension. The circuit delay time, which is proportional to the value of $CV/I_d$, is reduced by $1/\kappa$ since the capacitance is reduced by a factor of $1/\kappa$. Additionally, the power consumption, which is proportional to the $VI_d$ value, is reduced by a factor of $1/\kappa^2$. These are the most important conclusions of constant electric field scaling, which has been taken as a guiding principle for improving the performance of semiconductor technology.

On the other hand, the phenomenon in which the number of transistors placed on an integrated circuit doubles approximately every 2 years is known as Moore's Law [2]. This concept was proposed in 1965 and has been used in the semiconductor industry as a rule of thumb for increasing the number of transistors. In recent years, this law has also been used as a guideline for long-term planning and research and development goals to advance semiconductor technology [3].

The evolution of digital electronic devices has been achieved through breakthrough in LSI technology according to Moore's law and has facilitated the dramatic expansion of markets related to high-performance microprocessors and low-power applications. Even in today's highly information-oriented society, there remains a strong need for processors with high processing capability and high density integrated memories. To meet these demands, it is essential to further pursue scaling or equivalent performance improvements in MOSFETs.

## 2. Strained Si technology with epitaxial process
### 2.1 Mobility enhancement technology

As MOSFET device dimensions are scaled according to Moore's law, short channel effects such as tunnel current as a leakage current component due to a decrease in MOSFET threshold voltage and thinning of the gate insulating film cannot be ignored. Recognizing that it is difficult to improve the performance of MOSFETs by scaling alone when the gate length L is about 100 nm or less, research and development of device performance improvement technology that does not depend on scaling is being actively conducted. Strained Si technology has been put into practical use since the 90 nm node as a technique to improve the performance of planar bulk

MOSFETs without relying on scaling. This involves applying stress to cause lattice deformation (applying strain) to the Si crystal that serves as the channel. As seen in the drain current $I_d$ in Eq. (4.1), it is important to increase the carrier mobility μ of the channel material in order to increase $I_d$. Strained Si technology is a technology that applies strain to the Si crystal in the channel region to change the band structure and increase carrier mobility. The phenomenon in which electrical resistance changes when mechanical strain is applied to Si has been known as the piezoresistive effect for many years [4] and has been applied as a strained Si technology in advanced devices.

As summarized in Figs. 4.2 and 4.3, there are various ways to induce strain in the channel region of a MOSFET. Those methods are broadly classified into a method of forming a strained Si layer on a wafer and a method of locally applying stress only to the channel region. The former is called "substrate induced strain" and a strained Si layer is formed on the substrate. The latter is called "process induced strain" and strain is applied to the channel region by using several stressors in the device fabrication process.

The substrate induced strain technique utilizes a strained layer epitaxially grown on the substrate, as shown in Fig. 4.2a–c. A thin strained Si layer is formed on the substrate surface, in which in-plane biaxial stress is induced. Fig. 4.2a shows a bulk strained Si substrate with epitaxially grown Si layer on a thick strain relaxed buffer (SRB) SiGe layer formed on a Si substrate. Fig. 4.2b shows an SGOI (SiGe on Insulator) substrate in which a strained Si layer and a strain-relaxed SiGe layer are formed on a buried oxide (BOX) layer. The SiGe layer is formed on a Si substrate by growing a SiGe crystal layer in which the Ge composition is continuously or stepwise increased in the film thickness direction. As the thickness of the SiGe layer increases, SiGe strain relaxation proceeds through the introduction of misfit

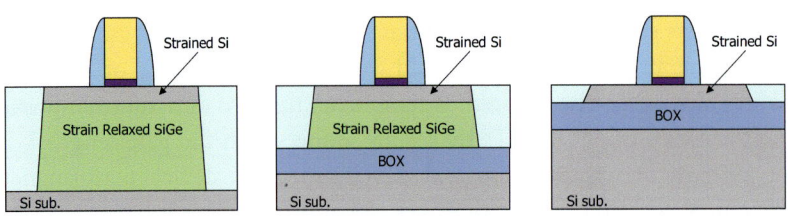

(a) Bulk strained Si substrate   (b) SGOI (SiGe on Insulator)   (c) SSOI (Strained Si on Insulator)

**Figure 4.2** Substrate induced strained Si technologies with virtual substrates.

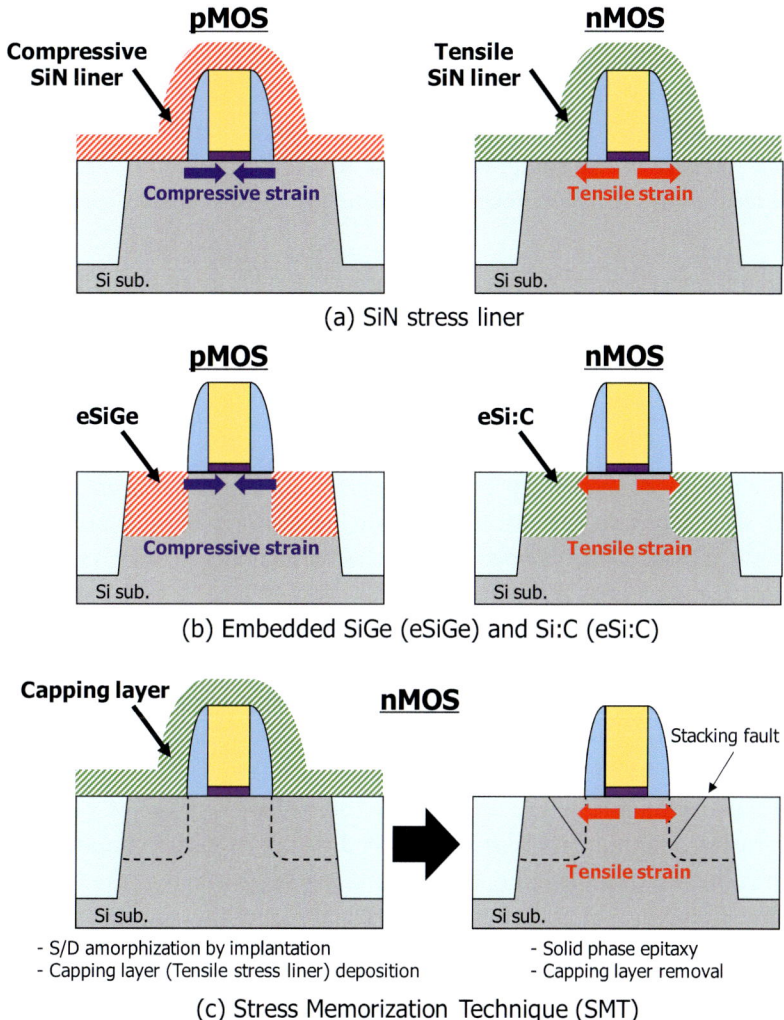

**Figure 4.3** Process induced strained Si technologies.

dislocations, which are confined in the Ge composition gradient region. The threading dislocation density in SiGe layers can be reduced by lateral propagation of misfit dislocations [5–7]. As a result, the SiGe layer is almost completely relaxed at the surface and its lattice constant approaches that of bulk SiGe. The SGOI substrate is formed by oxidizing the SiGe layer formed on the SOI (Silicon on Insulator) substrate, and the enrichment of Ge occurs due to the difference in the oxidation rate of Si and Ge [8,9]. It has been suggested that the strain relaxation of the SGOI layer is facilitated by the introduction of dislocations by Ge enrichment and the escape of misfit dislocations to the SOI/BOX interface [10,11]. When a thin Si layer is epitaxially grown on a strain-relaxed SiGe layer, tensile stress in the in-plane direction and compressive stress in the out-of-plane direction are induced in the Si layer since lattice constant of Si is smaller than that of SiGe.

Fig. 4.2c shows an SSOI (Strained Silicon on Insulator) substrate in which strain is applied to the SOI layer. The SSOI layer is obtained by bonding the strained Si layer of the bulk strained Si substrate fabricated by the method shown in Fig. 4.2a and the Si substrate with the BOX layer [12]. The strain applied to the SSOI layer remains even after removal of the thick strain-relaxed buffer SiGe layer and its underlying Si substrate due to the presence of atomically bonded interfacial contact between the BOX and SSOI layers. In general, for substrate induced strain, an in-plane biaxial stress is applied to the strained layer as the stress-inducing layer is interposed over the entire surface of the wafer.

Fig. 4.3a—c show a process induced strain technique that utilizes the application of stress from outside the channel region throughout the device fabrication process. Fig. 4.3a and b show methods using stress films such as SiN films and SiGe/Si:C (Carbon doped Si) layers embedded in the source/drain regions as stress sources. Local strain is induced into the Si channel region according to the strength of the internal stress in the SiN film. The internal stress in the SiN film is controlled by the deposition conditions. By increasing the film thickness, more strain can be obtained in the MOSFET channel region, which leads to further channel mobility enhancement. It is desirable to apply tensile stress to the nMOS channel region and compressive stress to the pMOS channel region, respectively, to enhance the electron and hole mobilities, respectively [13,14]. Therefore, a method of weakening unnecessary stress by ion implantation and a method of forming SiN films with the optimum polarity for both nMOS and pMOS (DSL: Dual Stress Liner) have been developed [15—17].

When SiGe is epitaxially grown on the source/drain regions, a compressive strain is applied along the channel direction to the Si channel region adjacent to SiGe because the lattice constant of SiGe is larger than that of Si. And the compressive strain enhances the hole mobility in the pMOS channel [18,19]. Techniques to increase stress transfer efficiency, such as optimizing the proximity of SiGe to the channel and the Ge concentration in SiGe in the source/drain regions, as well as the recess shape, are effective in improving device characteristics [20,21]. A similar principle can be applied to nMOS by taking advantage of the small lattice constant of Si:C and enhance electron mobility by applying tensile strain along the channel direction [22−24]. The structures shown in Fig. 4.3b are called embedded SiGe (eSiGe: embedded SiGe) and embedded Si:C (eSi:C: embedded Si:C) structures.

Fig. 4.3c shows a method of generating internal stress using ion implantation into the gate electrode and source/drain regions, which is called SMT (Stress Memorization Technique). After ion implantation, a capping layer (nitride or oxide) as a stress-inducing material is deposited over the device, followed by a thermal treatment to recrystallize the gate electrode and source/drain regions. As a result of recrystallization, a tensile strain is applied to the channel region and the strain is maintained after subsequent removal of the capping layer. It has been confirmed that tensile strain is caused by the presence of residual defects (stacking faults) at the channel-source/drain interface that occur during recrystallization [25−27]. Modulation of device performance using device isolation technologies such as local oxidation of silicon (LOCOS) and shallow trench isolation (STI) as methods of inducing external stress has been reported [28,29].

As described above, the process induced strain technique has advantages such as excellent compatibility with the device manufacturing process, less burden on the manufacturing cost, etc. In addition, different strains can be applied to nMOS and pMOS respectively by optimizing integration processes. Therefore, process induced strain Si technology is being introduced for mass production after the 90 nm node. However, on the other hand, it is known that the strain generated from stress-inducing materials such as SiN films, eSiGe, eSi:C, and STI induces complex strain distributions in device structures and exhibits layout dependence. This is because the stress varies depending on the distance between the channel region and the source/drain regions (or STI regions) and their geometry. The resulting strain distribution affects the modulation of device performance. Therefore, it is important to evaluate strain in localized regions of device structures in order to understand the effects on device properties and exploit them to improve device performance. In addition, increasing the amount of strain to improve mobility may also cause degradation of device characteristics due to the introduction of crystal defects. It is essential to understand the strain state of the device structure and then carefully design and control the strain in the device.

## 2.2 New device structure with three-dimensional structure

FinFET is a MOSFET with a three-dimensional device structure that is different from conventional planar MOSFETs [30−32]. Fin is a structure obtained by processing a Si substrate into a strip shape, and by covering it with a gate electrode, the top and both side surfaces of the surface Si are used as a channel. Unlike conventional planar MOSFETs that have a single-gate structure, they have a gate structure called a double-gate or tri-gate, which allows better control of the potential of the channel region by the gate electrode. Therefore, the punch-through resistance between S/D is high, and it is possible to reduce the leakage current between S/D (subthreshold leakage current) in the off state. The short channel effect can be suppressed even with shorter channel lengths, so the FinFET device structure is suitable for further device scaling. In addition, in planar MOSFETs, it was necessary to increase the impurity concentration for making the depletion layer thinner, but FinFETs are based on the principle of fully depleted operation due to their thin fin width, thus, the channel impurity concentration can be reduced. This reduces variations in device electrical characteristics due to variations in impurity concentration, and also suppresses the effect of impurity scattering on carriers, thereby a decrease in mobility can be suppressed. In this way, FinFET is a promising structure for realizing device scaling and higher performance.

## 3. Strain engineering for nanoscale strained SiGe FinFET

Rapid performance improvement of complementary metal-oxide-semiconductor (CMOS) technology has been achieved through scaling of the dimensions of the unit device [33]. However, as device dimensions shrink, it has become apparent that performance suffers due to short-channel effects, increased gate leakage, and increased parasitic resistance/capacitance. Therefore, new performance-enhancing elements have been incorporated state-of-the-art CMOS devices to mitigate these effects. The representative examples are introductions of new materials and novel device structures. The introduction of high-k/metal gate technology and strain engineering are key innovations for reducing gate leakage current and increasing carrier mobility [34−37]. Non-planar device structures such as FinFETs and horizontal/vertical Gate All Around (GAA) structures with superior gate to channel control have been proposed to enable continued scaling of high-performance CMOS devices [38–42]. Furthermore, improved performance of FinFET devices has been reported by implementing strained channel structures obtained by strain engineering such as source/drain epitaxial growth and stressed liner deposition [43,44].

It is necessary to explore performance improvement techniques for next generation CMOS technologies similar to the recent adoption of high mobility channel in 5 nm CMOS production FinFET technology [45]. In particular, strained SiGe FinFET with compressively strained SiGe in the channel has been considered as an attractive option for improving FinFET performance due to its high carrier mobility [46–49]. Although the strain induced in the SiGe channel is essential to achieve hole mobility enhancement [50], it is not yet clear whether Ge content or strain primarily has an effect on the mobility modulation. Thus, it is important to understand the effects of those physical parameters on the SiGe FinFET device performance. In addition, strain distributions within nanoscale non-planar device structures are more complex due to the nature of the 3D architecture. Because the strain modulation within 3D device structure is expected to become more pronounced with device scaling, its influence on the device characteristics is an issue to be concerned. In this section, the systematic decoupling of the effect of strain and Ge content on the device characteristics is reviewed. The investigation of the strain distributions within nanoscale strained SiGe FinFET device structures, its influence on the electrical characteristics, and techniques to mitigate the strain/electrical characteristic modulation are further discussed.

## 3.1 Strain engineering for SiGe FinFET

A high mobility channel material combined with FinFET device architecture which offers further gate length scaling by its superior electrostatics and gate control is an attractive option for improving device performance. Various research groups have reported performance advantages of SiGe channel FinFET over Si channel by introducing compressive strain in the SiGe channel region on either bulk Si or SOI substrates. One of the key metrics is integration process optimization including how to fabricate SiGe fin without crystalline defect introduction. Several approaches have been utilized to form SiGe fin, such as epitaxial growth of strained SiGe followed by lithographic patterning, cladding SiGe growth on Si FinFET structure, and Ge condensation technique with certain oxidizing and thermal processes typically for high Ge SiGe fin formation [48,49,51]. Also, the performance benefits of SiGe channel over Si pFETs have been achieved by interfacial layer engineering [52–54]. Recently, dual strained Si/SiGe channel FinFETs (tensile-strained Si nFET and compressively strained SiGe pFET) fabricated on a strain relaxed buffer (SRB) layer have been demonstrated to enhance both electron and hole mobility at the 7 nm technology ground rules [55,56]. Besides mobility enhancements, Si/SiGe dual channel configuration on SRB provides a common gate stack solution due to the higher valence band energy level of high-Ge-content (HGC) SiGe channel, which can reduce process/patterning complexity at a replacement metal gate (RMG) module. However, HGC SiGe pFETs fabricated on SRB have suffered from non-ideal gate stack formation due to intrinsically high interface trap charge (Dit) and relatively low hole mobility, limiting the performance of Si/SiGe dual channel CMOS. Since the channel strain of $Si_{1-x}Ge_x$ on SRB is different from $Si_{1-x}Ge_x$ directly grown on a Si substrate, a decoupled understanding of the electrical impact of strain and Ge content in the $Si_{1-x}Ge_x$ channel, especially within HGC $Si_{1-x}Ge_x$ channel, is required. Since most of the studies to date have been performed on strained SiGe structures which are fully strained to Si, no studies exist that have systematically investigated the effects of strain and Ge content in the SiGe channel over a wide range of Ge fractions. In this sub-section, the approach to control the amount of strain in SiGe channel and investigate the role of strain and Ge content in the SiGe channel are mainly reviewed.

### 3.1.1 Comparative study of strain and Ge content in $Si_{1-x}Ge_x$ channel

#### 3.1.1.1 Strained SiGe pFinFET formation on SRB

It is necessary to control strain amount in the strained SiGe FinFET in order to decouple the electrical impact of strain and Ge content in the $Si_{1-x}Ge_x$ channel. To control the strain amount, epitaxial $Si_{1-x}Ge_x$ layers ($0 \leq x \leq 0.4$) grown on Si substrate and $Si_{0.8}Ge_{0.2}$ strain relaxed buffer (SRB) virtual substrate were employed. Since $Si_{0.8}Ge_{0.2}$ SRB layer is fully relaxed, as confirmed by XRD Reciprocal Space Map (RSM) in Fig. 4.4, its lattice constant is 0.76% larger than that of Si [57], which results in inducing different strain amount in the epitaxially grown $Si_{1-x}Ge_x$ layers compared to those grown on Si substrate. Fig. 4.4 shows RSMs of the corresponding 224 reflections for the epitaxial $Si_{1-x}Ge_x$ layers grown on $Si_{0.8}Ge_{0.2}$ SRB virtual substrate. Since the position of the $Si_{1-x}Ge_x$ 224 diffraction peak along the in-plane direction ($Q_x$) is aligned to that of the $Si_{0.8}Ge_{0.2}$ SRB layer peak, the $Si_{1-x}Ge_x$ layers are fully strained, which means its lattice is matched to

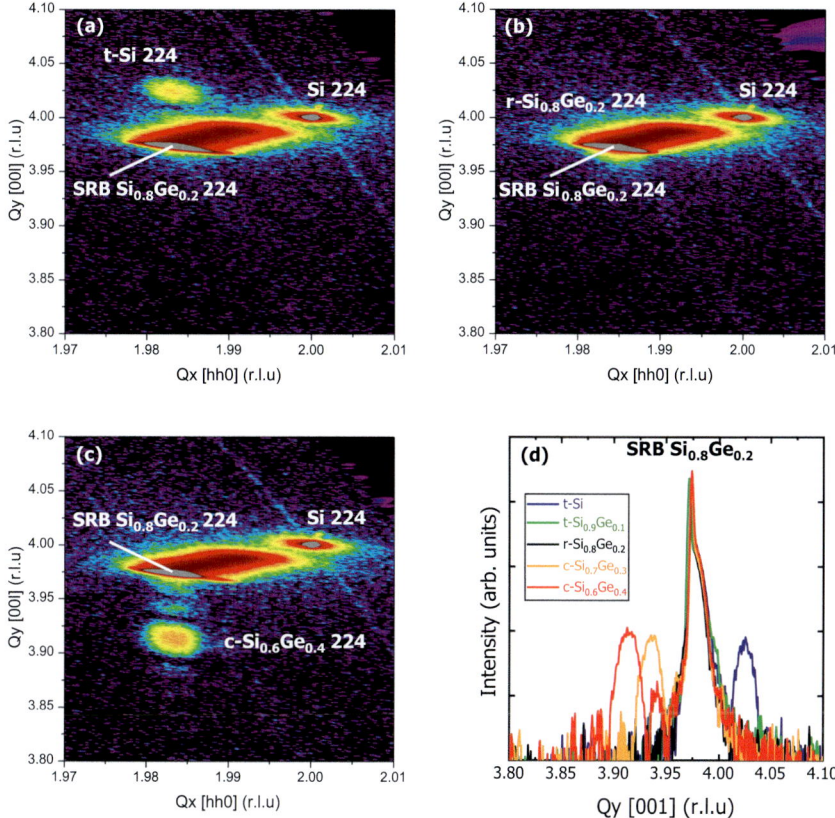

**Figure 4.4** Reciprocal space maps around the Si 224 and SiGe 224 for $Si_{1-x}Ge_x$ channels grown on the fully relaxed $Si_{0.8}Ge_{0.2}$ SRB layer. (a) Tensile strained Si, (b) relaxed $Si_{0.8}Ge_{0.2}$, and (c) compressively-strained $Si_{0.6}Ge_{0.4}$. (d) Cross-sectional profiles of the (224) diffraction peaks of $Si_{1-x}Ge_x$ channel epi and $Si_{0.8}Ge_{0.2}$ SRB layers extracted along $Q_y$ at $Q_x$ in 1.983 (r.l.u).

the underlying $Si_{0.8}Ge_{0.2}$ SRB layer. As shown in Fig. 4.4d the clear thickness fringes around the main $Si_{1-x}Ge_x$ peak indicate a high-quality $Si_{1-x}Ge_x$ layer with good crystallinity. The position of $Si_{1-x}Ge_x$ peak corresponds to the out-of-plane lattice parameter/strain in the $Si_{1-x}Ge_x$ layer. Growing an epitaxial $Si_{1-x}Ge_x$ layer with x < 0.2 or 0.2 < x on $Si_{0.8}Ge_{0.2}$ SRB virtual substrate means that the $Si_{1-x}Ge_x$ layer is under extension or compression along in-plane direction, respectively. Therefore, the $Si_{1-x}Ge_x$ layers can contain tensile or compressive in-plane strain depending on the Ge fraction. The in-plane strain of the $Si_{1-x}Ge_x$ layers on both SRB and Si obtained by the XRD is consistent with theoretical calculation as shown in

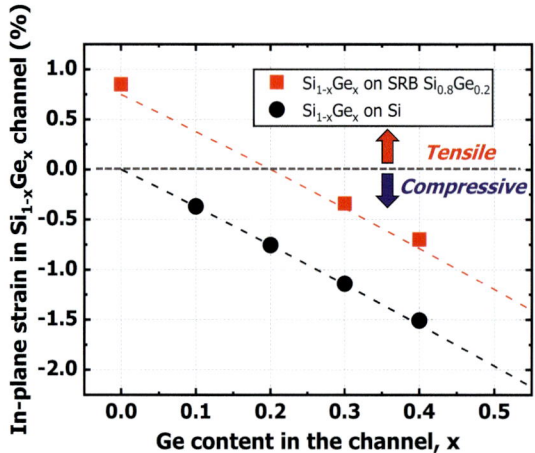

**Figure 4.5** In-plane channel strain in the $Si_{1-x}Ge_x$ channels on both Si and $Si_{0.8}Ge_{0.2}$ SRB estimated by XRD. The lines correspond to the in-plane strain obtained by theoretical calculation [57].

Fig. 4.5 [57]. Notice that the same channel strain with a different Ge content, −0.38% (in the case of $Si_{0.7}Ge_{0.3}$ layer on $Si_{0.8}Ge_{0.2}$ SRB and $Si_{0.9}Ge_{0.1}$ layer on Si substrate) and −0.78% (in the case of $Si_{0.6}Ge_{0.4}$ layer on $Si_{0.8}Ge_{0.2}$ SRB and $Si_{0.8}Ge_{0.2}$ layer on Si substrate), can be achieved.

Fig. 4.6 shows cross-sectional STEM images and EDX element maps of tensile-strained Si pFinFET fabricated on SRB and compressively strained $Si_{0.6}Ge_{0.4}$ pFinFET fabricated on SRB after M1 electrical test. The active fin height and width are 40 and 15 nm, respectively. There are no visible defects in the channel and channel/SRB interface. No Ge diffusion from SRB to the Si active fin is observed by EDX. One concern about HGC $Si_{1-x}Ge_x$ pFinFETs on SRB is that Ge pile-up happens at the $Si_{1-x}Ge_x$ fin surface during downstream thermal processing [56]. The Ge compositions were quantified in the $Si_{0.6}Ge_{0.4}$ pFinFET by EELS. Fig. 4.7 shows both vertical and lateral profiles of Si and Ge concentration in the active $Si_{0.6}Ge_{0.4}$ fin region on SRB, confirming that the variation of Ge content across the fin is less than 5% even after processing till the M1 level.

### 3.1.1.2 Impact of Ge content on SiGe FinFET

I−V curves of $Si_{1-x}Ge_x$ pFinFETs on SRB are shown in Fig. 4.8. A linear $V_{th}$ shift is observed as Ge content increases due to its valence band offset relative to Si. Fig. 4.9 shows $V_{th}$ shift for $Si_{1-x}Ge_x$ channel on $Si_{0.8}Ge_{0.2}$

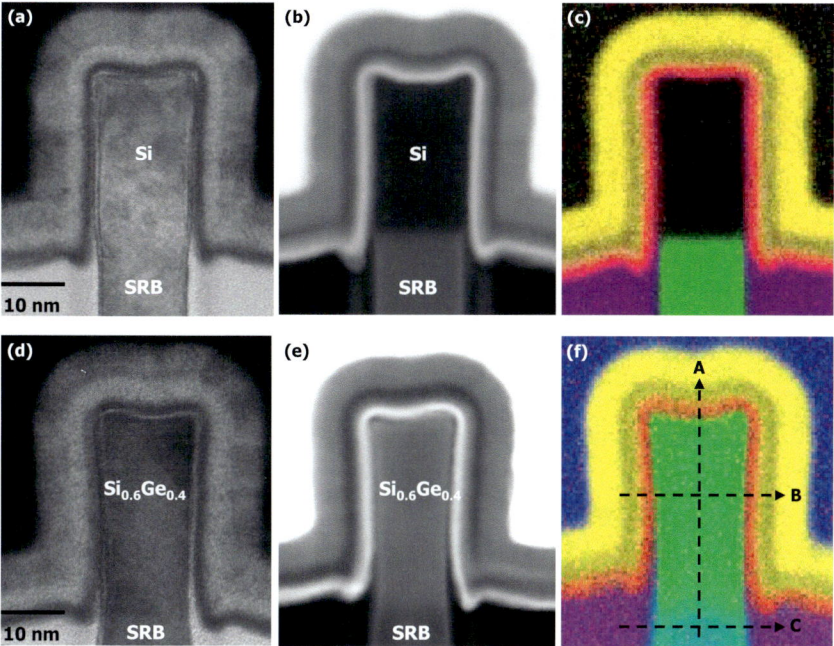

**Figure 4.6** HRTEM analysis of pFinFETs fabricated on $Si_{0.8}Ge_{0.2}$ SRB post M1 electrical test. (a) Cross-sectional STEM image, (b) HAADF STEM image, and (c) EDX profile of tensile-strained Si pFinFET, and (d) cross-sectional STEM image, (e) HAADF STEM image, and (f) EDX profile of compressively-strained $Si_{0.6}Ge_{0.4}$ pFinFET.

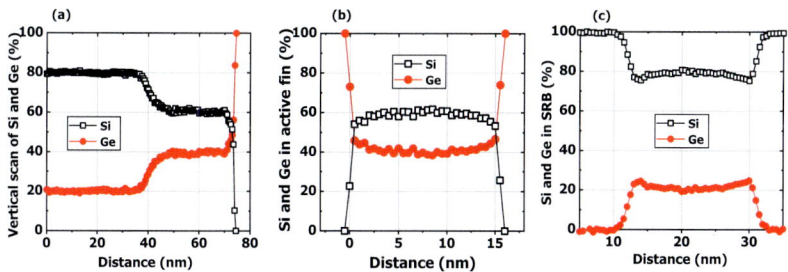

**Figure 4.7** EELS Ge and Si profiles in the $Si_{0.6}Ge_{0.4}$ fin region on $Si_{0.8}Ge_{0.2}$ SRB (post M1) along line A, C and B in Fig. 4.6. (a) Vertical scan of Si and Ge atomic concentration in the fin. (b) Lateral scan of Si and Ge atomic concentration in the $Si_{0.6}Ge_{0.4}$ fin. (c) Lateral scan of Si and Ge atomic concentration in the $Si_{0.8}Ge_{0.2}$ SRB.

SRB and Si substrate compared to a Si channel as a function of the Ge content. The $V_{th}$ shift in SiGe channel relative to Si pFET could be ascribed to both the channel strain and the valence band offset of SiGe relative to Si.

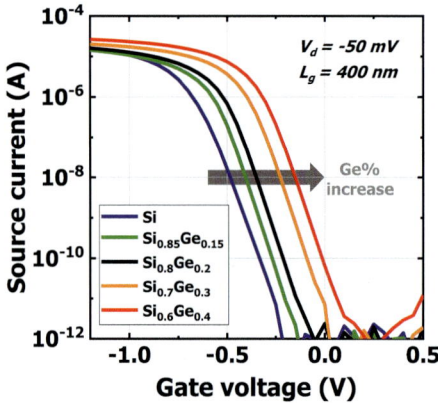

**Figure 4.8** I–V curves of $Si_{1-x}Ge_x$ ($0 \leq x \leq 0.4$) pFinFETs on $Si_{0.8}Ge_{0.2}$ SRB.

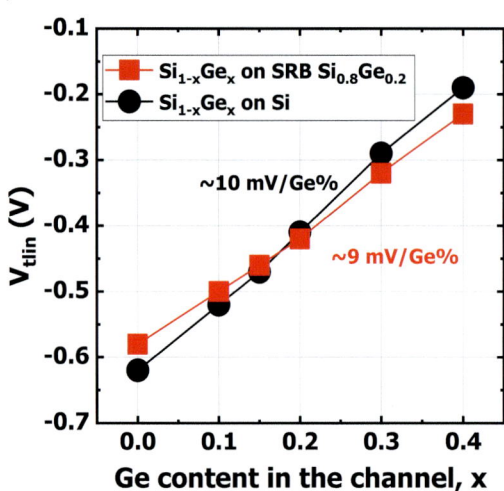

**Figure 4.9** $V_{th}$ shift in $Si_{1-x}Ge_x$ ($0 \leq x \leq 0.4$) channel relative to Si pFETs.

The same gate stacks (IL/HK/WFM) are used for all the devices to decouple the $V_{th}$ shift components in the $Si_{1-x}Ge_x$ channel. As shown in Fig. 4.9, a linear $V_{th}$ shift of ∼9 mV/Ge% in $Si_{1-x}Ge_x$ planar FETs on SRB is observed, which is very similar to $Si_{1-x}Ge_x$ planar FETs on Si (∼10 mV/Ge%), suggesting that the $V_{th}$ shift (corresponding to the higher valence band energy level) in $Si_{1-x}Ge_x$ FETs is mainly attributed to Ge content in the channel. The subthreshold slopes (SS) below 70 mV/dec is demonstrated for all the $Si_{1-x}Ge_x$ pFinFET devices and a slight increase of SS is observed as Ge content increases in the channel [58]. Additionally, it was reported that Dit value increased as the Ge content in the $Si_{1-x}Ge_x$ channel increased and is solely dominated by the Ge content in the $Si_{1-x}Ge_x$ channel, despite different channel strain of $Si_{1-x}Ge_x$ FETs on SRB and $Si_{1-x}Ge_x$ FETs on Si at a given Ge content [58].

One of the benefits of using pFET devices with SiGe channel is their superior Negative-Bias Temperature Instability (NBTI) performance over Si [59]. It was confirmed that the 10-year DC projected end-of-life (EOL) $V_{th}$ shift due to NBTI for SiGe channel formed on SRB and Si substrate follow almost the same trend [58]. As the Ge content in the SiGe channel increases, the NBTI performance improves regardless of the amount of channel strain. Since NBTI benefits of SiGe channel are attributed to the misalignment of hole trapping site in the IL and valence band, Ge content plays a critical role in NBTI improvement. In summary, the Ge content in the $Si_{1-x}Ge_x$ channel dominantly modulates the amount of $V_{th}$ shift, the value of Dit, and improvement of NBTI.

### 3.1.1.3 Impact of strain in SiGe channel FinFET

In this sub-section, the contribution of the channel strain and Ge content in the strained $Si_{1-x}Ge_x$ channel toward carrier transport is discussed. Fig. 4.10 shows extracted high field hole mobility as a function of the Ge content. Hole mobility of $Si_{1-x}Ge_x$ ($x \leq 0.2$) on SRB is lower than tensile-strained

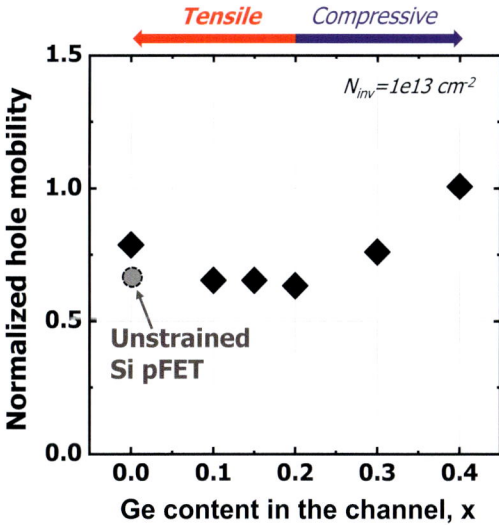

**Figure 4.10** Normalized high-field hole mobility at $N_{inv} = 1e13$ $cm^{-2}$ versus Ge content in $Si_{1-x}Ge_x$ ($0 \leq x \leq 0.4$) pFinFETs on $Si_{0.8}Ge_{0.2}$ SRB.

Si and is pretty similar to unstrained Si. On the other hand, hole mobility of $Si_{1-x}Ge_x$ ($0.2 \leq x$) on SRB is higher than unstrained $Si_{0.8}Ge_{0.2}$ on SRB, and $Si_{1-x}Ge_x$ channel with higher Ge content shows higher mobility. Fig. 4.11a shows normalized hole mobility of $Si_{1-x}Ge_x$ (x = 0.3 and 0.4) pFinFETs on SRB and $Si_{1-x}Ge_x$ (x = 0.1 and 0.2) pFinFETs on Si substrate as a function of inversion carrier density ($N_{inv}$). The in-plane compressive strain amount induced in the $Si_{1-x}Ge_x$ channel due to the lattice constant difference form the substrate is $-0.38\%$ and $-0.78\%$, respectively. Hole mobility in $Si_{1-x}Ge_x$ pFinFETs on SRB and $Si_{1-x}Ge_x$ pFinFETs on Si at a given channel strain exhibits a very similar behavior over a wide range of $N_{inv}$, although the Ge content within the $Si_{1-x}Ge_x$ channel is different. The mobility of $Si_{0.7}Ge_{0.3}$ pFinFETs on SRB is lower than that of $Si_{0.9}Ge_{0.1}$ pFinFETs on Si in the low-field region. This could be explained by Coulomb scattering due to charges trapped in the interface states since $Si_{1-x}Ge_x$ channel with higher Ge content showed relatively higher value of Dit in the $Si_{1-x}Ge_x$ channel, as mentioned in the previous sub-section. The impact of the channel strain on hole mobility in the $Si_{1-x}Ge_x$ channel is studied by comparing the high-field mobility at Ninv = 1e13 $cm^{-2}$ as a function of the channel strain (Fig. 4.11b). Regardless of Ge content in the channel, the hole mobility of $Si_{1-x}Ge_x$ pFinFET on SRB and $Si_{1-x}Ge_x$ pFinFET on Si is identical if the channel strain is aligned. This result also suggests that alloy scattering in $Si_{1-x}Ge_x$ FETs is not a significant limiting factor to carrier transport at room temperature unlike theoretical perspective, which speculated that scattering due to the random disorder potential

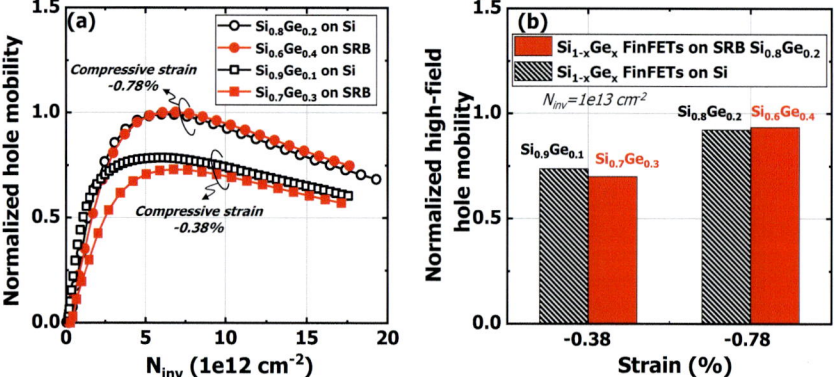

Figure 4.11 (a) Normalized hole mobility of $Si_{1-x}Ge_x$ pFinFETs on SRB and $Si_{1-x}Ge_x$ pFinFETs on Si as a function of $N_{inv}$. (b) Normalized high-field hole mobility at $N_{inv}$ = 1e13 $cm^{-2}$ versus channel strain.

in $Si_{1-x}Ge_x$ channel degrades hole mobility. This result indicates that the mobility enhancement in $Si_{1-x}Ge_x$ channel is mainly attributed to the compressive strain in the channel by the reduction of effective hole mass along the transport direction.

## 3.2 Local strain in nanoscale SiGe FinFET

Hole mobility in SiGe channel FinFET is predominantly modulated by channel strain as reviewed in the previous sub-section. Therefore, understanding strain variations within FinFET device structures is important especially in the nanoscale regime, as strain variations can be more complex in 3D device structures. Characterization of the local strain in nanoscale strained SiGe fin structures has been performed. Strain in the uniaxially stressed SiGe fin structure and the variations in local strain at the edge of the SiGe fin are also investigated. In addition, this indicates that the SiGe fin channel strain is modulated depending on the fin length and induces a local layout effect (LLE), which is expected to affect the device characteristics. Therefore, LLE in SiGe pFinFET is characterized through physical analysis (nano beam diffraction: NBD) and electrical characterization. In this subsection, techniques to mitigate LLE with full channel strain retention is further discussed.

### 3.2.1 Quantification of local strain distributions in nanoscale strained SiGe FinFET structures

To study the LLE, two types of samples were prepared: 1. SiGe fins on SOI(001) substrates and 2. SiGe fins on bulk Si(001) substrates. For type #1 samples, strained $Si_{1-x}Ge_x$ layers with a thickness of 20 nm were epitaxially grown on the SOI (001) substrates. The SOI layer thickness was 10 nm to form strained $Si_{1-x}Ge_x$/SOI fins with a height of 30 nm. The Ge fraction was x = 0.18, 0.28, and 0.38. For type #2 samples, a series of $Si_{1-x}Ge_x$ layers were epitaxially grown on the bulk Si(001) substrates. The Ge fractions and thicknesses of $Si_{1-x}Ge_x$ layers were systematically varied from x = 0.125 to 0.52 and from 5 to 60 nm, respectively. A patterning process was performed on the $Si_{1-x}Ge_x$/SOI and $Si_{1-x}Ge_x$/bulk Si layers to form SiGe fin structures with a fin width of 10 nm and a fin pitch of 42 nm. As shown in Fig. 4.12b, the longitudinal direction of the fin is defined to be along [110], and the fin width direction was defined to be parallel to $[1\bar{1}0]$, with the surface normal aligned parallel to [001]. The strain and lattice deformation in the SiGe fins was evaluated by high resolution X-ray

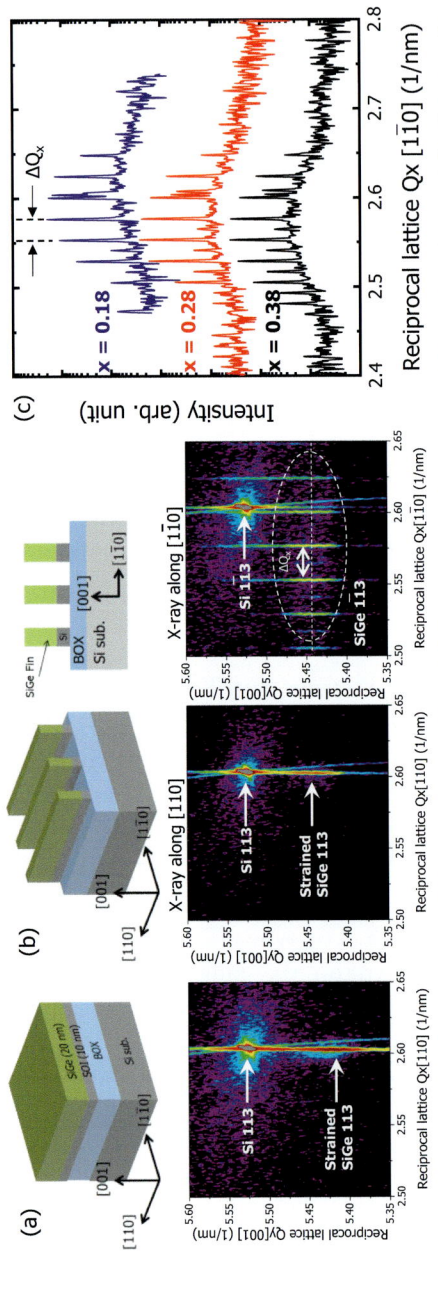

**Figure 4.12** RSMs around the 113 and (1$\bar{1}$3) reflections of Si and SiGe for (a) the $Si_{0.72}Ge_{0.28}$/SOI blanket film and (b) the $Si_{0.72}Ge_{0.28}$/SOI fin structures. X-ray incident direction was along [110] for blanket film, and X-ray incident direction of [110] and [1$\bar{1}$0] was used for fin structures (shown as schematic diagrams). The relationship between the X-ray incident direction and pattern geometry for SiGe fin structures is shown as schematic diagrams. (c) Cross-sectional profiles along $Q_x$ at $Si_{1-x}Ge_x$ 1$\bar{1}$3 diffraction peaks of $Si_{1-x}Ge_x$ fins (x = 0.18, 0.28, and 0.38).

diffraction (HRXRD) and Nanobeam Diffraction (NBD), which is a direct method to evaluate lattice deformation.

Fig. 4.12 shows RSMs of the corresponding 113 and $1\bar{1}3$ reflections of Si and SiGe for the $Si_{0.72}Ge_{0.28}$/SOI blanket film and $Si_{0.72}Ge_{0.28}$/SOI fin array structures with two orthogonal X-ray incidence directions ( [110] and $[1\bar{1}0]$). It is confirmed that the blanket SiGe film and SiGe fins along the fin direction ([110]) were fully strained since the $Q_x$ value of the SiGe 113 diffraction peak is aligned to that of the Si 113 substrate peak. In contrast, the SiGe $1\bar{1}3$ peak shifted to lower values in $Q_x$ compared to Si $1\bar{1}3$ peak indicating elastic relaxation along the fin width direction ($[1\bar{1}0]$). This result indicates that the biaxial stress in the blanket SiGe layer was converted to non-biaxial stress due to elastic strain relaxation along the fin width direction by patterning into fin structure. In addition, periodic streaks from the SiGe and SOI layers appeared in the spectra. The periodicity of the satellite peaks $\Delta Q_x$ corresponds to the fin pitch along $[1\bar{1}0]$. The measured $\Delta Q_x$ value was 0.0239 nm$^{-1}$ corresponding to 41.8 nm fin pitch in real space. Cross-sectional profiles along $Q_x$ of the $Si_{1-x}Ge_x$ $1\bar{1}3$ diffraction peaks of $Si_{1-x}Ge_x$ fins (x = 0.18, 0.28, 0.38) are shown in Fig. 4.12c. A shift of the SiGe diffraction peak to lower $Q_x$ value is more pronounced with increasing Ge fraction in the SiGe fins, indicating that the higher the Ge fraction, the larger the SiGe lattice spacing along $[1\bar{1}0]$ in SiGe fins after the elastic strain relaxation. Gaussian fits for both the broad SiGe diffraction peak and the satellite peaks on the cross-sectional profiles along $Q_x$ were performed to compute the in-plane and out-of-plane SiGe lattice constants. Fig. 4.13 shows the resulting in-plane and out-of-plane lattice deformation values of the $Si_{1-x}Ge_x$/SOI fins and $Si_{1-x}Ge_x$/bulk Si fins. The $[1\bar{1}0]$ in-plane lattice deformation ([001] out-of-plane lattice deformation) value is defined as the difference of $[1\bar{1}0]$ in-plane ([001] out-of-plane) SiGe lattice constant: $b_{SiGe}$ ($c_{SiGe}$) obtained from the broad SiGe diffraction peak position in RSM from the unstrained Si lattice constant: $a_{Si}$, and normalized by $a_{Si}$. Each line in the figure, annotated with specific Ge fractions, represents the theoretical in-plane and out-of-plane lattice deformation of the SiGe fins as a function of transverse in-plane stresses, which vary from a uniaxial stress state ($f = 0$) to a purely isotropic biaxial stress state ($f = 1$) for each fraction of Ge [60]. As the aspect ratio of the SiGe fins increase, it is observed that the lattice deformation values follow the expected trend (from $f = 1$ to $f = 0$) for their respective Ge fractions. For SiGe fins with

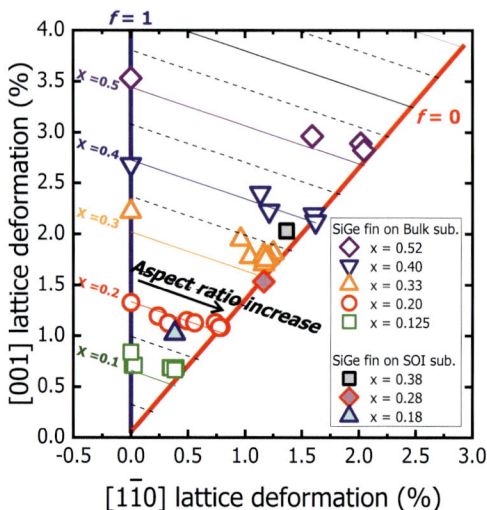

**Figure 4.13** Experimentally measured lattice deformation values for the $Si_{1-x}Ge_x$/SOI fins and $Si_{1-x}Ge_x$/bulk Si fins. The lattice deformation values are given by $(b_{SiGe}/a_{Si} - 1)$ for [1$\bar{1}$0] and $(c_{SiGe}/a_{Si} - 1)$ for [001]. Theoretical curves for the lattice deformation under the assumption of purely elastic behavior as a function of $f$ and Ge fraction $x$ are also plotted.

high aspect ratio, the lattice deformation values are close to those of the uniaxial stress state ($f = 0$). This result confirms that that the transverse in-plane stresses were almost fully, elastically relaxed in the SiGe fins due to patterning of the nanoscale fin structure. To further evaluate the lattice deformation in SiGe fins, NBD was performed on SiGe fin structure. Fig. 4.14 presents these NBD measurements, as a function of depth through the SOI and SiGe regions, for the [1$\bar{1}$0] in-plane and [001] out-of-plane lattice deformation profiles in a $Si_{0.62}Ge_{0.38}$/SOI fin structure. The average [1$\bar{1}$0] in-plane and [001] out-of-plane lattice deformation over the entire $Si_{0.62}Ge_{0.38}$ fin are 1.35% and 1.54%, respectively. When compared with the HRXRD results, which correspond to a volume-averaged lattice deformation, it is observed that the [1$\bar{1}$0] in-plane lattice deformation value (1.37%) matches well with the NBD measurements, while the [001] out-of-plane lattice deformation value is slightly higher (2.03%). This could possibly be due to strain relaxation along the longitudinal direction of the $Si_{0.62}Ge_{0.38}$ fin introduced by TEM sample preparation.

Another issue incumbent to on the spatial dependence of strain within the SiGe fins is related to the fin edges, where even the uniaxial stress state

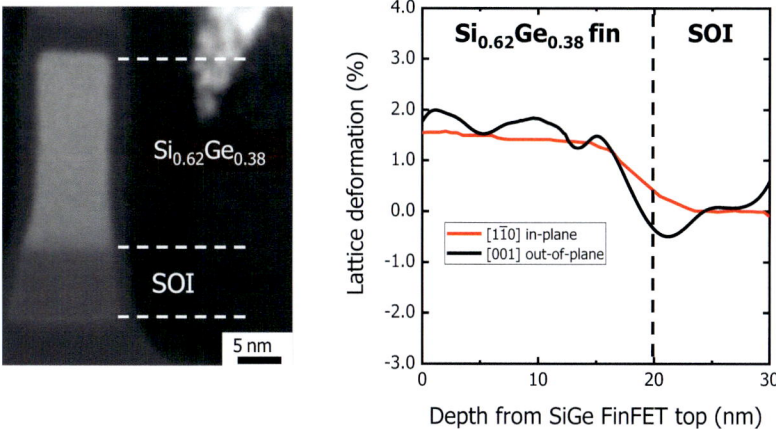

**Figure 4.14** Dark field cross-sectional TEM image and lattice deformation profiles in $Si_{0.62}Ge_{0.38}$/SOI fin structure. Both [1$\bar{1}$0] in-plane and [001] out-of-plane lattice deformation profiles are shown.

may change. The strain state near the $Si_{0.72}Ge_{0.28}$/SOI fin edges were investigated using NBD. Fig. 4.15a demonstrates a [110] lattice deformation contour map near the edge of this $Si_{0.72}Ge_{0.28}$ fin resulting from this investigation. No lattice deformation is observed at a lateral distance greater than approximately 300 nm from the $Si_{0.72}Ge_{0.28}$ fin edge but a significant increase in lattice deformation is confirmed at the $Si_{0.72}Ge_{0.28}$ fin edge particularly at its top corner. Fig. 4.15b plots the lattice deformation values extracted at $Si_{0.72}Ge_{0.28}$ fin top, middle, and bottom as a function of the

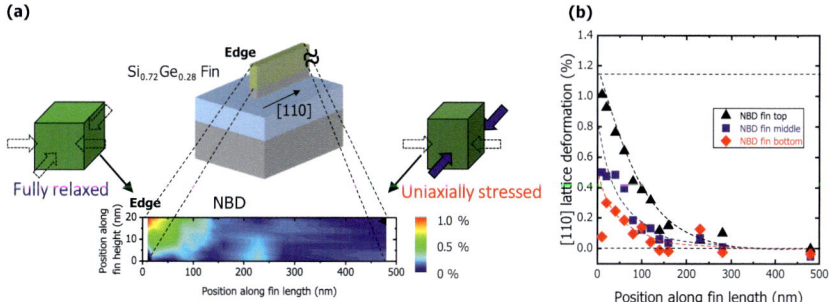

**Figure 4.15** (a) [110] lattice deformation contour map obtained by NBD near the edge of the 20 nm thick $Si_{0.72}Ge_{0.28}$ fin. (b) measured [110] lattice deformation profiles extracted at certain heights (top, middle, and bottom) in the SiGe fin. The dashed lines correspond the [110] lattice deformation values for $Si_{0.72}Ge_{0.28}$ with uniaxial stress state (0%) and fully relaxed stress state (1.15%).

position along the longitudinal direction of the fin. Positive value at the $Si_{0.72}Ge_{0.28}$ fin edge reflects the expansion of the SiGe lattice due to the presence of an additional traction-free surface along the longitudinal direction of the fin. Also, the lattice deformation at the $Si_{0.72}Ge_{0.28}$ fin top corner is close to the value of equilibrium $Si_{0.72}Ge_{0.28}$ (1.15%) indicating that SiGe is fully relaxed. The lattice deformation decreases laterally away from the edges as the $Si_{0.72}Ge_{0.28}$ fin transitions from a fully relaxed to a uniaxially stress state, and beyond a distance of 300 nm, this value is approximately 0. This indicates that the $Si_{0.72}Ge_{0.28}$ fin has same in-plane lattice constant along [110] as SOI underneath (fully strained $Si_{0.72}Ge_{0.28}$ fin). This result is also consistent with the strain state characterization near the fin edges obtained by synchrotron-based nanoXRD reported in Ref. [60]. Therefore, it is important to obtain depiction of the local and global elastic mechanical response within the 3D strained FinFET device structure from complementary strain mapping techniques using X-ray diffraction and electron diffraction, etc.

### 3.2.2 Local layout effect (LLE) in strained SiGe channel pFinFET and LLE mitigation

As reviewed in the previous sub-section, there is a potential variability in device performance when the current carrying regions of 20 nm tall FinFET devices include these sections 300 nm from the longitudinal fin edges [61]. Therefore, it is essential to consider this result in device design to avoid elastic relaxation effect at the edges and take advantage of the fully uniaxial stress in the SiGe fin. It is also important to minimize the strain non-uniformity within the SiGe fin through process techniques and device integration techniques such as reducing the height of the SiGe fin and introducing embedded SiGe (eSiGe) epitaxial structures in source/drain region. In this sub-section, the effect of the local strain distribution on the electrical characteristics of SiGe pFinFETs is investigated and a technique to mitigate this effect is examined.

SiGe pFinFETs with 20% Ge concentration were fabricated with optimized integration of SiGe fin for pFET into a standard 10 nm platform process flow [48]. With a fully compressively strained SiGe fin along [110] (channel direction), a 17% pFET $I_{eff}$ gain over Si fin channel was achieved at matched $I_{off}$, mainly attributed to 35% hole mobility enhancement by compressive stress along channel direction [48]. In order to evaluate strain modulation in SiGe pFinFETs with different fin length ($L_{fin}$), strain in a SiGe channel was measured using NBD [110]. Lattice deformation profile at the SiGe fin middle on $L_{\text{fin}} = 384$ nm baseline SiGe pFinFET is shown in Fig. 4.16a. The Lattice deformation changes from one end of the fin to the other. SiGe channel compressive strain is fully retained at the center active gate while the strain is relaxed at the fin ends. This is consistent with the results observed in the previous sub-section. Fig. 4.16b shows [110] lattice deformation profile at the SiGe fin middle on $L_{\text{fin}} = 128$ nm device, where channel strain is almost fully relaxed at the center active gate compared to $L_{\text{fin}} = 384$ nm case.

Electrical characteristics of LLE (delta $V_{\text{tlin}}$ and $I_{\text{odsat}}$ ratio as a function of $L_{\text{fin}}$) for baseline SiGe pFinFET are shown in Fig. 4.17a. Schematic illustrates device layouts to examine $L_{\text{fin}}$ dependence of electrical characteristics. $L_{\text{fin}}$ is a variable and is varied from two contacted poly pitch (CPP) ($L_{\text{fin}} = 128$ nm) to longer (maximum $L_{\text{fin}} = 1200$ nm), while keeping the other parameters constant. The active gate length is 20 nm with 64 nm CPP. Performance modulation of $I_{\text{odsat}}$ by 6% as well as $V_{\text{tlin}}$ shift by 30 mV is observed for the SiGe pFinFET with $L_{\text{fin}} = 128$ nm compared to the

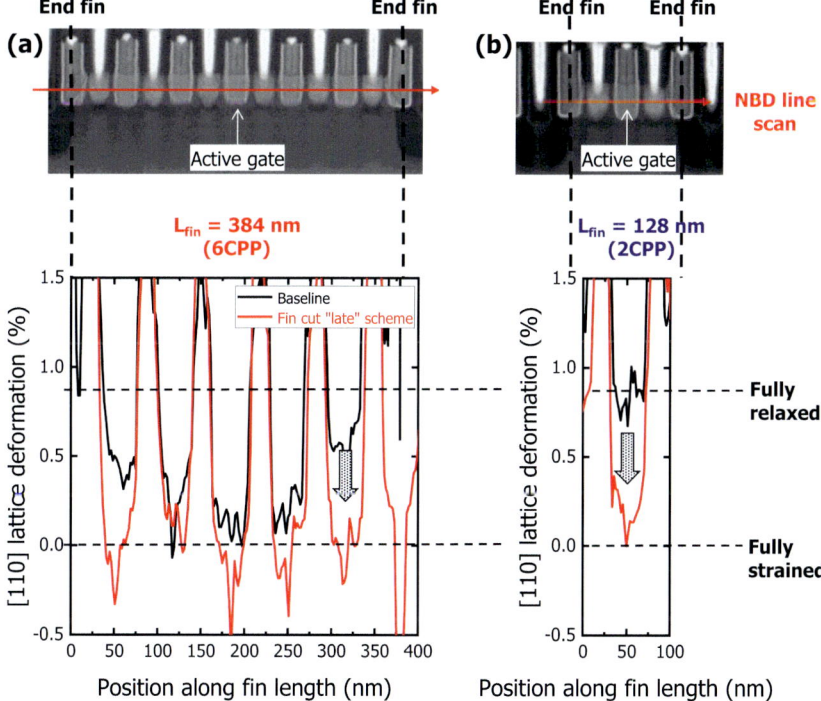

**Figure 4.16** [110] in-plane lattice deformation profiles at the SiGe pFinFET middle comparing baseline and fin cut "late" scheme. (a) $L_{\text{fin}} = 384$ nm, (b) $L_{\text{fin}} = 128$ nm.

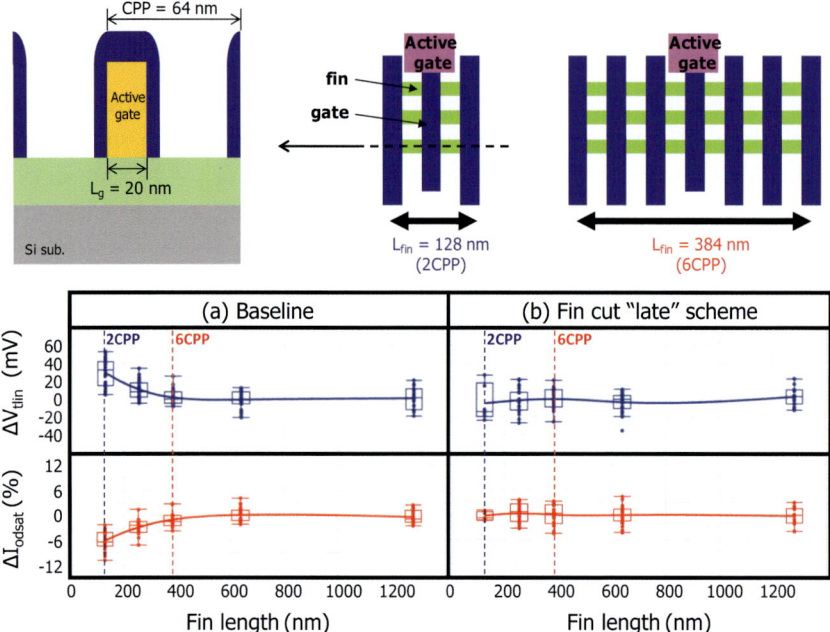

**Figure 4.17** Threshold voltage in linear region ($V_{tlin}$) and saturation current ($I_{odsat}$) of SiGe pFinFET relative to value at $L_{fin}$: 1200 nm as a function of $L_{fin}$. (a) Baseline, (b) method applying SiGe fin cut "late" scheme.

value at $L_{fin} = 1200$ nm. From this electrical characteristic result and above-mentioned strain distribution analysis, it may be inferred that hole mobility in SiGe FinFET with small $L_{fin}$ decreased due to the relaxation of the compressive strain in the channel direction, resulting in the degraded device performance as confirmed in Fig. 4.17a. The increase in the $V_{tlin}$ confirmed in Fig. 4.17a is also attributed to the increase in the SiGe bandgap due to the relaxation of the compressive strain in the channel direction.

As mentioned above, the effect of local strain distribution on the electrical characteristics of strained SiGe FinFETs have been clarified. However, the process of cutting the fin is unavoidable due to the need for element isolation in the actual device fabrication. Therefore, it is necessary to pay attention to a device design such as arranging nonfunctional "dummy" gates in order to avoid the influence of such LLE. Mitigation of LLE is critical to ensure minimum device dimensions and maximize the benefits of SiGe channel strain. Therefore, an integration solution to minimize LLE has been proposed and applied. The method is to perform cutting the SiGe fin downstream "late" [62], where major difference from baseline is to move SiGe fin cut process after spacer/source drain epi/interlevel dielectrics (ILD) formation. [110] Lattice deformation profile at the SiGe fin middle on $L_{fin}$ = 384 nm SiGe pFinFET with fin cut "late" scheme is shown in Fig. 4.16. It is clearly observed that channel strain is successfully retained with late fin cut scheme for both two $L_{fin}$ cases, $L_{fin}$ = 128 nm (Fig. 4.16a) and 384 nm (Fig. 4.16b). The electrical properties are also consistent with this strain result, showing completely suppressed LLE even with $L_{fin}$ = 128 nm case as shown in Fig. 4.17b.

Another possible method to mitigate LLE is to introduce an embedded SiGe (eSiGe) epitaxial structure in the source/drain (S/D) region. Although it becomes more challenging to induce channel strain from S/D epitaxy as technology advance due to reduction of S/D epitaxy volume and difficulty in epitaxial growth while maintaining decent crystallinity on complex 3D device structure, such as FinFET and GAA structure, it was confirmed that LLE is successfully mitigated by eSiGe process [62]. This is because the SiGe fin is patterned to a length shorter than that of $L_{fin}$ by the S/D recess, so that the $L_{fin}$ dependence practically disappears. In addition, the strain loss by elastic strain relaxation can be recovered by applying strain from S/D eSiGe, and as a result, a decrease in hole mobility in the SiGe fin channel can be prevented. These are encouraging techniques to mitigate LLE for enabling continued scaling further combined with strained SiGe FinFET fabrication technology.

## 4. Advanced source drain extension formation for scalded devices

Another issue that is caused by device scaling along vertical and horizontal directions is the increase of parasitic resistance in Source/Drain (S/D) region. The parasitic resistance in the S/D region increases by a factor of κ as the S/D length, width, and depth are shrunk by factor of κ, and the impact of the S/D parasitic resistance increase relatively in more scaled devices. The S/D region is formed by doping impurities, and there is some degree of spatial gradient in the dopant distribution in the S/D extension region (S/D junction region). It is also necessary to achieve a steeper dopant profile (junction profile) by a factors of κ to prevent degradation of the short channel performance due to the spreading of the S/D dopant into the channel region. S/D junction resistance and contact resistance are dominant factor in total series resistance of nanoscale CMOS transistors. S/D junction resistance can be reduced by improving abruptness of dopant profile in S/D extension region. Therefore, formation of the heavily doped low resistance S/D region with shallow and abrupt dopant profile in the S/D extension region is important to improve the performance of scaled devices.

## 4.1 Dopant diffusion control for device junction formation

Strain technology is a key element for improving the performance of state-of-the-art, high-performance metal-oxide-semiconductor field-effect transistors (MOSFETs). Recently, embedded carbon-doped Si (eSi:C) source/drain (S/D) technology has been adopted to improve nMOSFET performance [24,63—71]. Since the lattice constant of Si:C is smaller than that of Si, tensile strain is induced in the Si channel. This tensile strain enhances the electron mobility of nMOS devices [72,73]. Furthermore, an embedded Si:CP (eSi:CP) epitaxial process using in-situ phosphorus (P) doping is effective for reducing the resistance of the S/D region. For the formation of the eSi:CP layer, recessing the S/D region and growing eSi:CP with a high substitutional C concentration ($[C]_{sub}$) using selective epitaxial technology requires a complex crystal growth process and integration scheme [74—77]. In contrast, the approach of forming a Si:CP layer using P and C implantation and recrystallization heat treatment is a promising alternative that can achieve eSi:CP structures with relatively simple process integration [78,79]. However, the combination of ion implantation and recrystallization processes poses a challenge in realizing strained Si:CP films with high $[C]_{sub}$ (>1%) due to the low solid solubility limit of C in Si ($\sim 3 \times 10^{17}$ cm$^{-3}$), which is in thermodynamic equilibrium at high temperatures. Additionally, there is a trade-off between reducing sheet resistance and increasing strain because the activation of P and C to Si lattice substitution sites competes [80,81]. Therefore, in order to effectively utilize the strain effect and reduce the sheet resistance in devices implementing the S/D eSi:CP structure, it is essential to develop a technology to form strained Si:CP layers with high $[C]_{sub}$ and highly concentrated activated P.

On the other hand, diffusion of the S/D dopant into the channel region improves the linear current of the transistor but degrades the short channel characteristics, so forming the optimal S/D extension (SDE: source drain extension) is important in more scaled advanced device structures. A junction with a steep dopant profile must be formed to suppress short channel effects (SCE) at a given linear current. In other words, it is necessary to minimize the SDE sheet resistance (junction resistance) and maximize the linear current of the transistor while maintaining the targeted SCE characteristics through a given junction steepness.

To achieve highly strained S/D regions with steep dopant profiles at transistor junction regions, strained Si:CP layer formation technology by combining an in-situ P-doped Si (Si:P) epitaxial growth process and a C ion implantation process had been investigated. An amorphous layer is generated by implanting cluster C ions into the Si:P film. While monoatomic ion implantation with C ions ($^{12}C^+$) results in partial amorphization of the implanted region and nonuniform amorphous/crystalline (a/c) interface, an amorphous layer formed by cluster C ion implantation shows a uniform a/c interface [82–84]. The strained Si:CP layer is formed by recrystallizing this amorphous layer through thermal treatment. It has been demonstrated that the introduction of crystalline defects such as end of range defects (EOR) and stacking faults/dislocation loops can be reduced by low-temperature C ion implantation and reduction of the number of ion implantation steps. A method to control the activation of P and C by recrystallization thermal treatment conditions was demonstrated. At the same time, a steep profile (3 nm/decade) was achieved by the introduction of C suppressing P diffusion during thermal treatment (Fig. 4.18).

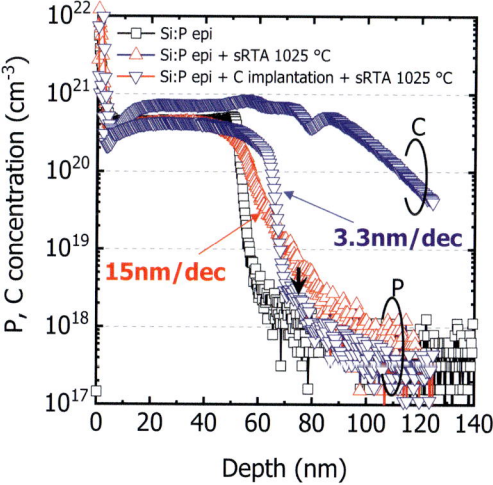

**Figure 4.18** Depth profiles of P and C in a Si:CP layer with a C concentration of 1.2% formed by C ion implantation and recrystallization thermal treatment (spike RTA 1025°C). P depth profiles of samples for Si:P epitaxial as-grown film and after thermal annealing (spike RTA 1025°C) are also shown.

## 4.2 Advanced epitaxial growth technique for source drain extension formation in scaled FinFET devices

Formation and optimization of the Source Drain Extension (SDE) is one of the key factors for aggressively scaled devices. It is necessary to minimize the sheet resistance of the SDE (junction resistance) while balancing the short channel effect (SCE) with an abrupt junction. Having a low resistive doped material with low diffusivity underneath the spacer is the most effective way to achieve high performance scaled devices.

SDE formation by an implantation technique and plasma doping have been reported [85–88]. However, extension doping for a 3D device structure, such as FinFET, is challenging due to the fin geometry. Therefore, SDE formed with selective doped epitaxial growth has been widely used for scaled FinFET devices due to its better sidewall coverage and higher active dopant concentration while eliminating crystalline damage introduction [89,90]. SDE is formed with dopant drive-in diffusion from the epitaxially grown S/D region. Phosphorus doped Si (Si:P) has been widely used as the epitaxial S/D material for nFET. However, degradation of SCE is a concern due to diffusion of phosphorus in Si during downstream processes. It is well known that carbon co-doping in Si:P (Si:CP) suppresses phosphorus diffusion, but the resistance of the film increases due to the presence of carbon.

SDE formation with epitaxially grown Arsenic doped Si (Si:As) is useful technique to achieve an abrupt profile without degrading film resistivity [91]. Fig. 4.19a and b show the hall carrier concentration and film resistivity

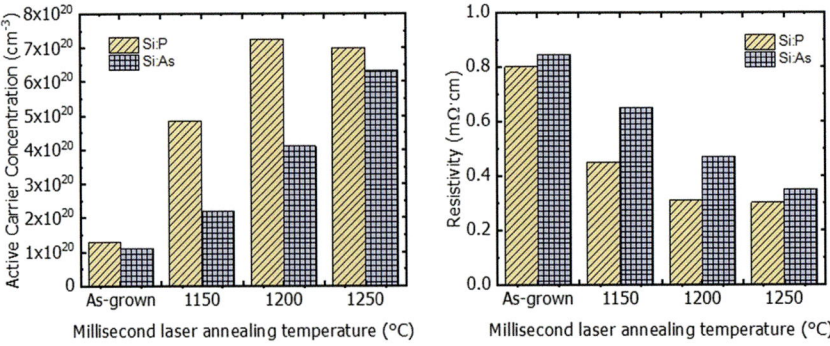

**Figure 4.19** (a) Hall carrier concentration of the P and As activated layers in Si:P and Si:As and film resistivity of the Si:P and Si:As layers as a function of additional milliseconds laser annealing temperature. The P and As concentrations in Si:P and Si:As epitaxial grown films were $1.2 \times 10^{21}$ atoms/cm$^3$.

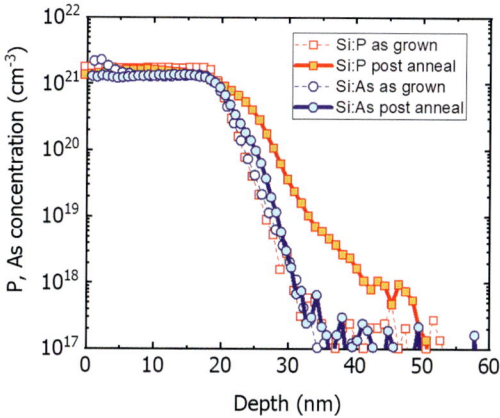

**Figure 4.20** Depth profiles for both P and As in the Si:P and Si:As layers before and after sRTA at 950°C.

of the P and As activated layers in Si:P and Si:As films as a function of additional milliseconds laser anneal temperature. With an increase in laser anneal temperature, the hall carrier concentration is increased. Active P and As concentration $>6 \times 10^{20}$ atoms/cm3 and film resistivity of Si:P and Si:As $< 0.4$ m$\Omega$ cm are achieved by laser anneal (1250°C). The depth profiles for both P and As in the Si:P and Si:As layers after spike rapid thermal annealing (sRTA) at 950°C are shown in Fig. 4.20. While a marked P diffusion is observed, As diffusion is minimal due to its low diffusivity in Si.

### 4.2.1 Si:As SDE formation on FinFET device

Fig. 4.21 shows the typical CMOS FinFET integration process and the SDE formation strategies with using 1) SDE formation with Si:As followed by Si:P S/D formation by in-situ doped selective epitaxial growth, and 2) S/D recess shape modification performed by S/D Fin recess with a RIE process followed by selective isotropic dry lateral etch process which etches the S/D region selective to the spacer material. Fig. 4.21 also shows the structural demonstration of a combination of the isotropic lateral recess and Si:As/Si:P selective epitaxial growth on a scaled FinFET device structure. Conformal Si:As layer growth at the SDE region underneath the spacer and Si:P layer growth at the S/D region is observed.

170  Handbook of Thin Film Deposition

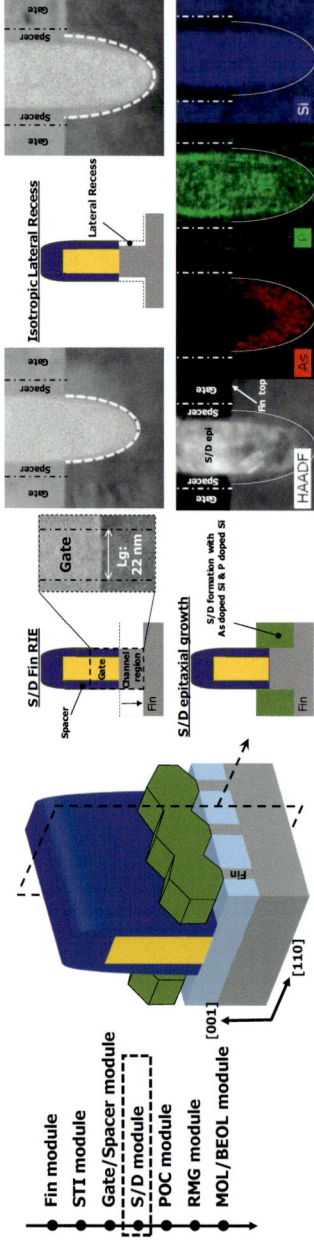

**Figure 4.21** Typical baseline integration process flow and the SDE formation strategies evaluated in this study. Lg in the short channel nFinFETs is 22 nm as shown in the inset. Cross-sectional TEM and EDX elemental mapping of the S/D structure formed by a combination of lateral recess and Si:As/Si:P selective epitaxy on device structure.

## 4.2.2 Electrical sensitivities to SDE material and lateral recess for device performance

To better understand the role of SDE formed with Si:As in short channel device performance, electrical readouts in nFinFETs with 5 nm S/D lateral recess are compared between Si:P SDE and Si:As SDE. The Si:As SDE process delivers around 10% Ron reduction and DIBL benefit of 14 mV due to less dopant diffusion into the channel region. Si:As SDE process gives 13% Ieff benefit at the same Ioff as shown by the $I_{off}$–$I_{eff}$ characteristics in Fig. 4.22. The SS–DIBL characteristics indicate that the Si:As SDE process gives a significant channel quality improvement as seen in the smaller SS degradation by suppression of dopant diffusion into channel region [91].

In addition, the electrical nFinFET device characteristic validation has been performed with different S/D recess shapes (lateral etch amount, depth) demonstrating that the shape highly impacts both junction abruptness and resistance. The two conventional methods for optimization of the junction are: (1) thermal process optimization and (2) spacer thickness optimization. Thermal process optimization, e.g., sRTA temperature and ramp up/down rates, controls dopant diffusion from the S/D region into the channel region. However, changes in the Ron–DIBL trade-off with respect to the thermal process can be expected due to DIBL degradation caused by modulation of the junction abruptness at a higher thermal budget. Spacer thickness optimization controls the distance between the S/D and the channel region. However, increases in parasitic capacitance, such as gate to S/D and gate to contact capacitance, can be expected as the

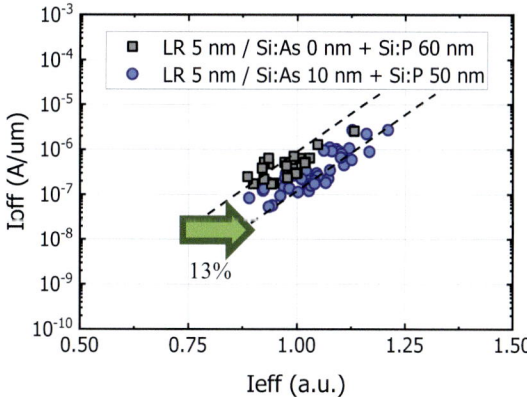

**Figure 4.22** $I_{off}$–$I_{eff}$ characteristics of the same sets of short channel (Lg = 22 nm) nFinFETs with SDE formed with Si:P and Si:As. The S/D lateral recess (LR) amount is 5 nm.

spacer becomes thinner. Adopting and further optimizing the S/D lateral recess process are promising option in order to avoid the problems mentioned above. The performance improvement of nFinFETs with various S/D lateral recess amounts with Si:As SDE is shown in Fig. 4.23. The data indicates that a lateral recess amount of 2 nm is optimal for this specific device structure (e.g., spacer thickness, thermal budget in the downstream processes). This is because of Ron–DIBL trade-off by varying the S/D lateral recess amount. DIBL degrades with more lateral recess amount for Si:As SDE due to dopant diffusion into the channel region [91]. Short channel device characteristics improvements by changing SDE from Si:P to Si:As are summarized in Fig. 4.24. The improvements from Si:As

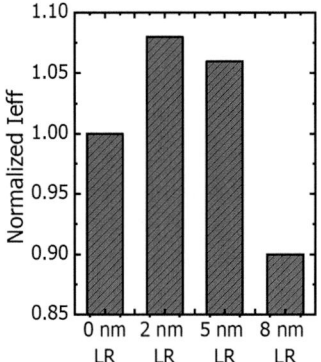

**Figure 4.23** Normalized Ieff values comparison with respect to the lateral recess amount with Si:As SDE.

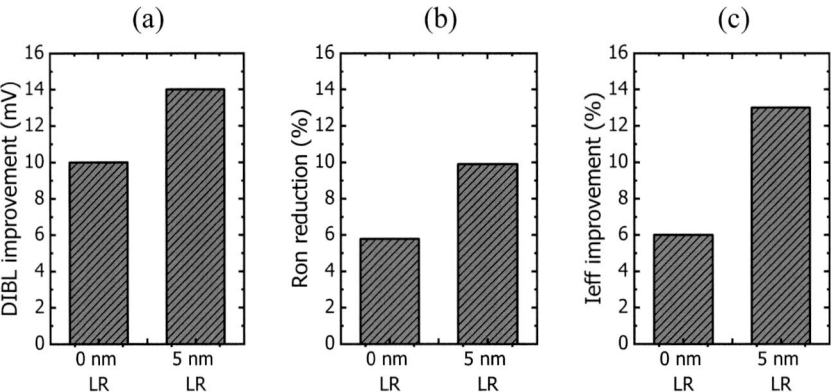

**Figure 4.24** Short channel device characteristics ($L_g = 22$ nm) improvements by changing SDE from Si:P to Si:As as a function of lateral recess amount. (a) DIBL improvement (mV), (b) Ron reduction (%), and (c) Ieff improvement (%).

SDE are more pronounced in the case of 5 nm lateral recess which has closer proximity structure compared to 0 nm lateral recess. These improvements prove the benefit of Si:As SDE in scaled/short channel device structures. S/D recess shape (lateral etch amount, depth), As concentration, and Si:As layer thickness play a critical role in modulating the SCE and device performance.

## 5. Performance enhancement techniques for gate-all-around (GAA) pFET device

The device scaling has been continued even since the introduction of FinFET technology at the 22 nm node. Horizontal gate-all-around (GAA) nanosheet (NS) devices are attractive candidates to replace FinFETs at the 3 nm technology node and beyond due to their excellent electrostatic properties and short-channel (SC) control [92–97]. GAA NS technology has multiple advantages compared to FinFET technology. GAA NS device offers increased drive current compared to FinFET technology, because the total effective width per active footprint is improved by ∼30% by stacking nanosheets compared to FinFET [97]. Also, GAA NS device enables gate length scaling thanks to its outstanding electrostatics and short channel control. Unlike FinFET, the active footprint is not quantized on NS. Therefore, flexible power/performance tuning can be achieved with variable nanosheet width. The industry has aligned on the fact that GAA NS structure is the architecture of choice beyond FinFET as we continue scaling.

However, pFET device performance degradation is a remaining concern, due to degradation of hole mobility by the transition from FinFET device structure to GAA device structure since the main channel surface changes from (110) to (001) plane. It is important to explore performance improvement techniques for next generation GAA CMOS technologies. Therefore, additional techniques are required to improve the performance of the pFET device and their development should be a continued focus of GAA NS technology.

One way to improve device performance is increasing channel carrier mobility. Inducing compressive strain into the channel region is effective in improving the hole carrier mobility, however it is becoming difficult to apply compressive strain in the channel region in more scaled, especially GAA device structures.

In this section, we discuss performance enhancement techniques that can be applied to GAA NS pFET device in order to increase hole mobility: (i) high hole mobility channel material, and (ii) channel surface orientation with high hole mobility.

## 5.1 Highly compressive strained SiGe channel application for GAA NS pFET

SiGe is one of the most promising materials for pFET channel. As mentioned in the previous section, the expected performance advantages of SiGe channel FinFET devices over Si FinFET devices have encouraged their application in GAA NS devices. Strained SiGe channel FinFET device shows intrinsic mobility gain compared to Si channel FinFET. Vt tunability is another benefit for high performance applications. Also, it shows superior NBTI performance. Because of above mentioned benefits, applying SiGe channel into GAA NS device has a great potential for further improvement of the pFET device performance. In addition, maintaining strain in the SiGe channel is important for performance improvement. Therefore, it is necessary to achieve a strained SiGe channel structure that is not affected by strain relaxation due to patterning processes. SiGe cladded NS channel formation through selective Si channel trimming and selective SiGe epitaxial growth post Si channel release has been proposed as an effective method for maintaining strain in the SiGe region [98,99]. The strain in SiGe is applied by the lattice constant difference between SiGe and Si, so there is no concern about strain relaxation of SiGe layer since there is no patterning step in the downstream process. It is also important to demonstrate that achieving the threshold voltage ($V_t$) modulation and reliability improvement of SiGe in a GAA device architecture is feasible. The electrical device characteristics of SiGe cladded NS channel devices with different Ge concentrations need to be compared to Si NS channel devices. Methods for providing a reduction in interface trap density (Dit) need to be evaluated for SiGe channel material.

### 5.1.1 SiGe cladded nanosheet structure formation

Fig. 4.25 shows proposed SiGe cladded NS channel formation strategies using typical GAA Si NS device integration process flow. Two major components of the flow are examined: (1) Si NS channel trimming and (2) SiGe cladding layer formation. The Si NS channel trimming is performed with a selective isotropic dry etch process which etches the Si channel region selective to the spacer and inner spacer materials. The SiGe cladding

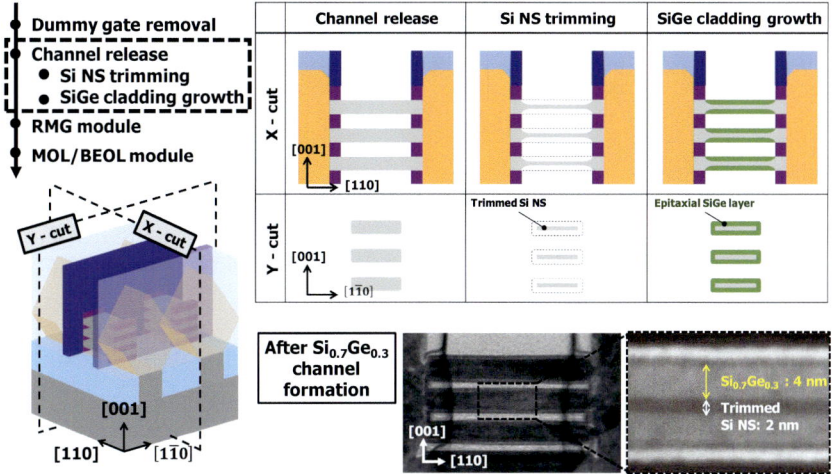

**Figure 4.25** Key integration processes for the SiGe NS channel device fabrication. Cros-sectional TEM images across the gate showing $Si_{0.7}Ge_{0.3}$ NS channel region.

layer formation is performed with a selective SiGe epitaxial growth on the trimmed Si NS channel. After the Si NS channel release, the Si NSs are thinned down to a thickness of 2 nm prior to selective SiGe epitaxial growth. SiGe layers can be epitaxially grown on trimmed Si NSs where Ge fractions and thicknesses of epitaxial SiGe layers can be controlled systematically by different process conditions. Furthermore, the additional growth of Si cap on SiGe layers can be realized as a method to improve channel interface characteristic. Fig. 4.25 contains cross-sectional TEM images across the gate after $Si_{0.7}Ge_{0.3}$ channel formation. No visible crystalline defects are observed in the 4 nm-thick $Si_{0.7}Ge_{0.3}$ layer grown on the 2 nm-thick trimmed Si NS, indicating superior crystallinity of $Si_{0.7}Ge_{0.3}$ layer. Since crystalline defects in the SiGe layer and at the interface cause deterioration of device characteristics, it is important to suppress the defects introduction by optimizing surface cleaning and epitaxial growth conditions.

HRTEM analysis of stacked SiGe NSs (Fig. 4.26) with sheet width (Wsheet) of 20 and 100 nm shows very flat, conformal 4 nm-thick $Si_{0.7}Ge_{0.3}$ layer growth on top and bottom of NSs of equivalent thickness. Uniform growth of SiGe layer is essential for reducing variations in device characteristics, and the SiGe growth conditions must be optimized.

One concern about SiGe channel device is high interface trap density (Dit). Therefore, an in-situ conformal Si cap epitaxy process to passivate

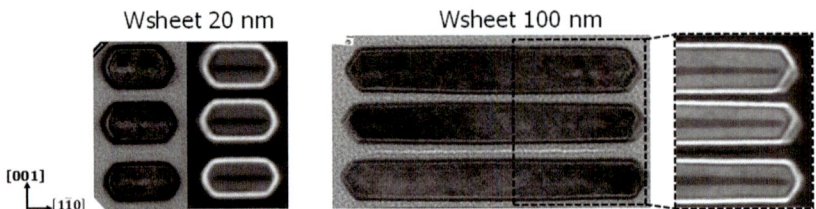

**Figure 4.26** Bright field and corresponding dark field cross-sectional HRSTEM images along the gate showing stacked $Si_{0.7}Ge_{0.3}$ NSs with sheet width (Wsheet) of 20 and 100 nm.

SiGe surface prior to interfacial layer (IL) formation is a useful technique. Fig. 4.27 shows a conformal 1 nm-thick defect-free epitaxial Si cap layer which is unaffected by the crystal growth plane orientation. Si cap thickness applied to the SiGe NS device can be designed so that most of the Si cap would be consumed during subsequent IL formation prior to high-k/metal gate processing. A cross-sectional STEM image and EDX element map of the $Si_{0.65}Ge_{0.35}$ channel structure after M1 electrical test are shown in Fig. 4.27. GAA structure is realized with the deposition of high-k gate dielectric and metal gate material into the space between SiGe NSs. The preservation of the SiGe layer and the absence of significant Ge diffusion after the thermal steps in downstream processes are also confirmed.

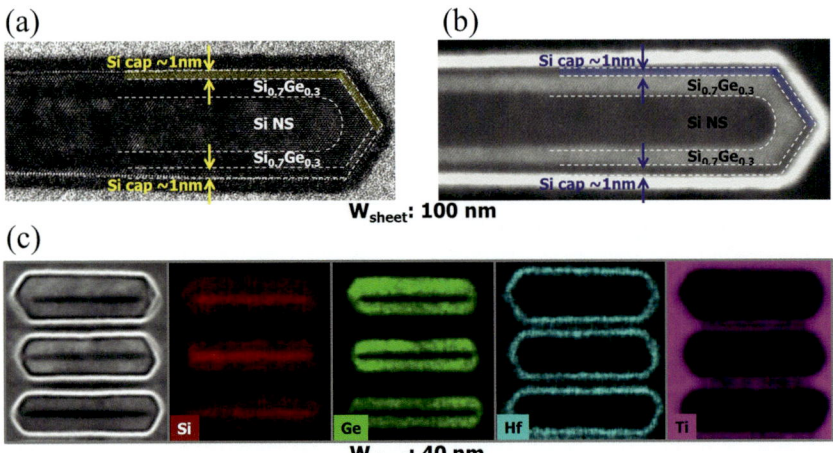

**Figure 4.27** (a) Bright field and (b) corresponding dark field cross-sectional HRSTEM images along the gate showing the epitaxial growth of the conformal 1 nm-thick Si cap layer on $Si_{0.7}Ge_{0.3}$ NSs with no visible defects. Wsheet = 100 nm. (c) Cross-sectional STEM image and EDX element map of stacked SiGe NSs channel with 4 nm-thick $Si_{0.65}Ge_{0.35}$ epitaxial growth along the gate post M1. Wsheet = 40 nm.

Precession Electron Diffraction (PED) characterization is a useful technique to investigate strain within nanoscale device structure by measuring lattice deformation in the crystalline material. Fig. 4.28 shows in-plane lattice deformation contour maps in the region of the stacked SiGe NSs channel with 4 nm-thick $Si_{0.7}Ge_{0.3}$ epitaxial growth obtained from both X-cut (across the gate) and Y-cut (along the gate), as shown in Fig. 4.25, for the SiGe NS structure with sheet width of 20 nm. Lattice deformation values are defined as the difference between in-plane lattice constants and the Si lattice constant, normalized by the Si lattice constant. The in-plane lattice deformation values were extracted from the middle of the SiGe NS channel as shown in Fig. 4.28. Preservation of half the amount of strain along the channel direction ([110]) is confirmed whereas strain along the NS width direction ([1$\bar{1}$0]) is found to be almost fully relaxed due to elastic relaxation. This results in the $Si_{0.7}Ge_{0.3}$ channel being uniaxially stressed. The compressive stress along the channel direction in the $Si_{0.7}Ge_{0.3}$ channel estimated from the [110] lattice deformation is ~1 GPa.

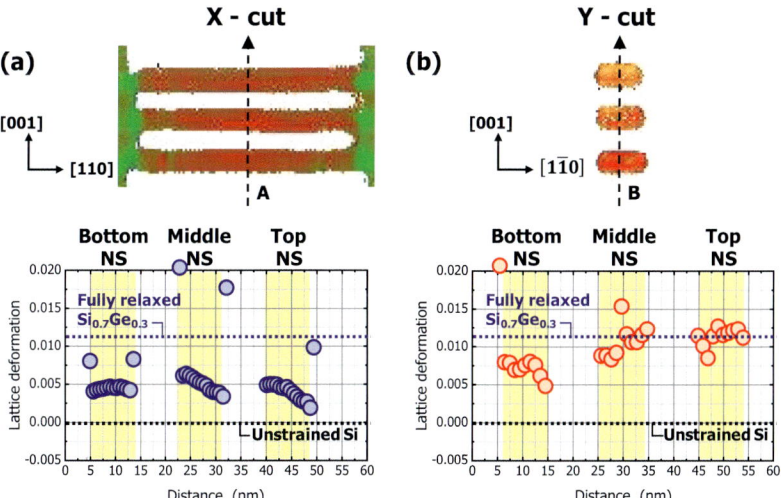

**Figure 4.28** Measured in-plane lattice deformation contour maps in the region of $Si_{0.7}Ge_{0.3}$ NS. (a) [110] Lattice deformation, (b) [1$\bar{1}$0] lattice deformation. Measured in-plane lattice deformation profiles extracted in the middle of the $Si_{0.7}Ge_{0.3}$ NSs are shown (profile A and profile B).

## 5.1.2 Effect of Ge fraction and SiGe thickness on device characteristics

Hole mobility in the $Si_{1-x}Ge_x$ channel NS pFET as a function of inversion carrier density (Ninv) is shown in Fig. 4.29. Fig. 4.29 also shows extracted peak hole mobility as a function of the $Si_{1-x}Ge_x$ epitaxial growth layer thickness. The $Si_{1-x}Ge_x$ channel with Si cap shows significantly higher hole mobility and the hole mobility increases for thicker $Si_{1-x}Ge_x$ films. Other carrier transport characteristics such as channel resistance ($R_{ch}$) and transconductance ($gm,max$) show similar trends [99]. The hole mobility of $Si_{1-x}Ge_x$ channel (x = 0.3 and 0.35) with Si cap is almost 100% higher than that of Si NS channel. The mobility benefit of $Si_{1-x}Ge_x$ channel is attributed to decreased phonon scattering for thicker channel bodies [100–102] and reduced effective hole mass along the transport direction caused by a high compressive strain in the $Si_{1-x}Ge_x$ channel [103,104]. Electrostatic such as subthreshold slope can also be improved when Si cap is applied, which is correlated with the gate stack interface characteristic improvement. Interface trap charge (Dit) reduction can be achieved by adding Si cap [99].

## 5.2 (110) versus (001) channel orientation GAA NS technology

Altering the channel orientation is another well-known technique for modulating carrier mobility independent of internal and external stressor elements. Hole mobility is higher for (110) planes than for (001) planes [105], and electron mobility is lower for (110) planes than for (001) planes. In this sub-section, pFET and nFET GAA Si channel NS devices fabricated

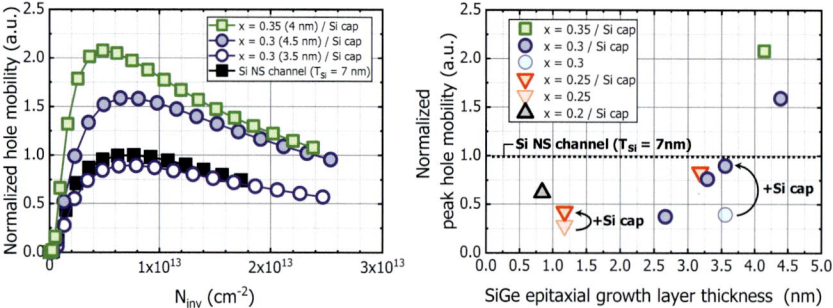

**Figure 4.29** Normalized hole mobility in the representative $Si_{1-x}Ge_x$ channel NS pFET as a function of inversion carrier density (Ninv). Wsheet = 20 nm. Normalized peak hole mobility for $Si_{1-x}Ge_x$ (x = 0.2, 0.25, 0.3, and 0.35) channel NS pFET without Si cap and with Si cap as a function of the $Si_{1-x}Ge_x$ epitaxial growth layer thickness. Wsheet = 20 nm.

on (001) and (110) bulk substrates are discussed. The same process integration can be utilized for both substrate types. Mobility is modulated by Si NS thickness ($T_{Si}$) due to quantum confinement and volume inversion, so the sensitivities of both (001) and (110) channel mobility and performance versus $T_{Si}$ should be understood.

### 5.2.1 Stacked Si NS pFET and nFET devices fabrication on (001) and (110) bulk Si substrates

The integration process flow for the GAA Si NS channel pFET and nFET devices fabrication on (001) and (110) bulk Si substrates is shown in Fig. 4.30. Stacked Si NS pFET and nFET devices are fabricated on (001) and (110) bulk Si substrates using the same integrated process flow. The thickness of the epitaxial Si layers in the SiGe/Si multilayer NS stack can be modulated to target $T_{Si}$. X-ray spectra for NS stacks grown on (001) and (110) bulk Si substrates are shown in Fig. 4.30. High-crystal quality for NS stacks of both orientations can be confirmed by the clear thickness fringes from the Si and SiGe layers. It is necessary to be careful about introducing crystalline defects when epitaxially growing SiGe on Si (110) substrate since SiGe critical thickness on Si (110) is much thinner than that on Si (001) substrate [106,107]. In order to grow SiGe with decent crystallinity, it is important to optimize not only epitaxial SiGe growth process condition but also cleaning process prior to the epitaxial growth. Fig. 4.31 shows schematics for crystallographic orientations for stacked GAA Si NS devices fabricated using (001) and (110) substrates. The transport direction is along <110> for both (001) and (110) NS devices. Si channel surface is (001) for (001) device, on the other hand, Si channel surface is (110) for (110) device, as shown here.

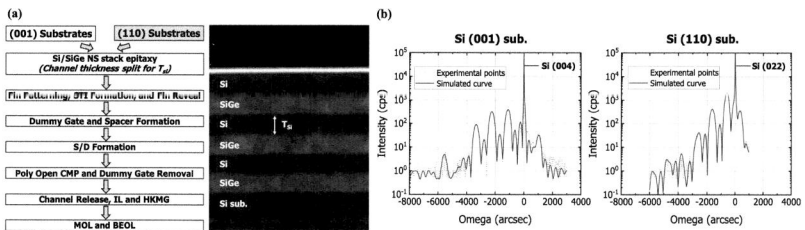

**Figure 4.30** (a) The integration process flow for the stacked GAA Si NS channel pFET and nFET devices fabrication on (001) and (110) bulk Si substrates. Cross-sectional TEM image showing epitaxially-grown SiGe/Si NS stack. (b) X-ray spectra for SiGe/Si NS stacks grown on (001) and (110) bulk Si substrates.

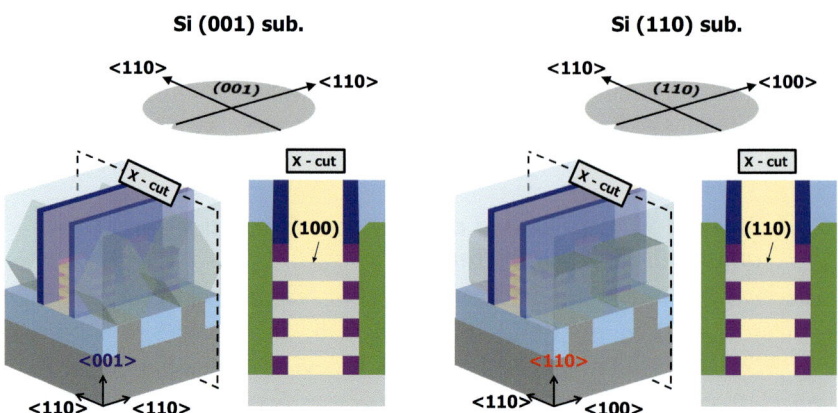

**Figure 4.31** Schematics showing crystallographic orientations for stacked GAA Si NS devices fabricated using (001) and (110) substrates.

### 5.2.2 Electrical sensitivities to $T_{Si}$ for device performance

Fig. 4.32 shows pFET and nFET hole and electron mobility for (001) and (110) long channel devices at $T_{Si} = 6$ nm versus inversion carrier density (Ninv) and compares extracted peak and high field mobility (Ninv@1E13cm$^{-2}$) for different $T_{Si}$. The (110) pFET hole mobility is significantly higher than (001) hole mobility for the $T_{Si}$ range of 4–8 nm. Less sensitivity of the (110) pFET hole mobility is observed with $T_{Si}$, while the (001) pFET hole mobility reduces with lower $T_{Si}$, similar to the $R_{ch}$ and $I_{odsat}$ trends [108]. Therefore, the hole mobility enhancement factor compared to the (001) pFET increases as $T_{Si}$ is reduced, which is consistent with UTBSOI planar MOSFETs [109]. The hole mobility enhancement factor is as high as 2.34 at $T_{Si} = 4$ nm. This is beneficial for device scaling while still maintaining decent electrostatics and performance for pFET devices. For (001) pFET NS, mobility degrades with decreasing $T_{Si}$ due to increased phonon scattering. Reduced (110) pFET mobility sensitivity to $T_{Si}$ can be explained by (1) rate of change of the (110) channel surface area to total channel surface area with changing $T_{Si}$ and (2) hole effective mass and scattering rate dependence on $T_{Si}$. It is known that increasing the ratio of the (110) channel surface area to the total surface area enhances hole mobility. This ratio change with $T_{Si}$ is 3.5× stronger for (001)-oriented devices than for (110)-oriented devices. By reducing $T_{Si}$ from 8 to 4 nm, there is a 15% change in the ratio for (110)-oriented devices compared to a 52% change in the ratio for (001)-oriented devices. The effective hole mass decreases and the scattering rate is nearly constant within the range of 8 to 4 nm $T_{Si}$ [110,111].

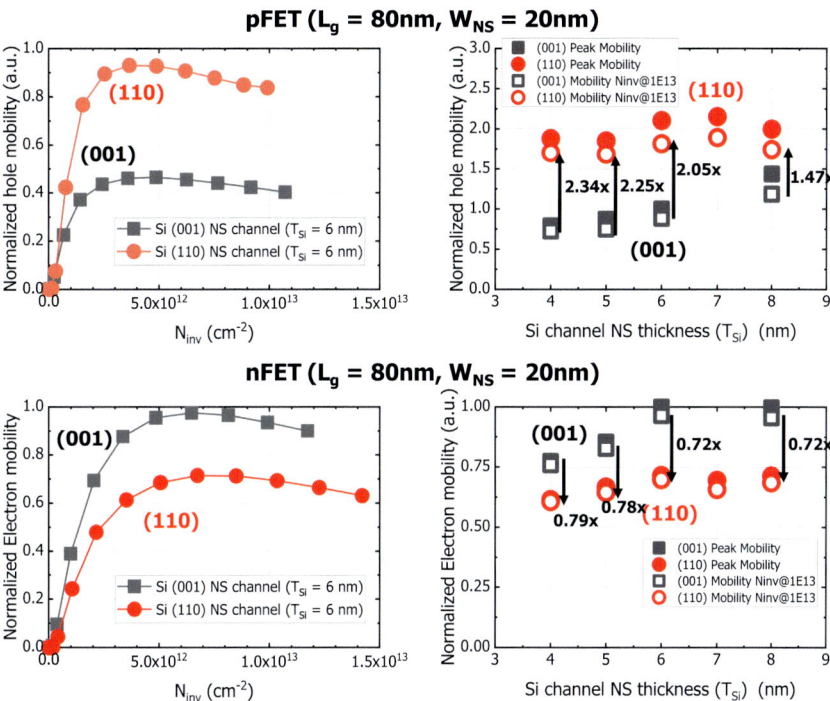

**Figure 4.32** pFET and nFET normalized hole and electron mobility versus $N_{inv}$ in long channel (001) and (110) GAA NS devices with $L_g = 80$ nm, $W_{NS} = 20$ nm and $T_{Si} = 6$ nm. Extracted peak and high field mobility comparison for different $T_{Si}$ are also shown.

On the other hand, the (110) nFET electron mobility is degraded relative to the (001) electron mobility as $T_{Si}$ varies from 4 to 8 nm. Both (110) and (001) show nFET electron mobility degradation for thinner $T_{Si}$. The nFET (110) electron mobility is degraded by 28% relative to (001) as $T_{Si}$ is reduced.

Fig. 4.33 compares nFET and pFET $R_{ch}$ and $I_{eff}$ for (110) and (001) short channel devices. The large impact on $R_{ch}$ with substrate orientation is observed, which is consistent with long channel devices [108]. (110) pFET $R_{ch}$ shows 15% improvement over (001) devices for $T_{Si} = 6$ nm. A slightly larger improvement is observed for lower $T_{Si}$ since the (110) hole mobility is less sensitive to interface scattering effects in thin silicon compared to (001). Opposite trends are observed for nFET (110) versus (001). (110) Rch shows 15% degradation compared to (001) Rch for $T_{Si} = 6$ nm. Even stronger degradation in $R_{ch}$ is observed for extremely thin $T_{Si}$. Better performance at $T_{Si} = 6$ nm for (110) pFET short channel devices with 22% higher $I_{eff}$ over (001), while the nFET (110) device shows 14% lower $I_{eff}$ than for (001).

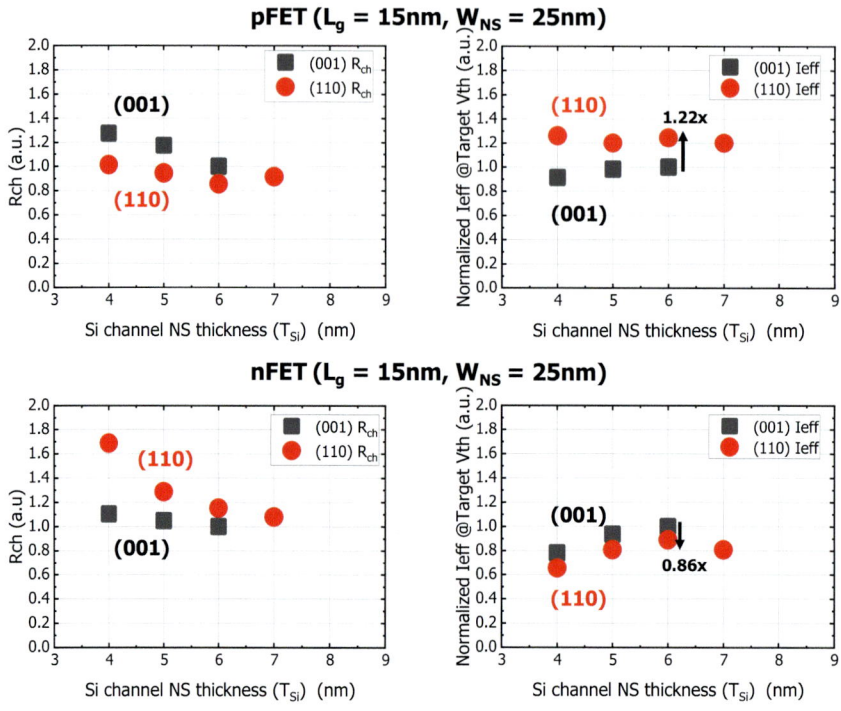

**Figure 4.33** Normalized $R_{ch}$ and $I_{eff}$ at target Vth versus $T_{Si}$ for pFET and nFET short channel (001) and (110) GAA NS devices with $L_g = 15$ nm and $W_{NS} = 25$ nm.

These results are directly related to the differences in channel orientation and electron/hole mobility. Similar to the long channel devices, the pFET (110) device performance improvement is less sensitive to $T_{Si}$. The net improvement in hole mobility and degradation of electron mobility for short channel devices is less than that seen in long channel devices. This can be explained by multiple factors that impact short channel devices, such as coulomb scattering, doping in the channel or external stressor effects [112]. Despite the lower improvement on short channel devices, NS technology fabricated on (110) substrates could have better performance tradeoff between nFET and pFET transistors over that built on (001) substrates.

# 6. Conclusions

Currently, FinFETs are already in practical use, and the possibility of further device scaling using new three-dimensional structure transistors has been demonstrated. Furthermore, with the prospect of practical application of extreme ultraviolet lithography, it will become possible to process to dimensions finer than 20 nm, which was difficult with conventional optical lithography technology, and the formation of ultra-fine device structures will become possible. Therefore, it is becoming more likely that horizontal/vertical gate all around (GAA) structures will be adopted as the next generation device structure to continue scaling. In addition, it is expected that structures will become even more complex in the future, such as the practical application of strained SiGe channels in FinFETs and the consideration of applying strain technology to GAA structures. In other words, it is becoming increasingly important to evaluate crystallinity and strain in actual device structures, rather than on a macroscopic scale, and to understand the mechanisms of their modulation and relaxation. In addition, there will be a strong need for engineering to evaluate, understand, and control in detail the influence of thermal load processes in device fabrication processes on crystallinity in local regions. The role of epitaxy technology in scaling devices is becoming increasingly important, not only in channel materials but also in SDE, S/D, and contact regions. The scaling of the device requires better process control over the thickness of the epitaxial grown thin film and geometrical constraints imposed by the three-dimensional device structure.

## References

[1] R.H. Dennard, F.H. Gaensslen, H.-N. Yu, V.L. Rideout, E. Bassous, A.R. LeBlanc, J. IEEE SC—9 (1974) 256.
[2] G.E. Moore, Electronics 38 (1965) 114.
[3] M. Bohr, IEEE IEDM, 2011, pp. 1.1.1—1.1.6.
[4] C.S. Smith, Phys. Rev. 94 (1954) 42.
[5] E.A. Fitzgerald, Y.-H. Xie, D. Monroe, P.J. Silverman, J.M. Kuo, A.R. Kortan, F.A. Thiel, B.E. Weir, J. Vac. Sci. Technol. B 10 (1992) 1807.
[6] E.A. Fitzgerald, S.B. Samavedam, Thin Solid Films 294 (1997) 3.
[7] F.K. LeGoues, B.S. Meyerson, J.F. Morar, P.D. Kirchner, J. Appl. Phys. 71 (1992) 4230.
[8] N. Sugii, S. Yamaguchi, K. Washio, J. Vac. Sci. Technol. B 20 (2002) 1891.
[9] T. Tezuka, N. Sugiyama, T. Mizuno, M. Suzuki, S. Takagi, Jpn. J. Appl. Phys. 40 (2001) 2866.
[10] N. Taoka, A. Sakai, S. Mochizuki, O. Nakatsuka, M. Ogawa, S. Zaima, T. Tezuka, N. Sugiyama, S. Takagi, Jpn. J. Appl. Phys. 44 (2005) 7356.
[11] M. Gunji, A.F. Marshall, P.C. McIntyre, J. Appl. Phys. 109 (2011) 014324.
[12] T.A. Langdo, M.T. Currie, A. Lochtefeld, R. Hammond, J.A. Carlin, M. Erdtmann, G. Braithwaite, V.K. Yang, C.J. Vineis, H. Badawi, M.T. Bulsara, Appl. Phys. Lett. 82 (2003) 4256.
[13] F. Ootsuka, S. Wakahara, K. Ichinose, A. Honzawa, S. Wada, H. Sato, T. Ando, H. Ohta, K. Watanabe, T. Onai, IEEE IEDM 575, 2000.

[14] S. Ito, H. Namba, K. Yamaguchi, T. Hirata, K. Ando, S. Koyama, S. Kuroki, N. Ikezawa, T. Suzuki, T. Saitoh, T. Horiuchi, IEEE IEDM 247, 2000.
[15] S. Pidin, T. Mori, K. Inoue, S. Fukuta, N. Itoh, E. Mutoh, K. Ohkoshi, R. Nakamura, K. Kobayashi, K. Kawamura, T. Saiki, S. Fukuyama, S. Satoh, M. Kase, K. Hashimoto, IEEE IEDM 213, 2000.
[16] H.S. Yang, R. Malik, S. Narasimha, Y. Li, R. Divakaruni, P. Agnello, S. Allen, A. Antreasyan, J.C. Arnold, K. Bandy, M. Belyansky, A. Bonnoit, G. Bronner, V. Chan, X. Chen, Z. Chen, D. Chidambarrao, A. Chou, W. Clark, S.W. Crowder, B. Engel, H. Harifuchi, S.F. Huang, R. Jagannathan, F.F. Jamin, Y. Kohyama, H. Kuroda, C.W. Lai, H.K. Lee, W.-H. Lee, E.H. Lim, W. Lai, A. Mallikarjunan, K. Matsumoto, A. McKnight, J. Nayak, H.Y. Ng, S. Panda, R. Rengarajan, M. Steigerwalt, S. Subbanna, K. Subramanian, J. Sudijono, G. Sudo, S.-P. Sun, B. Tessier, Y. Toyoshima, P. Tran, R. Wise, R. Wong, I.Y. Yang, C.H. Wann, L.T. Su, M. Horstmann, T. Feudel, A. Wei, K. Frohberg, G. Burbach, M. Gerhardt, M. Lenski, R. Stephan, K. Wieczorek, M. Schaller, H. Salz, J. Hohage, H. Ruelke, J. Klais, P. Huebler, S. Luning, R. van Bentum, G. Grasshoff, C. Schwan, E. Ehrichs, S. Goad, J. Buller, S. Krishnan, D. Greenlaw, M. Raab, N. Kepler, IEEE IEDM 1075, 2004.
[17] K. Uejima, H. Nakamura, T. Fukase, S. Mochizuki, S. Sugiyama, M. Hane, IEEE IEDM 220, 2007.
[18] S. Thompson, N. Anand, M. Armstrong, C. Auth, B. Arcot, M. Alavi, P. Bai, J. Bielefeld, R. Bigwood, J. Brandenburg, M. Buehler, S. Cea, V. Chikarmane, C. Choi, R. Frankovic, T. Ghani, G. Glass, W. Han, T. Hoffmann, M. Hussein, P. Jacob, A. Jain, C. Jan, S. Joshi, C. Kenyon, J. Klaus, S. Klopcic, J. Luce, Z. Ma, B. Mcintyre, K. Mistry, A. Murthy, P. Nguyen, H. Pearson, T. Sandford, R. Schweinfurth, R. Shaheed, S. Sivakumar, M. Taylor, B. Tufts, C. Wallace, P. Wang, C. Weber, M. Bohr, IEEE IEDM 61, 2002.
[19] T. Ghani, M. Armstrong, C. Auth, M. Bost, P. Charvat, G. Glass, T. Hoffmann, K. Johnson, C. Kenyon, J. Klaus, B. McIntyre, K. Mistry, A. Murthy, J. Sandford, M. Silberstein, S. Sivakumar, P. Smith, K. Zawadzki, S. Thompson, M. Bohr, IEEE IEDM 11.6.1, 2003.
[20] H. Ohta, Y. Kim, Y. Shimamune, Y. Sakuma, A. Hatada, A. Katakami, T. Soeda, K. Kawamura, H. Kokura, H. Morioka, T. Watanabe, J.O.Y. Hayami, J. Ogura, M. Tajima, T. Mori, N. Tamura, M. Kojima, K. Hashimoto, IEEE IEDM 247, 2005.
[21] N. Yasutake, A. Azuma, T. Ishida, N. Kusunoki, S. Mori, H. Itokawa, I. Mizushima, S. Okamoto, T. Morooka, N. Aoki, S. Kawanaka, S. Inaba, Y. Toyoshima, Symposium on VLSI Technology Digest of Technical Papers, 2007, p. 48.
[22] T.-Y. Liow, K.-M. Tan, D. Weeks, R.T.P. Lee, M. Zhu, K.-M. Hoe, C.-H. Tung, M. Bauer, J. Spear, S.G. Thomas, G.S. Samudra, N. Balasubramanian, Y.-C. Yeo, Symposium on VLSI Technology Digest of Technical Papers, 2008, p. 126.
[23] B. Yang, R. Takalkar, Z. Ren, L. Black, A. Dube, J.W. Weijtmans, J. Li, J.B. Johnson, J. Faltermeier, A. Madan, Z. Zhu, A. Turansky, G. Xia, A. Chakravarti, R. Pal, K. Chan, A. Reznicek, T.N. Adam, B. Yang, J.P. de Souza, E.C.T. Harley, B. Greene, A. Gehring, M. Cai, D. Aime, S. Sun, H. Meer, J. Holt, C. Theodore, S. Zollner, P. Grudowski, D. Sadana, D.-G. Park, D. Mocuta, D. Schepis, E. Maciejewski, S. Luning, J. Pellerin, E. Leobandung, IEEE IEDM 1, 2008.
[24] S. Narasimha, P. Chang, C. Ortolland, D. Fried, E. Engbrecht, K. Nummy, P. Parries, T. Ando, M. Aquilino, N. Arnold, R. Bolam, J. Cai, M. Chudzik, B. Cipriany, G. Costrini, M. Dai, J. Dechene, C. DeWan, B. Engel, M. Gribelyuk, D. Guo, G. Han, J. Habib, J. Holt, D. Ioannou, B. Jagannathan, D. Jaeger, J. Johnson, W. Kong, J. Koshy, R. Krishnan, A. Kumar, M. Kumar, J. Lee, X. Li, C.-H. Lin, B. Linder, S. Lucarini, N. Lustig, P. McLaughlin, K. Onishi, V. Ontalus,

R. Robison, C. Sheraw, M. Stoker, A. Thomas, G. Wang, R. Wise, L. Zhuang, G. Freeman, J. Gill, E. Maciejewski, R. Malik, J. Norum, P. Agnello, IEEE IEDM 3.3.1, 2012.
[25] K. Ota, K. Sugihara, H. Sayama, T. Uchida, H. Oda, T. Eimori, H. Morimoto, Y. Inoue, IEEE IEDM 27, 2002.
[26] C.-C. Liao, T.-Y. Chiang, M.-C. Lin, T.-S. Chao, IEEE IEDM 281, 2010.
[27] K.-Y. Lim, H. Lee, C. Ryu, K.-I. Seo, U. Kwon, S. Kim, J. Choi, K. Oh, H.-K. Jeon, C. Song, T.-O. Kwon, J. Cho, S. Lee, Y. Sohn, H.S. Yoon, J. Park, K. Lee, W. Kim, E. Lee, S.-P. Sim, C.G. Koh, S.B. Kang, S. Choi, C. Chung, IEEE IEDM 10.1.1, 2010.
[28] S. Tiwari, M.V. Fischetti, P.M. Mooney, J.J. Welser, IEEE IEDM 939, 1997.
[29] C. Gallon, G. Reimbold, G. Ghibaudo, R.A. Bianchi, R. Gwoziecki, S. Orain, E. Robilliart, C. Raynaud, H. Dansas, IEEE Trans. Electron Devices 51 (2004) 1254.
[30] D. Hisamoto, W.C. Lee, J. Kedzierski, E. Anderson, H. Takeuchi, K. Asano, T.J. King, J. Bokor, C. Hu, IEEE IEDM 1032, 1998.
[31] H.S.P. Wong, D.J. Frank, P.M. Solomon, IEEE IEDM 407, 1998.
[32] D. Hisamoto, W.C. Lee, J. Kedzierski, H. Takeuchi, K. Asano, C. Kuo, T.-J. King, J. Bokor, C. Hu, IEEE IEDM 2320, 2000.
[33] R.H. Dennard, F.H. Gaensslen, H.N. Yu, V. Leo Rideout, E. Bassous, A.R. Leblanc, Design of ion-implanted MOSFET's with very small physical dimensions, IEEE J. Solid State Circ. 9 (5) (1974), https://doi.org/10.1109/JSSC.1974.1050511.
[34] P.R. Chidambaram, C. Bowen, S. Chakravarthi, C. Machala, R. Wise, Fundamentals of silicon material properties for successful exploitation of strain engineering in modern CMOS manufacturing, IEEE Trans. Electron. Dev. 53 (5) (2006) 944−964, https://doi.org/10.1109/TED.2006.872912.
[35] S. Ito, H. Namba, K. Yamaguchi, T. Hirata, K. Ando, S. Koyama, S. Kuroki, N. Ikezawa, T. Suzuki, T. Saitoh, T. Horiuchi, Mechanical stress effect of etch-stop nitride and its impact on deep submicron transistor design, in: International Electron Devices Meeting 2000, IEEE, 2024, pp. 247−250. Technical Digest. IEDM (Cat. No.00CH37138).
[36] K. Ota, K. Sugihara, H. Sayama, T. Uchida, H. Oda, T. Eimori, H. Morimoto, Y. Inoue, Novel locally strained channel technique for high performance 55nm CMOS, in: Digest. International Electron Devices Meeting, IEEE, 2024, pp. 27−30.
[37] S.E. Thompson, M. Armstrong, C. Auth, S. Cea, R. Chau, G. Glass, T. Hoffman, J. Klaus, Z. Ma, B. Mcintyre, A. Murthy, B. Obradovic, L. Shifren, S. Sivakumar, S. Tyagi, T. Ghani, K. Mistry, M. Bohr, Y. El-Mansy, A logic nanotechnology featuring strained-silicon, IEEE Electron. Device Lett. 25 (4) (2004) 191−193, https://doi.org/10.1109/LED.2004.825195.
[38] H. Shang, L. Chang, X. Wang, M. Rooks, Y. Zhang, B. To, K. Babich, G. Totir, Y. Sun, E. Kiewra, M. Ieong, W. Haensch, Investigation of FinFET devices for 32nm technologies and beyond, in: 2006 Symposium on VLSI Technology, 2006. Digest of Technical Papers, IEEE, 2024, pp. 54−55.
[39] S. Barraud, V. Lapras, M.P. Samson, L. Gaben, L. Grenouillet, V. Maffini-Alvaro, Y. Morand, J. Daranlot, N. Rambal, B. Pre-vitalli, S. Reboh, C. Tabone, R. Coquand, E. Augendre, O. Rozeau, J.M. Hartmann, C. Vizioz, C. Arvet, P. Pimenta-Barros, N. Posseme, V. Loup, C. Comboroure, C. Euvrard, V. Balan, I. Tinti, G. Audoit, N. Bernier, D. Cooper, Z. Saghi, F. Allain, A. Toffoli, O. Faynot, M. Vinet, Vertically stacked-NanoWires MOSFETs in a replacement metal gate process with inner spacer and SiGe source/drain, in: 2016 IEEE International Electron Devices Meeting (IEDM), IEEE, 2016, pp. 17.6.1−17.6.4.
[40] D. Yakimets, G. Eneman, P. Schuddinck, T.H. Bao, M. Garcia Bardon, P. Raghavan, A. Veloso, N. Collaert, A. Mercha, D. Verkest, A.V.-Y. Thean, K.de Meyer, Vertical

GAAFETs for the ultimate CMOS scaling, IEEE Trans. Electron. Dev. 62 (5) (2015) 1433−1439, https://doi.org/10.1109/TED.2015.2414924.
[41] N. Loubet, T. Hook, P. Montanini, C.W. Yeung, S. Kanakasabapathy, M. Guillorn, T. Yamashita, J. Zhang, X. Miao, J. Wang, A. Young, R. Chao, M. Kang, Z. Liu, S. Fan, B. Hamieh, S. Sieg, Y. Mignot, W. Xu, S.C. Seo, J. Yoo, S. Mochizuki, M. Sankarapandian, O. Kwon, A. Carr, A. Greene, Y. Park, J. Frougier, R. Galatage, R. Bao, J. Shearer, R. Conti, H. Song, D. Lee, D. Kong, Y. Xu, A. Arceo, Z. Bi, P. Xu, R. Muthinti, J. Li, R. Wong, D. Brown, P. Oldiges, R. Robison, J. Arnold, N. Felix, S. Skordas, J. Gaudiello, T. Standaert, H. Jagannathan, D. Corliss, M.H. Na, A. Knorr, T. Wu, D. Gupta, S. Lian, R. Divakaruni, T. Gow, C. Labelle, S. Lee, V. Paruchuri, H. Bu, M. Khare, Stacked nanosheet gate-all-around transistor to enable scaling beyond FinFET, in: 2017 Symposium on VLSI Technology, IEEE, 2017, pp. T230−T231.
[42] H. Jagannathan, B. Anderson, C.W. Sohn, G. Tsutsui, J. Strane, R. Xie, S. Fan, K.I. Kim, S. Song, S. Sieg, I. Seshadri, S. Mochizuki, J. Wang, A. Rahman, K.Y. Cheon, I. Hwang, J. Demarest, J. Do, J. Fullam, G. Jo, B. Hong, Y. Jung, M. Kim, S. Kim, R. Lallement, T. Levin, J. Li, E. Miller, P. Montanini, R. Pujari, C. Osborn, M. Sankarapandian, G.H. Son, C. Waskiewicz, H. Wu, J. Yim, A. Young, C. Zhang, A. Varghese, R. Robison, S. Burns, K. Zhao, T. Yamashita, D. Dechene, D. Guo, R. Divakaruni, T. Wu, K.I. Seo, H. Bu, Vertical-transport nanosheet technology for CMOS scaling beyond lateral-transport devices, in: 2021 IEEE International Electron Devices Meeting (IEDM), IEEE, 2021, pp. 26.1.1−26.1.4.
[43] J. Kavalieros, B. Doyle, S. Datta, G. Dewey, M. Doczy, B. Jin, D. Lionberger, M. Metz, W. Rachmady, M. Radosavljevic, U. Shah, N. Zelick, R. Chau, Tri-gate transistor architecture with high-k gate dielectrics, metal gates and strain engineering, in: 2006 Symposium on VLSI Technology, 2006. Digest of Technical Papers, IEEE, 2024, pp. 50−51.
[44] F. Conzatti, N. Serra, D. Esseni, M. de Michielis, A. Paussa, P. Palestri, L. Selmi, S.M. Thomas, T.E. Whall, D. Leadley, E.H.C. Parker, L. Witters, M.J. Hytch, E. Snoeck, T.J. Wang, W.C. Lee, G. Doornbos, G. Vellianitis, M.J.H. van Dal, R.J.P. Lander, Investigation of strain engineering in FinFETs comprising experimental analysis and numerical simulations, IEEE Trans. Electron. Dev. 58 (6) (2011) 1583−1593, https://doi.org/10.1109/TED.2011.2119320.
[45] G. Yeap, X. Chen, B.R. Yang, C.P. Lin, F.C. Yang, Y.K. Leung, D.W. Lin, C.P. Chen, K.F. Yu, D.H. Chen, C.Y. Chang, S.S. Lin, H.K. Chen, P. Hung, C.S. Hou, Y.K. Cheng, J. Chang, L. Yuan, C.K. Lin, C.C. Chen, Y.C. Yeo, M.H. Tsai, Y.M. Chen, H.T. Lin, C.O. Chui, K.B. Huang, W. Chang, H.J. Lin, K.W. Chen, R. Chen, S.H. Sun, Q. Fu, H.T. Yang, H.L. Shang, H.T. Chiang, C.C. Yeh, T.L. Lee, C.H. Wang, S.L. Shue, C.W. Wu, R. Lu, W.R. Lin, J. Wu, F. Lai, P.W. Wang, Y.H. Wu, B.Z. Tien, Y.C. Huang, L.C. Lu, J. He, Y. Ku, J. Lin, M. Cao, T.S. Chang, S.M. Jang, H.C. Lin, Y.C. Peng, J.Y. Sheu, M. Wang, 5nm CMOS production technology platform featuring full-fledged EUV, and high mobility channel FinFETs with densest 0.021μm2 SRAM cells for mobile SoC and high performance computing applications, in: 2019 IEEE In-Ternational Electron Devices Meeting (IEDM), IEEE, 2019, pp. 36.7.1−36.7.4.
[46] P. Hashemi, K. Balakrishnan, S.U. Engelmann, J.A. Ott, K. Ali, A. Baraskar, M. Hopstaken, J.S. Newbury, K.K. Chan, E. Leobandung, R.T. Mo, D.-G. Park, First demon-stration of high-Ge-content strained-Si1−xGex (x=0.5) on insulator PMOS FinFETs with high hole mobility and aggressively scaled fin dimensions and gate lengths for high-performance applications, in: 2014 IEEE International Electron Devices Meeting, IEEE, 2014, pp. 16.1.1−16.1.4.

[47] H. Mertens, R. Ritzenthaler, A. Hikavyy, J. Franco, J.W. Lee, D.P. Brunco, G. Eneman, L. Witters, J. Mitard, S. Kubicek, K. Devriendt, D. Tsvetanova, A.P. Milenin, C. Vrancken, J. Geypen, H. Bender, G. Groeseneken, W. Vandervorst, K. Barla, N. Collaert, N. Horiguchi, A.V.Y. Thean, Performance and reliability of high-mobility Si0.55Ge0.45 p-channel FinFETs based on epitaxial cladding of Si Fins, in: 2014 Symposium on VLSI Technology (VLSI-Technology): Digest of Technical Papers, IEEE, 2014, pp. 1−2.
[48] D. Guo, G. Karve, G. Tsutsui, K.Y. Lim, R. Robison, T. Hook, R. Vega, D. Liu, S. Bedell, S. Mochizuki, F. Lie, K. Akarvardar, M. Wang, R. Bao, S. Burns, V. Chan, K. Cheng, J. Demarest, J. Fronheiser, P. Hashemi, J. Kelly, J. Li, N. Loubet, P. Montanini, B. Sahu, M. Sankarapandian, S. Sieg, J. Sporre, J. Strane, R. Southwick, N. Tripathi, R. Venigalla, J. Wang, K. Watanabe, C.W. Yeung, D. Gupta, B. Doris, N. Felix, A. Jacob, H. Jagannathan, S. Kanakasabapathy, R. Mo, V. Narayanan, D. Sadana, P. Oldiges, J. Stathis, T. Yamashita, V. Paruchuri, M. Colburn, A. Knorr, R. Divakaruni, H. Bu, M. Khare, FINFET technology featuring high mobility SiGe channel for 10nm and beyond, in: 2016 IEEE Symposium on VLSI Technology, IEEE, 2016, pp. 1−2.
[49] G. Tsutsui, R. Bao, K.-yong Lim, R.R. Robison, R.A. Vega, J. Yang, Z. Liu, M. Wang, O. Gluschenkov, C.W. Yeung, K. Watanabe, S. Bentley, H. Niimi, D. Liu, H. Zhou, S. Siddiqui, H. Kim, R. Galatage, R. Venigalla, M. Raymond, P. Adusumilli, S. Mochizuki, T.S. Devarajan, B. Miao, B. Liu, A. Greene, J. Shearer, P. Montanini, J.W. Strane, C. Prindle, E.R. Miller, J. Fronheiser, C.C. Niu, K. Chung, J.J. Kelly, H. Jagannathan, S. Kanakasa-bapathy, G. Karve, F.L. Lie, P. Oldiges, V. Narayanan, T.B. Hook, A. Knorr, D. Gupta, D. Guo, D. Rama, H. Bu, K. Mukesh, Technology viable DC performance elements for Si/SiGe channel CMOS FinFTT, in: 2016 IEEE International Electron Devices Meeting (IEDM), IEEE, 2016, p. 17., 4.1-17.4.4.
[50] K. Ikeda, M. Ono, D. Kosemura, K. Usuda, M. Oda, Y. Kamimuta, T. Irisawa, Y. Moriyama, A. Ogura, T. Tezuka, High-mobility and low-parasitic resistance characteristics in strained Ge nanowire PMOSFETs with metal source/drain structure formed by doping-free processes, in: 2012 Symposium on VLSI Technology (VLSIT), IEEE, 2012, pp. 165−166.
[51] P. Hashemi, K.-L. Lee, T. Ando, K. Balakrishnan, J.A. Ott, S. Koswatta, S.U. Engelmann, D.-G. Park, V. Narayanan, R.T. Mo, E. Leobandung, Demonstration of Record SiGe transconductance and short-channel current drive in high-Ge-content SiGe PMOS FinFETs with improved junction and scaled EOT, in: 2016 IEEE Symposium on VLSI Technology, IEEE, 2016, pp. 1−2.
[52] C.H. Lee, H. Kim, P. Jamison, R.G. Southwick, S. Mochizuki, K. Watanabe, R. Bao, R. Galatage, G. Guillaumet, T. Ando, R. Pandey, A. Konar, B. Lherron, J. Fronheiser, S. Siddiqui, H. Jagannathan, V. Paruchuri, Selective GeOx-scavenging from interfacial layer on Si1−xGex channel for high mobility Si/Si1−xGex CMOS application, in: 2016 IEEE Symposium on VLSI Technology, IEEE, 2016, pp. 1−2.
[53] C.H. Lee, R.G. Southwick, H. Jagannathan, Engineering the electronic defect bands at the Si1−xGex/IL interface: approaching the intrinsic carrier transport in compressively-strained Si1−xGex pFETs, in: 2016 IEEE International Electron Devices Meeting (IEDM), IEEE, 2016, pp. 31.1.1−31.1.4.
[54] P. Hashemi, T. Ando, S. Koswatta, K.L. Lee, E. Cartier, J.A. Ott, C.H. Lee, J. Bruley, M.F. Lofaro, S. Dawes, K.K. Chan, S.U. Engelmann, E. Leobandung, V. Narayanan, R.T. Mo, High performance and record subthreshold swing demonstration in scaled RMG SiGe FinFETs with high-Ge-content channels formed by 3D condensation and a novel gate stack process, in: 2017 Symposium on VLSI Technology, IEEE, 2017, pp. T120−T121.

[55] R. Xie, P. Montanini, K. Akarvardar, N. Tripathi, B. Haran, S. Johnson, T. Hook, B. Hamieh, D. Corliss, J. Wang, X. Miao, J. Sporre, J. Fronheiser, N. Loubet, M. Sung, S. Sieg, S. Mochizuki, C. Prindle, S. Seo, A. Greene, J. Shearer, A. Labonte, S. Fan, L. Liebmann, R. Chao, A. Arceo, K. Chung, K. Cheon, P. Adusumilli, H.P. Amanapu, Z. Bi, J. Cha, H.C. Chen, R. Conti, R. Ga-latage, O. Gluschenkov, V. Kamineni, K. Kim, C. Lee, F. Lie, Z. Liu, S. Mehta, E. Miller, H. Niimi, C. Niu, C. Park, D. Park, M. Raymond, B. Sahu, M. Sankarapandian, S. Siddiqui, R. Southwick, L. Sun, C. Surisetty, S. Tsai, S. Whang, P. Xu, Y. Xu, C. Yeh, P. Zeitzoff, J. Zhang, J. Li, J. Demarest, J. Arnold, D. Canaperi, D. Dunn, N. Felix, D. Gupta, H. Jagannathan, S. Kanakasabapathy, W. Kleemeier, C. Labelle, M. Mottura, P. Oldiges, S. Skordas, T. Standaert, T. Yamashita, M. Colburn, M. Na, V. Paruchuri, S. Lian, R. Divakaruni, T. Gow, S. Lee, A. Knorr, H. Bu, M. Khare, A 7nm FinFET technology Featuring EUV patterning and dual strained high mobility channels, in: IEEE International Electron Devices Meeting (IEDM), IEEE, 2016, pp. 2.7.1–2.7.4.

[56] Bae, Dong-il, G. Bae, K.K. Bhuwalka, S.-H. Lee, M.-G. Song, T.-soo Jeon, C. Kim, W. Kim, J. Park, S. Kim, U. Kwon, J. Jeon, K.-J. Nam, S. Lee, S. Lian, K.-ill Seo, S.-G. Lee, J.H. Park, Y.-C. Heo, M.S. Rodder, J.A. Kittl, Y. Kim, K. Hwang, D.-W. Kim, M.-song Liang, E.S. Jung, A novel tensile Si (n) and compressive SiGe (p) dual-channel CMOS FinFET Co-integration scheme for 5nm Logic applications and beyond, in: IEEE International Electron Devices Meeting (IEDM), IEEE, 2016, pp. 28.1.1–28.1.4.

[57] J.P. Dismukes, L. Ekstrom, R.J. Paff, Lattice parameter and density in Germanium-silicon alloys, The Journal of Physical Chemistry 68 (10) (1964) 3021–3027, https://doi.org/10.1021/j100792a049.

[58] C.H. Lee, S. Mochizuki, R.G. Southwick, J. Li, X. Miao, R. Bao, T. Ando, R. Galatage, S. Siddiqui, C. Labelle, A. Knorr, J.H. Stathis, D. Guo, V. Narayanan, B. Haran, H. Jagannathan, Comparative study of strain and Ge content in Si1−xGex channel using planar FETs, FinFETs, and strained relaxed buffer layer FinFETs, in: 2017 IEEE International Electron Devices Meeting (IEDM), IEEE, 2017, p. 37, 2.1-37.2.4.

[59] J.H. Stathis, M. Wang, R.G. Southwick, E.Y. Wu, B.P. Linder, E.G. Liniger, G. Bonilla, H. Kothari, Reliability challenges for the 10nm node and beyond, in: 2014 IEEE International Electron Devices Meeting, IEEE, 2014, p. 20.6, 1-20.6.4.

[60] S. Mochizuki, C.E. Murray, A. Madan, T. Pinto, Y.-Y. Wang, J. Li, W. Weng, H. Jagan-nathan, Y. Imai, S. Kimura, S. Takeuchi, A. Sakai, Quantification of local strain distributions in nanoscale strained SiGe FinFET structures, J. Appl. Phys. 122 (13) (2017) 135705, https://doi.org/10.1063/1.4991472.

[61] K. Cheng, A. Khakifirooz, N. Loubet, S. Luning, T. Nagumo, M. Vinet, Q. Liu, A. Reznicek, T. Adam, S. Naczas, P. Hashemi, J. Kuss, J. Li, H. He, L. Edge, J. Gimbert, P. Khare, Y. Zhu, Z. Zhu, A. Madan, N. Klymko, S. Holmes, T.M. Levin, A. Hubbard, R. Johnson, M. Terrizzi, S. Teehan, A. Upham, G. Pfeiffer, T. Wu, A. Inada, F. Allibert, B.Y. Nguyen, L. Grenouillet, Y. le Tiec, R. Wacquez, W. Kleemeier, R. Sampson, R.H. Dennard, T.H. Ning, M. Khare, G. Shahidi, B. Doris, High performance extremely thin SOI (ETSOI) hybrid CMOS with Si channel NFET and strained SiGe channel PFET, in: 2012 International Electron Devices Meeting, IEEE, 2012, pp. 18.1.1–18.1.4.

[62] G. Tsutsui, H. Zhou, A. Greene, R. Robison, J. Yang, J. Li, C. Prindle, J.R. Sporre, E.R. Miller, D. Liu, R. Sporer, B. Mulfinger, T. McArdle, J. Cho, G. Karve, F.L. Lie, S. Kanakasabapathy, R. Carter, D. Gupta, A. Knorr, D. Guo, H. Bu, SiGe FinFET for practical logic libraries by mitigating local layout effect, in: 2017 Symposium on VLSI Technology, IEEE, 2017, pp. T122–T123.

[63] E.R. Hsieh, S.S. Chung, Appl. Phys. Lett. 96 (2010) 093501.
[64] Z. Ren, G. Pei, J. Li, B. Yang, R. Takalkar, K. Chan, G. Xia, Z. Zhu, A. Madan, T. Pinto, T. Adam, J. Miller, A. Dube, L. Black, J.W. Weijtmans, B. Yang, E.C. Harley, A. Chakravarti, T. Kanarsky, R. Pal, I. Lauer, D.-G. Park, D. Sadana, Symp. VLSI Technology, 2008, p. 172.
[65] B. Yang, Z. Ren, R. Takalkar, L. Black, A. Dube, J.W. Weijtmans, J. Li, K. Chan, J. De Souza, A. Madan, G. Xia, Z. Zhu, J. Faltermeier, A. Reznicek, T.N. Adam, A. Chakravarti, G. Pei, R. Pal, B. Yang, E.C. Harley, B. Greene, A. Gehring, M. Cai, D. Sadana, D.-G. Park, D. Mocuta, D. Schepis, E. Maciejewski, S. Luning, E. Leobandung, ECS Trans. 16 (2008) 317.
[66] Y. Liu, O. Gluschenkov, J. Li, A. Madan, A. Ozcan, B. Kim, T. Dyer, A. Chakravarti, K. Chan, C. Lavoie, I. Popova, T. Pinto, N. Rovedo, Z. Luo, R. Loesing, W. Henson, K. Rim, Symp. VLSI Technology, 2007, p. 45.
[67] S.-M. Koh, X. Wang, K. Sekar, W. Krull, G.S. Samudra, Y.-C. Yeoa, J. Electrochem. Soc. 156 (2009) H361.
[68] T. Yamaguchi, Y. Kawasaki, T. Yamashita, N. Miura, M. Mizuo, J. Tsuchimoto, K. Eikyu, K. Maekawa, M. Fujisawa, K. Asai, Jpn. J. Appl. Phys. 50 (2011) 04DA02.
[69] S.-H. Dai, R. Liao, R.-M. Huang, L.-F. Chin, Y.-R. Liu, P. Kuo, C.-Y. Chen, K.-L. Chiu, C.-I. Li, C.-H. Tsai, C.-T. Tsai, C.-W. Liang, International Symposium on VLSI Technology Systems and Applications, 2011, 5872236.
[70] T.-Y. Liow, K.-M. Tan, D. Weeks, R.T.P. Lee, M. Zhu, K.-M. Hoe, C.-H. Tung, M. Bauer, J. Spear, S.G. Thomas, G.S. Samudra, N. Balasubramanian, Y.-C. Yeo, IEEE Trans. Electron Devices 55 (2008) 2475.
[71] M. Togo, J.W. Lee, L. Pantisano, T. Chiarella, R. Ritzenthaler, R. Krom, A. Hikavyy, R. Loo, E. Rosseel, S. Brus, J.W. Maes, V. Machkaoutsan, J. Tolle, G. Eneman, A.D. Keersgieter, G. Boccardi, G. Mannaert, S.E. Altamirano, S. Locorotondo, M. Demand, N. Horiguchi, A. Thean, Int. El. Devices. Meet, 18.2.1, 2012.
[72] N. Serra, D. Esseni, IEEE Trans. Electron Devices 57 (2010) 482.
[73] F. Conzatti, N. Serra, D. Esseni, M. De Michielis, A. Paussa, P. Palestri, L. Selmi, S.M. Thomas, T.E. Whall, D. Leadley, E.H.C. Parker, L. Witters, M.J. Hÿtch, E. Snoeck, T.J. Wang, W.C. Lee, G. Doornbos, G. Vellianitis, M.J.H. van Dal, R.J.P. Lander, IEEE Trans. Electron Devices 58 (2011) 1583.
[74] K.W. Ang, K.J. Chui, V. Blinznetsov, A. Du, N. Balasubramanian, M.F. Li, G. Samudra, Y.C. Yeo, Int. El. Devices. Meet, 2004, p. 1069.
[75] K.W. Ang, K.J. Chui, V. Blinznetsov, Y. Wang, L.Y. Wong, C.H. Tung, N. Balasubramanian, M.F. Li, G. Samudra, Y.C. Yeo, Int. El. Devices. Meet 497, 2005.
[76] K.J. Chui, K.W. Ang, H.C. Chin, C. Shen, L.Y. Wong, C.H. Tung, N. Balasubramanian, M.F. Li, G. Samudra, Y.C. Yeo, IEEE Electron. Device Lett. 27 (2006) 778.
[77] K.W. Ang, K.J. Chui, C.H. Tung, N. Balasubramanian, M.F. Li, G. Samudra, Y.C. Yeo, IEEE Electron. Device Lett. 28 (2007) 301.
[78] H. Itokawa, K. Miyano, Y. Oshiki, H. Onoda, M. Nishigoori, I. Mizushima, K. Suguro, in: 10th International Workshop on Junction Technology, IWJT-2010, IEEE, 2010 article 5475009.
[79] K. Yako, M. Fujiwara, H. Bu, in: International Workshop on Junction Technology, 2011, p. 77.
[80] W.Y. Woon, S.H. Wang, Y.T. Chuang, M.C. Chuang, C.L. Chen, Appl. Phys. Lett. 97 (2010) 141906.
[81] Z. Ye, Y. Kim, A. Zojaji, E. Sanchez, Y. Cho, M. Castle, M.A. Foad, Semicond. Sci. Technol. 22 (2007) 171.
[82] K. Sekar, W.A. Krull, ECS Trans. 28 (2010) 53.

[83] S.M. Koh, K. Sekar, D. Lee, W. Krull, X. Wang, G.S. Samudra, Y.C. Yeo, IEEE Electron. Device Lett. 29 (2008) 1315.
[84] A. Li-Fatou, A. Jain, W. Krull, M. Ameen, M. Harris, D. Jacobson, ECS Trans. 11 (2007) 125.
[85] M. Togo, Y. Sasaki, G. Zschätzsch, G. Boccardi, R. Ritzenthaler, J.W. Lee, F. Khaja, B. Colombeau, L. Godet, P. Martin, S. Brus, S.E. Altamirano, G. Mannaert, H. Dekkers, G. Hellings, N. Horiguchi, W. Vandervorst, A. Thean, VLSI Tech. Symp. Dig, IEEE, 2013, pp. T196−T197.
[86] Y. Sasaki, L. Godet, T. Chiarella, D.P. Brunco, T. Rockwell, J.W. Lee, B. Colombeau, M. Togo, S.A. Chew, G. Zschaetszch, K.B. Noh, A. De Keersgieter, G. Boccardi, M.S. Kim, G. Hellings, P. Martin, W. Vandervorst, A. Thean, N. Horiguchi, 2013 IEEE International Electron Devices Meeting, IEEE, 2013, pp. 542−545.
[87] G. Zschätzsch, Y. Sasaki, S. Hayashi, M. Togo, T. Chiarella, A.K. Kambham, J. Mody, B. Douhard, N. Horiguchi, B. Mizuno, M. Ogura, W. Vandervorst, 2011 International Electron Devices Meeting, IEEE, 2011, pp. 841−844.
[88] Y. Sasaki, R. Ritzenthaler, A. De Keersgieter, T. Chiarella, S. Kubicek, E. Rosseel, A. Waite, J. del Agua Borniquel, B. Colombeau, S.A. Chew, M.S. Kim, T. Schram, S. Demuynck, W. Vandervorst, N. Horiguchi, D. Mocuta, A. Mocuta, A.V-Y. Thean, VLSI Tech. Symp. Dig, IEEE, 2015, p. T30.
[89] T. Chiarella, S. Kubicek, E. Rosseel, R. Ritzenthaler, A. Hikavyy, P. Eyben, A. De Keersgieter, L.-Å. Ragnarsson, M.S. Kim, S.-A. Chew, T. Schram, S. Demuynck, M. Cupák, L. Rijnders, M. Dehan, N. Horiguchi, J. Mitard, D. Mocuta, A. Mocuta, A. V-Y Thean, 46th European Solid-State Device Research Conference, ESSDERC 2016, IEEE, 2016, pp. 131−134.
[90] G. Tsutsui, R. Bao, K.-Y. Lim, R. Robison, R. Vega, J. Yang, Z. Liu, M. Wang, O. Gluschenkov, C.W. Yeung, K. Watanabe, S. Bentley, H. Niimi, D. Liu, H. Zhou, S. Siddiqui, H. Kim, R. Galatage, R. Venigalla, M. Raymond, P. Adusumilli, S. Mochizuki, T.S. Devarajan, B. Miao, B. Liu, A. Greene, J. Shearer, P. Montanini, J. Strane, C. Prindle, E. Miller, J. Fronheiser, C.C. Niu, K. Chung, J. Kelly, H. Jagannathan, S. Kanakasabapathy, G. Karve, F. Lie, P. Oldiges, V. Narayanan, T. Hook, A. Knorr, D. Gupta, D. Guo, R. Divakaruni, H. Bu, M. Khare, 2016 IEEE International Electron Devices Meeting (IEDM), IEEE, 2016, pp. 456−459.
[91] S. Mochizuki, B. Colombeau, L. Yu, A. Dube, S. Choi, M. Stolfi, Z. Bi, F. Chang, R.A. Conti, P. Liu, K.R. Winstel, H. Jagannathan, H.-J. Gossmann, N. Loubet, D.F. Canaperi, D. Guo, S. Sharma, S. Chu, J. Boland, Q. Jin, Z. Li, S. Lin, M. Cogorno, M. Chudzik, S. Natarajan, D.C. McHerron, B. Haran, 2018 IEEE International Electron Devices Meeting (IEDM), IEEE, 2018, pp. 811−814.
[92] S. Bangsaruntip, G.M. Cohen, A. Majumdar, Y. Zhang, S.U. Engelmann, N.C.M. Fuller, L.M. Gignac, S. Mittal, J.S. Newbury, M. Guillorn, T. Barwicz, L. Sekaric, M.M. Frank, J.W. Sleight, 2009 IEEE International Electron Devices Meeting (IEDM), IEEE, 2009, pp. 297−300.
[93] I. Lauer, N. Loubet, S.D. Kim, J.A. Ott, S. Mignot, R. Venigalla, T. Yamashita, T. Standaert, J. Faltermeier, V. Basker, M.A. Guillorn, VLSI Tech. Symp. Dig, IEEE (2015) 140−141.
[94] S.D. Kim, M. Guillorn, I. Lauer, P. Oldiges, T. Hook, M.H. Na, 2015 IEEE SOI-3D-Subthreshold Microelectronics Technology Unified Conference (S3S), IEEE, 2015, pp. 1−3.
[95] H. Mertens, R. Ritzenthaler, A. Chasin, T. Schram, E. Kunnen, A. Hikavyy, L.-Å. Ragnarsson, H. Dekkers, T. Hopf, K. Wostyn, K. Devriendt, S.A. Chew, M.S. Kim, Y. Kikuchi, E. Rosseel, G. Mannaert, S. Kubicek, S. Demuynck, A. Dangol, N. Bosman, J. Geypen, P. Carolan, H. Bender, K. Barla, N. Horiguchi,

D. Mocuta, 2016 IEEE International Electron Devices Meeting (IEDM), IEEE, 2016, pp. 524−527.
[96] G.J. Bae, D.-I. Bae, M. Kang, S.M. Hwang, S.S. Kim, B. Seo, T.Y. Kwon, T.J. Lee, C. Moon, Y.M. Choi, K. Oikawa, S. Masuoka, K.Y. Chun, S.H. Park, H.J. Shin, J.C. Kim, K.K. Bhuwalka, D.H. Kim, W.J. Kim, J. Yoo, H.Y. Jeon, M.S. Yang, S.-J. Chung, D. Kim, B.H. Ham, K.J. Park, W.D. Kim, S.H. Park, G. Song, Y.H. Kim, M.S. Kang, K.H. Hwang, C.-H. Park, J.-H. Lee, D.-W. Kim, S-M. Jung, H.K. Kang, 2018 IEEE International Electron Devices Meeting (IEDM), IEEE, 2018, pp. 656−659.
[97] N. Loubet, T. Hook, P. Montanini, C.-W. Yeung, S. Kanakasabapathy, M. Guillom, T. Yamashita, J. Zhang, X. Miao, J. Wang, A. Young, R. Chao, M. Kang, Z. Liu, S. Fan, B. Hamieh, S. Sieg, Y. Mignot, W. Xu, S.-C. Seo, J. Yoo, S. Mochizuki, M. Sankarapandian, O. Kwon, A. Carr, A. Greene, Y. Park, J. Frougier, R. Galatage, R. Bao, J. Shearer, R. Conti, H. Song, D. Lee, D. Kong, Y. Xu, A. Arceo, Z. Bi, P. Xu, R. Muthinti, J. Li, R. Wong, D. Brown, P. Oldiges, R. Robison, J. Arnold, N. Felix, S. Skordas, J. Gaudiello, T. Standaert, H. Jagannathan, D. Corliss, M.-H. Na, A. Knorr, T. Wu, D. Gupta, S. Lian, R. Divakaruni, T. Gow, C. Labelle, S. Lee, V. Paruchuri, H. Bu, M. Khare, Symposium on VLSI Technology, IEEE, 2017, pp. T230−T231.
[98] S. Mochizuki, B. Colombeau, J. Zhang, S.C. Kung, M. Stolfi, H. Zhou, M. Breton, K. Watanabe, J. Li, H. Jagannathan, M. Cogorno, T. Mandrekar, P. Chen, N. Loubet, S. Natarajan, B. Haran, Symposium on VLSI Technology, IEEE, 2020, pp. 1−2.
[99] S. Mochizuki, M. Bhuiyan, H. Zhou, J. Zhang, E. Stuckert, J. Li, K. Zhao, M. Wang, V. Basker, N. Loubet, D. Guo, B. Haran, H. Bu, 2020 IEEE International Electron Devices Meeting (IEDM), IEEE, 2020, pp. 19−22.
[100] K. Uchida, J. Koga, R. Ohba, T. Numata, S. Takagi, International Electron Devices Meeting. Technical Digest, IEEE, 2001, pp. 633−636.
[101] D. Esseni, M. Mastrapasqua, G.K. Celler, F.H. Baumann, C. Fiegna, L. Selmi, E. Sangiorgi, International Electron Devices Meeting 2000. Technical Digest, IEEE, 2000, pp. 671−674.
[102] C.W. Yeung, J. Zhang, R. Chao, O. Kwon, R. Vega, G. Tsutsui, X. Miao, C. Zhang, C.W. Sohn, B.K. Moon, A. Razavieh, J. Frougier, A. Greene, R. Galatage, J. Li, M. Wang, N. Loubet, R. Robison, V. Basker, T. Yamashita, D. Guo, 2018 IEEE International Electron Devices Meeting (IEDM), IEEE, 2018, pp. 652−655.
[103] D. Guo, G. Karve, G. Tsutsui, K-Y. Lim, R. Robison, T. Hook, R. Vega, D. Liu, S. Bedell, S. Mochizuki, F. Lie, K. Akarvardar, M. Wang, R. Bao, S. Burns, V. Chan, K. Cheng, J. Demarest, J. Fronheiser, P. Hashemi, J. Kelly, J. Li, N. Loubet, P. Montanini, B. Sahu, M. Sankarapandian, S. Sieg, J. Sporre, J. Strane, R. Southwick, N. Tripathi, R. Venigalla, J. Wang, K. Watanabe, C.W. Yeung, D. Gupta, B. Doris, N. Felix, A. Jacob, H. Jagannathan, S. Kanakasabapathy, R. Mo, V. Narayanan, D. Sadana, P. Oldiges, J. Stathis, T. Yamashita, V. Paruchuri, M. Colburn, A. Knorr, R. Divakaruni, H. Bu, M. Khare, Symposium on VLSI Technology, IEEE, 2016, pp. 1−2.
[104] C.H. Lee, S. Mochizuki, R.G. Southwick, J. Li, X. Miao, R. Bao, T. Ando, R. Galatage, S. Siddiqui, C. Labelle, A. Knorr, J.H. Stathis, D. Guo, V. Narayanan, B. Haran, H. Jagannathan, 2017 IEEE International Electron Devices Meeting (IEDM), IEEE, 2017, pp. 820−823.
[105] J. Chen, T. Saraya, T. Hiramoto, Symposium on VLSI Technology, IEEE, 2009, pp. 90−91.
[106] R. Hull, J.C. Bean, L. Peticolas, D. Bahnck, Appl. Phys. Lett. 59 (1991) 964.

[107] S. Saito, Y. Sano, T. Yamada, K.O. Hara, J. Yamanaka, K. Nakagawa, K. Arimoto, Mater. Sci. Semicond. Process. 113 (2020) 105042.
[108] S. Mochizuki, N. Loubet, P. Mirdha, C. Durfee, H. Zhou, G. Tsutsui, J. Frougier, R. Vega, L. Qin, N. Felix, D. Guo, H. Bu, 2023 International Electron Devices Meeting (IEDM), IEEE, 2023, pp. 1–4.
[109] G. Tsutsui, M. Saitoh, T. Hiramoto, IEEE Electron Device Lett. 26 (11) (2005) 836–838.
[110] M. Poljak, V. Jovanović, T. Suligoj, IEEE 2011 International SOI Conference, IEEE, 2011.
[111] L. Donetti, F. Gámiz, N. Rodriguez, F. Jiménez-Molinos, J.B. Roldán, Solid-State Electronics 54 (2) (2010) 191–195.
[112] G. Sun, Y. Sun, T. Nishida, S.E. Thompson, J. Appl. Phys. 102 (2007) 084501.

# CHAPTER 5

# Equipment and manufacturability issues in chemical vapor deposition processes

Loren A. Chow
SVT Associates, Los Altos, CA, United States

## 1. Introduction

The ever more powerful functionality expected of today's microprocessors such as accurate spatial computing, distributed ledger technology, and generative AI has placed increasing demands on device scaling. This in turn has led to updated film requirements, the development of novel materials, and changes in architecture. Chemical vapor deposition (CVD) technology, in keeping pace, continues to undergo significant improvements in capability and equipment design. Taken as a whole, CVD offers an array of deposition technologies with a broad range of capabilities. For example, with its capability to deposit conformal films one layer at a time, atomic layer deposition (ALD) is a strong candidate for 3D architectures such as gate-all-around transistors, where a thin film that wraps around the transistor channel is needed. For applications where a trench fill is required, high-density plasma CVD (HDPCVD), which offers a sputter component during deposition, can be a solution. In a manufacturing environment, low-pressure CVD (LPCVD), which normally runs in a reaction-limited regime, enables epitaxial Group IV films to be grown with high uniformity, even in batch systems. With its ability to deposit a range of different materials with sharp interfaces, metal-organic CVD (MOCVD) has enabled use of III–V materials, which are advantageous for their high mobilities and direct bandgap. The list of other films and their respective properties that can be delivered by CVD goes on: low-resistivity interconnect barriers, low-dielectric constant intermetal dielectrics, strained transistor channel, dielectrics offering high etch selectivity, etc. What all of these film characteristics have in common with each other is that they enable device scaling with every technology node, and CVD has the ability to deliver such sought-after film properties in a manufacturing environment.

This chapter surveys the capability of the CVD technologies most commonly used today in semiconductor manufacturing, such as ALD, LPCVD, MOCVD and HDPCVD. For CVD, in general, there are excellent resources for explaining the thermodynamics, kinetics, and applications of CVD. However, the goal of this chapter is different: It is to offer an explanation of the suite of CVD technologies as they exist in manufacturing today, their advantages, and their drawbacks.

The ideal audience for this chapter is an engineer searching for the CVD technology offering the highest probability of success in depositing a given material for a specific application. This chapter will also discuss new films and their requirements from front-end-of-line (FEOL) through interconnects. Also covered are common metrology technologies associated with CVD, tool selection criteria, and commercial considerations such as cost of ownership parameters.

## 2. Basic principles of CVD

For all the sophisticated tasks CVD films are called upon to perform, the deposition process, at least in concept, is rather simple. At a molecular level, the process can be described in three parts: arrival of the precursor molecule, the surface reaction, which incorporates the desired reactant atoms, and removal of by-products. The process begins with a precursor molecule entering a reaction chamber by forced flow or, in the case of solid and liquid precursors, by a carrier gas. Then, by diffusion, the precursor molecule drifts to the wafer surface.

There are a number of possible flow regimes near the substrate surface as the incoming gas washes over the wafer. That is, the flow can be turbulent, laminar, or a mix of both. The regime is given by the Reynolds number, a dimensionless quantity given by Ref. [1]:

$$Re = U\rho L/\mu$$

where $U$ is the bulk velocity, $\rho$ the fluid density, $L$ the characteristic length (the surface length in the direction of the flow in this case), and $\mu$ the fluid viscosity.

The Reynolds number is usually about a few hundred in CVD [2]. A Reynolds number less than 1100 corresponds to laminar flow deposition.

(It is noted here that for a showerhead design the flow can be modeled as flowing outward radially toward the edge of the wafer.) The laminar gas flow across the wafer leads to a velocity profile above the wafer. That is, at the substrate surface, the flow velocity is zero due to the viscosity of the fluid, and the velocity parallel to the wafer increases with distance perpendicular from the surface until at some point the flow velocity is the same as that before reaching the substrate. This region where there is a marked impact on flow velocity due to substrate effects is called the boundary layer (Fig. 5.1). More precisely, the upper limit on boundary layer flow velocity is customarily defined as 99% of the maximum flow velocity. Assuming flow across a plate, its thickness, $\delta(x)$, is given by Ref. [2]:

$$\delta(x) = 5x / (Re_x)^{1/2}$$

where $x$ is the distance along the plate. Integrating over the length of the plate, one finds that the average boundary layer thickness is $(10/3)\ L/(ReL)^{1/2}$.

Whether the boundary layer has an impact on growth depends on the reaction rate at the surface. Because of the lower flow velocity within the layer, the influx of new precursor molecules is smaller compared to that outside the layer. Then, if the reaction rate at the wafer is high, the precursor consumption can exceed the arrival of new reactants. The boundary layer would be a gap depleted of reactants through which fresh precursor gases need to pass to sustain growth. In this case, growth would be controlled by transport to the substrate and is known as "transport-limited" growth. In such conditions, it is crucial that flows and reactor design enable arrival of reactants to ensure uniform film growth. However, at low

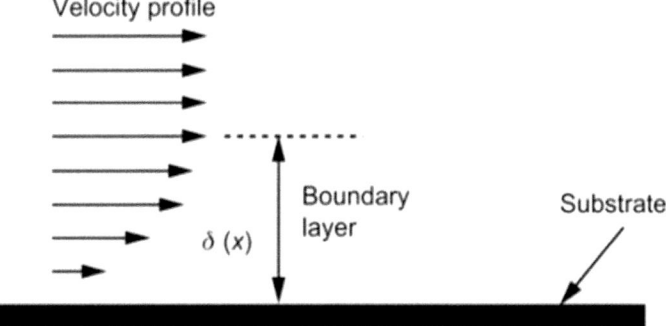

Figure 5.1 The boundary layer for laminar flow.

temperatures or when the surface reaction of precursor species does not keep pace with the incoming flux of reactant species, the existence of the boundary layer has no significant effect on growth. That is, the reaction rate is so slow that there is a build-up of species waiting to react. There is no depletion of precursor in the boundary layer in this case.

This treatment of the boundary layer assumes continuous flow and a high degree of interaction between molecules in the flow. At pressures below about a millitorr, the molecular mean free path is roughly of the order of the chamber dimensions. In this case, there is no continuous transport or fluid flow as pertinent to the above discussion. At such a long mean free path, the precursor molecules are independent entities having limited interaction with each other [3] and, as such, no boundary layer is formed.

When the precursor molecule makes its way through the boundary layer and arrives at the substrate, it is ideally still in its original, stable form, with the atom of interest—the one that will eventually be incorporated into the film—still attached to the rest of the precursor molecule. This arrival and subsequent sticking to the surface is called adsorption. While on the substrate, which is usually hot—substrate temperatures in CVD normally range from 400°C to 1000°C—the precursor molecule breaks apart due to the bond-breaking heat from the wafer. (Section 6.6 in this chapter will review plasma-enhanced CVD (PECVD), where bonds are broken by energetic electrons, possibly enabling a lower deposition temperature.) With the bonds of the desired atom no longer satisfied, it becomes reactive. The heat from the wafer serves another important purpose: it provides energy to the reactant atom for surface diffusion. As will be seen later in this chapter, surface diffusion is especially important for epitaxial growth, where the atom will need enough energy to diffuse along the surface to find a kink or ledge to incorporate itself. Energetically, in CVD, it is preferable for an atom to bond to other like atoms, such as that found in a kink, rather than begin a process of heterogeneous nucleation on the wafer surface. While the desired atom becomes a part of the film, the remainder of the original precursor molecule desorbs as a gaseous by-product of the reaction, to be removed from the chamber as exhaust.

This raises a third purpose (in addition to precursor volatilization and surface diffusion) for a suitably high substrate temperature: thermal desorption of contaminants such as oxygen, carbon, and hydrogen. Considering the silane reaction for silicon epitaxy:

$$SiH_4 \rightarrow Si + 2H_2$$

The hydrogen by-product can terminate—and passivate—the silicon surface, halting growth. Hence, the presence of hydrogen, which may originate from silanes or arrive simply as a chamber impurity, sets a lower bound for growth of silicon [4] at around 500°C [5]. Although this example is specific to silicon growth with silane as a precursor, it illustrates a fundamental property required of any practical precursor: that it breaks apart at the substrate with the by-products desorbing as a gas (hydrogen in this example), leaving behind only the species intended for integration into the film (silicon).

Depending on the material to be deposited, the precursors used, and the targeted material properties, there may exist trade-offs between substrate temperature, film quality, and deposition rate. A high temperature may be required to crack the precursor into its reactive constituents, enhancing the deposition rate. A high temperature can also aid surface diffusion, which would promote terrace-ledge growth for monocrystalline deposition. However, a high temperature can also lead to thermal desorption of the reactant. If precursor decomposition is not a limitation, low-temperature deposition at a low deposition rate can lead to highly uniform films with refined grain size [6].

In selecting a precursor for CVD, perhaps the two most important technical considerations are vapor pressure and cracking, or decomposition, temperature. A precursor with a vapor pressure less than, say, 0.76 Torr (1 mbar) at room temperature is not practical for CVD, as the minuscule quantity of vapor that could be brought to the substrate would lead to a vanishingly small deposition rate. Although the vapor pressure can be increased through heating, this can lead to precursor decomposition. This leads to a second consideration for precursor selection: thermal stability. Precursor compounds that decompose at or below 100°C can easily decay in storage and lead to irreproducible results [7].

As will be discussed in Section 6, precursor selection will also be dependent on the specific CVD technology to be utilized. Although CVD techniques can be broadly sorted by pressure: atmospheric pressure CVD (APCVD: 760 Torr and slightly below), reduced pressure CVD (RPCVD: 1–100 Torr), LPCVD (100 m Torr–1 Torr), and ultrahigh vacuum CVD (UHVCVD: $10^{-2}$ to $10^{-4}$ Torr) [8], such a categorization greatly oversimplifies the various CVD approaches available. Section 3 summarizes the history of CVD, how the aforementioned pressure regimes evolved, and the rise—and in some cases the fall—of prominent CVD technologies.

## 3. A brief history of CVD equipment

The value of CVD as a coating technology was recognized nearly 130 years ago by the lighting industry. To coat a lamp filament, gasoline vapor was introduced into a filament chamber. The heated filament cracked hydrocarbon molecules in the vapor, resulting in the deposition of a layer of graphite on the filament surface. This process optimized the filament resistance and emissivity [9]. In 1925, Anton Eduard van Arkel and Jan Hendrik de Boer developed a process, bearing their names, which purified metals initially containing contaminants such as nitrogen and oxygen. Volatile metal iodides are formed through a reaction with iodine, leaving impurities behind. The iodides, in turn, decomposed when heated by a tungsten filament leading to the deposition of a pure metal—in van Arkel's case, titanium, hafnium, and zirconium [10]. Even today, CVD remains an important technology for extraction of metals and the production of hard coatings (e.g., titanium nitride, titanium carbide, and boron carbide). However, it was the electronics industry, with the invention of the transistor in 1947 and the integrated circuit in 1958 that drove—and continues to drive—the expansion of research and development of CVD, which has enabled the deposition of thin, high-purity films so essential to integrated circuit manufacturing [10].

The year 1960 introduced the acronym "CVD" for chemical vapor deposition to semiconductor fabrication [11], where a variation of the aforementioned iodine process deposited epitaxial silicon. The dominant transport mechanism was thermal convection [12]. The ability to deposit high-quality epitaxial silicon enabled well-controlled doping levels at the device layer regardless of the doping in the underlying substrate.

Through the 1960s, commercial production used vertical (bell jar), barrel, and horizontal reactors for epitaxial deposition. The vertical reactor's bell jar design (Fig. 5.2) encourages circulating flow, with the idea that mixing spent gases with fresh reactants will lead to a uniform environment in the chamber, resulting in film uniform in thickness and electrical properties [13]. However, such a design made it difficult to control growth, whether to form abrupt interfaces, grade composition, or selectively deposit a film.

The original embodiment of the horizontal reactor involved so-called plug flow or displacement flow, where incoming gases push away downstream gases, leading to limited mixing but tighter control over growth compared with mixed flow. Because this deposition was often performed at

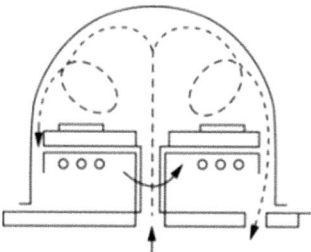

**Figure 5.2** The bell jar design. (*Source: Taken from Silicon Deposition by Chemical Vapor Deposition, p. 55, Handbook of Thin-Film Deposition Processes and Techniques, Krishna Seshan (editor).*)

atmospheric pressure, deposition was transport limited (for reasons that will be explained later in this chapter). That is, precursor depletion as the gas flowed from the front part of the wafer to the back required that the substrate be tilted to optimize film thickness uniformity.

The advent of LPCVD in the mid-1970s [14], however, enabled epitaxial deposition as a reaction-limited process, meaning a nonuniformity in precursor flux to the wafer still permitted uniform growth. The horizontal flow reactor with the tilted susceptor saw its commercial use drop for silicon epitaxy from greater than 90% in the early 1970s to less than 5% by the end of the 1980s [13]. In an LPCVD environment, a number of wafers could be arranged perpendicular to the axis of the tube with a pitch designed to maximize throughput.

In fact, well before the invention of LPCVD, such a "tube," which simultaneously processed multiple wafers neatly arranged along an axis, was a relatively mature product. Available since the early 1950s, diffusion furnaces had been in use for batch processing and by the late 1960s they were capable of reaching temperatures, with precision, exceeding 1000°C [15]. Diffusion furnaces were so named because they drove dopant diffusion for semiconductor junctions, filling a critical need, especially before the first ion implantation systems were commercially available in the late 1960s [16]. Even after the proliferation of implant systems, diffusion furnaces retained their use as dopant drivers (though the diffusion length was of a much smaller scale since profiles were largely defined by the implant step) as well as for dopant activation. Through the 1980s, however, scaling requirements demanded ever tighter control of dopant profiles and, in turn, shorter annealing times. By the late 1980s, single-wafer rapid thermal processing (RTP) came to prominence [17] and featured temperature ramps of the order of 200°C/s [18]. Today, single-wafer millisecond anneals are used to meet junction abruptness requirements, activate ultrashallow junctions, and form silicides [19]. The point is, for its original purpose of driving dopants for junction formation, batch furnaces have been, to a large degree, displaced by single-wafer chambers. Yet, "diffusion furnace" is a moniker that still exists today for tube furnaces even though the term fails to indicate the varied roles it plays for CVD batch processing.

At the time of LPCVD's rise in the mid-to late 1970s, diffusion furnaces in the horizontal configuration were mainstream. Most commonly, in a manufacturing environment, the tubes were stacked in groups of three or four. Hence, LPCVD was initially adopted by existing horizontal diffusion furnaces. However, through the 1980s and into the 1990s, diffusion reactors in the vertical configuration proved advantageous. High-volume manufacturing evolved to ultimately utilize, for tube reactors, vertical furnaces for batch processing. Compared with horizontal furnaces, the vertical configuration is more compact for a given batch size, hence offering a better economy of floor space. It also offers better uniformity because the wafer boat can be rotated during processing and can more easily accommodate automation requirements. So, today, virtually all diffusion furnaces have the vertical configuration in an automated manufacturing environment.

The diffusion furnace is a so-called hot-wall reactor. The furnace is heated using electrical coils arranged in zones—often a central zone covering the center portion of the tube at the top and bottom of the vertical furnace to compensate for heat loss out the ends of the tube. Lining the inside of the furnace, covering the heating elements and acting as a diffuser to evenly distribute heat to the boat, is a silicon carbide wall. With the wafer boat surrounded by the carbide wall, in the equilibrium state, the wafers and tube will reach the temperature of the wall itself. Indeed, temperature uniformity, and therefore growth uniformity, is an advantage of a hot-wall reactor.

A vertical diffusion furnace (Fig. 5.3) can process more than 100, 300-mm wafers at once. As alluded to in the discussion of rapid thermal annealing, the thermal mass associated with such a large boat entails long process times for temperature ramp-up (of the order of $15°C/min$) and cool-down (roughly $4°C/min$). A long duration at high process temperatures, however, is often incompatible with the requirements of manufacturing scaled transistors, where abrupt—and often thin—junction profiles are needed. To get around the high thermal mass associated with batch processing, single-wafer RTP using arc lamps began development in the early 1980s [20]. While the original intent of RTP was to anneal the silicon substrate after ion implantation, RTP technology was used with CVD ("rapid thermal chemical vapor deposition" or RTCVD) by the late

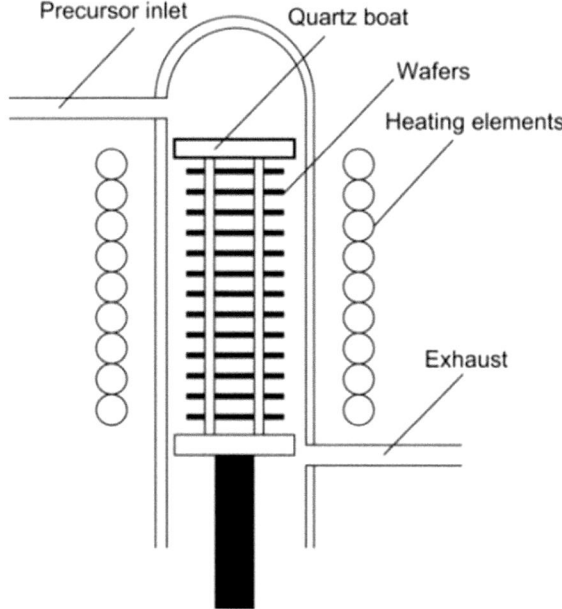

**Figure 5.3** A vertical diffusion furnace.

1980s [21]. This gave the engineer another parameter that could be used for process optimization. Instead of simply turning on and off gas flows, reactions could begin and end with rapid heating, up to 300°C/s [22], and cooling (the cooling rate is dependent on a number of factors, including radiative properties of the film and annealing temperature, but roughly this would be of the order of 100°C/min [23]).

In RTP, heat from the arc lamp is transmitted by radiative heat transfer through a quartz window or liner. Even before the advent of RTP, quartz had long been the material of choice to surround the wafer in furnace processing, whether in the form of a tube or jar. Quartz is ideal for this application due to its strength, high purity (lowering the risk of contamination), low coefficient of thermal expansion (allowing survival of thermal shocks), compatibility with chlorine (allowing, for example, selective deposition) and, of course, transparency. It is this transparency that allows for so-called cold-wall deposition, which limits deposition on the chamber wall, which in turn minimizes film spalling and particle generation. For this purpose, certain RTCVD chambers utilize a water-cooled stainless steel wall, with a quartz window between the lamp array and the wafer.

Because the entire wafer surface must be exposed to the lamp, all RTP processes for 300 mm utilize single-wafer chambers. This has a number of advantages. These include lower thermal mass (less thermal budget impact), lower cycle time (process does not need to wait for a full boat), greater process control (better probability of uniformity given smaller volume and that each wafer is exposed to the same environment), and precursor flexibility (due to smaller chamber volume).

The RTP process, however, does not come without challenges. Some processes are not compatible with cold-wall chambers such as the silicon nitride process using dichlorosilane with ammonia. Ammonium chloride deposits as a fine white powder in cooler areas of the chamber, creating unwanted particles and possibly line clogging [24]. Also, because the arc lamp can reach temperatures of up to 6000 K, there is a risk of temperature overshoot, especially when the emissivity of the absorbing film is not properly considered. In a hot-wall reactor, the temperature of the wafer does not exceed the temperature of the wall (also known as a thermal diffusion plate, often made of SiC).

While the 1970s was a time of rapid evolution for CVD in furnaces (LPCVD in particular), it was also a period of marked progress for low-temperature CVD alternatives. PECVD, which volatilizes precursors at low temperatures, was developed by Reinberg in 1971 and enabled silicon nitride deposition at 350°C [25]. This enabled the deposition of a strong, conformal barrier against sodium without compromising the existing aluminum metalization. Today, PECVD is a mainstay for low-temperature dielectric deposition, valued especially for its ability to deliver reasonably conformal nitride films whose stress can be controlled and low-$k$ interlayer dielectrics.

A variation of the PECVD theme is HDPCVD. Although early HDPCVD work was involved with amorphous silicon carbide deposition in the 1980s, for silicon microelectronics it was first used for shallow trench isolation [26] and subsequently for interlayer dielectric applications [27]. The defining feature of HDPCVD is its ability to fill trenches. During trench isolation fill, growth and sputter occur concurrently, with the highest growth rate occurring at the horizontal plane, the lowest on the vertical plane, and the highest etch rate at a 45 degrees angle as shown in Fig. 5.4. This works in favor of trench fill as the sputter component maintains an entry at the trench opening, etching the dielectric at the trench corners. This combination of sputtering at the trench corner and low growth rate from the vertical walls promotes bottom-up growth in the trench. Bottom-up growth would result in a fill free of a center seam, which would exist for an oxide grown from the sidewalls in a thermal

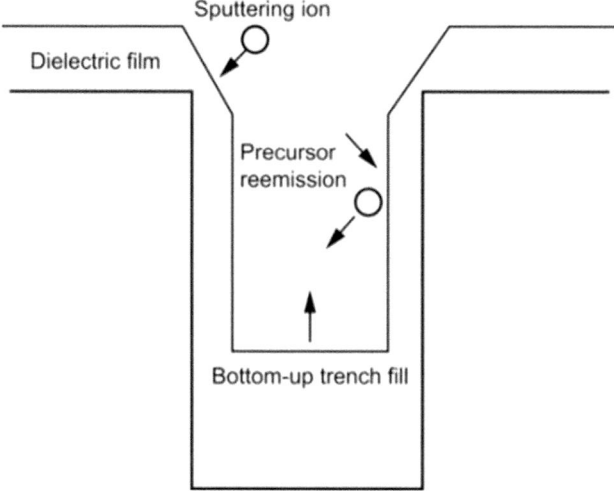

**Figure 5.4** The bottom-up fill of high-density plasma chemical vapor deposition.

process. The higher the aspect ratio of the trench, the smaller the deposition/sputtering ratio must be to avoid crowning (depositing a film that covers the trench but does not actually fill it) and voids. Hence, a higher aspect-ratio trench will result in a lower wafer throughput.

With the ability to incorporate fluorine during interlayer dielectric deposition [28], hence lowering the dielectric constant of the oxide, HDPCVD saw use as an interlayer dielectric (ILD) deposition technology for the subtractive aluminum process, which required oxide gap fill. However, with the logic industry moving to a copper damascene process, ILD deposited by a trench fill process was no longer needed. So, HDP tooling was largely replaced for ILD deposition by less expensive CVD technologies such as PECVD. HDP is still used today, however, for interlayer dielectric deposition in magnetoresistive random-access memory (MRAM) after transistor and bit line formation [29]. Moreover, because it is credited for depositing a denser protective nitride film than PECVD, HDPCVD has been found to offer a stronger defense against MRAM device performance degradation due to subsequent interconnect processing [30].

Dielectric deposition by plasma-based CVD at low temperature is usually simplified by the fact that no oxide removal is required. That is, the underlying surface for the plasma-deposited film is usually an oxide, as thin as that oxide layer might be, due to exposure to ambient or even oxygen from the previous process step. However, what if it was desired to deposit a

Group IV (such as Si, SiGe, or SiC) epitaxial film—requiring an oxide-free surface—at low temperature? Low temperatures during epitaxial growth may be needed, for example, to avoid dopant diffusion, creation of defects, or strain relaxation. To address this, Meyerson in 1986 reported the use of UHVCVD [31] for homoepitaxial silicon growth. The idea here is to have a base pressure low enough, say, 1-5E-9 [32], that the surface is prevented from oxidizing. There is a correlation between the partial pressure of $H_2O$ and $O_2$ and the growth temperature for maintaining a clean silicon surface required for epitaxial growth [33]. That is, the higher the background pressure in the chamber, the higher the substrate temperature must be for clean epitaxial growth. Indeed, up until the early 1980s, epi silicon involved temperatures greater than 1100°C to bake off the oxide and greater than 1000°C to keep an oxide from returning [34]. At a base pressure of 5E-9 Torr, however, defect-free growth of silicon without carbon or oxygen incorporation has been reported as low as 600°C [35], a temperature at which there is virtually no dopant diffusion [32]. Moreover, in addition to maintaining an abrupt interface, restricted dopant diffusion can also enable dopant concentrations even beyond the solubility limit. Low-temperature epitaxial growth, UHVCVD's advantage, is also one of its big challenges in a manufacturing environment, however. Because growth is reaction limited, growth rates by UHVCVD become exceedingly small at 600°C and below [36].

Just as the 1960, 1970s, and 1980s saw the development of new CVD technologies for depositing Group IV materials and dielectrics, it was also during this period that MOCVD rose to prominence for depositing compound semiconductors. By the early 1960s, gallium arsenide had been known for its ability to enable current oscillations at microwave frequencies [37] as well as to emit light [38], hence attracting attention as a material for high-frequency and optoelectronic devices. It was realized shortly thereafter that it would be desirable to grow GaAs layers on insulating substrates, providing enhanced electrical isolation. For optoelectronics, preference was given to a transparent substrate material, such as sapphire [39]. In 1968, Manasevit, who gave MOCVD its name, reported growing GaAs on sapphire, spinel, beryllium oxide, and thorium oxide substrates [40]. The "metal" emphasized the metal component and "metal-organic" applied to metal alkyl compounds—often the Group III precursor—at the time [41]. While the Group III precursors were commonly metal alkyls, the Group V were hydrides, and so it went: just as $AsH_3$ led to GaAs, use of phosphine, $PH_3$, led to GaP. Mixtures of the hydrides led to ternary compounds such as GaAsP. But it was Manasevit's plumbing of ammonia, $NH_3$, that delivered what is today the compound most commonly deposited by MOCVD for its use in light emitting diodes (LEDs) and, to a smaller extent, power management devices: gallium nitride.

The theme throughout this brief history of CVD is that requirements rooted in device scaling in a mass production environment drove the creation and survival of certain CVD technologies while others were left behind. Section 4 summarizes materials—metals, dielectrics, and semiconductors—deposited by CVD that enabled scaling.

## 4. CVD applications and their impact on scaling

Scaling prompts new requirements for materials and the manner in which they are deposited. As transistor density increases, for example, the critical dimensions of the components of the transistor as well as the surrounding shallow trench isolation regions shrink. For materials that can be deposited by CVD, this can mean newer gate dielectrics to minimize gate leakage, a resulting change in the gate metal or an increase in germanium content in the source and drain to enhance channel mobility.

In the metal layers, interconnects are drawn ever closer by scaling, hence increasing the parasitic capacitance, which leads to increased cross talk, power consumption, and RC (resistive-capacitive) time delay. This drives the need for ILD materials with reduced dielectric constants such as carbon-doped oxides deposited by CVD. Moreover, because of the increasing interconnect density, there exists a need for copper barrier layers that are thin, highly conductive, and can prevent diffusion. As such, CVD is seen as a suitable candidate for barrier layers.

What follows, then, is a review of metals, dielectrics, and semi-conductors deposited today by CVD to address scaling. A timeline summary of such enhancements is found in Fig. 5.5.

### 4.1 CVD metals

Since the early 1990s, tungsten deposited by CVD has been used as a contact plug material [56]. Tungsten has a low resistivity, is strongly resistant to electromigration, and is an effective diffusion barrier against copper. Pure tungsten cannot, however, be electroplated from an aqueous solution [57]. Also, because of the tight critical dimensions of contact vias, W sputtering does not meet step coverage requirements. Tungsten by CVD, however, not only delivers step coverage as a contact plug but can

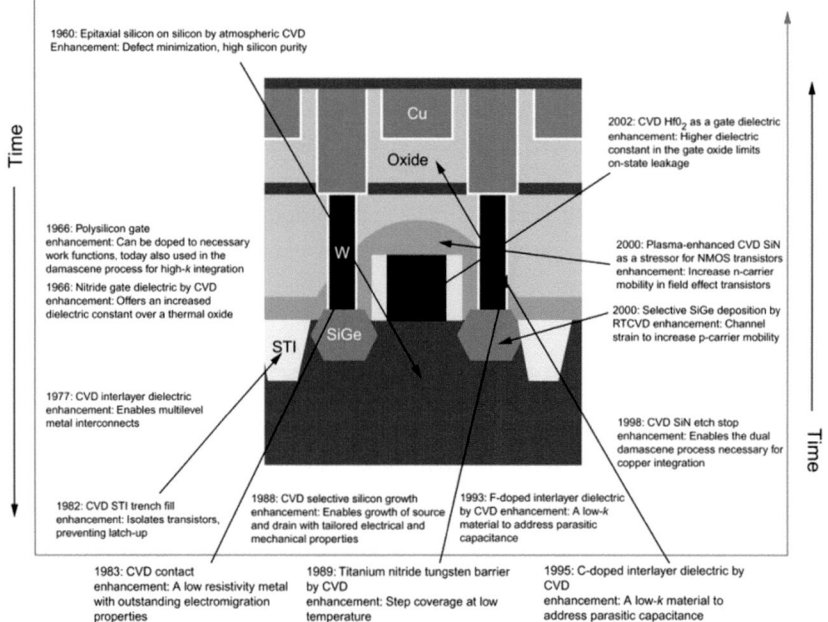

**Figure 5.5** First report of chemical vapor deposition materials that enabled device scaling [42–55].

also be a selective process, depositing only on metal silicides without lithography [58].

Titanium and titanium nitride are commonly deposited together as an adhesion and barrier layer for tungsten. Titanium is deposited first. Due to its high reaction activity, Ti is a getter for interfacial impurities and effectively reduces metal oxides, hence providing a sound ohmic contact [59] and reducing contact resistance [60]. Although titanium silicide was commonly used to reduce contact resistance through the late 1990s, the sheet resistance of $TiSi_2$ is very sensitive to thermal conditions and increases as the linewidth is made narrower [61]. Hence, the industry subsequently moved to cobalt and, more recently, nickel silicides [62,63]. The growth of NiSi, however, is inhibited or even prevented altogether by the presence of a native oxide. However, deposition of titanium after nickel deposition (which itself is usually not deposited by CVD as the precursors lead to either carbon incorporation or poor step coverage [64]) serves to not only protect the underlying nickel from oxygen contamination but the Ti also reduces the silicon oxide by diffusing through the Ni film, hence enabling the formation of NiSi [65].

Titanium nitride is a diffusion barrier to prevent the $WF_6$ from reacting with the underlying Ti, forming volcanoes [66]. TiN also acts as an adhesion layer for W [67]. Although originally deposited by physical vapor deposition, titanium nitride deposition transitioned to CVD for its step coverage while meeting adhesion and barrier requirements [68]. Sputtering of TiN can be performed at substrate temperatures below 300°C with deposition rates as high as 1 μm/min; however, sputtered TiN has a sticking coefficient close to unity, precluding it from filling high aspect ratio vias [69]. Likewise, Ti can also be sputtered and, indeed, sputtered Ti has long survived sputtered TiN. However, conjecture is that the usefulness of sputtered Ti is nearing an end, again because of ever-increasing contact-hole aspect ratios [70].

## 4.2 Metals more commonly deposited by non-CVD equipment

It is worthwhile to look briefly at some metals not commonly deposited by CVD as it reveals some limitations of CVD.

Tantalum nitride is a common choice as a copper diffusion barrier, as sputtered TaN has an amorphous structure that permits a thinner effective barrier than TiN [71]. CVD growth of TaN is largely a precursor challenge. Growth of TaN by CVD using $Ta(NMe_2)_5$, one of the most commonly used precursors for TaN growth, and ammonia led to tetragonal phase $Ta_3N_5$, which is a dielectric material with high resistivity [72]. MOCVD precursors are volatile, enabling a suitable deposition temperature but they have limited thermal stability and are sensitive to air and moisture [73]. Alternatively, halide precursors can be utilized for tantalum nitride, but they result in by-products that are corrosive to either Cu or $SiO_2$ [74]. As a barrier, TaN is often used with Ta, in which case TaN acts as both a barrier and adhesion layer for tantalum [75].

Copper can be deposited by CVD, but with poor adhesion to common barrier metals [76]. However, electroplating offers excellent gap fill characteristics [77] and, because copper electroplating is done at room temperature, it is attractive from a thermal budget perspective. Electroplating has a larger grained microstructure [78] resulting in superior electromigration properties to CVD copper [79]. Finally, because no vacuum needs to be maintained and the deposition rate typically reaches 1 μm/min, electroplating is cost-effective compared to CVD [80].

## 4.3 Dielectrics

For logic manufacturing, thermal budget must be considered to prevent dopant diffusion and maintain abrupt junctions in the transistor.

Plasma-enhanced vapor deposition enables film synthesis at temperatures less than 400°C, the approximate maximum temperature for back-end processing [81]. Such films include interlayer dielectrics with low dielectric constants (3.0 and below). The plasma volatilizes the precursor, eliminating the need to use a high temperature for precursor cracking.

The effort in depositing a low-$k$ dielectric film can be lost, however, if the etch stop layer has a high dielectric constant, which would compromise the effective dielectric constant of the dielectric stack. The etch stop acts as a barrier to further etching of the interlayer dielectric in the damascene process for metalization. An etch stop should itself have a low dielectric constant, act as a diffusion barrier against copper, and offer a high etch selectivity compared with the interlayer dielectric. Silicon nitride has been used as an etch stop material, as it offers good barrier properties and etch selectivity. However, the dielectric constant is high, and scaling requirements prompt the use of an alternative material. Deposition of etch stop films remains in the domain of CVD, with materials such as amorphous SiC, SiCO, and SiCN as candidates [82,83].

At the device level, scaling has pushed the source and drain of the transistor close together, leading to short channel effects partially addressed by channel doping. Such doping, however, in turn causes carrier-mobility degradation. Yet, the transistor current must be maintained or even increased from one technology node to the next to drive interconnects and meet performance expectations while keeping the general trend of supply voltage, $V_{dd}$, low.

To enhance carrier mobility in logic devices, a number of strain mechanisms have been implemented for both p-type metal-oxide-semiconductor (PMOS) and n-type metal-oxide-semiconductor (NMOS) transistors by CVD. For PMOS devices, compressive strain increases hole mobility, while carrier mobility in NMOS devices is enhanced by tensile strain [84]. One method for imposing a uniaxial compressive strain on PMOS devices is to synthesize a source and drain through selective, epitaxial deposition of silicon germanium (SiGe) [85]. Because the lattice parameter of the SiGe is larger than that for Si, the source and drain would impose a compressive stress on the channel in between. The process involves a recess etch step to provide a pocket for the subsequent SiGe deposition [86].

For NMOS devices, mobility enhancement has been achieved for several years through a silicon nitride layer deposited over the transistor. Originally used as a stopping layer for the contact etch, silicon nitride had long been known to have the ability to be deposited as a highly stressed film. Indeed, silicon nitride films deposited by PECVD can, depending on process conditions, impose a stress of order of gigapascals, tensile or compressive [87]. Reports show enhanced performance for both planar [88] and nonplanar transistors [89]. However, gate pitch scaling compromises the strain benefit of the nitride layer, and careful process optimization, possibly involving post-treatment [90], will be needed to have usefulness as a stressor.

Transistor scaling also drives the need for novel dielectric materials in the gate stack. Short channel effects associated with gate-length reduction has mandated that gate oxide be thinned to maintain control of the channel [91]. However, as the gate oxide is thinned linearly, as has been the case from one technology node to the next, there is an exponential increase in gate leakage. Hence, high dielectric constant oxides are needed to enable smaller effective oxide thickness for gate control, but thicker physical thickness to prevent leakage. For this, a hafnium oxide [92] gate has been implemented.

From a gate-dielectric deposition standpoint, it is required that the technology deliver a gate dielectric of highly uniform thickness across the wafer (for device performance uniformity), offer precise control of film thickness (for predictable performance from wafer to wafer), and deposit a film that is smooth at the atomic level (to avoid charge trapping) [93]. As will be discussed later in this chapter, ALD meets all three requirements and, as such, is the technology most commonly used in a logic production environment for high-$k$ gate-dielectric deposition. To avoid depletion and to screen out undesirable phonon effects associated with a polysilicon gate, metal electrodes, one type for NMOS, another for PMOS, are integrated with the high-$k$ film. Such films may also be deposited by ALD [94].

High-$k$ dielectrics are also playing an increasingly important role in dynamic random-access memory (DRAM). From one technology node to the next, the minimum capacitance per cell must be maintained at $\sim 10-15$ fF to provide adequate sensing margin and data retention time [95]. The challenge, then, is scaling the cell size while keeping the capacitance fixed. Historically, this was addressed by high aspect ratio silicon-insulator-silicon capacitor trenches using $SiO_2$ as the insulator. As scaling proceeded, however, the $SiO_2$ film could not be thinned further without deeply impacting the data retention time, motivating implementation of high-$k$ dielectric materials. Complementary metal-oxide-semiconductor (CMOS) gate dielectric criteria such as bandgap and impact on channel mobility are not considerations for a DRAM capacitor. This allows, compared with logic, greater flexibility in materials selection. In the early 2000s, DRAM makers switched to tantalum pentoxide, aluminum oxide, or a mixture of both [96]. Zirconium oxide found its way into DRAM cells in 2007 [97].

DRAM trench capacitors possess aspect ratios approaching 100:1 [98], thus requiring a highly conformal insulator deposition (greater than 90%). A dielectric film with poor conformality will risk either current leakage or poor capacitance (if the film needs to be grown thicker to ensure coverage everywhere in the trench) or both. Also, because of capacitance loss caused by silicon depletion, the industry moved to metal electrodes. With its chemical stability and low resistance, titanium nitride is the consensus material used as both top and bottom electrodes. Due to such stringent conformality requirements, especially of the insulator, the films are deposited by CVD, commonly by ALD. Indeed, DRAM manufacturers were the first to use atomic ALD for their high aspect ratio structures. Deposition of the electrodes by ALD may also be mixed with pulsed CVD to increase throughput [99].

## 4.4 Semiconductors

Originally chosen for its wide bandgap relative to germanium, silicon has been the transistor material of choice for more than 70 years. Due to scaling requirements, silicon's survival in logic devices has been dependent on mobility enhancements due to strain. Such strain has been imposed by CVD-deposited silicon nitride stressors for NMOS devices and selective deposition by CVD for PMOS. For the latter, the source and drain areas are etched away followed by a selective deposition of silicon germanium. Due to its larger lattice parameter compared to silicon, the epitaxial silicon germanium source and drain squeeze the channel between, causing a compressive strain, which in turn increases hole-carrier mobility. To maintain or increase transistor performance with decreasing transistor size, the germanium content in SiGe source drains has increased steadily since their introduction. Moreover, the source and drain can be doped in situ during the growth process [100]. There exists, of course, a limit to the germanium concentration in an epitaxially deposited source and drain. This realization has led to research into alternative channel materials and transistor architectures.

For example, a germanium quantum well, compressively strained by a SiGe upper and lower barrier, demonstrated a hole mobility quadruple that of the current strained silicon [101]. The quantum well stack was grown on a silicon substrate by rapid thermal CVD.

In summary, device scaling has led to short channel effects in transistors. Implemented to address such effects were measures such as channel doping and gate oxide thinning. Also, increasing transistor densities have prompted interconnects to be manufactured with tighter pitches, increasing the risk of cross talk and RC time delay. As discussed in the aforementioned examples, CVD continues to enable scaling through the deposition of films that are crystalline and noncrystalline, create tensile stress and compressive stress, are conformal and blanket, and in the form of metals, semiconductors, and dielectrics.

In addition to new CVD materials, however, scaling also drives ever-tightening requirements for contamination and metrology. Section 5 discusses contamination requirements and protocols to minimize contamination. Accepted metrology technologies are also covered to characterize contamination and the composition, mechanical and electrical properties, and thickness of films grown in a CVD reactor.

## 5. Contamination and metrology
### 5.1 Contamination

With device scaling comes tighter etch features and in turn ever-increasing restrictions on contamination which can have a significant impact on wafer yields. Indeed, the arrival of a single particle less than 100 nm in diameter on the front side of the wafer during the manufacturing process can destroy the functionality of an entire chip [102]. Moreover, with the complete wafer flow often needing more than 100 process steps, the wafer is handled extensively and hence backside particles are also a concern. Such particles can lead to photolithography problems by either distorting the flatness [103] of the wafer or lifting the wafer out of the depth of focus during exposure.

A primary source of particles in a CVD process is the reactor wall, especially for hot-wall chambers, where the high surface temperature can promote film growth and, eventually, flaking. Although the wafer can undergo a cleans process subsequent to deposition, this does not address flaking or contamination during growth. Periodic chamber cleans using a dry etch process such as that using $SF_6$ are frequently utilized to confront film build-up. However, this comes with drawbacks. The first wafer entering the chamber after the clean will be exposed to a clean reactor environment, whereas subsequent wafers will not. This is known as the "first wafer effect." Hence, not only will reactor time be spent on cleaning the chamber, but it will be also spent on nonproduction growth. So, both processes will compromise throughput. Another effect of the clean is that species from the dry etch, such as fluorine or a compound, can remain on the surface of the reactor only to leave the chamber wall, in one form or another, during deposition [104].

In addition to particles, contamination can also take the form of metal atoms. Namely, copper is of paramount concern. Displacing aluminum as an interconnect metal, copper exceeds aluminum in conductivity and electromigration properties. However, copper diffuses readily through silicon and dielectric materials. For interlayer dielectrics at the interconnect level, copper diffusion can cause dielectric breakdown [105]. In silicon, copper has a low solubility. Hence, it can form stress-inducing compounds at the surface. The $Cu_3Si$ precipitate can induce stress at the silicon surface and is highly resistive, degrading device performance [106], and the copper atom on its own is a deep-level trap, reducing carrier lifetime. In fact, transition metals, in general, such as nickel [107], iron [108], and gold [109], likewise produce deep-level traps in silicon.

To prevent the deleterious effects of copper contamination, equipment in a manufacturing environment is segregated to isolate wafers that have been exposed to copper or even exposed to tools that have been exposed to copper. That is, noncopper systems are forbidden from accepting any wafer that has been exposed to copper in its lifetime. Moreover, before a wafer lot is accepted into a noncopper tool, a check is performed to detect whether copper or other contamination elements are present. Total reflection X-ray fluorescence (TXRF) is perhaps the most common method for checking metal contamination [110]. In the literature, the acceptable threshold for copper contamination before the wafer can be introduced into equipment varies and depends on the process concerned. For critical FEOL processing, the upper limit for copper contamination suggested is 5E9 atoms/$cm^2$ [111], while for back-end-of-line processing, a figure for the upper limit is 5E10 atoms/$cm^2$ [112].

In fact, a 5E9–1E10 atoms/$cm^2$ limit for the front end has also been recommended by the ITRS road map for certain transition metals. Unlike mobile metals such as sodium and potassium, which can be easily removed, transition metals such as Ni, Cr, Co, Cu, and Fe can dissolve in silicon or form unwanted silicides [113]. In addition to the aforementioned deep-level traps, which can reduce minority carrier lifetime, transition metals also have high diffusion coefficients leading to frontside contamination even if their point of origin is the backside of the wafer and have a strong solubility dependence on temperature, possibly resulting in precipitates upon wafer cooling [108].

## 5.2 Metrology

After the wafer is processed by a CVD system, it is normally characterized for composition and mechanical and electrical properties. This is especially the case in the research and development stages. Although device results are the ultimate indicator of the success or failure of a CVD process, there are a number of metrology tools the engineer utilizes to enable fast cycling between determining film quality and the ensuing CVD experiment to optimize film properties. What follows is a description of wafer-level analytical techniques often used in the semiconductor industry.

### 5.2.1 Sample imaging and film thickness: atomic force microscopy, Nomarski, transmission electron microscopy, scanning electron microscopy, and optical microscopy

Whatever the CVD process, an image of the film can indicate film performance, whether it be, for example, to determine step coverage, gap fill, mechanical strength, film thickness, epitaxial growth, or etch selectivity. Although the process engineer will often need electron microscopy to refine a process, for a new CVD process, the first characterization is made by the human eye. For oxides, thickness uniformity can be determined by the chromatic uniformity across the wafer by visual inspection. The eye can also detect gross defects of several microns or greater, depending on contrast [114]. Improved resolution can be obtained by using an optical microscope, which, due to diffraction effects, has a resolution limit of half the wavelength of visible light, or about 200 nm. Another optical microscopy technique, reflected light differential interference contrast (DIC) microscopy, a variation of the classic Nomarski imaging, offers topographical information. Topographical features on the surface of the film create optical path differences for light glancing off the sample. The optical path differences are then transformed by reflected light DIC microscopy into intensity differences that can reveal features such as hillocks [115], cross-hatched patterns related to lattice mismatch [116], and cracks [117], and to verify substrate quality prior to deposition [118]. An important advantage of reflected light DIC microscopy over Nomarski imaging is that the sample can be opaque and therefore no sample preparation is needed.

Topographical information is also imaged by atomic force microscopy (AFM), which involves use of a small cantilever with a sharp tip at the end of the beam. Roughly speaking, the cantilever itself is about 250 µm long, 10 µm thick, and 35 µm wide [119]. As the sharp tip underneath the end of the beam glides over the surface of the film, it interacts with the film's topographical features, which affect the deflection of the beam. The amount of deflection, in turn, is characterized by light reflecting from the top of the beam to a bank of photodiodes. The lateral resolution of the AFM is less than 10 Å, while vertical resolution of AFM is under 1 Å [120]. AFM can determine surface roughness, which is an indicator of film quality and in turn, because of surface roughness scattering, carrier mobility [121]. Moreover, AFM can detect surface defects such as hillocks [122] and threading defects [123], both of which can arise from CVD growth.

Electron microscopy is to be credited for the high-resolution black-and-white photos seen with virtually every new semiconductor—product introduction. Virtually all recent scaling enhancements in the semiconductor industry, such as low-$k$ dielectrics, copper interconnects, SiGe source and drain, nitride stressor, high-$k$ dielectric gate oxide, metal gate, and fin field effect transistors (FinFETs), made their public debut in photos using electron microscopy. Scanning electron microscopy (SEM) is used to determine film thickness, though contrast is needed between film layers. (To avoid charge accumulation in the SEM chamber, dielectric needs to be coated with metal beforehand.) Passing SEM inspection is often the first requirement for CVD film applications requiring step coverage or gap fill.

Although SEM has an advantage over transmission electron microscopy (TEM) due to its simple sample preparation, TEM is needed to resolve features of less than 1 nm. Although there are broad, simple-to-execute measurements to gauge film quality such as X-ray diffraction (XRD), Hall measurements, and even SEM, TEM because of its fine resolution (capable of less than 1 Å [124]) is often the method offering the highest confidence of understanding the success or failure of a CVD process. Indeed, TEM is frequently used as final validation of other metrology techniques. From a TEM image, the engineer can visually determine if a given film is epitaxially deposited, what defects exist (e.g., stacking faults, twins, and threading dislocations) in the film and the film thickness. Dark-field TEM can, in some cases, be used to more easily highlight crystalline defects in the film [125].

There are two notable downsides to TEM in addition to the fact that it is a destructive procedure. Both are due to the labor-intensive nature of TEM. First, photos for a single sample can run into thousands of dollars. Second, turnaround time for a TEM sample is often of the order of days, especially for cross-sectional photos—photos that are often of most interest. Hence, other techniques are generally used first to characterize CVD films to enable faster experimental turnaround.

### 5.2.2 Composition: XRD, secondary mass spectrometry, mass spectroscopy, and Fourier transform infrared

When studying crystalline films, a versatile analytical tool is XRD. From XRD spectra, one can infer crystalline structure (from peak pattern), composition (by peak location), degree of crystallinity (from peak width), grain or particulate size (diffraction angle and peak width at half maximum [126]), and film strain and relaxation (deduced from comparing the lattice parameter in the film with the known lattice parameter of the film in its relaxed state [127]). Data collection for XRD is nondestructive, can be performed in minutes, and can even be performed in situ [128].

Like XRD, Fourier transform infrared (FTIR) spectroscopy is a potentially fast, contactless, nondestructive technique offering compositional information. By sending light of a given wavelength through a sample, the molecules in the film can undergo bond stretching, rocking, or asymmetric deformation. In other words, light can be absorbed. If the wavelength of the absorbed light is known, it can be compared against known values for given molecular bonds. Although FTIR can offer much information quickly, there are drawbacks. For certain molecules, it is not a sensitive technique. A sample can contain 1%—2% water and still not have it appear in an FTIR scan. Also, FTIR cannot detect homonuclear diatomic molecules such as $O_2$ or $N_2$.

A destructive technique—but one that is highly sensitive—offering compositional information is secondary mass spectrometry (SIMS). By sputtering the surface of the film and measuring ejected secondary ions by mass spectrometry, SIMS is extremely surface sensitive. By ion milling into the film or film stack, accurate compositional information can be obtained as a function of depth. Hence, SIMS can be used to determine composition, doping profiles, interface abruptness, impurity concentration, and film thickness.

X-ray photoelectron spectroscopy (XPS) is heavily utilized for CVD-film characterization due to the wealth of composition information it provides, the fast turnaround time (scans can be as fast as 10 min), and the relatively simple sample preparation involved. XPS, like SIMS, is surface sensitive, providing elemental information of the top 5—6 nm [129] of the sample. Unlike SIMS, however, XPS is generally nondestructive [130]. In

XPS, an X-ray beam of known wavelength incident on the sample surface causes core electrons to emit from the sample. By measuring the kinetic energy of the photoelectrons ejected from the sample and knowing the photon energy of the X-ray beam, binding energies of emitted photoelectrons can be inferred, leading to identification of elements in the sample. Also, the number of counts for a given energy peak is related to the elemental concentration. Because the binding energy of the electron depends in part on the oxidation state of the atom and the local chemical environment, differentiation can be made regarding the speciation of the atom (such as aluminum in a pure Al sample versus that in $Al_2O_3$ [131,132]).

The surface sensitivity of this technique also makes it possible to detect contamination on the surface of the film or wafer. Because the yield of photoelectrons from hydrogen and helium are below XPS detection limits, XPS can detect only elements with atomic number 3 and higher [131].

Information regarding thermal stability, composition, and moisture content can be offered by thermogravimetric analysis. Here, a small portion of the post-deposition wafer is placed on sensitive balance, which is then heated. Alternatively, the film can be scraped off the substrate directly onto the balance. Materials desorb during the temperature ramp, and the mass of the thermally desorbed material is plotted as a function of temperature [133]. For low-$k$ dielectrics, this technique is especially sensitive in detecting moisture absorption, which can have a significant impact on the film dielectric constant.

### 5.2.3 Mechanical: stud pull, wafer bowing, and nanoindentation

Mechanical properties of CVD films can come into play at the transistor level when attempting to predict the mobility enhancement of a stressor film. At the interconnect level, mechanical properties are especially important for the ILD, where the dielectric constant, generally, varies with mechanical strength. That is, films with a lower dielectric constant typically have lower mechanical robustness. This in turn can lead to cracking at the packaging level [134] and delamination [135].

In addition to the aforementioned XRD, film stress can be determined using a system that measures the wafer's radius of curvature before and after film deposition. Compared with XRD, such a system is simpler, capable of performing its measurement more quickly, and less expensive. Stoney's equation can then be applied, assuming the thickness of the film is much less than that of the substrate. The elastic modulus of the film can be found from nanoindentation. In this procedure, force and penetration depth are recorded as a diamond-tipped indenter presses into a film. Determination of the elastic modulus requires, however, a knowledge of the film's Poisson's ratio.

A gross estimate of the film's adhesion can be inferred from a stud pull test. Here, multiple studs are glued to the surface of the film. After the adhesive is fully cured, a specialized machine pulls the stud with a force perpendicular to the wafer surface. There is normally a wide statistical spread in adhesion values for this test, so a minimum of 20 pulls is performed for a given film.

### 5.2.4 Electrical: hall, reciprocal space mapping, and capacitance−voltage dot measurement

For the characterization of an interlayer dielectric, capacitance−voltage (CV) dot measurements indicate capacitance as it varies with frequency, known as "frequency dispersion." A typical range is, say, 10 kHz to 1 MHz. A high-frequency dispersion is indicative of high bond polarization [136]. Sample preparation is simple: the dielectric is deposited on a metal film. Then, aluminum CV dots are deposited onto the dielectric through a mask. Using one probe of the capacitance meter contacts a CV dot, while using the other, the metal underlayer.

In the context of gate dielectrics, the dielectric is sandwiched between a metal (or doped polysilicon) and a doped semiconductor. As the voltage to the top metal ($V_g$) is varied, a curve can be traced from accumulation to depletion to inversion. By applying an alternating current and varying the frequency, one can deduce dielectric characteristics as well as interface properties such as Fermi-level pinning [137] and trap density [138].

Hall experiments can independently measure the mobility and carrier density inside a channel [139] and are even capable of detecting defects and interface traps [140]. For a typical doped semiconductor sample at room temperature, mobility should climb significantly as temperature is decreased due to reduced phonon scattering. This inverse relationship between temperature and mobility holds until at some point, around 77 K or so, mobility decreases due to scattering from ionized impurities. Indeed, for Si, Ge, and GaAs, the room temperature mobility is roughly one-seventh that at 77 K [141]. In other words, if the Hall mobility at 77 K is not substantially higher than that at room temperature, it could indicate that impurities and defects are abundant, compromising the carrier mobility [142]. Although the Hall measurement is a destructive process, a Hall system is compact enough to fit on a typical lab counter with each data point taking roughly a few minutes.

The versatility and need for CVD moving forward will also make itself apparent later in this chapter in a discussion regarding prospective materials under consideration to meet scaling requirements. What follows next, however, is a review of CVD technologies prominent in the semiconductor industry, and their capabilities and applications.

## 6. Summary of CVD technologies

### 6.1 ALD

For its ability to deposit highly conformal films and smooth films with extraordinary thickness uniformity, ALD has gained strong interest for applications such as, for example, copper barriers [143], liners [144], and gate dielectrics [145]. ALD takes advantage of the difference between the binding energy for physical adsorption and that for chemisorption. For ALD, only one precursor is present in the chamber during deposition. The process is self-limiting. That is, once the reactant has completely covered the substrate, there are no more reactions at the substrate. So, for a film that is a binary compound, deposition occurs half a monolayer at a time. The remaining precursor would then be evacuated from the chamber, usually flushed out with an inert gas such as argon or nitrogen.

An example of an ALD process is the deposition of $Al_2O_3$ and is given by Fig. 5.6. Water vapor is introduced into the chamber, leading to hydroxyl groups forming bonds with the silicon surface. Any excess water is then purged from the reactor. The purge is followed by the introduction of the aluminum precursor source (Fig. 5.6a), trimethyl aluminum (TMA), which reacts with the adsorbed hydroxyl groups (Fig. 5.6b). The metal atom bonds with the oxygen; the hydrogen atom, meanwhile, leaves the surface as a reaction by-product in methane, which is pumped out of the chamber (Fig. 5.6c). TMA continues to react with any remaining adsorbed hydroxyl groups. Since TMA does not react with itself and only with the adsorbed hydroxyl groups, growth is terminated after one layer. The remaining unreacted TMA in the chamber is then purged. The cycle then returns to $H_2O$ introduction (Fig. 5.6d), but this time the oxygen atom attaches itself to the aluminum atom, displacing a dangling methyl group (Fig. 5.6e). A hydrogen atom from the original $H_2O$ precursor combines with the released methyl group to form methane. Any extra $H_2O$ molecules in the reactor will not react with the hydroxyl group attached to the

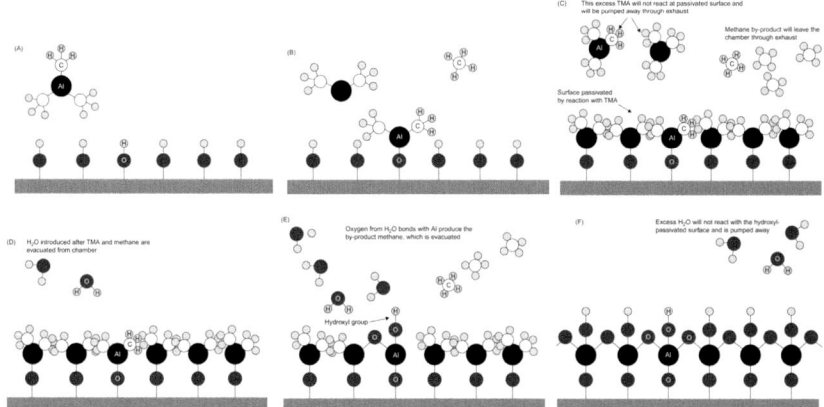

**Figure 5.6** The atomic layer deposition process. (a) Introduction of the aluminum precursor trimethyl aluminum (TMA). (b) Reaction of TMA with hydroxyl groups. (c) TMA surface passivation. (d) Introduction of water, the oxygen precursor. (e) oxygen combines with aluminum, displacing methyl groups. (f) With the H2O reactions with aluminum running their course, hydroxyl groups again passivate the surface; introduction of TMA begins the next cycle.

aluminum atom and will instead seek dangling methyl groups to react with. Hence, hydroxyl groups form a passivation layer (Fig. 5.6f). The cycle repeats itself, half monolayer by half monolayer, until the desired thickness is achieved. It is noted that at the heart of the formation of $Al_2O_3$ by ALD is the strong AlO bond [146], making it thermodynamically favorable for either the OH or Al-methyl bond at the film surface to be broken when fresh precursor (TMA in the former case and $H_2O$ for the latter) is introduced at the beginning of a cycle.

The periodic, self-limiting nature of ALD brings with it certain advantages. Because the precursors are pulsed into the chamber at separate intervals, the likelihood of gas-phase nucleation between separate precursors is minimized. This enables high chamber volumes associated with batch reactors, although a larger chamber volume will of course have a longer purge time. Also lending itself to batch reactors is the self-limiting aspect of ALD. That is, deposition stops on each wafer and within each wafer throughout the reactor after a fresh layer has been deposited, leading to highly uniform films independent of wafer spacing or distance from the precursor inlet.

In filling a trench using conventional CVD, for example, for a feature with a high aspect ratio, there can be preferential growth on the top corners of the trench, simply because reactants will be transported to the top of the trench before the sidewalls and trench bottom. This can lead to runaway growth at the top of the trench, leading to crowning or keyholing. The layered growth characteristic of ALD, however, can bring excellent results for growth on features requiring sidewall coverage, with conformal coating on structures with aspect ratios as high as 1000:1 [147]. Moreover, the half-monolayer-at-a-time deposition permits precise control over film thickness with outstanding uniformity with thickness variation within a few angstroms across a 300-mm wafer [148]. Although gate dielectrics such as $HfO_2$ have been sputtered, physical vapor deposition does not grant the thickness control offered by ALD. Such uniformity (1 sigma <1% for 300-mm wafers is common [149]) is especially important for films that directly impact device electrical characteristics. Even a slight variation in the gate dielectric thickness, for example, will significantly broaden the device performance range throughout the wafer.

One disadvantage often cited with conventional ALD is the slower deposition rate associated with each growth cycle, as the entire process is a series of discrete, time-consuming steps: precursor introduction, subsequent purge, introduction of a second precursor, and subsequent purge. To address this, there are ALD variations that attempt to enhance the overall deposition rate. One is plasma-assisted ALD. Here, the use of a plasma can, with respect to the conventional thermal ALD, increase the reaction rates at the surface [150] as well as enhance the removal of product molecules and enable the reduction of substrate temperature during growth [151]. Another ALD variant, pulsed CVD, uses the same sequential introduction of precursors, but with shortened purge times. Then, multiple precursor gases may be simultaneously present in the reactor. This offers the possibility of gas-phase reactions between precursors and deposition thicknesses of greater than one monolayer per cycle, increasing the deposition rate but at the risk of compromised conformality [152].

In the aforementioned description of atomic layer deposition (ALD), the wafer is stationary while the chamber environment changes with time as precursors are injected sequentially. This model of ALD is known as "temporal ALD." A significant drawback of temporal ALD is the need for a chamber purge to assure a given precursor is evacuated before the next deposition step. Over the past decade, however, there has been rapid development, industry acceptance and proliferation of so-called "spatial ALD." Here, it's the precursor environment that remains static and the wafer is in motion.

In one configuration of spatial ALD, two sets of two stations (say, one for oxygen deposition and another for aluminum), for four deposition stations total, simultaneously deposit self-limiting monolayers of material as with temporal ALD. Each station is dedicated to depositing a single material. Instead of a purge step after deposition, however, a wafer with a subsequent layer to be deposited is rotated through a nitrogen curtain to the next station. The removal of a purge step (which would be needed for temporal ALD) greatly enhances the wafer throughput, especially for thin ALD layers. In other words, spatial ALD can make sense for a process where the purge step would otherwise occupy a significant portion of the process time. For thick films, it could still be beneficial from a throughput standpoint to use temporal ALD in, say, a batch furnace.

### 6.2 Subatmospheric ACVD

Although the term "subatmospheric" CVD (SACVD) can mean chemical vapor deposition at a pressure less than 760 Torr, it is frequently acknowledged in the literature to fit in the window between LPCVD and APCVD or roughly 100–600 Torr [153]. Such is the assumed SACVD pressure range for the purposes of this chapter.

This pressure range, while it does permit the deposition of silicon and germanium, is not a popular one for Group IV deposition. In this range, the benefits of true LPCVD (such as thickness uniformity as permitted by reaction-limited growth and the minimization of impurities in the reactor) and APCVD (which enables high temperature, high deposition rate growth without the need for vacuum pumps) are compromised. However, for dielectric deposition, especially for step coverage, SACVD has strengths in areas where other CVD technologies fall short. For example, although PECVD enables oxide deposition at a temperature range safe for back-end processing, it has nonconformal step coverage [154]. As for APCVD, though it is credited for having a high throughput [155], the step coverage capability for oxide deposition is marginal [156], probably due to its operating in the transport-limited regime. SACVD is even mentioned as being preferable to APCVD for back-end blanket dielectric applications for its better particle performance and, because more wafers can be processed between chamber cleans, better throughput [157].

Use of SACVD for oxide deposition where step coverage is required often involves tetraethoxysilane (TEOS). On its own, temperatures exceeding 700°C are required to volatilize TEOS [158], making it impractical for back-end processing. However, when used in combination with ozone, the process temperature can be reduced significantly, to as low as 200°C [159]. Silane can be used to deposit silicon dioxide in a temperature range common to back-end processing [160]. However, while silane is pyrophoric, TEOS is stable and can be handled easily. In terms of performance, the conformality of a TEOS/$O_3$ process exceeds that of silane-based oxides, since the TEOS/$O_3$ combination leads to an intermediate precursor with high surface mobility [161]. As for LPCVD, TEOS with ozone was tried at low pressure, but results included films that were porous [162] and with high moisture content, high stress, and film shrinkage that can lead to cracking [163].

For TEOS and ozone deposition of $SiO_2$, at fixed reactant ratios, the deposition rate first increases with temperature, as one would expect, but it then achieves a maximum and decreases. The decrease is believed to be due to parasitic gas-phase reactions which in turn limit precursor flux [153].

The need for low-temperature conformal oxide growth was largely driven by back-end dielectrics in a subtractive aluminum process. But, with the move to copper interconnects, the implementation of which involves a damascene process, the need for a conformal interlayer dielectric went the way of aluminum. However, a recent development for SACVD is its potential use for through-silicon-vias (TSVs), which connect stacked chips [164]. The manufacturing process involves etching a deep trench to be followed by dielectric formation along the trench sidewall. A low-resistivity metal would then fill the core. Because the conductive core must extend through the entire depth of the trench, a bottom-up dielectric fill such as HDP would not be acceptable, and because the substrate would contain interconnects and devices, the dielectric deposition temperature would have to be low, eliminating the possibility of thermal oxide growth.

## 6.3 LPCVD

In the early 1960s researchers demonstrated that reduced pressures enabled highly uniform films and, in the early 1970s, it was realized that closely spaced substrates positioned vertically could be processed at low pressure deposition without sacrificing uniformity [165]. With its low-pressure process regime, deposition in an LPCVD reactor is often reaction limited rather than transport limited. If for approximation purposes the pressure in an APCVD reactor is 1000-times greater than that for LPCVD, then, all other parameters being equal, the diffusion constant increases by 1000 and the boundary layer thickness over the wafer surface increases by about 30. Hence, transport to the substrate increases by roughly 30 compared with

APCVD, making deposition reaction limited at low temperatures. Because deposition in this regime is independent of transport and therefore independent of the vagaries of gas flow and wafer orientation, LPCVD can enable a highly uniform film and is therefore the technology commonly selected for epitaxial deposition of silicon and germanium.

For reasons similar to what was discussed earlier about UHVCVD, an advantage of LPCVD is that potential contaminants such as $H_2$, $O_2$, and out-diffused dopants are quickly pumped out of the system. This reduces particulate generation and unwanted oxidation. Moreover, as-grown oxides deposited by LPCVD do not incorporate defects in the underlying silicon as thermally grown oxides might. Also, because there is an inverse relationship between mean free path and pressure, collisions are minimized in a low-pressure environment and the likelihood of gas-phase nucleation of particles is reduced compared to an atmospheric pressure chamber. This, in turn, can enable precursor flexibility. Similarly, another advantage of LPCVD is that due to the low pressure in the chamber, pyrophorics such as silane can be used.

Deposition rate is heavily dependent on temperature, as in the reaction-limited regime the growth process is limited by the reaction rate, which is in turn governed by an Arrhenius relationship (the rate constant is proportional to $\exp(-E/RT)$, where $E$ is the activation energy and $R$ the universal gas constant). Hence, the substrate temperature must be highly uniform within the wafer for uniform film growth. As an aside, it is noted that, while the deposition rate has an exponential dependence on temperature in the reaction-limited regime, there is a $T^{1/2}$ dependence for a transport-limited process. Moreover, with all other parameters constant, a lower chamber pressure leads to a higher temperature at which the process transitions from reaction limited to transport limited, allowing the process to remain on the exponential curve. In theory, a lower chamber pressure leads to a higher deposition rate at high temperatures. Texts and course presentations often show this as a figure with the natural logarithm of growth on the $y$-axis and $1/T$ on the $x$-axis and lower pressures leading to ever higher growth rate curves. This can be misleading. What must also be considered is that at sufficiently low pressure the reactant flux to the substrate surface is not high enough to sustain the predicted growth rate.

Because precursor depletion is not of concern in the reaction-limited regime, film uniformity is not heavily dependent on gas flow dynamics. So, LPCVD can also be used in a batch processing system where slotted wafers are stacked in groups of up to 200 wafers, leading to a throughput advantage over single-wafer systems. From a throughput perspective, this can be attractive in a manufacturing environment, especially for processes involving low temperatures (and hence low deposition rates) or thick films.

Hole-mobility enhancement by imposing a compressive strain on a silicon channel is accomplished by deposition of monocrystalline silicon germanium in the source and drain of PMOS devices. This process is performed by selective epitaxial growth (SEG), where crystalline SiGe growth occurs only on exposed areas of silicon. Either a mask layer or a patterned dielectric prevents growth of, in this case, SiGe on all other parts of the wafer. Because of this topography, a low-pressure process is needed. A process at, say, atmospheric pressure has poor step coverage and is not capable of depositing a well-controlled layer at the bottom of the trench formed by the mask. Moreover, because of dopants that are introduced during growth, a low-temperature process is required [166]. That is, in a low-temperature, high-pressure process reactants will have a limited surface diffusion distance before a high flux of reactants rains down on the substrate, hindering crystalline growth.

## 6.4 APCVD

Chemical vapor deposition at atmospheric pressure is as old as CVD itself and, indeed, up until the late 1970s virtually all epitaxial silicon deposition was done using APCVD [167]. APCVD is characterized by film deposition in the transport-limited regime, high gas flows, and, because vacuum equipment is not required, low equipment cost.

Because of the lower diffusion constant associated with higher pressures, APCVD enables deposition at high temperatures, in excess of 1000°C, for crystalline growth. That is, a reactant landing on the substrate has enough thermal energy to diffuse on the surface and enough time to find a site, or ledge, for crystalline growth before other precursor species arrive to block its path. Also, processing in this temperature range results in desorption of the native oxide and the maintenance of an oxide-free surface [32].

For epitaxial growth, APCVD can be used at low temperatures as well. A silicon wafer in a hydrogen ambient maintains nearly complete hydrogen coverage at 600°C, whereas in a vacuum at the same temperature hydrogen is almost completely desorbed [168]. With hydrogen adsorbed on the silicon surface, unwanted oxidation is prevented both prior to and during deposition. Hence, clean epitaxial silicon growth has been demonstrated down to as low as 500°C, leading to lower atomic diffusion compared with high temperature growth, avoiding, say, autodoping, impurity redistribution, wafer slip, and segregation (e.g., that of Ge during SiGe growth). Also, recently, epitaxial growth of silicon by atmospheric pressure plasma chemical vapor deposition (AP-PCVD) has been demonstrated at temperatures as low as 570°C with a deposition rate of 0.2 μm/min, greater than LPCVD and an order of magnitude higher than the rate for APCVD—for the same temperature [169].

As a dielectric deposition technology, APCVD has given ground since the 1970s to other CVD techniques, but nevertheless continues to be in use today for its high throughput and ability to process at low temperatures. In a common configuration, a continuous stream of wafers is introduced into the chamber by conveyor belt, with process gases confined in the reactor by a nitrogen laminar-flow shroud. Dielectric deposition by APCVD produces more particles than technologies operating at a lower pressure, however. Hence, chamber cleans need to be implemented more frequently, negating part or all of throughput gains by batch APCVD.

Because deposition in an atmospheric pressure chamber is normally transport limited, temperature uniformity is not as critical as that for a low-pressure system. In the transport-limited regime, after all, the deposition rate varies roughly with the square root temperature rather than temperature in the exponential term as with a reaction-limited regime. However, configuring the chamber to enable uniform flux of reactants is critical to thickness uniformity. So, although batch processing is possible with APCVD, stacking wafers along an axis with a tight pitch as in a vertical diffusion furnace would lead to nonuniform growth, since reactants would arrive at the outer edge of the wafer first, depleting reactants before they can reach the wafer center.

Numerous APCVD configurations exist to promote uniform reactant flux to the substrate. A classic design, used mostly in a laboratory environment, features a tilted substrate (Fig. 5.7) in a tube. A simple calculation considering conservation of mass and the shrinking flow area downstream

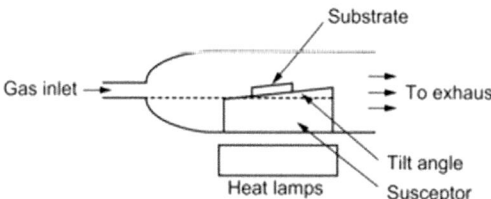

**Figure 5.7** A substrate in an atmospheric chemical vapor deposition chamber is tilted to enable uniform mass transfer. (*Source: Taken from MOCVD Technology and Equipment, p. 189, Handbook of Thin-Film Deposition Processes and Techniques, Second edition, Krishna Seshan (editor).*)

would show an increase in gas velocity. The increase in flow rate (and hence increase in precursor delivery) compensates for the reactant depletion upstream. Production chambers usually involve wafer movement to address any nonuniformities in precursor delivery. This includes the aforementioned conveyor reactor as well as a single-wafer chamber with a rotating susceptor.

APCVD generally has poor step coverage [170], stemming from the fact that deposition rate is limited by mass transport. That is, that part of a structure receiving the most reactant species will have the most growth. A structure such as a trench, for example, has features that have different acceptance angles. Either corner at the top of a trench, for example, will have an acceptance angle of 270 degrees, whereas the concave corners at the trench bottom will each have a 90 degrees acceptance angle. Then, unless the sticking coefficient of the reactant is very low (say, less than 0.01), different parts of a structure will have different growth rates. Moreover, the high deposition rates often associated with APCVD may lead to trench crowning, preventing reactants from reaching the trench.

Whether the film is polycrystalline or epitaxial, the deposition rate will be determined, in part, by the flux of atoms to the surface. It is here that APCVD holds an advantage over CVD technologies utilizing a lower pressure. That is, increasing both temperature (for reaction kinetics) and pressure (for flux of reactants) increases the deposition rate. At high pressure, however, the risk of gas-phase nucleation runs high, especially for volatile precursors. Then, for high-deposition-rate epitaxial silicon, a precursor such as dichlorosilane or trichlorosilane is utilized. Silane, for example, volatilizes at or even below 600°C [171] and is therefore commonly avoided in a high-temperature APCVD process.

## 6.5 MOCVD

As will be discussed later in this section, today there is a strong interest in compound semiconductors such as InGaAs as channel materials due to their high mobilities. Nitrides likewise continue to gain attention for optoelectronics, radio frequency (RF) applications, and power management. AlN, GaN, and InN have bandgap energies (of 6.2, 3.4, and 0.7 eV) covering from ultraviolet to the entire visible spectrum. Also, the strong III-N bond makes the nitride very stable, resistant to large avalanche breakdown fields and high temperatures [172]. Moreover, they have high thermal conductivities and large high-field electron drift velocities [173]. Such III-V materials can be deposited by molecular beam epitaxy. However,

MOCVD, with its ability to run batch processes and comparatively high deposition rates enabled by processes at or near atmospheric pressure, is prominent in a production environment. Moreover, MOCVD has the ability to deposit conformal films, can deposit uniformly over large wafers (including the largest Si substrates available today), can handle source materials with high vapor pressures, and can produce multilayer and graded composition layers [174].

When used to deposit epitaxial films, MOCVD is also known as MOVPE (vapor phase epitaxy). The motivation leading to MOCVD was heteroepitaxial deposition of high electron mobility and optoelectronic compound materials, such as GaAs [175]. Deposition of such III–V (or II–VI) compound semiconductors by MOCVD is enabled by the fact that many metal-organic compounds have a high vapor pressure and can therefore be transported by a carrier gas to the reactor. Also, such precursors volatilize at moderate temperatures inside the reactor [176].

As mentioned in Section 3, Manasevit first tried to deposit GaAs on crystalline insulators. He used triethylgallium (TEG) as the Ga source as metal alkyls were a subject of his graduate studies and the precursor could be readily procured [177]. Once commercially available, he subsequently used the more volatile trimethylgallium (TMG) [178]. With both precursors being volatile compounds, necessary quantities of either could flow to the reaction chamber using a bubbler at room temperature. Today, TMG remains perhaps the most commonly used precursor for gallium compounds deposited by MOCVD. Its chief competitor is still TEG, which reportedly can lead to lower carbon contamination compared to TMG [179]. Like many metal alkyls, TMG meets the basic requirements of an MOCVD precursor: it is stable enough and has a suitably high vapor pressure for transport to the substrate. At the wafer surface it decomposes cleanly, leaving the gallium atom as a part of the film with its gaseous side products readily removed from the chamber. Although TMG is pyrophoric and sensitive to water, it is relatively nontoxic.

Hydrides such as arsine and phosphine historically have been common sources for the Group V components arsenic and phosphorus, respectively. They decompose readily during MOCVD growth [180]. Moreover, they are gaseous and can therefore be delivered from a simple, cylinder-based system. However, while TMG has gained long-term acceptance as a Ga source (as with its trialkyl counterparts for In and Al, trimethylindium and trimethylaluminum), the high toxicity of the Group V hydrides, especially arsine and phosphine, continues to motivate development and proliferation of alternative precursors. For replacement of arsine, a promising candidate is tertiarybutylarsine (TBA).

The TBA molecule is the arsine molecule with one of its hydrogen functions replaced by a heavier butyl group. Compared to arsine, TBA has a lower vapor pressure and toxicity [181]. Likewise, a safer alternative to phosphine is tertiarybutylphosphine (TBP). TBA is 2- to 3-times less toxic than arsine, and TBP is several orders of magnitude less toxic than phosphine. Moreover, both TBP and TBA are liquids, presenting much lower speeds of exposure compared to their gaseous counterparts [182]. Also, the decomposition efficiency of TBA and TBP is reported to be higher than that for their hydride counterparts [183], leading to less source consumption, which helps address environmental concerns. Drawbacks with TBA and TBP include a history of leaving behind oxygen impurities in films [184] and the fact that they are expensive, at times prohibitively so [185].

Going one step further away from the gaseous hydrides is the use of a solid precursor such as solid arsenic, whose vapor pressure is high enough to be useable for MOCVD but low at room temperature [186]. Its storage is simple and does not carry the leak hazard of the high-pressure cylinders used to handle arsine. However, one challenge is that nonarsine sources are known to leave behind oxygen (a deep-level trap compromising mobility) and carbon (a dopant) impurities [187]. Another is maintaining a precise flow rate of the precursor. That is, with all solid precursors, the flux leaving the solid is dependent on exposed surface area, which can change with time as the precursor is depleted.

The most common nitrogen source for MOCVD deposition is $NH_3$. Because of the stability of $NH_3$, however, high temperatures are needed to volatilize it. For InN, growth temperatures can exceed 550°C, and above 1000°C for GaN and AlN [188]. As such, reactor components must be able to withstand high temperatures. Graphite, due to its chemical inertness and thermal stability, is a common material choice for the wafer susceptor and can withstand growths in excess of 1100°C [189]. Alternatively, nitrogen can be sourced from metal-organic precursors. However, resulting films are known to incorporate carbon impurities [190].

For a liquid precursor, the delivery system often involves a bubbler. To maintain an adequate vapor pressure, the bubbler is immersed in a temperature-controlled bath, and to avoid precursor condensation inside the lines leading to the chamber, the lines are heated all the way from the delivery system to the chamber. Condensation on the wall of the delivery line can in turn lead to particle generation and deterioration of deposition reproducibility [191]. The lines must not be so hot, however, as to lead to precursor decomposition. To minimize the risk of a cold spot—and to minimize the costs associated with line heating—the precursors are brought as close to the chamber as possible. The combined need for precursors to be close to the reactor—they are often located on the manufacturing floor—and their associated baths leads to MOCVD systems having a footprint much larger than a conventional CVD system.

Gas-phase interactions between Group III (or Group II) organometallics and Group V (or Group VI) organometallics can produce a low-vapor-pressure polymer that can condense inside precursor lines supplying the reactor or the chamber wall itself [192]. It is therefore often desirable to separate the Group III and Group V precursors until, ideally, just before the wafer surface is reached. Broadly speaking, the precursor gases can be introduced into the chamber by one of two methods: through a side injector or through a showerhead. Both designs exist for MOCVD. For most cases in a production environment, the wafer holder rotates for precursor flux and temperature uniformity. Depending on chamber design, precursors can be introduced into the reactor from either a side inlet or a showerhead. While side inlets have demonstrated worthiness for conventional Group IV deposition, the compound semiconductor devices deposited by MOCVD often require abrupt interface layers and film composition control. A showerhead design enables precursor introduction close to the wafer surface, hence reducing the residence time which promotes abrupt interfaces and reduces likelihood of gas-phase nucleation.

Film deposition can occur in a reaction-limited or transport-limited regime. As with high throughput epitaxial silicon (see Section 6.4), the deposition rates are highest in transport-limited growth and this is therefore frequently used in commercial reactors [193]. Critical to thickness uniformity in the transport-limited regime is a uniform precursor flux to the substrate. As such, showerhead configurations are often favored for their demonstrated thickness and composition uniformity during transport-limited growth [194,195] provided the wafer is rotating with adequate speed (in the 1000s of revolutions per minute).

The pressure range for MOCVD deposition is similar to that of conventional CVD [196], ranging from atmospheric down to UHV (background pressure 1E-8 Torr) [197]. Broadly stated, consistent with principles discussed earlier in this chapter, a low-pressure regime is often characterized by reaction-limited growth and conformal deposition [198]. As alluded to above, MOCVD performed at atmospheric pressure offers comparatively higher deposition rates [199].

Layered structures requiring abrupt interfaces as with superlattices and heterostructures are of interest for both their mechanical and electrical properties [200–202]. Moreover, abruptness in composition between layers is becoming ever more important for device performance as devices scale. Composition profile, in turn, is strongly dependent on precursor residence time in the reactor during deposition. That is, the longer the residence time, the greater the smearing, or dispersion, of the compositional gas front [203]. Contributors to longer residence time can be recirculation cells inside the gas chamber, moving large volumes of precursor gas at slow speeds and dead space in valves and tubing [204]. Hence, both equipment and process can be engineered to minimize residence times: Reduction of dead space in the delivery systems, reactor design to eliminate recirculation and deposition process at low pressure process and high flow rates.

## 6.6 PECVD

PECVD utilizes an electrode to volatilize precursors. This offers the ability to deposit at low temperatures, often below 400°C, which is especially important in processes following transistor fabrication where junction depth, junction abruptness, and gate performance can all be negatively affected by subsequent high-temperature processes. Although capable of depositing metals and silicides, and even epitaxial silicon (though not at deposition rates competitive with APCVD at high temperature) [205], PECVD is primarily used in semiconductor processing where low temperatures are required, such as for back-end dielectrics deposition of silicon dioxide (and doped variations thereof), silicon nitride, and silicon oxynitride. In these films, due to precursor volatilization, hydrogen is normally present. Moreover, process tuning is normally required to achieve a stoichiometric film if such a film is desired. Usually, PECVD silicon nitride films tend to be nonstoichiometric; that is, $SiN_x$ with $x \neq 4/3$ [206].

In PECVD, an electrode at high frequency (often 13.65 MHz) delivers bond-breaking, electron-stripping energy to inflowing gases, producing a soup of volatile species such as radicals, ions, and free electrons. The ions and radicals are reactive and play a role in film growth, while the free electrons are needed to maintain the plasma. For the process to sustain itself, a low chamber pressure is needed, since the mean free path of the electrons must be large enough to obtain the necessary energy for disassociation or ion creation. The electrons move much faster than the atomic and molecular species, so the electrons are whipping around and the atoms and molecules are essentially staying still. Associated with this process is a plasma glow, where free accelerated electrons excite orbital electrons, which give off light upon relaxation.

As an example of a PECVD process, silane can be used to deposit silicon. A silane glow discharge can contain $SiH$, $SiH_2$, $SiH_3$, $H$, and $H_2$ as well as positive ions $SiH_3^+$, $SiH_2^+$, $SiH^+$, $H^+$, and $Si_2H_2^+$ [207]. Because the substrate is submerged in the plasma, there is a flux of ionized species incident on the substrate [208]. So, with increasing power delivered to the plasma, the deposition rate initially increases due to precursor volatility but then levels out or even decreases due to ion sputtering. Then, there are two drawbacks of conventional PECVD: lack of control over what species exist in the reactor and unintended ion implantation or bombardment.

Use of a remote, or downstream, plasma, however, can address both concerns. Remote plasma-enhanced CVD (RPCVD or REPECVD) utilizes a plasma discharge away from the substrate surface. If the plasma is far from the substrate (a "far" distance is dependent on the flow rate of the excited species [208], but a reasonable estimate is of the order of tens of centimeters, with the distance inversely proportional to the flow rate), only long lifetime radical species [209] can reach the film formation region. Hence, specific radicals desired for film synthesis can be selected while avoiding substrate damage due to ion implantation. From a process recipe standpoint, the use of a remote plasma decouples plasma generation from film deposition. Each process can then be independently optimized. Increasing the power delivered to the plasma to generate a higher radical density, for example, will not damage the substrate. A shortcoming of remote plasma is that particles in excited states generated in the discharge area can recombine before reaching the wafer. Hence, the deposition rate can be up to an order of magnitude lower than for standard PECVD [210].

The ability of PECVD to deposit at temperatures low enough to maintain dopant profiles in the underlying transistors has made it the technology of choice for depositing low-$k$ dielectrics. A film's dielectric constant is determined by its density and the polarization in its bonds. Everything else being equal, the more porous the film, the lower the dielectric constant. However, a balance must be struck between porosity and the mechanical needs of the film. The polarization issue can be addressed by the addition of carbon, which makes a bond with silicon more covalent and less ionic compared with the SiO bond. PECVD demonstrates the ability to deliver on both counts: porosity control and carbon doping of silicon dioxide [211]. For porosity, an inverse relationship between deposition rate and film density exists. Also, the plasma conditions can be tuned to prevent complete disassociation of the precursor molecule, leading to a more porous film [212]. The deposition rate, meanwhile, increases with increasing RF power [213]. On a similar note, a decrease in the gas flow rate leads to a decrease in the deposition rate.

The plentitude of volatile species produced by PECVD can enable the process engineer to control residual film stress. An example is silicon nitride, which is used as an etch stop and stressor for mobility enhancement. The stress in the nitride film is largely dictated by hydrogen content, which in turn is influenced by temperature (the higher the temperature, the lower the hydrogen content), pressure (lower pressure leads to lower hydrogen content), and plasma conditions (the details of the complicated relationship between the plasma power and the corresponding hydrogen concentration can be found in Ref. [214]). Silicon nitride films with low hydrogen content correspond to films with compressive stress, as silicon-centered tetrahedrons, the basic building blocks of the nitride are joined to each other, forming a dense film (which would have, by the way, a low etch rate). A hydrogen atom, however, can position itself at the corner of a tetrahedron, acting as a terminator, preventing the linkage with a neighboring tetrahedron, leading to a less dense film [215]. In addition, the stress in the film is also determined by the NH to SiH ratio for a given hydrogen content. Due to the electronegativities of Si, N, and H atoms, for H and N attached to the same Si atom, the SiH bond decreases the strength of the SiN bond. This increases the length of the weakened SiN bond, hence making it more tensile [216].

There are a number of hybrid CVD approaches that combine two or more of the technologies discussed, hence taking advantage of the strengths of each. For instance, we saw that PECVD has been combined with APCVD to achieve low temperature, epitaxial, high growth rate silicon films. MOCVD has been combined with APCVD, known as AP-MOCVD, and is reported to offer, compared to low-pressure MOCVD, superior mobility for GaN [217] and ZnO [218] in addition to high growth rates. As a third example, plasma has been combined with ALD. Plasma-enhanced ALD, or PEALD, is now commercially available and, compared with conventional ALD, has been credited for higher growth rate per cycle (especially at very low temperatures such as 65°C) and a more efficient surface reaction, decreasing the impurity concentration [219]. Although PEALD is reported to be more conformal than physical vapor deposition (PVD) [220], it does not quite match the conformality of conventional ALD [221].

## 7. CVD tool selection for research and manufacturing

A number of factors are to be considered when selecting a CVD tool. The first question to be asked, of course, is "What is the CVD tool for?" The criteria for research will be different from those for a manufacturing environment. For research, flexibility is paramount, since the deposition system will be used to study feasibility for a variety of materials, a range of recipes for a single material, or both. Such flexibility includes the ability to deliver an array of precursors to be used in a chamber equipped to deliver a suitably wide range for process conditions such as temperature, pressure, and flow rate. Since research is a proof-of-concept stage of development, processing can be done at the single-coupon level. If there is any wafer size requirement, it would be dictated by upstream and downstream processing requirements.

As mentioned earlier in this chapter, a single-wafer chamber offers a number of advantages over batch processing for CVD processing in a research environment. The first is a cost and environmental sustainability issue. Compared to a batch system, a single-wafer chamber will use less process chemicals, by volume, per process run due to its smaller chamber size. This is especially the case when several recipe splits are run per wafer lot, which is often the case for research. Moreover, a batch reactor will usually require so-called dummy wafers to be processed along with the wafer of interest in order to mimic the performance in a manufacturing environment. Second, due to the volume of a batch reactor, gas-phase nucleation is a risk. This can impose restrictions on precursor selection, as a volatile precursor molecule more easily cracks and its constituent can react before arriving at the wafer surface. Finally, with a batch reactor, especially a tube reactor, there can be precursor depletion since reactants will first arrive at wafers closest to the inlet. This can be addressed through process parameters such as adjusting the inlet flows or reactor pressure, or use of a temperature gradient. However, in research, one is interested in the limitations of the film, not the reactor. In other words, one wants as wide a process window as possible to determine feasibility and to optimize film properties. If it happens that a batch reactor is ideal for a given film, so be it. However, arriving at this conclusion should be incidental, not forced because of the reactor one happens to be using when trying to establish a concept.

If the answer to this section's original question is that the tool is to be used for manufacturing or possibly development in preparation for manufacturing, the criteria change from prioritizing materials flexibility to emphasizing factors such as cost, film uniformity, and defects (or particles). By this point, research has delivered a film that has a high confidence of meeting manufacturing, integration, and product requirements.

The target film and its associated requirements will be the key criteria for deposition equipment selection. A supplier can differentiate itself on film quality through, to name a few, CVD deposition technology (e.g., LPCVD vs APCVD), chamber design, chamber clustering (i.e., having multiple chambers per platform, hence enabling an in situ deposited film), precursor delivery capability, precursor intellectual property, and process know-how. No matter what approach is taken by a supplier to stand out, however, determining the best technical candidate will usually come down to film characterization and integration.

For example, ultralow-$k$ interlayer dielectrics with dielectric constants as low as 2.0 are attractive from a parasitic capacitance standpoint, and indeed, such films can be deposited by CVD [222]. However, integration of such films is perhaps the greatest challenge facing low-$k$ dielectrics. Because ultralow-$k$ dielectrics are necessarily of low density, they are mechanically weak and, often, exhibit poor adhesion. Since film quality is often tied to the equipment supplier (due to, for example, chamber design or process-related intellectual property), selection of manufacturing hardware, then, will be dependent on not only the properties of the film itself but also on whether the film can be properly integrated into the chip during the manufacturing process. Also critical is the performance of the device after integration of this film—that is, whether the device performance improved with respect to a baseline due to the integration of this new film. This determination—probability of integration and overall performance—will usually represent the bulk of the time and resources involved with equipment selection.

For manufacturing, a second important consideration, in addition to the technical criteria mentioned above, is capital cost, where a number of factors come into play. Throughput, as determined by availability, run rate, and preventive maintenance (PM) frequency will affect the total number of tools needed and therefore the total capital cost. Also affecting throughput are necessary processes related to the hardware. For example, as discussed earlier, a hot-wall chamber brings the advantage of temperature uniformity. However, it can also lead to the occurrence of particles due to deposition on the chamber walls. To mitigate the particle risk, intermittent clean processes are used, which affect throughput.

When considering a batch system in a manufacturing environment, cost of ownership is perhaps the most important motivation. With diffusion furnaces able to process as many as 200 wafers in a single boat, the throughput benefits can indeed be attractive. However, when calculating the cost benefit of such a system, failure consequences and queue time should be considered. Regarding the former, a system failure or power interruption of a furnace processing eight lots of wafers is far more substantial than that for a single-wafer system with one wafer per process chamber. As far as wafer throughput is concerned, before that batch of 200 wafers can be processed, the tool will need to wait until eight full lots of 25 wafers have arrived at the loading station. A single-wafer system, however, can begin processing with the arrival of a single lot. A similar consideration should be given for the output of a tool that processes scores of wafers at once. While the next process step may also be a multi-lot process, inevitably a single-wafer processing step will be involved (e.g., lithography) downstream. At that point, one lot can be processed but the other seven lots will wait for processing, adding to cycle time (the total time the wafer spends in the fab—this is typically of the order of weeks to months).

Another consideration when deciding between multi- and single-wafer processing is the level of process flexibility afforded by each [223], especially if consecutive in-situ steps are involved. A large-batch furnace can contain up to two tubes, with an in-situ transfer from one tube to the next. In this example, the available processing is constrained to that which is available in a furnace. A single-wafer cluster platform, however, can contain four or more chambers, each conceivably with different functions (Fig. 5.8). For example, three process chambers, one each for preclean, gate oxide, and gate material, can all be on the same platform, with the wafer moving from one chamber to the next without breaking vacuum. Also, the number of chambers for any given step can be tuned to match throughput between steps. Use of multiple consecutive in-situ wafer processing steps has the added benefit of minimizing the handling outside of a vacuum environment and the need for particle-cleans steps.

Scaling often requires novel hardware to be introduced, whether in the form of precursor delivery, enabling new chemistries in the chamber, or even particle reduction. In any case, such a novel introduction is intellectual property and the supplier can charge for its use accordingly. Also, with research and development costs not yet amortized, novel equipment can also carry a pricing premium for those who adopt early in the product cycle.

When considering cost, one should also consider clean room space, which has roughly doubled over 10 years to roughly \$8500/ft$^2$ [224]. The

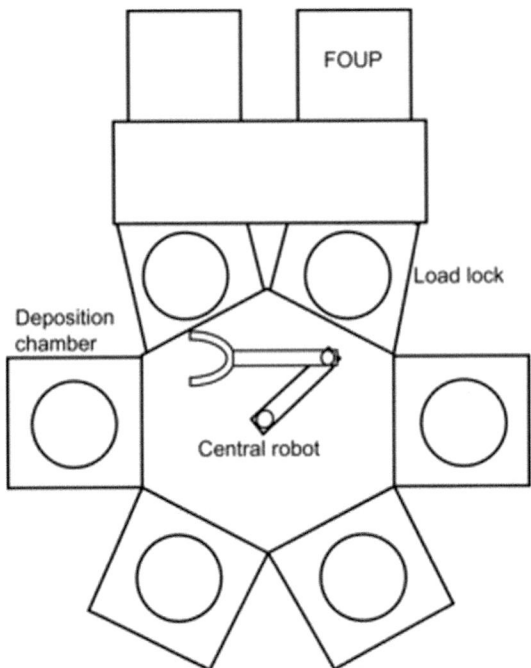

**Figure 5.8** A cluster platform with two load lock chambers and four process chambers.

footprint of a 300-mm CVD cluster tool is, say, 20 m$^2$ or about 200 ft$^2$. One must also factor in the area occupied by support equipment such as pumps, abatement, and the gas box, though frequently one or more of these components can reside in the less costly sub-fab underneath.

## 8. CVD trends and projection

For more than six decades, the chip industry has manufactured planar metal-oxide-semiconductor field-effect-transistors (MOSFETs) as the basic building blocks of computing. This transistor architecture works well for a channel length that is long compared to its thickness, where the gate has functional control over charge carriers in the underlying channel. With ever-shrinking channel lengths from one technology node to the next, however, the gate in a planar transistor correspondingly possesses less control over the channel. Hence, either the off-state leakage either increases or, if the drain voltage is decreased to minimize off-state leakage, the on-state current is diminished.

Non-planar, or three-dimensional (3D) MOSFET architecture helps to address the issue of gate control with additional gate coverage around the periphery of the channel compared to that of a planar MOSFET. The first 3-D MOSFET, known as a "FinFET," became commercially available in 2012 [225]. In this configuration, the gate bounds three sides of the channel, which is shaped as a fin, to maximize gate control and minimize the area of the fourth side of the channel not covered by a gate.

A next step in the evolution of gate control, then, is the so-called "gate all-around" (GAA) FET or GAAFET. Here, the gate completely surrounds the channel, leaving no surface of the channel exposed between the transistor source and drain. The first commercially available GAAFETs were manufactured in 2022 [226] using an industry-consensus stacked-channel configuration. Such an arrangement has alternating layers of channel, gate oxide and gate metal.

Most commonly, the manufacturing of the GAAFETs involves first depositing alternating epitaxial layers of Si and SiGe by LPCVD [227] between the source and drain. This is followed by a selective etch to remove the SiGe layer, leaving behind Si bridges connecting the source and drain. ALD, with its excellent conformality, is used to deposit both the gate oxide and the metal gate to fill the gaps left behind by the SiGe etch.

Prior to the SiGe removal, a slight recess is made in each of the SiGe layers on both ends (the source end and drain end) for a spacer dielectric as a sidewall against the source and drain. That is, the SiGe is etched away and replaced by the gate dielectric and metal. This spacer, often SiN, is a dielectric layer preventing shorting between the eventual metal gate and the source and drain. Because the spacer deposition involves filling of a shallow vertical cavity, CVD tooling—whether LPCVD, PECVD or ALD–is the deposition category of choice for this purpose [228].

Taking the ribbon-FET to the extreme is a thin or so-called "2D" channel. Off-state leakage is nearly eliminated for field-effect transistors made from 2D materials because charge carriers are confined in an atomically thin channel and, hence, uniformly controlled by the gate voltage [229]. Successful commercial implementation of 2D materials, in turn, is dependent on affordable growth of continuous wafer-scale single-crystal films [230]. Transition metal dichalcogenides are of interest as materials for 2D transistor materials for their bandgap, which is tunable by composition and number of layers. Two such materials $MoS_2$ and $WSe_2$ have been used to demonstrate current switching ratios up to $10^8$ and a subthreshold slope of 60 mV/decade [231]. MOCVD's ability to control deposition at the monolayer level combined with its capacity for volume production makes it uniquely suited for the commercial synthesis of 2D transistor materials [232].

With gate pitch scaling degrading the effectiveness of silicon nitride stress layers, there is motivation to find alternative methods of increasing carrier mobility in NMOS devices. One approach is to utilize a lattice mismatched material in the source and drain region. The reader will recall this has been done using SiGe in the source and drain for PMOS devices in high-volume production since the 90-nm technology node (2003). For NMOS devices, however, instead of creating compressive strain, the idea is to create tensile strain using silicon carbon as the source and drain material. For this, SiC, which possesses a lattice parameter smaller than for Si, is a candidate [233]. As with SiGe source and drain, the SiC is deposited epitaxially and selectively. Significant obstacles exist to integrating SiC successfully, however. There is a vast lattice mismatch between diamond and Si, and C has a low solubility in Si. Hence, to grow metastable films beyond the critical thickness without $Si_{1-y}C_y$ precipitation requires nonequilibrium growth conditions—namely low temperature and high growth rate [234]. Quite unfortunately, for reasons covered in this chapter—precursor thermal decomposition among them—growth rate falls sharply with decreasing temperature, compromising manufacturability.

As mentioned earlier in this chapter, the search for mobility for NMOS logic has also led to quantum well devices featuring III—V compound semiconductors such as indium antimonide (InSb) and indium gallium arsenide (InGaAs) [235]. Although molecular beam epitaxy is a common research tool for building quantum well stacks for its material flexibility and control, the deposition rate is slow, of the order of 1 μm/h. However, MOCVD can offer a deposition rate a few times higher for the same material. Moreover, even though such devices demand abrupt interfaces and can have layers with thicknesses of the order of 1 nm [236], such control can be offered by MOCVD [237,238]. If logic is to be committed to silicon substrates moving forward, compound devices will likely require virtual substrates involving thick (greater than 1 μm [239]) buffer layers and, hence, deposition rate will be critical in a manufacturing environment.

For PMOS transistors, channel strain through implementation of SiGe has provided hole-mobility enhancements in successive process nodes since the 90-nm manufacturing technology, more than 20 years ago [86]. Indeed, even today, SiGe is under strong consideration as a stressor material for gate-all-around devices [240]. However, such enhancements will run their course, and a new materials system will be required. Prominent among all PMOS channel material candidates is germanium. The techniques most commonly used for epitaxial germanium deposition are molecular beam epitaxy and some form of CVD, whether it be by LPCVD, ALD, or even low-energy PECVD [241]. Although molecular beam epitaxy (MBE) offers excellent heterointerface control [242] and can be even used for selective Ge deposition [243], its low deposition rate is not suitable for a high-volume manufacturing environment. Moreover, a strained Ge quantum-well device stack deposited by reduced-pressure exhibited a hole mobility of $4.3 \times 10^6$ cm$^2$/V-s, or roughly twice the highest electron mobility obtained by strained Si [244].

In the domain of back-end interconnects, while Cu has a bulk conductivity second only to silver across all elemental metals, scaling Cu as an interconnect material has led to increased electron scattering at the conductor surfaces and grain boundaries, negatively impacting the resistivity of the Cu interconnect. Such scattering can cause the resistivity of 10-nm-wide Cu lines to be approximately an order of magnitude larger than bulk copper [245]. Another drawback of copper is the associated diffusion barrier and liner, both of which compromise effective interconnect conductivity. As alluded to earlier, copper interconnects require diffusion barriers because Cu can otherwise easily diffuse into the interlayer dielectric to form physical spikes and deep level traps [246]. Tantalum nitride is commonly selected as a diffusion barrier. Because TaN has poor adhesion to copper, however, a liner to promote Cu adhesion such as Ta is often used. The diffusion barrier and liner, though necessary for Cu integration, increase the effective resistivity of the interconnect ($\rho_{Cu} < 6.5$ μΩ-cm [247], $\rho_{Ta} = 13$ μΩ-cm [248], $\rho_{TaN} > 200$ μΩ-cm [249]). Also, the barrier and liner layers do not scale in step with manufacturing nodes, as reduced thicknesses negatively affect dielectric breakdown and electromigration properties [250]. Specifically, due to the inherent roughness of the interlayer dielectric trench wall and thickness non-uniformity of the barrier layer itself, a minimum barrier thickness must be targeted to minimize risk of holes in the barrier. So, for all other factors unchanged, the thickness of the barrier material occupies a greater proportion of the overall width of the trench or via as pitch shrinks with each successive technology node.

The aforementioned undesirable barrier effects can, of course, be mitigated by replacement with a low resistivity material. A candidate barrier material such as ruthenium possesses a resistivity much lower than that of Ta ($\rho_{Ru} = 7.1$ μΩ-cm vs. $\rho_{Ta} = 13$ μΩ-cm [248]), excellent wettability, negligible solubility and strong adhesion to electroplated copper [248]. Ruthenium as a barrier layer can be deposited using PVD, CVD and ALD. Deposition by PVD, however, is widely known to result in strong shadow effects. Deposition by CVD, on the other hand, has demonstrated highly conformal Ru films with excellent purity [248].

To summarize, interconnect scaling has led to (1) increased electron scattering at conductor surfaces and grain boundaries [251] and (2) the diffusion barrier taking a larger proportion of the interconnect width. Both of these factors lead to increased line resistance for Cu scaling. This motivates a search for a material that, while perhaps having a bulk resistivity higher than that of Cu, may nevertheless have a lower effective resistivity as an interconnect metal. Such a move away from Cu would first involve metal layers M0 and M1—the local interconnect layers—as they have the finest line widths.

For a narrow interconnect, classical transport models suggest that a material's resistivity scales with the product of $\rho_0$ and $\lambda$ [252]. Where, $\rho_0$ is the bulk resistivity of the material and $\lambda$ is its bulk electron mean free path. All other parameters being equal, a lower bulk resistivity implying a lower resistivity, even for narrow interconnects, is intuitive. As for the electron mean free path, a smaller $\lambda$ implies a lower probability of the charge carrier being subject to size-dependent scattering by surfaces or grain boundaries, with grain-size tending to be proportional to the line width [252].

A possible bonus in moving away from Cu is obviating the need for a diffusion barrier and liner. A feature of the latter is its limiting of electromigration. The cohesive energy of a metal is closely related to its melting temperature and its ability to form vacancies. Hence, an inverse relationship is expected between melting temperature and degree of electromigration. In other words, melting temperature can be used as a proxy for electromigration [253]. Similarly, vacancy scales with interconnect metal diffusion which, in turn, can lead to dielectric breakdown [254]. Hence, cohesive energy can also be an indication of the need for a diffusion barrier.

The $\rho_0$-$\lambda$ product and melting temperature, then, can be used as a filter to find candidate replacement materials for copper. Metals with a lower $\rho_0$-$\lambda$ and higher melting temperature than copper include platinum group metals—Pt, Rh, and Ru—and refractory metals Mo, Ir, Os and Nb. Cobalt is also a candidate and has the advantage of being extensively studied. However, Co requires a diffusion barrier [255]. Despite its higher $\rho_0$-$\lambda$ product, tungsten continues to receive consideration due to its high melting temperature and decades-long use as a contact and via fill material [70].

Rhodium and Iridium are more expensive than gold and exhibit poor dielectric adhesion [256]. Ruthenium has itself long been under strong consideration as a copper barrier [257], hence as an interconnect material it would not need a diffusion barrier [258]. Also, being a noble metal, ruthenium is resistant to oxidation. However, ruthenium is expensive, roughly three orders of magnitude more so than copper.

The fabrication process for copper interconnects requires a damascene process because of the lack of a volatile etch by-product [259]. Hence, for certain non-Cu interconnects, a subtractive metalization process, one form of which is a so-called "semi-damascene" process, where a dielectric layer is etched, with trenches overfilled with an interconnect metal until a full metal layer if formed over the dielectric [256]. This metal layer, which theoretically would not require CMP, would then be etched. Note possible advantages of this process. The grain-boundary size of a blanket metal layer, as opposed to one where the metal is filling a pre-defined geometry, can be more easily controlled [260], hence potentially minimizing charge-carrier scattering. Such a process can, in theory, also avoid the need for metal CMP [261].

The subtractive metalization process forms an empty space between metal interconnects. This can be utilized to form an air gap as the interlayer dielectric, potentially reducing the interconnect parasitic capacitance compared to a conventional dielectric material. For example, an airgap can be made in the trench by tailoring a plasma CVD recipe to result in an incomplete gapfill [262], leaving a keyhole in the trench.

Although other candidate materials exist, including the aforementioned, the semi-damascene process has been demonstrated to the public most commonly with ruthenium as the interconnect metal [263]. When gapfill is a necessity for semi-damascene, ruthenium deposition by ALD allows for conformal coverage [264]. The overfill layer can be deposited by CVD or PVD [265].

Regarding affordability considerations, transistor scaling has led to chip design and manufacturing costs increasing with each technology node [266]. A leading-edge fabrication facility can cost more than $10B [267]. Larger chip sizes can include more transistors and therefore more functionality. However, larger chips are prone to more defects and therefore decreased yield. Moreover, functionality expectations demand more transistors than can be packed onto even the largest chips that can be processed by today's semiconductor manufacturing equipment [268]. Finally, such designs at the chip level are costly, with skews difficult to implement.

Heterogeneous integration is the incorporation of separate components into a higher-level assembly and is a pathway to address the aforementioned challenges of system-on-chip. Such components can be DRAM, compute die, cache, flash memory, etc. and together they form a so-called system in package (SIP). A leading approach for HI is three-dimensional-stacking, for which through-silicon-vias (TSVs) have emerged as the key technology to enable communication between components in the 3D system [269]. In TSV formation, interconnects traverse through the thickness of the substrate. Hence, TSVs allow direct signal routing between stacked chips, not only enhancing signal speed, but obviating the need for wires on the periphery of individual components or wire bonding to an interposer [270].

Through-silicon-via formation involves a deep etch into the Si substrate leaving a trench to be filled with a conducting material, usually copper. Electrical isolation of the conducting material is needed to prevent coupling with the Si substrate, especially at high device frequencies [271]. $SiO_2$ is the most common material for this purpose, although use of $Si_3N_4$ [269] has also been documented. Growth of this insulating layer, or "liner," requires a conformal deposition on the wall of the via, which can have an aspect ratio of a challenging 10:1 or even 15:1 [272]. A number of different techniques in the CVD family can be deployed for this task. Indeed, CVD technologies having demonstrated TSV liner deposition include SACVD [271], PECVD [273], LPCVD [273], ALD [274] and Hot-Filament Assisted CVD [272]. (This last technology utilizes a heated zone to volatilize incoming precursors upstream of deposition to minimize temperature impact to the wafer.) The CVD technology one chooses depends on factors such as thermal budget, step coverage and the dielectric properties of the liner.

## References

[1] S.P. Krumdieck, CVD reactors and delivery system technology, in: A.C. Jones, M.L. Hitchman (Eds.), Chemical Vapour Deposition: Precursors, Processes and Applications, RSC Publishing, London, UK, 2009, pp. 37–92.
[2] M. Ohring, Mater Sci Thin Films, Academic Press, San Diego, CA, 1992.
[3] D.M. Dobkin, M.K. Zuraw, Principles of Chemical Vapor Deposition, Kluwer Academic Publishers, Dordrecht, The Netherlands, 2003.
[4] K. Kolasinski, Surface Science Foundations of Catalysis and Nanoscience, John Wiley and Sons, West Sussex, England, 2008.
[5] J. Holm, J.T. Roberts, Sintering, coalescence and compositional changes of hydrogen-terminated silicon nanoparticles as a function of temperature, J. Phys. Chem. 113 (2009) 15955–15963.
[6] D.J. Devlin, I.O. Usov, Report on CVD Processing of Mo Tubing, LA-UR-14-26755, Los Alamos Laboratory, 2014.

[7] E. Woelk, D.V. Shenai-Khatkhate, R.L. DiCarlo, A. Amamchyan, M.B. Power, B. Lamare, J. Cryst. Growth 287 (2006) 684–687.
[8] J.D. Cressler, SiGe and Si Strained-Layer Epitaxy for Silicon Heterostructure devicesBoca Raton, FL, CRC Press, 2006.
[9] W.D. Kingery, Ceramic materials science in society, Annu. Rev. Mater. Sci. 19 (1989) 1–21.
[10] M. Allendorf, From bunsen to VLSIInterface, Spring, 1998, pp. 1–3.
[11] H.O. Pierson, Handbook of Chemical Vapor Deposition: Principles, Technology and Applications, Noyes Publications, Norwich, New York, USA, 1992.
[12] J.E. May, Kinetics of epitaxial silicon deposition by a low pressure iodide process, J. Electrochem. Soc. 112 (7) (1965) 710–713.
[13] W. O'Mara, R.B. Herring, L.P. Hunt, Handbook of Semiconductor Silicon Technology, Noyes Publications, Park Ridge, New Jersey, U.S.A, 1990.
[14] R.J. Gieske, J.J. McMullen, L.F. Donaghey, Low pressure chemical vapor deposition of polysilicon, in: Proceedings of the 6th International Conference on Chemical Vapor Deposition, Atlanta, GA, October 10–13, 1977.
[15] VLSI Research, Diffusion and Oxidation, 1991, p. 4.4.1.4.
[16] C.B. Yarling, History of industrial and commercial ion implantation 1906–1978, J. Vac. Sci. Technol. A 18 (4) (2000) 1746–1750.
[17] J. Nakos, J. Shepard, The expanding role of rapid thermal processing in CMOS manufacturing, Mater. Sci. Forum 573–574 (2008) 3–19.
[18] J.B. Guibe, J.M. Dilhac, B. Dahhou, Adaptive control of a rapid thermal processor using two long-range predictive methods, J. Process Control 2 (1) (1992) 3–8.
[19] J.P. Lu, Y. He, H. Chen, Millisecond anneal for ultra-shallow junction applications, in: International Workshop Junction Technology, 2010, pp. 1–4.
[20] A. Gat, Heat-pulse annealing of arsenic-implanted silicon with a CW arc lamp, IEEE Electron Device Lett EDL. 2 (4) (1981) 85–87.
[21] R.S. Gyurcsik, T.J. Riley, F.Y. Sorrell, A model for rapid thermal processing: achieving uniformity through lamp control, IEEE Trans. Semicond. Manuf. 4 (1) (1991) 9–13.
[22] M.C. Ozturk, D.T. Grider, J.J. Wortman, M.A. Littlejohn, Y. Zhong, D. Batchelor, Rapid thermal chemical vapor deposition of germanium on silicon and silicon dioxide and new applications of ge in ULSI technologies, J. Electron. Mater. 19 (10) (1990) 1129–1134.
[23] A.R. Londergan, G. Nuesca, C. Goldberg, G. Peterson, A.E. Kaloyeros, B. Arkles, Interlayer mediated epitaxy of cobalt silicide on silicon (100) from low temperature chemical vapor deposition of cobalt formation mechanisms and associated properties, J. Electrochem. Soc. 148 (1) (2001) C21–C27.
[24] M.C. Ozturk, F.Y. Sorrell, J.J. Wortman, F.S. Johnson, D.T. Grider, Manufacturability issues in rapid thermal chemical vapor deposition, IEEE Trans. Semicond. Manuf. 4 (2) (1991) 155–165.
[25] R.K. Waits, Evolution of integrated circuit vacuum process: 1959–1975, J. Vac. Sci. Technol. A 18 (4) (2000) 1736–1745.
[26] T. Gocho, Y. Morita, J. Sato, Trench isolation technology for 0.35-μm device by bias ECR CVD, in: Symposium on VLSI Technology, 1991, pp. 87–88.
[27] S. Krishnan, S. Nag, Assessment of Charge-Induced Damage from High Density Plasma (HDP) Oxide Deposition 1st International Symposium on Plasma Process-Induced Damage, 1996, pp. 67–70.
[28] K. Koyanagi, K. Kishimoto, T.-C. Huo, A. Matsumoto, N. Okada, N. Sumihiro, Stability and application to multilevel metallization of fluorine-doped silicon oxide by high-density plasma chemical vapor deposition, Jpn. J. Appl. Phys. 39 (2000) 1091–1097.

[29] J. Lee, H. Kim, H. Kim, S. Kim, K. Lee, S. Lee, Reduction of plasma-induced damage during HDP-CVD oxide deposition in the inter layer dielectric (ILD) process, Microelectron. Eng. 88 (8) (2011) 2489−2491.
[30] K. Suemitsu, Y. Kawano, H. Utsumi, H. Honjo, R. Nebashi, S. Saito, Improvement of thermal stability of magnetoresistive random access memory device with SiN protective film deposited by high-density plasma chemical vapor deposition, Jpn. J. Appl. Phys. 47 (4) (2008) 2714−2718.
[31] B.S. Meyerson, Low-temperature silicon epitaxy by ultrahigh vacuum/chemical vapor deposition, Appl. Phys. Lett. 48 (12) (1986) 797−799.
[32] B.S. Meyerson, UHV/CVD growth of Si and Si:Ge alloys: chemistry, physics, and device applications, Proc. IEEE 80 (10) (1992) 1592−1608.
[33] F.W. Smith, G. Ghidini, Reaction of oxygen with Si(1 1 1) and (1 0 0): critical conditions for Growth of $SiO_2$, J. Electrochem. Soc. 129 (6) (1982) 1300−1306.
[34] D.L. Harame, B.S. Meyerson, The early history of IBM's SiGe mixed signal technology, IEEE Trans. Electron. Dev. 48 (11) (2001) 2555−2567.
[35] T.N. Adam, S. Bedell, A. Reznicek, D.K. Sadana, R.J. Murphy, A. Venkateshan, Low-temperature epitaxial Si, SiGe, and SiC in a 300 mm UHV/CVD reactor, ECS Trans. 33 (6) (2010) 149−154.
[36] C. Rosenblad, T. Graf, J. Stangl, Y. Zhuang, G. Bauer, J. Schulze, Epitaxial growth at high rates with LEPCVD, Thin Solid Films 336 (1998) 89−91.
[37] J.B. Gunn, Microwave oscillations of current in III-V semiconductors, Solid State Commun. 1 (4) (1963) 88−91.
[38] R.N. Hall, G.E. Fenner, J.D. Kingsley, T.J. Soltys, R.O. Carlson, Coherent light emission from GaAs junctions, Phys. Rev. Lett. 9 (9) (1962) 366−368.
[39] G.W. Turner, H.K. Choi, B.-Y. Tsaur, Microwave MESFET's fabricated in GaAs layers grown on SOS Substrates, IEEE Electron Device Lett EDL. 8 (10) (1987) 460−462.
[40] H.M. Manasevit, A.C. Thorsen, Heteroepitaxial GaAs on aluminum oxide I: early growth studies, Metall. Trans. A 1 (1970) 623−628.
[41] J.-H. Ryou, R. Kanjolia, R.D. Dupuis, CVD of III-V compound semiconductors, in: A.C. Jones, M.L. Hitchman (Eds.), Chemical Vapour Deposition: Precursors, Processes and Application, RSC Publishing, London, UK, 2009, pp. 272−319.
[42] J.E. May, Kinetics of epitaxial silicon deposition by a low pressure iodide process, J. Electrochem. Soc. 12 (7) (1965) 710−713.
[43] C.H. Fa, T.T. Jew, The poly-silicon insulated-gate field-effect transistor, IEEE Trans. Electron. Dev. 13 (2) (1966) 290−291.
[44] E. Kooi, A. Schmitz, Springer, Germany, 2005.
[45] J. Peters, U.S. Patent 4,419,385, which references A. Amick, G.L. Shnable, J.L. Vossen, Deposition techniques for dielectric films on semiconductor devices, J. Vac. Sci. Technol. 14 (5) (1977) 1053−1063.
[46] R. Rung, H. Momose, Y. Nagakubo, Deep trench isolated CMOS devices, in: IEDM Technical Digest, 1982, pp. 237−240.
[47] T. Moriya, S. Shima, Y. Hazuki, M. Chiba, M. Kashiwagi, A planar metallization process—Its application to trilevel aluminum interconnection, in: International Electron Devices Meeting, 1983, pp. 550−553.
[48] T.R. Yew, O. Kenneth, R. Reif, Erratum: silicon epitaxial growth on (100) patterned oxide wafers at 800°C by ultralow-pressure chemical vapor deposition, Appl. Phys. Lett. 52 (24) (1988) 2061−2063.
[49] N. Yokoyama, K. Hinode, Y. Homma, LPCVD TiN as barrier layer in VLSI, J. Electrochem. Soc. 136 (3) (1989) 882−883.
[50] T. Usami, K. Shimokawa, M. Yoshimaru, Low dielectric constant interlayer using fluorine-doped silicon oxide, Jpn. J. Appl. Phys. 33 (1994) 408−412.

[51] A. Nara, H. Itoh, Low dielectric constant insulator formed by downstream plasma CVD at room temperature using TMSiO$_2$, Jpn. J. Appl. Phys. 36 (1997) 1477–1480.
[52] J. Yota, J. Hander, A.A. Saleh, A comparative study on inductively-coupled plasma high-density plasma, plasma-enhanced, and low pressure chemical vapor deposition silicon nitride films, J. Vac. Sci. Technol. A 18 (2000) 372–376.
[53] S. Gannavaram, N. Pesovic, C. Ozturk, Low temperature ($\leq$800°C) recessed junctionselective silicon–germanium source/drain technology for sub-70 nm CMOS, in: IEDM Technical Digest International, 2000, pp. 437–440.
[54] S. Ito, H. Namba, K. Yamaguchi, T. Hirata, K. Ando, S. Koyama, Mechanical stress effect of etch-stop nitride and its impact on deep submicron transistor design, in: IEDM Technical Digest, 2000, pp. 247–250.
[55] S.J. Lee, T.S. Jeon, D.L. Kwong, R. Clark, Hafnium oxide gate stack prepared by in situ rapid thermal chemical vapor deposition process for advanced gate dielectrics, J. Appl. Phys. 92 (2002) 2807–2809.
[56] R.V. Joshi, A new damascene structure for submicrometer interconnect wiring, IEEE Electron. Device Lett. 14 (3) (1993) 129–132.
[57] M.A.M. Ibrahim, S.S. Abd El Rehim, S.O. Moussa, Electrodeposition of non-crystalline cobalt tungsten alloys from citrate electrolytes, J. Appl. Electrochem. 33 (2003) 627–633.
[58] T. Tsutsumi, H. Kotani, J. Komori, S. Nagao, A selective LPCVD tungsten process using silane reduction for VLSI appications, IEEE Trans. Electron. Dev. 37 (3) (1990) 569–576.
[59] J.K. Lan, Y.L. Wang, K.Y. Lo, C.P. Liu, C.W. Liu, J.K. Wang, Integration of MOCVD titanium nitride with collimated titanium and ion metal plasma titanium for 0.18-μm logic process, Thin Solid Films 398–399 (2001) 544–548.
[60] A.C. Westerheim, J.M. Bulger, C.S. Whelan, T.S. Sriram, L.J. Elliott, J.J. Maziarz, Integration of chemical vapor deposition titanium nitride for 0.25 μm contacts and vias, J. Vac. Sci. Technol. B 16 (5) (1998) 2729–2733.
[61] T. Morimoto, T. Ohguro, S. Momose, T. Iinuma, I. Kunishima, K. Suguro, Self-aligned nickel-mono silicide technology for high-speed deep submicrometer logic CMOS ULSI, IEEE Trans. Electron. Dev. 42 (5) (1995) 915–922.
[62] H. Iwai, T. Ohguro, S. Ohmi, NiSi sailicide technology for scaled CMOS, Microelectron. Eng. 60 (2002) 157–169.
[63] R.T.P. Lee, L.-T. Yang, T.-Y. Liow, K.-M. Tan, A.E.-J. Lim, K.-W. Ang, Nickel-silicide: carbon contact technology for N-channel MOSFETs with silicon-carbon source/drain, IEEE Electron. Device Lett. 29 (1) (2008) 89–92.
[64] Z. LiR, R.G. Gordon, V. Pallem, H. Li, D.V. Shenai, Direct-liquid-injection chemical vapor deposition of nickel nitride films and their reduction to nickel films, Chem. Mater. 22 (2010) 3060–3066.
[65] W.L. Tan, K.L. Pey, S.Y.M. Chooi, J.H. Ye, T. Osipowicz, Effect of a titanium cap in reducing interfacial oxides in the formation of nickel silicide, J. Appl. Phys. 91 (5) (2002) 2901–2909.
[66] G.C. D'Couto, G. Tkach, K.A. Ashtiani, L. Hartsough, E. Kim, R. Mulpuri, In situ physical vapor deposition of ionized Ti and TiN thin films using hollow cathode magnetron plasma source, J. Vac. Sci. Technol. B 19 (1) (2001) 244–249.
[67] S. Panda, J. Kim, B.H. Weiller, D.J. Economou, D.M. Hoffman, Low temperature chemical vapor deposition of titanium nitride films from tetrakis (ethylmethylamido) titanium and ammonia, Thin Solid Films 357 (1999) 125–131.
[68] J. Zhao, E.G. Garza, K. Lam, C.M. Jones, Comparison study of physical vapor-deposited and chemical vapor-deposited titanium nitride thin films using X-ray photoelectron spectroscopy, Appl. Surf. Sci. 158 (2000) 246–251.

[69] J.N. Musher, R.G. Gordon, Atmospheric pressure chemical vapor deposition of titanium nitride from tetrakis (diethylamido) titanium and Ammonia, J. Electrochem. Soc. 143 (2) (1996) 736—744.
[70] T. Luoh, C.-T. Su, T.-H. Yang, K.-C. Chen, C.-Y. Lu, Advanced tungsten plug process for beyond nanometer technology, Microelectron. Eng. 85 (2008) 1739—1747.
[71] R.G. Gordon, J. Barton, S. Suh, Chemical Vapor Deposition (CVD) of tungsten nitride for copper diffusion barriers, Mater. Res. Soc. Symp. Proc. 714E (2001) L8.10.1—L8.10.6.
[72] M.H. Tsai, S.C. Sun, H.T. Chiu, C.E. Tsai, S.H. Chuang, Metal organic chemical vapor deposition of tantalum nitride by tertbuylimidotris (diethylamido) tantalum for advanced metallization, Appl. Phys. Lett. 67 (8) (1995) 1128—1130.
[73] T. Chen, C. Xu, T.H. Baum, G.T. Stauf, J.F. Roeder, A.G. DiPasquale, New tantalum amido complexes with chelate ligands as metalorganic (MO) precursors for chemical vapor deposition (CVD) of tantalum nitride thin films, Chem. Mater. 22 (2010) 27—35.
[74] Z. Li, R.G. Gordon, D.B. Farmer, Y. Lin, J. Vlassak, Nucleation and adhesion of ALD copper on cobalt adhesion layers and tungsten nitride diffusion barriers, Electrochem. Solid State Lett. 8 (7) (2005) G182—G185.
[75] C. Zhao, et al., Failure mechanisms of PVD Ta and ALD TaN barrier layers for Cu contact applications, Microelectron. Eng. 84 (11) (November 2007) 2669—2674.
[76] J. Chae, H.-S. Park, S. Kang, Atomic layer deposition of nickel by the reduction of preformed nickel oxide, Electrochem. Solid State Lett. 5 (6) (2002) C64—C66.
[77] P.C. Andricacos, C. Uzoh, J.O. Dukovic, J. Horkans, H. Deligianni, Damascene copper electroplating for chip interconnections, IBM J. Res. Dev. 42 (5) (1998) 567—574.
[78] S.-C. Chang, J.-M. Shieh, B.-T. Dai, M.-S. Feng, Y.-H. Li, The effect of plating current densities on self-annealing behaviors of electroplated copper films, J. Electrochem. Soc. 149 (9) (2002) G535—G538.
[79] C. Ryu, K.-W. Kwon, A.L.S. Loke, V.M. Dubin, R.A. Kavari, G.W. Ray, et al., Electromigration of submicron Damascene copper interconnects, in: Symposium on VLSI Technology Digest of Technical Papers, 1998, pp. 156—157.
[80] W. Ruythooren, K. Attenborough, S. Beerten, P. Merken, J. Fransaer, E. Beyne, Electrodeposition for the synthesis of microsystems, J. Micromech. Microeng. 10 (2000) 101—107.
[81] K. Barmak, C. Cabral, K.P. Rodbell, H.M.E. Harper, On the use of alloying elements for Cu interconnect applications, J. Vac. Sci. Technol. B 24 (2006) 2485—2498.
[82] Z. Chen, K. Prasad, C.Y. Li, S.S. Su, D. Gui, P.W. Lu, Characterization and performance of dielectric diffusion barriers for Cu metallization, Thin Solid Films 462—463 (2004) 223—226.
[83] Y.H. Wang, M.R. Moitreyee, R. Kumar, L. Shen, K.Y. Zeng, J.W. Chai, A comparative study of low dielectric constant barrier layer, etch stop and hardmask films of hydrogenated amorphousSi-(C,O, N), Thin Solid Films 1—2 (460) (2004) 211—216.
[84] Y.C. Yeo, Enhancing CMOS transistor performance using lattice-mismatched materials in source/drain regions, Semicond. Sci. Technol. 22 (2007) S177—S182.
[85] S. Gannavaram, Electron Devices Meeting, IEDM Technical Digest International, 2000, pp. 437—440.
[86] T. Ghani, M. Armstrong, C. Auth, M. Bost, P. Charvat, G. Glass, et al., A 90nm high volume manufacturing logic technology featuring novel 45nm gate length strained

silicon CMOS transistors, in: Electron Devices Meeting IEDM '03 Technical Digest, 2003, pp. 11.6.1–11.6.3.
[87] S. Ito, H. Namba, T. Hirata, T. Hirata, K. Ando, S. Koyama, Effect of mechanical stress induced by etch-stop nitride: impact on deep-submicron transistor performance, Microelectron. Reliab. 42 (2) (2002) 201–209.
[88] S.E. Thompson, M. Armstrong, C. Auth, M. Alavi, M. Buehler, R. Chau, A 90-nm logic technology featuring strained-silicon, IEEE Trans. Electron. Dev. 51 (11) (2004) 1790–1797.
[89] J. Kavelieros, B. Doyle, S. Datta, G. Dewey, M. Doczy, B. Jin, et al., Tri-gate transistor architecture with high-k Gate dielectric, in: Metal Gates and Strain Engineering, VLSI Technology, Digest of Technical Papers, 2006, pp. 50–51.
[90] J. Tian, B. Zuo, W. Lu, M. Zhou, L.C. Hsia, Stress modulation of silicon nitride film by initial deposition conditions for transistor carrier mobility enhancement, Jpn. J. Appl. Phys. 49 (2010).
[91] S.E. Thompson, P. Packan, M. Bohr, MOS transistors: scaling challenges for the 21$^{st}$ century, Intel Technol. J. Q3 (1998) 1–19.
[92] K. Mistry, C. Allen, C. Auth, B. Beattie, D. Bergstrom, M. Bost, A 45nm logic technology with high-k + Metal Gate transistors, strained silicon, 9 Cu interconnect layers, 193nm dry patterning, and 100% Pb-free packaging, in: Electron Devices Meeting, IEDM, 2007, pp. 247–250.
[93] M.T. Bohr, R.S. Chau, T. Ghani, K. Mistry, The high-k solution, IEEE Spectrum 44 (10) (2007) 29–35.
[94] J. Pan, C. Woo, C.-Y. Yang, U. Bhandary, S. Guggilla, N. Krishna, Replacement metal-gate NMOSFETs with ALD TaN/EP-Cu, PVD Ta, and PVD TaN electrode, IEEE Electron. Device Lett. 24 (5) (2003) 304–305.
[95] J.E. Jang, S.N. Cha, Y.J. Choi, D.J. Kang, T.P. Butler, D.G. Hasko, Nanoscale memory cell based on a nanoelectromechanical switched capacitor, Nat. Nanotechnol. 3 (2008) 26–30.
[96] M. McCoy, Forging the way to high-k dielectrics, Chem. Eng. News 83 (26) (2005) 26–29.
[97] D. Scansen, DRAM Gets More Exotic, EE Times, 2008.
[98] S. Franssila, Introduction to Microfabrication, John Wiley and Sons, West Sussex, United Kingdom, 2010.
[99] M. Verghese, J.W. Maes, N. Kobayashi, Atomic layer deposition goes mainstream in 22 nm logic technologies, Solid State Technol. 53 (10) (2010) 18–21.
[100] S. Takehiro, M. Sakuraba, T. Tsuchiya, J. Murota, High Ge fraction intrinsic SiGe-heterochannel MOSFETs with embedded SiGe source/drain electrode formed by in-situ doped selective CVD epitaxial growth, Thin Solid Films 517 (1) (2008) 346–349.
[101] R. Pillarisetty, B. Chu-Kung, S. Corcoran, G. Dewey, J. Kavelieros, H. Kennel, High mobility strained germanium quantum well field effect transistor as the p-channel device option for low power (Vcc = 0.5V) III–V CMOS architecture, in: IEEE International Electron Devices Meeting, 2010, pp. 6.7.1–6.7.4.
[102] V.A. Andreev, E.M. Freer, J.M. de Larios, J.M. Prausnitz, C.J. Radke, Silicon-wafer cleaning with aqueous surfactant-stabilized gas/solids suspensions, J. Electrochem. Soc. 158 (1) (2011) H55–H62.
[103] A. Carlson, T. Le, Correlation of wafer backside defects to photolithography hot spots using advanced macro inspection, in: 31$^{st}$ International Symposium, Microlithography, 2006.
[104] G. Cunge, B. Pelissier, O. Joubert, R. Ramos, C. Maurice, New chamber walls conditioning and cleaning strategies to improve the stability of plasma processes, Plasma Sources Sci. Technol. 14 (2005) 599–609.

[105] B. Li, T.D. Sullivan, T.C. Lee, D. Badami, Reliability challenges for copper interconnects, Microelectron. Reliab. 44 (2004) 365−380.
[106] H.C. Chung, C.P. Liu, Effect of crystallinity and preferred orientation of Ta$_2$N films on diffusion barrier properties for copper metallization, Surf. Coat. Technol. 200 (2006) 3122−3126.
[107] B.M. Wang, Y.S. Wu, Using phosphorus-doped α-si gettering layers to improve NILC poly-Si TFT performance, J. Electron. Mater. 39 (2) (2010) 157−161.
[108] A.A. Istratov, Iron contamination in silicon technology, Appl. Phys. A 70 (2000) 489−534.
[109] M.A. Cappelletti, Theoretical study of neutron effects on PIN photodiodes with deep-trap levels, Semicond. Sci. Technol. 24 (2009).
[110] C. Sparks, J. Barnett, D.K. Michelson, C. Gondran, S.-C. Song, A. Martinez, Advanced TXRF analysis: background reduction when measuring high-k materials and mapping metallic contamination, Solid State Phenom. 134 (2008) 285−288.
[111] A. Daniel, N. Cabuil, T. Lardin, D. Despois, M. Veillerot, C. Geoffroy, Comparison of direct-total-reflection X-ray fluorescence, sweeping-total-reflection X-ray fluorescence and vapor phase decomposition-total-reflection X-ray fluorescence applied to the characterization of metallic contamination on semiconductor wafers, Spectrochim. Acta B. 63 (12) (2008) 1375−1381.
[112] W.Y. Chou, B.-Y. Tsui, C.-W. Kuo, T.-K. Kang, Optimization of back side cleaning process to eliminate copper contamination, J. Electrochem. Soc. 152 (2) (2005) G131−G137.
[113] International Technology Roadmap for Semiconductors, Front End Processes, 2001, 2001, p. 8.
[114] D.R. Williams, J. Porter, G. Yoon, A. Guirao, H. Hofer, L. Chen, How far can we extend the limits of human vision? in: R.R. Krueger, R.A. Applegate (Eds.), Wavefront Customized Visual Corrections: The Quest for Super Vision II SLACK Incorporated, Thorofare, NJ, 2004, p. 22.
[115] G. Dhanaraj, Y. Chen, H. Chen, D. Cai, H. Zhang, M. Dudley, Chemical vapor deposition of silicon carbide epitaxial films and their defect characterization, J. Electron. Mater. 36 (4) (2007) 332−339.
[116] Z.C. Feng, H.C. Lin, J. Zhao, T.R. Yang, I. Ferguson, Surface and optical properties of AlGaInP films grown on GaAs by metalorganic chemical vapor deposition, Thin Solid Films 498 (1−2) (2006) 167−173.
[117] H.F. Liu, W. Liu, S.J. Chua, Epitaxial growth and chical lift-off of GaInN/GaN heterostructures on c- and r-sapphire substrates employing ZnO sacrificial templates, J. Vac. Sci. Technol. A 28 (2010) 590−594.
[118] S.E. Saddow, T.E. Schattner, J. Brown, L. Grazulis, K. Mahalingam, G. Landis, Effects of substrate surface preparation on chemical vapor deposition growth of 4H-SiC epitaxial layers, J. Electron. Mater. 30 (3) (2001) 228−234.
[119] M. Kopycinska-Muller, R.H. Geiss, D.C. Hurley, Contact mechanics and tip shape in AFM-based nanomechanical measurements, Ultramicroscopy 106 (2006) 466−474.
[120] F. Giessibl, Atomic resolution of the silicon (111)-(7X7) surface by atomic force microscopy, Science 267 (5194) (1995) 68−71.
[121] B. Liu, Y.W. Lu, G.R. Jin, Y. Zhao, X.L. Wang, Q.S. Zhu, Surface roughness scattering in two dimensional electron gas channel, Appl. Phys. Lett. (2010) 97.
[122] J.C. Moore, J.E. Ortiz, J. Xie, H. Morkoç, A.A. Baski, Study of leakage defects on GaN films by conductive atomic force microscopy, J. Phys. Conf. Ser. 61 (2007) 90−94.
[123] K. Gradkowski, T.C. Sadler, L.O. Mereni, V. Dimastrodonato, P.J. Parbrook, G. Huyet, Crystal defect topography of Stranski−Krastanow quantum dots by atomic force microscopy, Appl. Phys. Lett. 97 (2010) 191106.

[124] D.B. Williams, C.B. Carter, The Transmission Electron Microscope, Springer, New York, NY, 2009, p. 6.
[125] G.F. Iriarte, Using transmission electron microscopy (TEM) for chemical analysis of semiconductors, Microscopy: Science, Technology, Applications and Education (2010) 1888–1896.
[126] K. Tao, D. Zhang, J. Zhao, L. Wang, H. Cai, Y. Sun, Low temperature deposition of boron-doped microcrystalline Si:H thin film and its application in silicon based thin film solar cells, J. Non-Cryst. Solids 356 (2010) 299–303.
[127] H. Nitta, J. Tanabe, M. Sakuraba, J. Murota, Carbon effect on strain compensation in $Si_{1-x-y}Ge_xC_y$ films epitaxially grown on Si(100), Thin Solid Films 508 (2006) 140–142.
[128] M.I. Richard, M.J. Highland, T.T. Fister, A. Munkholm, J. Mei, S.K. Streiffer, In situ synchrotron x-raystudies of strain and composition evolution during metal-organic chemical vapor deposition of InGaN, Appl. Phys. Lett. 96 (2010).
[129] J.H. Kim, V.A. Ignatova, J. Heitmann, L. Oberbeck, Deposition temperature effect on electrical properties and interface of high-k $ZrO_2$ capacitor, J. Phys. D Appl. Phys. 41 (2008).
[130] G.C. Smith, A.K. Livesay, Maximum entropy: a new approach to non-destructive deconvolution of depth profiles from angle-dependent XPS, Surf. Interface Anal. 19 (1–12) (1992) 175–180.
[131] P.A.W. van der Heide, X-Ray Photoelectron Spectroscopy: An Introduction to Principles and Practices, John Wiley and Sons, Inc., Hoboken, NJ, 2012.
[132] K. Norrman, S. Cros, R. de Bettignies, M. Firon, F.C. Krebs, Lifetime and Stability Studies Polymer Photovoltaics, SPIE, Bellingham, WA, 2008.
[133] K. Xi, H. He, D. Xu, R. Ge, Z. Meng, X. Jia, Ultra low dielectric constant polysilsesquioxane films using $T_8(Me_4NO)_8$ as porogen, Thin Solid Films 518 (17) (2010) 4768–4772.
[134] A. Yeoh, M. Chang, C. Pelto, T.-L. Huang, S. Balakrishnan, G. Leatherman, Copper die bumps (first level interconnect) and low-K dielectrics in 65nm high volume manufacturing, in: Electronic Components and Technology Conference, 2006, pp. 1611–1615.
[135] E. Andideh, T. Scherban, B. Sun, J. Blaine, C. Block, et al., Interfacial adhesion of copper-low k interconnects, in: Proceedings of the IEEE 2001 International Interconnect Technology Conference, 2001, pp. 257–259.
[136] K. Maex, M.R. Backlanov, D. Shamiryan, F. Iacopi, S.H. Brongersma, Z.S. Yanovitskaya, Low dielectric constant materials for microelectronics, J. Appl. Phys. 93 (11) (2003) 8793–8839.
[137] H.C. Lin, W.-E. Wang, G. Brammertz, M. Meuris, M. Heyns, Electrical study of sulfur passivated $In_{0.53}Ga_{0.47}As$ MOS capacitor and transistor with ALD $Al_2O_3$ as gate insulator, Microelectron. Eng. 86 (7–9) (2009) 1554–1557.
[138] P.D. Ye, B. Yang, K.K. Ng, J. Bude, G.D. Wilk, S. Halder, GaN metal-oxide-semiconductor high-electron-mobility-transistor with atomic layer deposited $Al_2O_3$ as gate dielectric, Appl. Phys. Lett. 86 (2005) 063501.
[139] S. Dhar, S. Haney, L. Cheng, S.-R. Ryu, A.K. Agarwal, L.C. Yu, Inversion layer carrier concentration and mobility in 4H–SiC metal-oxide-semiconductor field-effect transistors, J. Appl. Phys. 108 (2010) 054509.
[140] V. Tilak, K. Matocha, G. Dunne, F. Allerstam, E.O. Sveinbjornsson, Trap and inversion layer mobility characterization using hall effect in silicon carbide-based MOSFETs with gate oxides grown by sodium enhanced oxidation, IEEE Trans. Electron. Dev. 56 (2) (2009) 162–169.
[141] S.K. Tewksbury, Semiconductor materials, in: J.C. Whitaker (Ed.), The Electronics Handbook, CRC Press, Salem, MA, 1996, p. 119.

[142] L.P. Nguyen, C. Fenouillet-Beranger, G. Ghibaudo, T. Skotnicki, S. Cristoloveanu, Mobility enhancement by CESL strain in short-channel ultrathin SOI MOSFETs, Solid State Electron. 54 (2) (2010) 123–130.
[143] S.H. Kim, K.T. Kim, S.-S. Kim, D.-J. Lee, K.-S. Kim, H.-M. Kim, A bilayer diffusion barrier of ALD-Ru/ALD-TaCN for direct plating of Cu, J. Electrochem. Soc. 155 (8) (2008) H589–H594.
[144] C.K. Hu, L. Gignac, E. Liniger, S. Grunow, J.J. Demarest, B. Redder, Comparison of electromigration in Cu interconnects with atomic-layer- or physical-vapor-deposited TaN liners, J. Electrochem. Soc. 154 (9) (2007) H755–H758.
[145] M.L. Green, M.-Y. Ho, B. Busch, G.D. Wilk, T. Sorsch, T. Conard, Nucleation and growth of atomic layer deposited HfO2 gate dielectric layers on chemical oxide (Si–O–H) and thermal oxide ($SiO_2$ or Si–O–N) underlayers, J. Appl. Phys. 92 (12) (2002) 7168–7174.
[146] S.M. George, B. Yoon, A.A. Dameron, Surface chemistry for molecular layer deposition of organic and hybrid organic–inorganic polymers, Chem. Rev. 110 (2010) 111–131.
[147] J.W. Elam, D. Routkevitch, P.P. Mardilovich, S.M. George, Conformal coating on ultrahigh-aspect-ratio nanopores of anodic Alumina by Atomic layer deposition, Chem. Mater. 15 (18) (2003) 3507–3517.
[148] M.Y. Ho, H. Gong, G.D. Wilk, B.W. Busch, M.L. Green, P.M. Voyles, Morphology and crystallization kinetics in $HfO_2$ thin films grown by atomic layer deposition, J. Appl. Phys. 93 (2003) 1477–1481.
[149] Y. Okuyama, C. Barelli, C. Tousseau, S. Park, Y. Senzaki, Batch process for atomic layer deposition of hafnium silicate thin films on 300-mm-diameter silicon substrates, J. Vac. Sci. Technol. A 23 (3) (2005) L1–L3.
[150] S.M. Rossnagel, A. Sherman, F. Turner, Plasma-enhanced atomic layer deposition of Ta and Ti for interconnect diffusion barriers, J. Vac. Sci. Technol. B 18 (4) (2000) 2016–2020.
[151] T.O. Kaariainen, S. Lehti, M.-L. Kaariainen, D.C. Cameron, Surface modification of polymers by plasma-assisted atomic layer deposition, Surf. Coat. Technol. 205 (Suppl. 2) (2011) S475–S479.
[152] A.K. Roy, W.A. Goedel, Control of thickness and morphology of thin alumina films deposited via pulsed chemical vapor deposition (pulsed CVD) through variation of purge times, Surf. Coat. Technol. 205 (2011) 4177–4182.
[153] I.A. Shareef, G.W. Rubloff, M. Anderle, W.N. Gill, J. Cotte, D.H. Kim, Subatmospheric chemical vapor deposition ozone/TEOS process for $SiO_2$ trench filling, J. Vac. Sci. Technol. B 13 (4) (1995) 1888–1892.
[154] S. Mani, T.M. Saif, Mechanism of controlled crack formation in thin-film dielectrics, Appl. Phys. Lett. (2005) 86.
[155] M. Yin, L. Zhao, X. Xu, W. Wang, Atmospheric pressure plasma enhanced chemical vapor deposition of borophosphosilicate glass films, Jpn. J. Appl. Phys. 47 (4) (2008) 1735–1739.
[156] J.K. Lan, Y.-L. Wang, C.G. Chao, K. Lo, Y.L. Cheng, Effect of substrate on the step coverage of plasma-enhanced chemical-vapor deposited tetraethylorthosilicate films, J. Vac. Sci. Technol. B21 (2003) 1224–1229.
[157] C. Leung, E. Ong, Silicon-based dielectrics, in: S.P. Murarka, M. Eizenberg, A.K. Sinha (Eds.), Interlayer Dielectrics for Semiconductor Technologies, Elsevier, London, UK, 2003, p. 124.
[158] A.C. Adams, C.D. Capio, The deposition of silicon dioxide films at reduced pressure, J. Electrochem. Soc. 126 (1979) 1042–1046.

[159] S. Nguyen, D. Dobuzinsky, D. Harmon, R. Gleason, S. Fridman, Reaction mechanisms of plasma- and thermal assisted chemical vapor deposition of tetraethylorthosilicate oxide, J. Electrochem. Soc. 137 (7) (1990) 2209—2215.
[160] C.S. Tan, R. Reif, Silicon multilayer stacking based on copper wafer bonding, Electrochem. Solid State Lett. 8 (1) (2005) G1—G4.
[161] C. Chang, T. Abe, M. Esashi, Trench filling characteristics of low stress TEOS/ozone oxide deposited by PECVD and SACVD, Microsyst. Technol. 10 (2004) 97—102.
[162] W.N. Gill, S. Ganguli, Gas phase and surface reactions in subatmospheric chemical vapor deposition of tetraethylorthosilicate-ozone, J. Vac. Sci. Technol. B 15 (4) (1997) 948—954.
[163] M. Matsuura, Y. Hayashide, H. Kotani, H. Abe, Film characteristics of APCVD Oxide using organic silicon and ozone, Jpn. J. Appl. Phys. 30 (7) (1991) 1530—1538.
[164] H. Kikuchi, Y. Yamada, A.M. Ali, J. Liang, T. Fukushima, T. Tanaka, Tungsten through-silicon via technology for three-dimensional LSIs, Jpn. J. Appl. Phys. 47 (4) (2008) 2801—2806.
[165] W. Kern, G.L. Schnable, Low-pressure chemical vapor deposition for very large-scale integration processing—a review, IEEE Trans. Electron. Dev. ED-26 (4) (1979) 647—657.
[166] M. Racenelli, D.W. Greve, Low-temperature selective epitaxy by ultrahigh-vacuum chemical vapor deposition from SiH4 and GeH4/H2, Appl. Phys. Lett. 58 (19) (1991) 2096—2098.
[167] M. Hammond, Silicon epitaxy by chemical vapor deposition, in: K. Seshan (Ed.), Handbook of Thin Film Deposition, Deposition Processes and Techniques, Noyes publications. William Andrew Publishing, Norwich, NY, USA, 2002, pp. 45—110.
[168] T.O. Sedgwick, J.N. Burghartz, D.A. Grutzmacher, Low temperature pressure chemical vapor deposition for epitaxial growth of SiGe bipolar transistors, semiconductor Silicon 1994, Proc Seventh Int Symp Silicon Mater Sci Technol (1994) 298.
[169] T. Ohnishi, Y. Kirihata, H. Ohmi, H. Kakiuchi, K. Yasutake, In situ doped si selective epitaxial growth at low temperatures by atmospheric pressure plasma CVD, ECS Trans. 25 (8) (2009) 309—315.
[170] M. Madou, Fundamentals of Microfabrication: The Science of Miniaturization, second ed., CRC Press, Boca Raton, FL, 2002, p. 150.
[171] H. Fritzsche, C.C. Tsai, Porosity and oxidation of amorphous silicon films prepared by evaporation, sputtering and plasma-deposition, Sol. Energy Mater. 1 (5—6) (1979) 471—479.
[172] X.L. Nguyen, T.N.N. Nguyen, B.T. Chau, M.C. Dang, The fabrication of GaN-based light emitting diodes(LEDs), Adv. Nat. Sci. Nanosci. Nanotechnol. (2010) 1.
[173] D.A. Neumayer, J.G. Ekerdt, Growth of group III nitrides. A review of precursors and techniques, Chem. Mater. 8 (1996) 9—25.
[174] P.J. Wright, M.J. Crosbie, P.A. Lane, D.J. Lane, A.C. Jones, D.J. Williams, Metal organic chemical vapor deposition (MOCVD) of oxides and ferroelectric materials, J. Mater. Sci. Mater. Electron. 13 (2002) 671—678.
[175] S.O. Kasap, P. Capper, Springer Handbook of Electronic and Photonic Materials, Springer, New York, NY, 2006.
[176] W. Richter, Physics of metal organic chemical vapor deposition, Adv. Solid State Phys. 26 (1986) (1986) 335—359.
[177] H.M. Manasevit, The beginnings of metalorganic chemical vapor deposition (MOCVD), in: R. Feigelson (Ed.), 50 Years Progress in Crystal Growth: A Reprint Collection, Elsevier B.V., Amsterdam, The Netherlands, 2004, pp. 217—220.
[178] H.M. Manasevit, W.I. Simpson, The use of metal-organics in the preparation of semiconductor materials, J. Electrochem. Soc. 116 (1969) 1725—1732.

[179] A. Saxler, D. Walker, P. Kung, X. Zhang, M. Razeghi, J. Solomon, Comparison of trimethylgallium andtriethylgallium for the growth of GaN, Appl. Phys. Lett. 71 (22) (1997) 3272−3274.
[180] C.R. Abernathy, W.S. Hobson, Carbon-impurity incorporation during the growth of epitaxial group III-V materials, J. Mater. Sci. Mater. Electron. 7 (1996) 1−21.
[181] J. Derluyn, K. Dessein, G. Flamand, Y. Mols, J. Poortmans, G. Borghs, Comparison of MOVPE grown GaAs solar cells using different substrates and group-V precursors, J. Cryst. Growth 247 (2003) 237−244.
[182] S.P. Denbaars, A.L. Holmes, M.E. Heimbuch, Compressively strained 1.55-um InxGa1-xAsyP1-y/InP quantum well laser diodes grown by MOCVD with tertiarybutylarsine (TBA) and tertiarybutylphosphine (TBP), SPIE 2148 (1994) 179−188.
[183] A. Moto, S. Tanaka, T. Tanabe, S. Takagishi, GaInP/GaAs and mechanically stacked GaInAs solarcells grown by MOCVD using TBAs and TBP as V-precursors, Sol. Energy Mater. Sol. Cells 66 (2001) 585−592.
[184] G. Chen, G. Chen, D. Cheng, R.F. Hicks, A.M. Noori, S.L. Hayashi, Metalorganic vapor-phase epitaxyof III/V phosphides with tertiarybutylphosphine and tertiarybutylarsine, J. Cryst. Growth 270 (2004) 322−328.
[185] T.C. Hsu, Y. Hsu, G.B. Stringfellow, Effect of P precursor on surface structure and ordering in GaInP, J. Cryst. Growth 193 (1998) 1−8.
[186] M.G. Arellano, R.C. Ojeda, R.P. Sierra, S.M. Moreno, Growth of $Al_xGa_{1-x}As$/GaAs structures for single quantum wells by solidarsenic MOCVD system, Rev. Mexic. Fisica 53 (6) (2007) 441−446.
[187] J. Diaz-Reyes, M. Galvan-Arellano, R.S. Castillo-Ojedo, R. Pena-Sierra, Characterization of $Al_xGa_{1-x}As$ layers grown on (100) GaAs by metallic-arsenic-based-MOCVD, Vacuum 84 (2010) 1182−1186.
[188] D.P. Norman, L.W. Tu, S.Y. Chiang, P.H. Tseng, P. Wadekar, S. Hamad, Effect of temperature and V/III ratio on the initial growth of indium nitride using plasma-assisted metal-organic chemical vapor deposition, J. Appl. Phys. 109 (2011).
[189] A. Gupta, D. Paramanik, S. Varma, C. Jacob, CVD growth and characterization of 3C-SiC thin films, Bull. Mater. Sci. 27 (5) (2004) 445−451.
[190] R.M. Guerrero, J.R.V. Garcia, Growth of AlN films by chemical vapor deposition, Superficies y Vacio. 9 (1999) 82−84.
[191] C.S. Hwang, H.-I. Yoo, Metal-organic chemical vapor deposition of high dielectric (Ba, Sr) $TiO_3$ Thin films for dynamic random access memory applications, in: J.H. Park, T.S. Sudarshan (Eds.), Chemical Vapor Deposition, ASM International, Materials Park, OH, 2001, pp. 205−242.
[192] Zilko J. Metal organic chemical vapor deposition: technology and equipment. In: Seshan K, editors. Handbook of Thin Film Deposition, second ed., p. 151−204.
[193] H. Li, Mass transport analysis of a showerhead MOCVD reactor, J. Semiconduct. 32 (2011) 3.
[194] S.C. Warnick, M.A. Dahleh, Feedback control of MOCVD growth of submicron compound semiconductor films, IEEE Trans. Control Syst. Technol. 6 (1) (1998) 62−71.
[195] R. Zuo, Q. Xu, H. Zhang, An inverse-flow showerhead MOVPE reactor design, J. Cryst. Growth 298 (2007) 425−427.
[196] A. Brevet, P.M. Peterlé, L. Imhoff, M.C. Marco de Lucas, S. Bourgeois, Initial stages of $TiO_2$ thin films MOCVD growth studied by in situ surface analyses, J. Cryst. Growth 275 (1−2) (2005) 1263−1268.
[197] J.-P. Lu, R. Raj, Ultra-high vacuum chemical vapor deposition and in situ characterization of titanium oxide thin films, J. Mater. Res. 6 (9) (1991) 1913−1918.

[198] J. Lee, H.J. Yang, J.H. Lee, J.Y. Kim, W.J. Nam, H.J. Shin, Highly conformal deposition of pure Co films by MOCVD using $Co_2(CO)_8$ as a precursor, J. Electrochem. Soc. 153 (6) (2006) G539–G542.

[199] R.D. Dupuis, H. Temkin, L.C. Hopkins, InGaAsP/InP double heterostructure lasers grown by atmospheric-pressure MOCVD, Electron. Lett. 21 (2) (1985) 60–62.

[200] X. Wu, Gradient and lamellar heterostructures for superior mechanical properties, MRS Bull. 46 (2021) 244–249.

[201] H. Yi, *Crossover from Ising- to Rashba-type superconductivity in epitaxial $B_2Se_3$/monolayer $NbSe_2$* heterostructures, Nat. Mater. 21 (2022) 1366–1372.

[202] J. Wang, Electrical properties and applications of graphene, hexagonal boron nitride (h-BN), and graphene/H-BN heterostructures, Mater. Today Phys. 2 (September 2017) 6–34.

[203] C.A. Wang, et al., MOVPE Growth of LWIR AlInAs/GaInAs/InP Quantum Cascade Lasers: Impact of Growth and Material Quality on Laser Performance, Massachusetts Institute of Technology, 2017.

[204] J.J. Coleman, Metalorganic chemical vapor deposition, Proc. IEEE 85 (11) (November 1997) 1715–1729.

[205] C. Rosenblad, H.R. Deller, A. Dommann, T. Meyer, P. Schroeter, H. von Känel, Silicon epitaxy by low-energy plasma enhanced chemical vapor deposition, J. Vac. Sci. Technol. A 16 (5) (1998) 2785–2790.

[206] H. Huang, K.J. Winchester, A. Suvorova, B.R. Lawn, Y. Liu, X.Z. Hu, Effect of deposition conditions on mechanical properties of low-temperature PECVD silicon nitride films, Mater. Sci. Eng., A 435–436 (2006) 453–459.

[207] A. Matsuda, T. Tanaka, Plasma spectroscopy—glow discharge deposition of hydrogenated amorphous silicon, Thin Solid Films 92 (1–2) (1982) 171–187.

[208] R.J. Markunas, R. Hendry, R.A. Rudder, Patent: Remote Plasma Enhanced CVD Method and Apparatus for Growing an Epitaxial Semiconductor Layer, 1993.

[209] M. Nakamura, S. Kato, T. Aoki, L. Sirghi, Y. Hatanaka, Formation mechanism for $TiO_x$ thin film obtained by remote plasma enhanced chemical vapor deposition in $H_2$-$O_2$ mixture gas plasma, Thin Solid Films 401 (1–2) (2001) 138–144.

[210] S.E. Alexandrov, M.L. Hitchman, Plasma enhanced chemical vapour deposition processes, in: A.C. Jones, M.L. Hitchman (Eds.), Chemical Vapour Deposition: Precursors, Processes and Applications, RSC Publishing, London, UK, 2009, pp. 494–534.

[211] A. Grill, Plasma enhanced chemical vapor deposited SiCOH dielectrics: from low-k to extreme low-k interconnect materials, J. Appl. Phys. 93 (3) (2003) 1785–1790.

[212] A. Grill, Low and Ultralow Dielectric Constant Films Prepared by Plasma-Enhanced Chemical Vapor depositionDielectric Films for Advanced Microelectronics, John Wiley & Sons, Ltd, West Sussex, England, 2007.

[213] J. Batey, E. Tierney, Low-temperature deposition of high-quality silicon dioxide by plasma-enhanced chemical vapor deposition, J. Appl. Phys. 60 (9) (1986) 3136–3147.

[214] B.F. Hanyaloglu, E.S. Aydil, Low temperature plasma deposition of silicon nitride from silane and nitrogen plasmas, J. Vac. Sci. Technol. A 16 (5) (1998) 2794–2803.

[215] M.K. Gunde, M. Macek, The relationship between the macroscopic properties of PECVD silicon nitride and oxynitride layers and the characteristics of their networks, Appl. Phys. Mater. Sci. Process 74 (2) (2002) 181–186.

[216] R. Arghavani, Z. Yuan, N. Ingle, K.-B. Jung, M. Seamons, S. Venkataraman, Stress management insub-90-nm transistor architecture, IEEE Trans. Electron. Dev. 51 (10) (2004) 1740–1743.

[217] J. Dai, H. Liu, W. Fang, L. Wang, Y. Pu, Y. Chen, Atmospheric pressure MOCVD growth of high-quality ZnO films on GaN/$Al_2O_3$ templates, J. Cryst. Growth 283 (1–2) (2005) 93–99.

[218] Y.C. Huang, Z.-Y. Li, H. Chen, W.-Y. Uen, S.-M. Lan, S.-M. Liao, Characterizations of gallium-doped ZnO films on glass substrate prepared by atmospheric pressure metal-organic chemical vapordeposition, Thin Solid Films 517 (18) (2009) 5537–5542.

[219] C. Detavernier, J. Dendooven, D. Deduytsche, J. Musschoot, Thermal versus plasma-enhanced ALD: growth kinetics and conformality, ECS Trans. 16 (4) (2008) 239–246.

[220] G. Vellianitis, M.J.H. van Dal, L. Witters, G. Curatola, G. Doornbos, N. Collaert, Gatestacks for scalable high-performance FinFETs, in: IEEE International Electron Devices Meeting (IEDM 2007), 2007, pp. 681–684.

[221] J. Dendooven, D. Deduytsche, J. Musschoot, R.L. Vanmeirhaeghe, C. Detavernier, Conformality of $Al_2O_3$ and AlN deposited by plasma-enhanced atomic layer deposition, J. Electrochem. Soc. 157 (4) (2010) G111–G116.

[222] A. Grill, D.A. Neumayer, Structure of low dielectric constant to extreme low dielectric constant SiCOH films: Fourier transform infrared spectroscopy characterization, J. Appl. Phys. 94 (10) (2003) 6697–6707.

[223] S. Ikeda, K. Nemoto, M. Funabashi, T. Uchino, H. Yamamoto, N. Yabuoshi, Process integration of single-wafer technology in a 300-mm fab, realizing drastic cycle time reduction with high yield and excellent reliability, IEEE Trans. Semicond. Manuf. 16 (2) (2003) 102–110.

[224] Yole Group, The 2023 Global Fab Landscape: Opportunities and Obstacles, Nov, 2022. https://www.yolegroup.com/strategy-insights/the-2023-global-fab-landscape-opportunities-and-obstacles/.

[225] L. Mari, What is a FinFET?, EE Power, October 23, 2020. https://eepower.com/technical-articles/what-is-a-finfet/#.

[226] R. Smith, Samsung Starts 3 nm Production: The Gate-All-Around (GAAFET) Era Begins, June 30, 2022. https://www.anandtech.com/show/17474/samsung-starts-3nm-production-the-gaafet-era-begins.

[227] J. Jung, et al., Process steps for high quality Si-based epitaxial growth at low temperature via RPCVD, Materials 14 (13) (July 2021) 3373.

[228] J. Li, et al., Study of silicon nitride inner spacer formation in process of gate-all-around nano-transistors, Nanomaterials 10 (4) (2020) 793.

[229] M. Chhowalla, et al., Two-dimensional semiconductors for transistors, Nat. Rev. Mater. 1 (2016) 16052, https://doi.org/10.1038/natrevmats.2016.52.

[230] K. Momeni, A computational framework for guiding the MOCVD-growth of wafer-scale 2D materials, Npj Comput. Mater. 8 (2022) 240, https://doi.org/10.1038/s41524-022-00936-y.

[231] T. Roy, et al., Field-effect transistors built from all two-dimensional material components, ACS Nano 8 (6) (2014) 6259–6264.

[232] M. Heuken et al., Recent progress in large-area CVD growth of 2D materials, 2021 ECS Meeting Abstract, MA-2021-02 606.

[233] K.W. Ang, K.J. Chui, V. Bliznetsov, A. Du, N. Balasubramanian, M.F. Li, Enhanced performance in 50 nm N-MOSFETs with silicon-carbon source/drain regions, International Electron Devices Meeting, in: IEDM Technical Digest, 2004, pp. 1069–1071.

[234] M. Bauer, B. Machkaoutsan, C. Arena, Highly tensile strained silicon–carbon alloys epitaxially grown into recessed source drain areas of NMOS devices, Semicond. Sci. Technol. 22 (2007) S183–S187.

[235] Q. Wang, et al., Integrated fabrication of a high strain InGaAs/GaAs quantum well structure under variable temperature and improvement of properties using MOCVD technology, Opt. Mater. Express 11 (8) (2021) 2378–2388.

[236] M. Radosavljevic, et al., Non-planar, multi-gate InGaAs quantum well field effect transistors with high-K gate dielectric and ultra-scaled gate-to-drain/gate-to-source separation for low power logic applications, in: IEDM Technical Digest, 2010, pp. 6.1.1–6.1.4.
[237] J.H. Zhao, X.H. Tang, T. Mei, B.L. Zhang, G.S. Huang, MOCVD growth of InGaAsP/InGaAs multi-step-quantum well structure for QWIP application by using TBA and TBP in N2 ambient, J. Cryst. Growth 268 (3–4) (2004) 432–436.
[238] F. Heinrichsdorff, A. Krost, D. Bimberg, A.O. Kosogov, P. Werner, Self organized defect free InAs/GaAs and InAs/InGaAs/GaAs quantum dots with high lateral density grown by MOCVD, Appl. Surf. Sci. 123/124 (1998) 725–728.
[239] M.K. Hudait, et al., Heterogeneous integration of enhancement mode $In_{0.7}Ga_{0.3}As$ quantum well transistor on silicon substrate using thin ($\leq 2$ μm) composite buffer architecture for high-speed and low-voltage (0.5 v) logic applications, Int. Electron Devices Meet. (IEDM) (2007) 625–628.
[240] E. Mohapatana, et al., Design and optimization of stress/strain in GAA nanosheet FETs for improved FOMs at sub-7 nm nodes, Phys. Scripta 98 (2023) 065919.
[241] M. Bosi, G. Attolini, Progress in crystal growth and characterization of materials, Prog. Cryst. Growth Char. Mater. 56 (3–4) (2010) 146–174.
[242] T.H. Loh, H.S. Nguyen, R. Murthy, M.B. Yu, W.Y. Loh, Selective epitaxial germanium on silicon-on-insulator high speed photodetectors using low-temperature ultrathin $Si_{0.8}Ge_{0.2}$ buffer, Appl. Phys. Lett. 91 (7) (2007).
[243] Q. Li, S.M. Han, S.R.J. Brueck, S. Hersee, Y.-B. Jiang, Selective growth of Ge on Si(100) through vias of $SiO_2$ nanotemplate using solid source molecular beam epitaxy, Appl. Phys. Lett. 83 (24) (2003) 5032–5034.
[244] M. Myronov, et al., Holes outperform electrons in group IV semiconductor materials, Small Sci. 3 (4) (April 2023).
[245] D. Gall, et al., The resistivity bottleneck: the search for new interconnect materials, in: 2020 International Symposium on VLSI Technology, Systems and Applications (VLSI-TSA), September 22, 2020.
[246] H. Kim, et al., Cu diffusion into the glass under bias temperature stress condition for through glass vias (TGV) applications, in: International Symposium on Microelectronics (2018), 2018 (1).
[247] A. Pyzyna, et al., Resistivity of copper interconnects beyond the 7 nm node, in: Symposium on VLSI Technology, 2015.
[248] Z. Li, et al., Materials 13 (21) (November 2020) 5049.
[249] N. Arshi, et al., Effects of nitrogen composition on the resistivity of reactively sputtered TaN thin films, Surf. Interface Anal. 47 (2015) 154–160.
[250] L.G. Wen, et al., Ruthenium metallization for advanced interconnects, IITC Conference Paper (May 2016) 34–36.
[251] D. Gall, Metals for low-resistivity interconnects, in: IEEE Interconnect Technology Conference (IITC), 2018, pp. 157–159.
[252] D. Gall, The search for the most conductive metal for narrow interconnect lines, J. Appl. Phys. 127 (5) (February 7, 2020).
[253] C. Adelmann, et al., Alternative metals for advanced interconnects. IITC Conference Paper, IEEE, May 2014, pp. 173–175.
[254] S. Davis, Can binary or ternary compounds beat Cu in future interconnect applications? Semicond. Digest (October 7, 2022).
[255] Cadence website, https://community.cadence.com/cadence_blogs_8/b/breakfast-bytes/posts/iedm18-interconnect.
[256] Z. Tokei, Scaling the BEOL–A Toolbox Filled With New Processes, Boosters and Conductors, September 8, 2019, in: https://www.imec-int.com/en/imec-magazine/imec-magazine-september-2019/scaling-the-beol-a-toolbox-filled-with-new-processes-boosters-and-conductors.

[257] R. Bernasconi, L. Megagnin, Review—ruthenium as diffusion barrier layer in electronic interconnects: current literature with a focus on electrochemical deposition methods, J. Electrochem. Soc. 166 (2019) D3219.
[258] G. Murdoch, et al., First demonstration of two metal level semi-damascene interconnects with fully self-aligned vias at 18MP, in: 2022 IEEE Symposium on VLSI Technology and Circuits, June 2022.
[259] M. Khan and M. S. Kim, http://classweb.ece.umd.edu/enee416/GroupActivities/Damascene%20Presentation.pdf.
[260] S.J. Yoon, et al., Large grain ruthenium for alternative interconnects, IEEE Electron. Device Lett. 40 (1) (January 2019).
[261] G. Murdoch, Z. Tokei, IMEC Demonstrates Semi-Damascene Interconnects With Fully Self-Aligned Vias at 18 nm Metal Pitch, June 23, 2022. https://www.imec-int.com/en/articles/imec-demonstrates-semi-damascene-interconnects-fully-self-aligned-vias-18nm-metal-pitch.
[262] S. Nguyen, et al., Pinch off plasma CVD process and material for nano-device air gap/spacer formation, ECS J. Solid State Sci. Technol. 7 (10) (2018).
[263] H.W. Kim, et al., Recent trends in copper metallization, Electronics 11 (18) (2022) 2914 (September 2022).
[264] A. Rogozhin, et al., Plasma enhanced atomic layer deposition of ruthenium films using $Ru(EtCp)_2$ precursor, Coatings 11 (2) (2021) 117.
[265] Y.L. Chen, et al., Grain-boundary/interface structures and scatterings of ruthenium and molybdenum metallization for low-resistance interconnects, Appl. Surf. Sci. 629 (August 30, 2023) 157440.
[266] S.K. Mah, et al., *A feasible Alternative to FDSOI and FinFET: Optimization of $W/La_2O_3/Si$ planar PMOS with 14-nm gate length*, Materials 14 (2021) 5721.
[267] bloomberg.com, TSMC Plans for First German Chip Fab with Cost Up to €10 Billion, Jillian Deutsch and Alberto Nardelli, May 3, 2023.
[268] New York Times, US Focuses on Invigorating 'Chiplets' to Stay Cutting Edge in Tech, Don Clark, May 11, 2023.
[269] W.W. Shen, K.N. Chen, Three-dimensional integrated circuit (3D IC) key technology: through-silicon via (TSV), Nanoscale Res. Lett. 12 (2017). Article number: 56.
[270] S.L. Burkett, et al., Tutorial on forming through-silicon vias, J. Vacuum Sci. Technol. A 38 (2020) 031202.
[271] M. Lisker, et al., Sub-atmospheric chemical vapor deposition of SiO2 for dielectric layers in high aspect ratio TSVs, in: Meeting Abstract, MA 2011-01, 1129, 2011.
[272] V. Jousseaume, et al., Conformal isolation of high-aspect-ratio TSVs using a low-κ dielectric deposited by filament-assisted CVD, Microelectron. Eng. 167 (January 5, 2017) 80–84.
[273] L. Hofmann, et al., Study on TSV isolation liners for a via Last approach with the use in 3D-WLP for MEMS, Microsyst. Technol. 22 (2016) 1665–1677.
[274] D. Zhang, et al., Room temperature ALD oxide liner for TSV applications, July 2015, in: Proceedings—Electronic Components and Technology Conference, 2015, pp. 59–65.

CHAPTER 6

# CMP: Scaling down and stacking up: How the trends in semiconductors are affecting chemical-mechanical planarization

Wei-Tsu Tseng
IBM Semiconductor Technology Research, Albany, NY, United States

## 1. Introduction

Ever since the invention of integrated circuits (IC) in the early 1960s, Si based semiconductors have been progressing according to Gordon Moore's observation in 1965, which stated that *"the number of transistors that can be packed into a given unit of space will roughly double every 2 years"* [1]. What started out as a proposal back then soon became a law that governs the advancement of Si semiconductors for over 5 decades. Examples of device scaling are provided in Fig. 6.1. Quite often, such scaling and enhanced performance are accompanied by the introduction of new materials and/or new manufacturing process. Since its invention in the early 1980s, chemical-mechanical planarization (CMP) has been one of the enabling process technologies contributing to the realization of Moore's law for decades.

Tremendous technical challenges begin to emerge when the transistor geometry approaches nanometer regime. Meanwhile, longer and more complex process flow plus difficulties in meeting the ever-stringent requirement gradually delay the introduction of new technology nodes. The net result is that the industry began to "slow down" and deviated from Moore's law. The industry began to debate and contemplate over what the next evolution can be that will drive the industry forward in the more-than-Moore and beyond–Moore era.

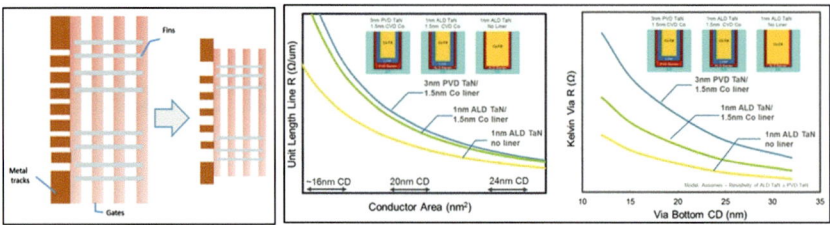

**Figure 6.1** Examples of areal scaling: CMOS (*left*) and Cu interconnects (*right*) [2]. *(Source: Applied Materials.)*

Nevertheless, a new paradigm arose over the past 2 decades that promises to continue the trend of packing higher device density while delivering improved power and speed performance. Instead of continuing the planar shrinkage in geometry, devices are being *stacked up* on top of each other to increase density and boost functionality. 3D devices such as nanosheets [3] (as shown in Fig. 6.2) and 3D NAND flash memories are some of the innovative solutions.

In addition, by combining chips with different process nodes and technologies, heterogeneous integration (HI) [4] opens up a broad avenue for continued increase in functional density and decrease in cost per function required to maintain the progress in cost and performance of microelectronics. HI refers to the integration of separately manufactured components into a higher-level assembly (System in Package—SiP). In parallel, wafer-level bonding technologies such as hybrid bonding vertically connects die-to-wafer (D2W) or wafer-to-wafer (W2W) via closely spaced copper pads. The development of such wafer-level bonding technologies

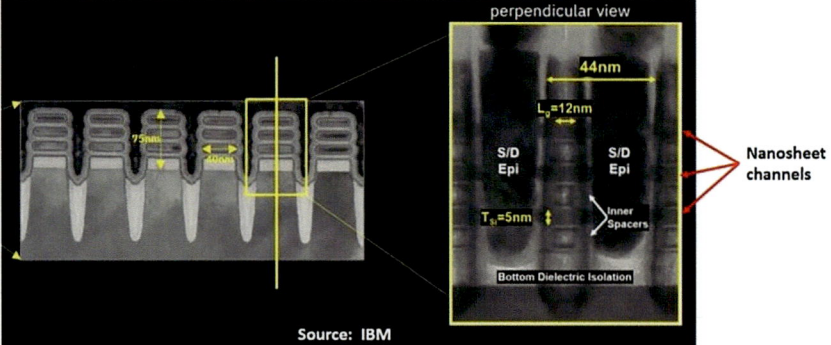

**Figure 6.2** An example of 3D device: 2 nm transistors with fully planarized and stacked nanosheets [3]. *(Source: IBM.)*

further enables heterogeneous integration, providing a powerful and flexible means to deliver increased power and improved device speed efficiency. Once again, CMP is the pivotal process step for wafer bonding as exemplified in Fig. 6.3 [5].

In summary, Si based semiconductor device is continuing its ground rule scaling horizontally while growing its functionality vertically. Impacts of such aggressive change in device integration scheme on CMP will be deliberated in the next session.

## 2. CMP challenges
### 2.1 More CMP steps

Inevitably, the ever-shrinking device geometry plus the advent of 3D devices warrant more processing steps, including CMP to complete the manufacturing cycle. As shown in Fig. 6.4, for advanced logics as well as memories, the increasing CMP steps result mainly from FEOL processing. Additionally, higher density and complexity of device structures require more levels of metalization and local interconnection, hence more steps of W or Co CMP. Specifically for 3D NAND memory devices, increasing number of stack layers call for repetitive ILD CMP steps to planarize the high accumulated topography after dielectric deposition.

Coming along with the increasing numbers of CMP steps is the need to polish novel materials, or to polish existing materials with new and more challenging requirements. For example, implementation of SiGe, III-V compound semiconductors [7] for advanced logics, and the reliance of flowable chemical vapor deposited (FCVD) [8,9] dielectrics to enhance gap fill require additional CMP steps to polish new materials. Meanwhile, new

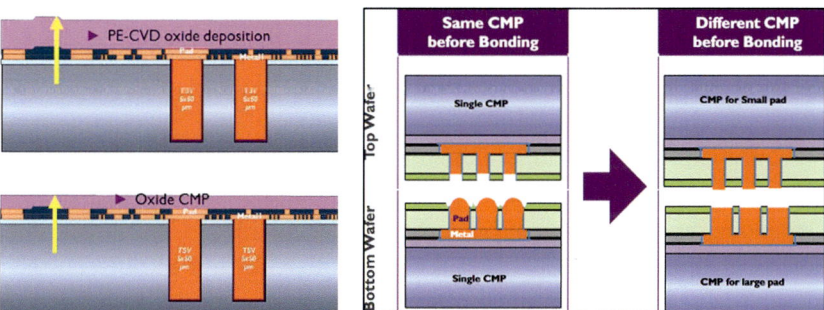

**Figure 6.3** Oxide CMP (*left*) and Cu CMP (*right*) for hybrid wafer bonding [5].

**Advanced logic**

| # CMP Process | | | | | | | |
|---|---|---|---|---|---|---|---|
| 19 | | | | | | | 10-11 Cu |
| 18 | | | | | | | Co IM (3) |
| 17 | | | | | | | HM (4-12) |
| 16 | | | | | 10-13 Cu (*) | | W Plugs |
| 15 | | | | | Co IM (0-2) | | W-TS |
| 14 | | | | | HM (2-6) | | Co-TS |
| 13 | | | | 8-12 Cu (*) | W Plugs | | HM |
| 12 | | | | HM (2-4) | W-TS | | SAC2 |
| 11 | | | | W-Plugs | SAC2 | | SAC1 |
| 10 | | | | W-TS | SAC1 | | Co Gate |
| 9 | | | | SAC | W Gate | | HM |
| 8 | | | | W-Gate | HM | | POP |
| 7 | | | 9-10 Cu (*) | HM | POP | | ILD0 |
| 6 | | 9-10 Cu | W-Plugs | POP | ILD0 | | Gate Poly |
| 5 | | 7-8 Cu | W | W-TS | ILD0 | Gate Poly | III-V |
| 4 | 7-8 Cu | W | Al / W Gate | Al/W Gate | Gate Poly | SiGe | SiGe |
| 3 | W | Al Gate | POP | POP | HM | HM | HM |
| 2 | ILD | ILD | ILD | ILD | STI2 | STI2 | STI2 |
| 1 | STI | STI | STI | STI | STI1 | STI1 | STI1 |
| # CMP Layer | 10 | 12 | 14 | 15 | 18-25 | 24-30 | 25-34 |
| Technology Node | 65nm Planar | 45nm | 28nm HKMG | 20nm | 16/14/10nm | 7nm FinFETs | 5nm GAA |

**Memory**

| # CMP Process | | | | | | | |
|---|---|---|---|---|---|---|---|
| 16 | | | | | | | Cu (X) |
| 15 | | | | | | | W via (X) |
| 14 | | | | | | | W |
| 13 | | | | ILD | | Cu (X) | W (X) |
| 12 | | | ILD | Cu (X) | | W via (X) | W (X) |
| 11 | | | Cu | W via (X) | | W | HM (X) |
| 10 | | | W via (X) | TiN | | W (X) | ILD4 |
| 9 | | Cu | TiN | ILD4 | | ILD3 | ILD3 |
| 8 | | W via (X) | ILD4 | ILD3 | Cu | SoP (X) | SoP (X) |
| 7 | | W | ILD3 | W (X) | W (X) | Poly (X) | Poly (X) |
| 6 | | ILD3 | W (X) | ILD2 | ILD | SoN (X) | SoN (X) |
| 5 | | Ox Buff (X) | ILD2 | HM Buff (X) | W | HM Buff (X) | HM Buff (X) |
| 4 | | ILD2 | Poly | Poly | ILD | ILD2 | ILD2 |
| 3 | | Poly | ILD1 | ILD1 | Poly | ILD1 | ILD1 |
| 2 | | ILD1 | Ox Buff | Ox Buff | Ox Buff (X) | Ox Buff (X) | Ox Buff (X) |
| 1 | | STI | STI | STI | STI | STI | STI |
| # CMP Layer | | 10 | 14 | 17 | 10 | 14-26 | 17-32 |
| Node Technology | | 3Xnm | 2Xnm DRAM | 1Xnm | 2X - 1Xnm Planar | 3D 32-36L NAND | 3D 48-64L |

**Figure 6.4** Increase in CMP steps versus advances in technology nodes for logic (*top*) and memories (*bottom*) devices [6].

integration scheme such as self-aligned contact (SAC) [10] for middle-of-the-line (MOL) calls for a CMP process with (reverse) high nitride-to-oxide selectivity.

Lastly, the introduction of hybrid bonding and heterogeneous integration add a few more steps of CMP to the entire manufacturing process. Depending upon the integration approaches, these added CMP steps can be polishing Si, polyimide, or various flavors of oxide and nitride. Furthermore, backside grinding and polishing becomes integral part of process flow.

## 2.2 Sub-nm wafer-level uniformity

By nature, when device geometries shrink progressively with each new technology nodes, the tolerance for variability and non-uniformity also scales down accordingly. For example, Fig. 6.5 exhibits the correlation between the specification of line height and its within-wafer standard deviation (1-σ) to achieve a process potential capability [11], $C_p$, of 1.33.

As illustrated in Fig, 6.5, 1σ falls below 1 nm (10 Å) when line height is reduced below 80 nm. This would apply to RMG gate height, W/Co via height, and thin Cu wires for 5 nm logics technology node and beyond.

It should be note that, the total variability of manufacturing process must take into account contributions from within-wafer (wiw), wafer-to-wafer (wtw), and lot-to-lot (LtL):

$$1\sigma_{\text{total}} = \sum (1\sigma_{\text{wiw}} + 1\sigma_{\text{wtw}} + 1\sigma_{\text{LtL}})$$

As a consequence, sub-nm or angstrom level standard deviation is needed from each component in order to achieve the target $C_p$ performance in Fig. 6.5.

The above sub-nm control requirements impose critical challenges to metrology as one can anticipate. In fact, as shown in Fig. 6.6, currently the typical metrology uncertainty is already above the uncertainty requirements of a few angstroms for advanced nodes.

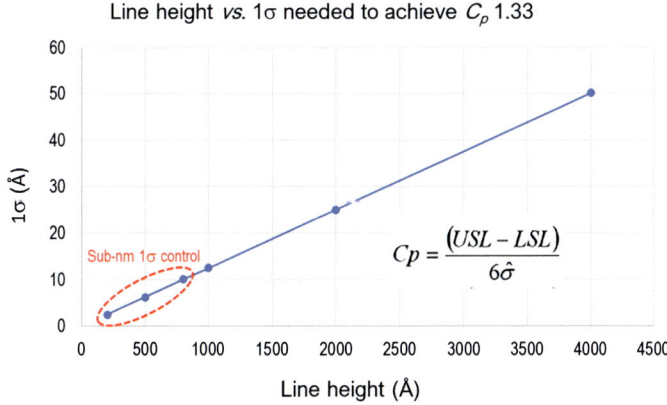

**Figure 6.5** Line height versus 1σ for achieving $C_p$ 1.33. Assumption: (USL−LSL) = Spec ± 5%.

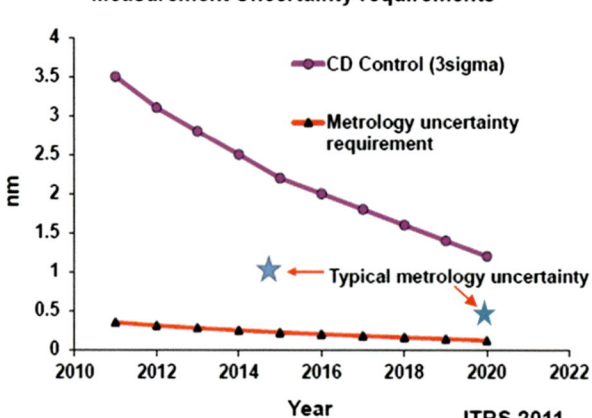

**Figure 6.6** CD control and measurement uncertainty requirements [12,13].

## 2.3 Planarity and surface roughness

Concurrent with the above variability requirement is the stringent planarity and roughness requirement at the same scale. In the context of CMP, this means the tolerance for dishing and erosion within the die would be reduced to nm and sub-nm as well. For example, for 500 Å line height with 2% variability, the tolerance for amount of dishing and erosion would be 10 Å everywhere within the die, regardless of the line width, pattern density, and dimension of the macros. Such within-die thickness and topography uniformity requirement is arguably the greatest challenge to CMP for advanced technology nodes and has been a subject of growing R&D efforts in recent years [14–17]. Additionally, the within-die planarity requirements become even more stringent at critical layers patterned by extreme ultraviolet (EUV) lithography due to its low depth-of-focus.

Surface roughness can be considered another form of topography variation at short length scale. As the uniformity and planarity requirements enter the sub-nm regime, atomic-level surface roughness becomes imperative for CMP. This is especially the case in the era of wafer-to-wafer bonding where surface finishing is critical to bonding strength [18,19].

## 2.4 Even lower tolerance for defectivity

Another natural consequence of scaling is that impacts of defects are amplified with shrinking device geometry. As illustrated in Fig. 6.7., a post-CMP polish residue (PR) or foreign material (FM) of the same size will

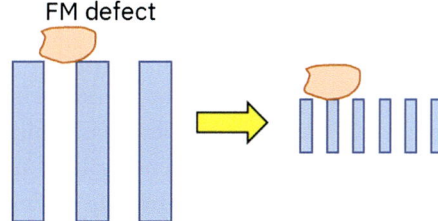

**Figure 6.7** Size effect of defects on shrinking line geometry. An FM particle of the same size would lead to elevated impacts on arrays of lines when both width and line height are reduced by 50%.

shadow more lines or patterns in the x-y plane when the line width is reduced by 50%. In addition, such a defect blocks a higher proportion of space along the z-direction, when the line height is shortened. As a result, it presents a more pronounced problems on subsequent processing steps such as deposition and lithography, leading to a higher kill ratio [20]. Similarly, the tolerance for CMP scratch is lowered significantly as the impacts of a CMP scratch of the same length, width, and depth are heightened on macros of finer geometry.

Not only is the severity of defects elevated, the quantity of defects is also surging. As shown in Fig. 6.8., higher resolution optics reveals smaller defects in significantly larger quantity, according to a few studies [21,22]. This would impose even greater challenge to CMP defect reduction and mitigation.

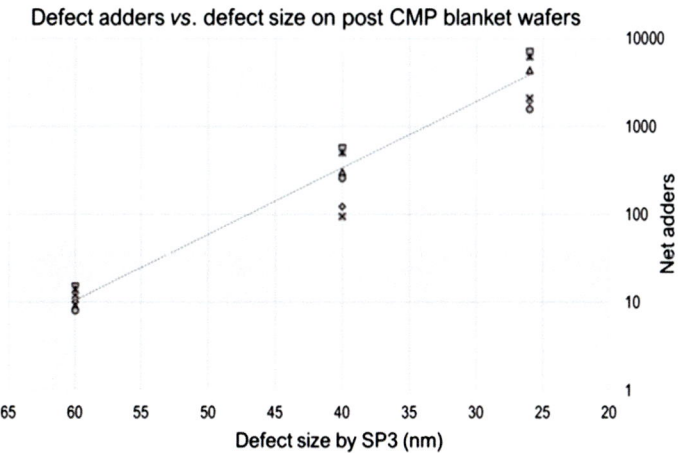

**Figure 6.8** Logarithmic correlation between inverse of defect size and net defect adders (NADD) on blanket wafers [21].

## 2.5 Post CMP metal loss in fine patterns

Metal corrosion caused by CMP is a well-known and extensively researched subject. Depending upon the integration scheme and chemistry involved, the corrosion can take place in form of chemical etching, galvanic attack, photo-corrosion, and electrostatic charge built up [23]. The extent of CMP corrosion is exacerbated with shrinking geometry to the degree that segment of Co liner on top can be missing [24] and even half of metal vias/wires can be lost as shown in Fig. 6.9.

Previous partition experiments suggest each of the fluids involved in CMP, including slurries, clean chemical, and even DIW plays certain role to the observed liner and conductor metal loss [25]. The implications of such findings are likely to impose additional challenges to slurry and clean chemical formulation as well as to design of CMP equipment and facility.

# 3. The path forward

Over the past 4 decades, advancements in CMP equipment and consumables have been enabling new integration schemes and elevating performance with each new technology nodes. In the era of nanometer geometry and 3-dimensional integration scheme, continuous innovations are even more critical in order to meet the aforementioned challenges. Trends in CMP equipment and consumables are projected and discussed in this session.

## 3.1 Equipment

Nowadays the most advanced 300 mm wafer CMP polishers are evolving in multiple areas.

**3.1.1 Multiplaten configuration**: multiple platens are incorporated into the polisher to increase throughput and process flexibility. Most

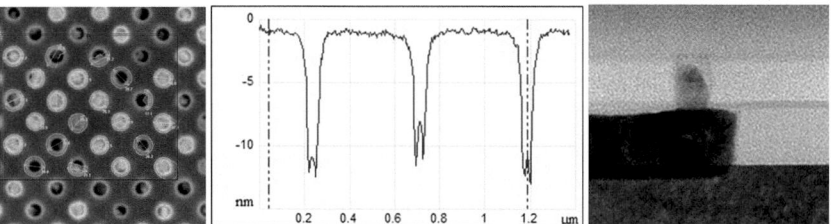

**Figure 6.9** Post CMP Cu metal loss observed from top-down SEM inspection (*left*) and AFM scan (*center*), and confirmed by cross-sectional TEM (*right*) [25].

commonly, their configuration is based on the assumption of 2-platen process, where the first platen clears the overburden ("bulk polish") while the second platen trims down the remaining layers to target thickness ("buff polish"). Configurations of a 300 mm CMP polisher are exemplified in Fig. 6.10. To meet the needs of heterogeneous integration, the option of adding one platen for backside thinning and polish is being offered as well.

**3.1.2 Carriers with multiple zones**: The number of pressure zones in 300 mm carriers is trending higher and higher in order to provide more granularity and tuneability for within-wafer uniformity control. Carriers with up to 11 pressure zones (excluding retaining ring) are being provided these days. This is usually accompanied by advanced algorithms to enable closed-loop control (CLC) of zone pressures in situ during end-pointing process. An example of carrier pressure zoning design is illustrated in Fig. 6.11 [26].

**3.1.3 End-pointing**: optical (e.g., white light or laser), electrical (e.g., eddy current), and mechanical (e.g., motor torque) end-pointing techniques, already in use extensively in the industry, continue to progress with multiple sensors to provide extended wafer coverage for more robust end-point signals. For metal CMP, end-pointing the barrier polish step (buff polish) to a target dielectric thickness remains a challenge though. Yet, this can be compensated by metrology-based APC control, as will be described in the section below.

**3.1.4 Metrology**: on-board metrology combined with advanced process control (APC) algorithm to correct polish time has been proven effective in reducing wafer-to-wafer and run-to-run variability. Specifically, spectral reflectometry technology to measure optical critical

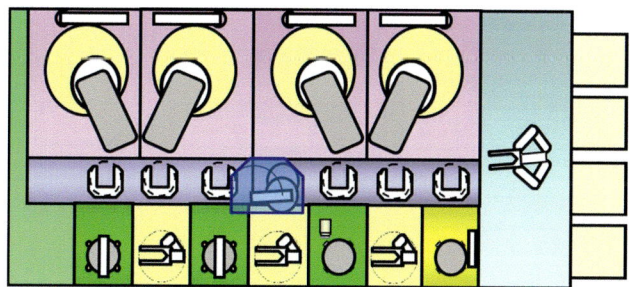

**Figure 6.10** Configuration of a 300 mm CMP polisher. *(Source: Ebara.)*

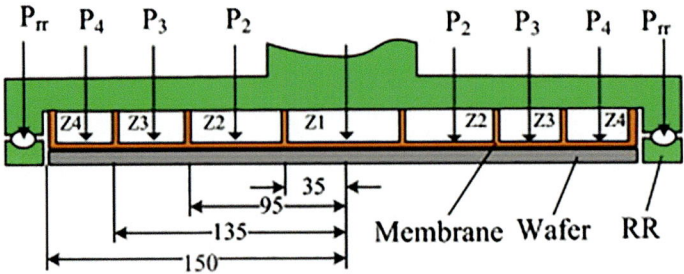

**Figure 6.11** An example of CMP carrier pressure zoning design [26].

dimension (OCD) has been integrated into CMP polishers and utilized extensively in high volume manufacturing (HVM) environment [27,28]. One such example is displayed in Fig. 6.12 for W gate CMP. Additionally, advanced optics such as polarized DUV has enabled precision measurement of fine features and 3D structures.

**3.1.5 Machine learning (ML) and artificial intelligence (AI)**: Machine learning and more recently, artificial intelligence have been applied to semiconductor manufacturing for process control [29–31] and predictive modeling [32]. Innovations in cloud storage and new processors like graphics processing unit (GPU) enable higher data processing capacity, further advancing the applications of ML and AI. Removal rates projection, incoming variability analysis, and multi-variable process optimization are just a few examples of opportunities for ML and AI to play active roles in CMP.

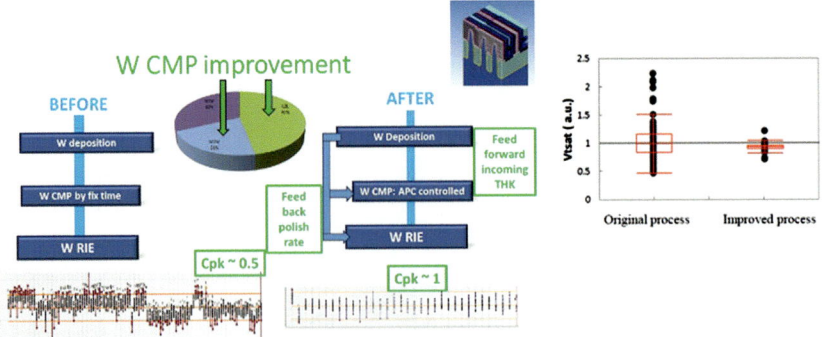

**Figure 6.12** Run-to-run W gate height uniformity improvement by APC through feed-forward from prior processes and feedback to the CMP process [28].

**3.1.6 Wafer handling and tool ambient control**: The exacerbated metal corrosion resulting from CMP in fine geometries prompts the need for ambient control and shorter wafer transient time. Exposure of metal wafers to oxygen and/or DIW plus residual slurry or chemical should be kept at minimum to reduce the extent of corrosion. From that perspective, tool ambient control (e.g., purging oxygen with nitrogen) and application of functional water [33], instead of DIW to CMP operation provide potential countermeasures against CMP metal loss issue. Robotics design with advanced algorithms to reduce idle and transient time can be helpful too.

**3.1.7 In-line chemical monitoring**: Besides on-board metrology for APC control, in-line chemical metrology to monitor slurries and chemicals in situ is another emerging trend. Up to this point, chemical characteristics such as pH, reflective index (RI), conductivity, % solid, % $H_2O_2$, and zeta-potential are mostly measured off-line at the facility/sub-fab level. The capability to monitor slurries and clean chemicals *real time* at or near point-of-use (POU) would be beneficial from process debug and/or control point of view.

## 3.2 Slurry

In many cases, the aggressive trend of scaling down and stacking up calls for the need to remove thicker overburden while stopping on thinner stop layer or barrier materials. This leads to the continuing drive for even higher selectivity in slurries. Additionally, high removal rates and high planarization efficiency are required for 3D devices and heterogeneous integration schemes.

Among the greatest challenge is the staircase inter-level dielectric planarization (ILP) CMP for 3D NAND as illustrated in Fig. 6.13. For example, to remove 4 μm of oxide and stop on 20 nm of nitride in a 2-min CMP process, an oxide-to-nitride selectivity of >200:1 is required to ensure enough process window. Also required in this case is the tolerance of a few 10 s of nm of oxide erosion over a planarization length of a few mm (1000 s of μm).

Likewise, stringent requirements for metal/liner selectivity and within-die planarity exist in MOL and BEOL CMP processes. As an example, a selectivity (e.g., metal/barrier or metal/dielectric) of >100:1 is needed in order to remove 2000 Å of W or Cu overburden and stop on 3 nm of liner/barrier layer in a 90 s bulk polish process. Dishing and erosion should

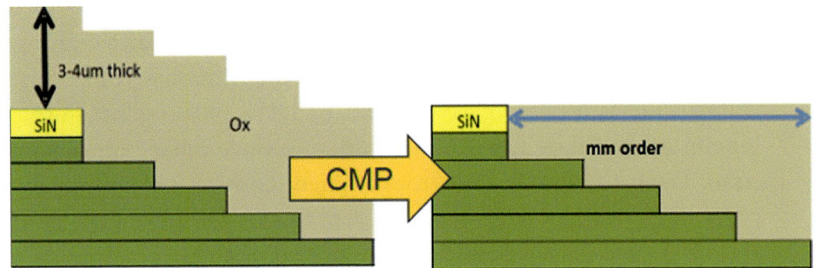

**Figure 6.13** Staircase ILP CMP for 3D NAND [6].

be controlled down to ≤10 nm across the repeating devices or macros a few mm to 10 s of mm distance away within a die.

Last but not least, defect reduction is also a critical criterion for slurry formulation in the era of shrinking geometry and 3D integration as highlighted before.

Ceria slurries continue to be adopted extensively for FEOL CMP owing to their built-in oxide/nitride selectivity by nature [34]. Meanwhile, colloidal silica slurries are used for oxide, polysilicon, and metal CMP. For concerns over defectivity, repeatability, and reliability, both ceria and silica abrasives are trending toward high-purity smaller particles (i.e., <50 nm) with tighter particle size distribution. Larger particle colloidal silica slurry slurries also find their applications in TSV, HI, and wafer bonding CMP applications where large amount of materials removal is required. Alumina slurries, despite their shrinking volume, remain in the industry for their unique characteristics that find themselves in certain niche CMP applications.

One of the main drawbacks of ceria slurry is its high scratch defectivity compared with colloidal silica slurry. Nowadays, innovations in processing technology have given birth to colloidal ceria particles with circular morphology for defect reduction. The introduction of nano ceria (∼5 nm) presents further opportunities for scratch reduction [35]. Smaller and circular abrasive particles also offer advantages of lower surface roughness. However, as particle size decreases, the surface charges (or bonding sites) to volume ratio increases inevitably, rendering them more difficult to remove from wafer surface during post CMP clean process. In addition, detection and characterization of such nano-size particles pose new challenges to metrology.

Besides abrasive particles, additives are integral part of slurry formulation serving all varieties of functions such as oxidizer, tuning selectivity booster, planarization agent, complexion agents, surfactants, and corrosion inhibitor. For example, the introduction of nano particles would require surfactants or dispersants to prevent particle agglomeration. By prohibiting oxide formation on nitride surface, amino acids such as lysine and proline are common slurry additives for elevating oxide/nitride selectivity [36]. The effectiveness of these selectivity boosters depends on pH of slurry. It and can vary with the size, phase, morphology, and crystallinity of abrasives as shown in Fig. 6.14 [37].

Dishing and erosion can be addressed through slurry formulation too. The common practice is to add specific polymers in the slurry that would absorb onto the down features on wafer surface, protecting it from chemical reactions with slurry and contact with abrasive particles [38,39]. On the other hand, the additives on the up features would be abraded away by abrasives through pad/abrasive contact, leading to planarization of the up features as shown in Fig. 6.15.

The above planarization mechanism by slurry additives would work effectively during ILP or bulk polish, where large amount of varying topography of the same overburden material exists on incoming wafer surface. During the metal buff steps (e.g., W barrier, or Cu barrier polish), however, two layers of dissimilar materials (metals or dielectric) would be exposed and polished simultaneously throughout the duration of buff polish, namely, conductor plus barrier, conductor plus hardmask, and conductor plus insulator, as depicted in Fig. 6.16. In this case, the ideal scenario to achieve minimum dishing and erosion is to be able to polish all these metals and dielectrics *non-selectively* until reaching the target thickness. Once again, surfactant or additives can be formulated into slurry in order to enhance or slow down the removal of certain material for more effective dishing and erosion control [39,40].

Corrosion inhibitor is another critical component in metal CMP slurries. Benzotriazole (BTA) and its derivatives are common additives to Cu

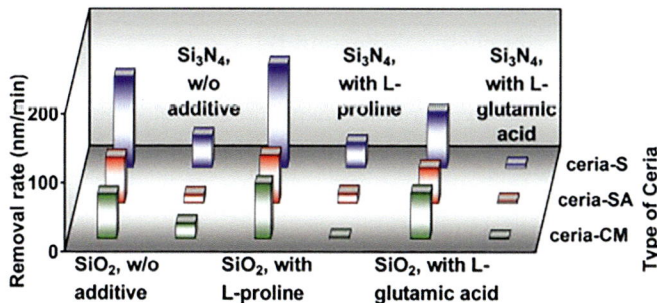

**Figure 6.14** Interactions between type of ceria abrasives and slurry additives and their resulting $SiO_2/Si_3N_4$ removal rate selectivity [37].

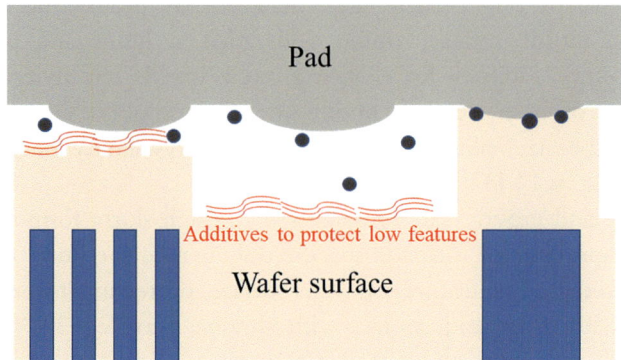

**Figure 6.15** Polishing and planarizing the up features while protecting/passivating the down features through slurry additives.

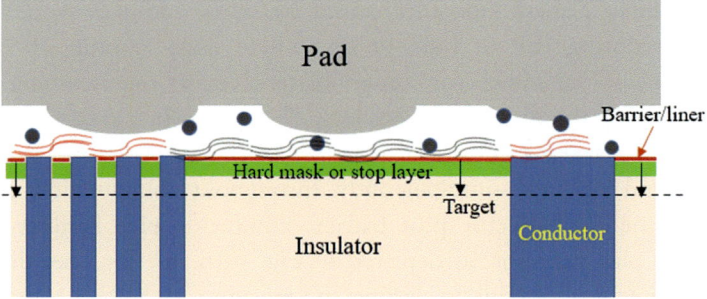

**Figure 6.16** Metal buff (metal barrier) CMP step. Conductor and insulator are metal and dielectric, respectively. Hardmask or stop layer can be either metal or dielectric. More than one slurry additives may be needed to enable non-selective polish of dissimilar materials at same the removal rate.

and Co slurries to mitigate corrosion [41,42]. The post CMP metal loss in advanced interconnects [24,25], though on the order of nanometers only, would necessitate the search for more effective corrosion inhibitors in slurries and post clean chemicals beyond BTA.

Besides the needs for more effective additives to meet the selectivity, topography, and defectivity requirements, the implementation of novel materials such polymers for hybrid bonding [43] and Mo for MOL metalization [44] provide opportunities for new slurry formulation.

### 3.3 Pads and conditioners

Together pad and conditioner form another vital element that plays vital roles in CMP planarization, defectivity, and uniformity. The majority of CMP pads are made of thermoplastic or thermoset polyurethane (PU).

Their porosity, pore size, the height and density of surface asperities, and mechanical characteristics of top pads are the main attributes would affect CMP performances. Meanwhile, the aggressiveness of diamond tips on conditioners is the dominating factor that brings about pad surface texture. As a consequence, the selection and pairing of pad and conditioner, plus the optimization of conditioning process are the key to maximize CMP performance. A simplified illustration of pad/conditioner interactions, and wafer/slurry/pad interfaces is displayed in Fig. 6.17 [45].

Depending on hardness and elasticity, PU CMP pads can be roughly divided into "*hard pads*" (hardness Shore D 50–80) and "*soft pads*" (hardness Shore A ~ 20). The hardness quoted here applies to the top pad only. The sub pad underneath can modulate the long-range mechanical properties (e.g., compliance) that also affect polish and planarization performance. "Hard pads" are used predominantly for bulk polish and ILP for their relatively higher planarization efficiency. "Soft pads" are used mainly for metal buff polish due to their lower defectivity performance. Pore structures of typical hard and soft pads are shown in Fig. 6.18.

Pore size, porosity, and PU resin hardness of pads are critical to CMP performances. In general, softer polyurethane resin helps reduce defectivity. In addition, removal rates can be increased with increasing percent porosity for fixed pore size and resin hardness [47]. Additionally, pad surface groove width and pattern design can modulate removal rates and uniformity as well [48]. From pad material perspectives, constant porosity, uniform pore size with tight distribution along the width and diameter of top pad should be the key to uniformity within-wafer, wafer-to-wafer, and run-to-run. For poromeric soft pads, longer and straighter pores would provide more area to transport the polish debris away, leading to more consistent removal rate and defect performance [21,49].

The density, height, and height distribution of pad surface asperities depend strongly on aggressiveness of the type of conditioner used. In general, coarse diamond tips impart rough texture to pad surface. Displayed

**Figure 6.17** (*Left*) Pad surface texturing by conditioner; (*right*) Pad surface asperities in contact with slurry and wafer [45].

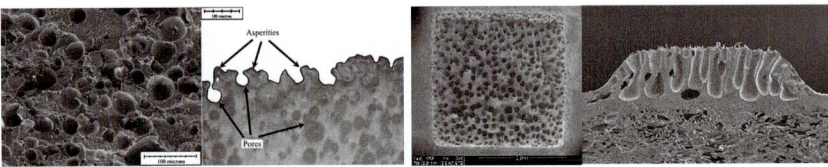

**Figure 6.18** Pores and asperities of a hard pad (*left*) [46]; cross sectional SEM pictures showing pore structures of a hard pad (*center*) and a poromeric soft pads (*right*) [47].

in Fig. 6.19 are surface textures on pads of the same hardness but conditioned by two different types of conditioners [50].

Low roughness or high density of fine asperities provides more sites for active wafer-abrasive-asperity contacts. In the case of chemical reaction driven slurries such as ceria, this results in elevated $Ce^{3+}/Ce^{4+}$ redox reactions and hence higher removal rates as shown in Fig. 6.20a. The high contact area resulting from high density of pad asperities also provides more mechanical support across macros of varying pitches and pattern densities on wafer surface, leading to lower dishing and enhanced planarization efficiency as shown in Fig. 6.20b.

It should be noted that for silica or other slurries where material removal is more driven by mechanical abrasion, higher density of asperity contacts would result in lower contact pressure and hence lower removal rates, assuming the same pore structure and mechanical properties of the top pad.

The above example illustrates the complicated nature of pad-conditioner interactions that can generate different pad surface textures with varying removal rate, topography, and defectivity. Consequently, in addition to slurry selection, pad-conditioner pairing should be an essential part of CMP process development and optimization.

As suggested above, more uniform distribution of pad asperities in their width, spacing, and height provides the opportunities to improve uniformity within-wafer and wafer-to-wafer, plus potential benefits of defect

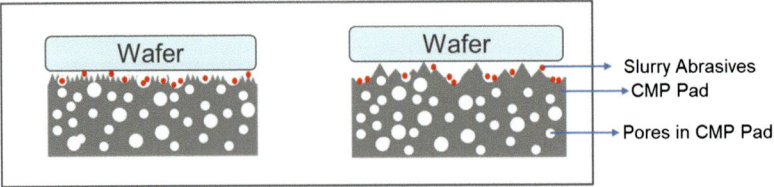

**Figure 6.19** Surface textures on pads of the same hardness after conditioning by two different conditioners. The one on the left shows lower surface roughness (i.e., higher density of pad asperities) than the one on the right [50].

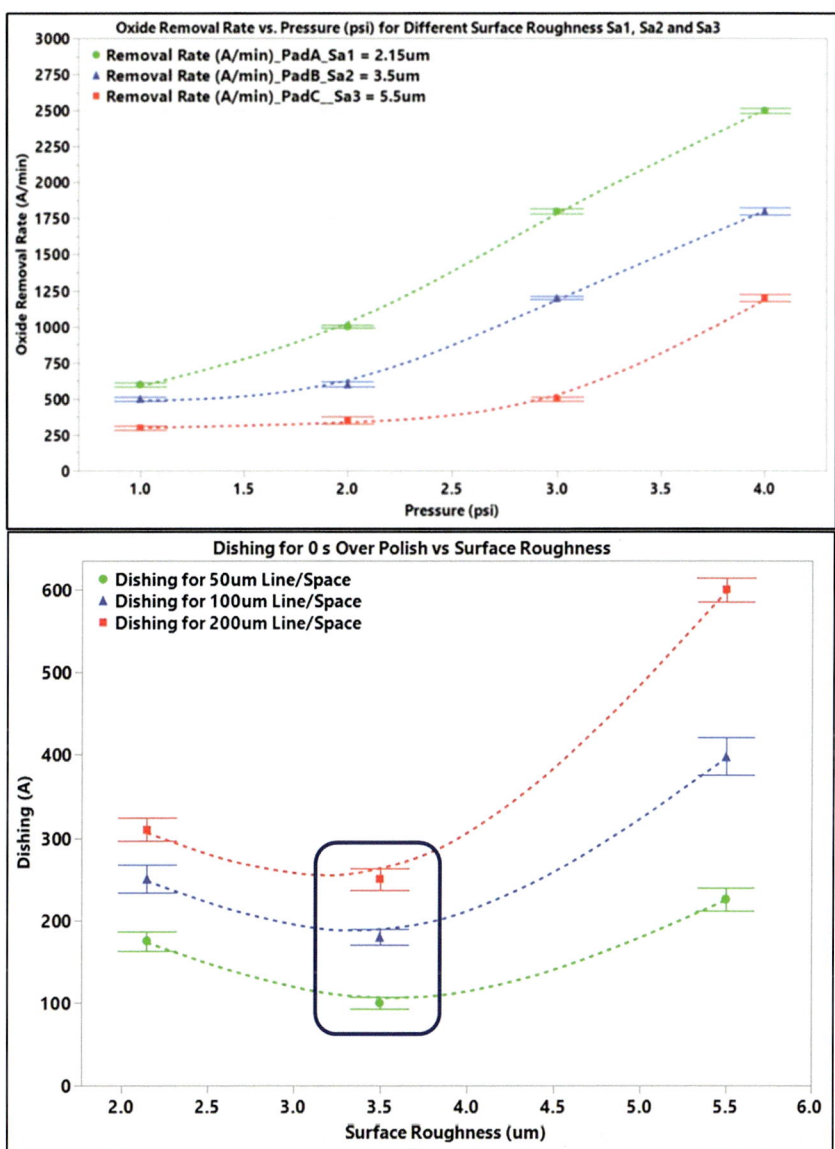

**Figure 6.20** (a): Effects of pad surface roughness on oxide removal rates with ceria slurry. (b): Effects of pad surface roughness on dishing (topography) [50].

reduction and better planarization efficiency. Likewise, uniform distribution of diamond tips with fixed shape and morphology can help generate more uniform and repeatable pad surface textures. In fact, such design concept is becoming an emerging trend in both pad and conditioner, enabled by innovations in manufacturing technologies. The differences between traditional pads and pads with engineered asperities ("EA pads") are illustrated in Fig. 6.21 [51].

### 3.3.1 Pads with engineered asperities (EA pads)

Compared with traditional pads, the asperities on EA pads are manufactured into regulated shapes. They are arranged in long-range order with designed patterns and tunable size and density to tailor for varying CMP performance requirement. Usually, no conditioning with diamond tips is needed as this would damage the regulated patterns on pad surface. Instead, "conditioning" with bristle brush in-situ and/or ex-situ is often required to maintain polish performance. Reported benefits of EA pads include superior planarization efficiency, reduced defectivity, improved within-wafer uniformity, and reduced slurry utilization [14,51,52].

Overall, EA pads as a group represents a major paradigm shift in pad design, from complete *randomness* to regularity and symmetry on surface asperities. In this case, asperities are already pre-designed and *"built-in"* during manufacturing process. Consequently, no diamond tip conditioning is required to bring up and re-generate the surface textures needed for traditional pads. The contact mode in wafer-slurry-asperity tri-body layer shifts from random point contacts in traditional pads to regulated surface contact in EA pads. The true contact area is generally higher with EA pads, leading to their superior planarization efficiency. Additionally, the long-range regulated surface patterns may have given rise to more efficient slurry transport and abrasive delivery, as suggested by the observed reduced slurry utilization and potential for defect reduction.

By nature, the CMP performance of EA would depend strongly on the geometry, distribution, and materials properties of surface asperities. For example, contact area would be determined by the shape and pattern

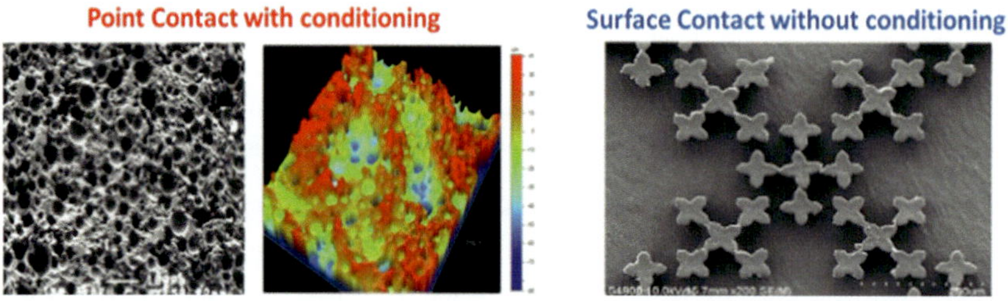

**Figure 6.21** Top-down view of a traditional PU pad (*left*) versus a pad with engineered asperities (*right*) [51].

density of asperities. Co-planarity of surface patterns over large area would be critical to process stability and repeatability. Similarly, groove geometry and the "valley" between asperities would play even greater roles in modulating the delivery of fresh slurry and transport of polish debris than in traditional pads [53]. All the above parameters would affect RR, planarity, and defectivity, and lend themselves to the precision manufacturing process required for EA pads.

For both traditional and EA pads, the trend toward uniformity and regularity in pores and asperities open up broad new opportunities to tailor CMP performance. For instance, the dilemma between high RR and low defectivity for hard pads, or between high planarization efficiency and low defectivity for soft pads can be alleviated through pad surface design and materials engineering [54]. Meanwhile, such a paradigm shift could also signify a fundamental change in the hydrodynamics and tribology states, for example, from partial lubrication to hydrodynamic lubrication, or even hydroplaning [55]. Furthermore, for reactive ceria slurries, with increase in contact area, the extent of surface reactions that are vital to oxide RR could be enhanced or inhibited, depending on whether the pad surface chemistry could facilitate the $Ce^{3+}/C^{4+}$ redox reactions. In general, more R&D efforts and fundamental understanding are needed to explore the potentials of these new pad designs in order to meet the ever-stringent CMP requirements ahead.

Besides engineering surface patterns through innovative manufacturing processes, novelties in materials themselves provide another avenue to explore for new CMP pads. Non-porous materials, thermoset versus thermoplastic PU, and non-PU polymers are all being investigated as potential candidates for next-generation pads.

Similar to EA pads, a new group of conditioners are gaining tractions in the industry. Collectively known as chemical-vapor deposited (CVD) conditioners, they are characterized by chemical vapor deposited diamond or diamond-like surface coating with tips of controlled shapes arranged in long-range regulated patterns. Examples of a conventional versus CVD conditioners are shown in Fig. 6.22 for comparison.

The regular patterns and tight control of tip surface heights of CVD conditioners lead to larger pad-conditioner contact area and more even wearing over pad surface. The net results are, compared with conventional conditioners, CVD conditioners offer lower and more consistent pad cut rates with lower pad surface roughness [58] as shown in Fig, 6.23. They also

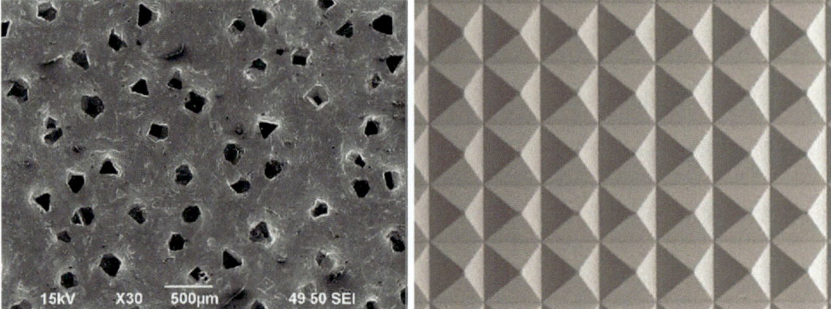

**Figure 6.22** Top-down SEM showing diamond tips of a conventional diamond conditioner (*left*) [56] versus a CVD conditioner (*right*) [57].

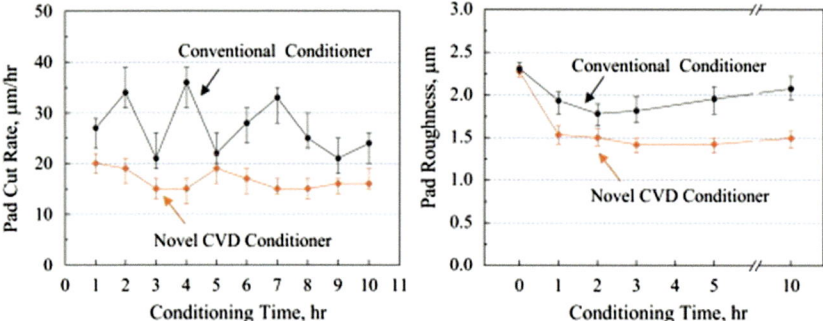

**Figure 6.23** Comparison of pad cut rate and pad surface roughness between a conventional versus CVD diamond conditioners [58].

lead to tighter height distribution of pad asperities like the ones illustrated in Fig. 6.24 [59].

The more uniform and moderate pad wear rates of CVD conditioners make them suitable for conditioning on soft pads. Their highly ordered and repeatable tips and patterns translates to more effective removal of polish debris, and opportunities of reducing scratches [60]. Their more gentle and effective conditioning can translate to extended pad service life. In addition, the lower surface roughness and tighter asperity height distribution can present opportunities for planarity improvement too. For pad break-in or conditioning on hard pads, however, the moderate pad wear rates of CVD conditioners can become a challenge.

In general, effective conditioning is a subtle balance between pad wear rate and conditioner cut rate, as shown in Fig. 6.25. The extent of conditioning needed to bring out the surface texture for optimal performance

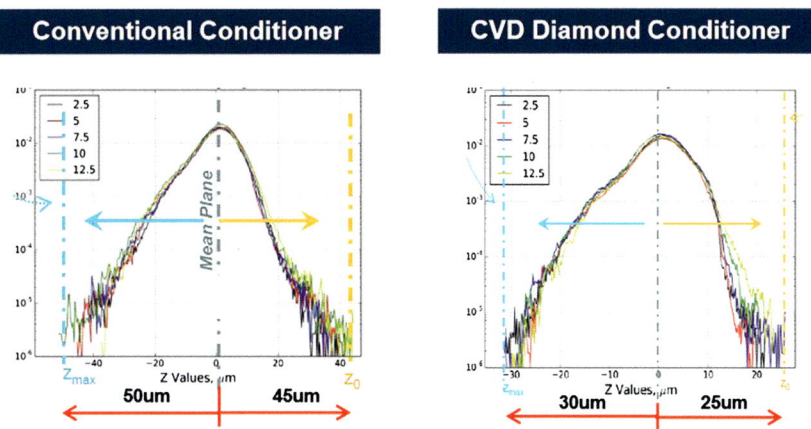

**Figure 6.24** Comparison of surface height probability density functions between a conventional versus CVD diamond conditioners [59].

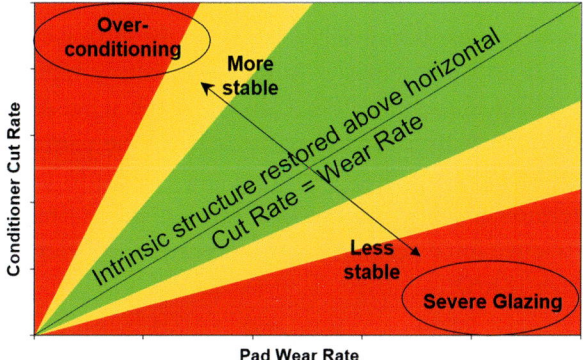

**Figure 6.25** Balance between pad wear rates and conditioner cut rates [61].

of RR, defectivity, and planarity also depends heavily on the mechanical characteristics of pads, the aggressiveness of conditioners, and to a lesser degree, the type of slurries too. Considering all these factors, conventional conditioners remain as the majority conditioners for hard pads, despite the aforementioned advantages of CVD conditioners.

### 3.3.2 Post CMP clean

Similar to slurries, pads, and conditioners, post CMP clean chemicals are facing their own set of challenges in the era of nanometer geometry, 3D structures, and HI. The trends toward smaller abrasive particles in slurries suggest the debris and residues from wafer and pad surface would become smaller proportionately. Besides, the additives from slurries, many of them being organic in nature, need to be cleaned off the wafer surface. The vulnerability of fine metal patterns to corrosive attack by DIW and any CMP chemicals calls for an additional requirement for the post clean process to passivate post polish metals while removing smaller debris or particles through more effective surfactants, stronger charge repulsion and/ or bond-breaking mechanisms. Meanwhile, the integration of porous and/ or hydrophobic dielectric materials drives the need for effective surfactants during cleaning and repulsion of moistures from wafer surface afterward. Last not least, the shrinking CD and pitch down to nanometer regime also suggests decreasing tolerance for metal or other mobile ions in the device and insulator. On the regard, post CMP cleaning process is also responsible for extracting and removing ionic species such as Na, Ca, Mg, Fe, Zn, Al, Cr, Cl, and S — impurities originating from slurries, pads, conditioner disks, prior wafer processing steps, components such as carrier and retaining ring, or even the ambient inside the CMP tool.

The cleaning mechanism usually involves slight surface etching and lift-off of particles from wafer surface through electrostatic repulsion between particles and surface as shown in Fig. 6.26 [62]. This is assisted by the sheer force of rotating PVA brush scrubbers to remove the particles away from wafer surface.

From intermolecular force point of view, as particle size decreases, the combined attractive force per unit area $F_a$ increases and is dominated by van der Waals force [21,63].

$$F_a = C/r$$

where r is particle radius, C is a parameter associated with process and consumables. The natural consequence is that debris, residues, and particles of

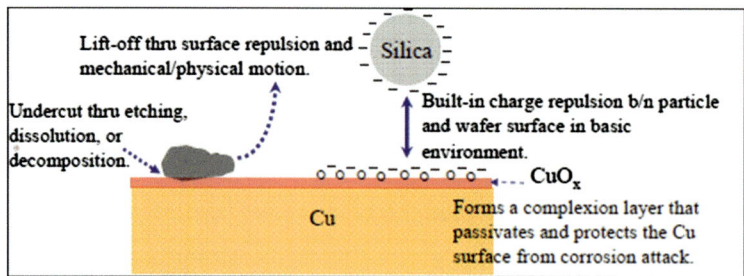

**Figure 6.26** Undercutting and charge repulsion mechanisms for particle removal during post CMP cleaning [62].

smaller size would experience stronger van der Waals attraction with wafer surface and hence are more difficult to remove during post CMP cleaning. A stronger chemical etching or undercut component could help break the van der Waals force and lift off the smaller particles as shown in Fig. 6.26. However, such aggressive chemical etching usually induces higher surface roughness, leading to reliability concern. Therefore, dissolution or charge repulsion mechanisms would be the preferable mechanisms for cleaning device of fine geometries.

In the semiconductor industry, APM (ammonia/peroxide mixture, SC1) and HPM (hydrochloric/peroxide mixture, SC2) solutions have been used extensively to remove organics/particles and metal ions, respectively. Diluted HF (DHF) is commonly adopted for oxide or nitride clean as well. However, provided the ever-stringent criteria, the commodity chemicals above are showing deficiency in cleaning performance. Search of new chemistries is showing promises for more effective post CMP cleaning. One such example is shown in Fig. 6.27 for various chemicals to clean ceria CMP particles, relative to SC1 [64].

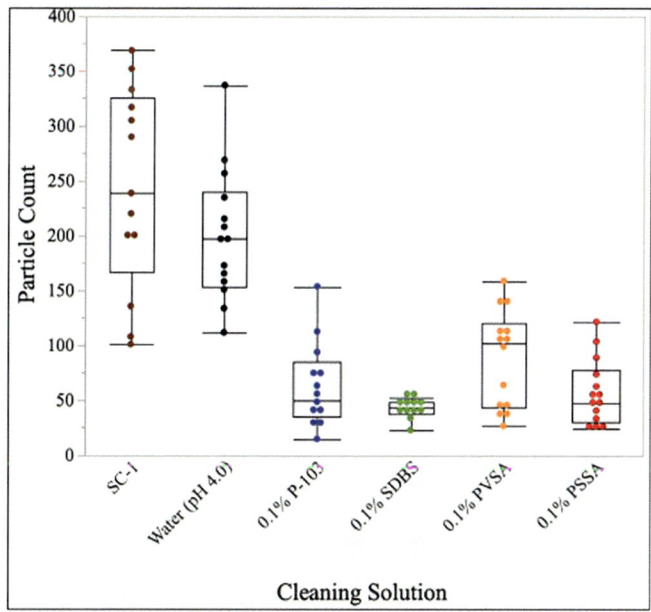

**Figure 6.27** Effectiveness of various clean chemicals in removing ceria particles from TEOS oxide surface with brush scrubbers [64]. *P-103*, pluronic p103; *PSSA*, poly(4-styrene sulfonate, ammonium salt); *PVSA*, poly(vinylsulfonic acid sodium salt); *SDBS*, sodium dodecylbenzenesulfonate.

In general, similar to slurries, post CMP clean chemicals are trending toward specialty formulated complex mixtures of additives, each of them playing specific roles for the eventual goal of *cleaning and protecting* polished wafer surface. For examples, surfactants and/or chelating agents (e.g., amines or organic acids) are added to help dissociate or solubilize the residues [65]. Additionally, chelating or complexing agents other than BTA are being explored as more effective corrosion inhibitors for Cu and Co [66].

In general, similar to slurries, post CMP clean chemicals are trending toward specialty formulated complex mixtures of additives, each of them playing specific roles for the eventual goal of *cleaning and protecting* polished wafer surface. For examples, surfactants and/or chelating agents (e.g., amines or organic acids) are added to help dissociate or solubilize the residues [65]. Additionally, chelating or complexing agents other than BTA are being explored as more effective corrosion inhibitors for Cu and Co [66].

Besides chemical clean through specialized formulation, the physical force involved to assist particle removal is critical to cleaning efficiency. As shown in Fig. 6.28, this is often accomplished through a rotating PVA brush, either a roll type or pencil type, in contact with spinning wafer while clean chemical is being spread onto wafer surface.

A recent study suggests cleaning efficiency is found to exhibit a strong inverse correlation between wafer-level defects and shear force [68]. The latter was shown to decrease with solution availability at the brush-wafer interface which in turn was shown to depend on brush porosity and the diffusion rate of the solution through the pores, as exhibited in Fig. 6.29.

Besides pore size and pore structures, dimensional stability and consistent coefficient of friction (COF) of brushes during service would suggest minimum fluctuation of brush-wafer contact-pressure and better cleaning efficiency [70]. Recent years have also seen growing efforts in conditioning roller brushes to reduce contamination and prevent pore clogging [67,69,71].

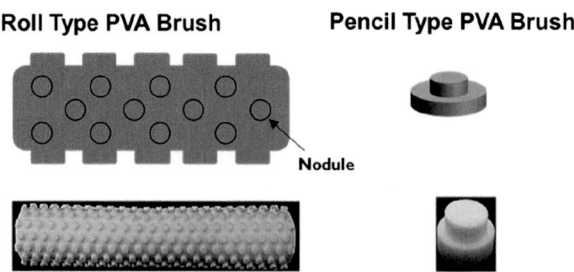

**Figure 6.28** A schematic diagram showing a roll type and pencil type brushes [67].

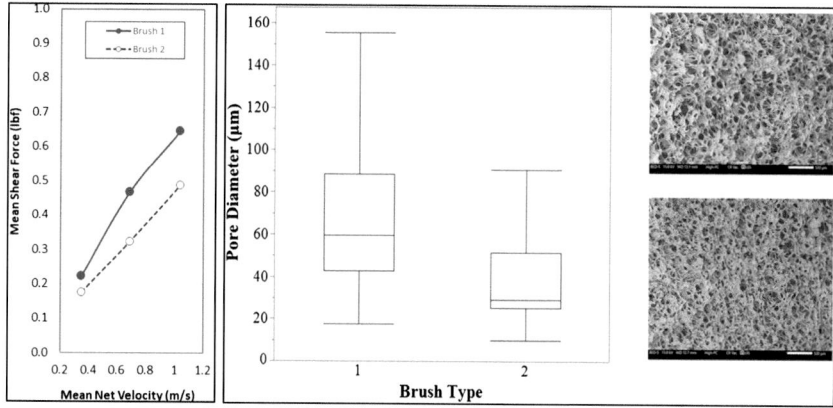

**Figure 6.29** Mean shear force (*left*) and pore diameter (*right*) of two different types of PVA brushes [68]. The brush with smaller pores, i.e., brush 2, was found to exhibit higher cleaning efficiency.

Sonication such as megasonic is another common physical means to assist wafer cleaning. The mechanical force of bubble cavitation generated by megasonic power can improve residue removal and enhances the mass transfer rate. Early on, it was widely adopted for post CMP cleaning as well. However, concerns over surface roughening or even device damage have halted the development and adoption of megasonic as part of post CMP cleaning process step, despite continuing research activities on the subject [70]. Nevertheless, known methods from other cleaning technologies provide the alternatives to help boost particle removal efficiency for post CMP applications. Thermal energy, waterjet, and supercritical carbon dioxide ($CO_2$) are some of the examples that could be explored for semiconductor cleaning. However, the balance or dilemma between using high-intensity energy form for particle removal and the induced irreversible damage on wafer surface (e.g., roughening and oxidation) could be the challenge and limitation.

In general, CMP in the post-Moore and HI/3D era is becoming more *control-intensive* in nature in order to meet the critical challenges. On one hand, it is trending toward shorter, more *moderate and chemical* in both polish and post clean for devices of nanometer geometry. On the other hand, it is tasked with large μm-level material removal down to nanometer precision for HI and wafer bonding. Therefore, tunability, flexibility, and adaptability in equipment, consumables, process design, and logistics deployment are required in the R&D as well as manufacturing environments.

## 4. Conclusions

As Moore's law approaches its physical limitation, the advent of heterogeneous integration and wafer-to-wafer bonding enables the continuous quest for increasing speed and power saving of Si-based semiconductors. Consequently, device geometries are scaling down while more layers of them are being built on wafer front side and backside as well. The aggressive shrinking and stacking trends present new challenges to CMP that can be summarized as follows:

- More CMP steps. These include addition of existing processes such as ILP and W, plus insertion of new steps such as backside and dielectric CMP for wafer-to-wafer bonding.
- CMP of bizarre new materials such as Mo and polymers.
- Nano to Angstrom scale control in uniformity & topography within-die, within-wafer, wafer-to-wafer, and run-to-run.
- Much reduced defect tolerance in terms of size and density.

To meet the above challenges, CMP equipment and consumables are advancing into new grounds proactively.

- **Equipment**: becoming more complicated and intelligent for precision process control. More granularity in carrier zoning and end-pointing; repetitive cleaning stations; extensive utilization of APC with on board metrology; and in-situ chemical monitoring.
- **Slurry**: trending toward smaller & more circular abrasives with higher purity and tighter particle size distribution; more specialized additives to tailor for tunable RR, selectivity, and planarity.
- **Pad and conditioner**: engineering in asperity/tip design; morphing from randomness to regularity with long-range order; innovation in materials manufacturing; and co-optimization of slurry and pad surface chemistries.
- **Post CMP clean**: trending from cleaning only to *cleaning and protecting* polished wafer surface; optimized additives for charge repulsion and (metal) surface passivation; better understanding of brush/chemical/residue interactions; innovations in brush design and material engineering.

Last but not least, the application of ML/AI to assist and expedite process development and control is becoming the "next big thing" in CMP, as well as in most other semiconductor manufacturing processes.

## Acknowledgments

I would like to express my deep gratitude to colleagues in IBM Semiconductor Technology Research in Albany, NY, USA for their assistance at work plus the idea exchange that has paved the way to the content of this chapter directly or indirectly. The list includes but not limited to Donald Canaperi, Pinlei Chu, Matthew Malley, Emiko Motoyama, James Cuerdon, Jeff Lang, Daniel LaGuerre, Atharv Jog, Claire Silvestre, Takeshi Nogami, Somnath Ghosh, Matthew Shoudy, Sang-Kee Eah, Juan-Manuel Gomez, and Sam Choi.

I am also obliged to peers from CMP community, in both industry and academia for the joint development projects we executed, for the academic studies we carried out, and for the many comments I received and ideas we exchanged during meetings and conferences that have been inspiring me throughout the years. Without them, this chapter would not be possible.

## References

[1] G. Templeton, What is Moore's Law? ExtremeTech (2015). July 29, https://www.extremetech.com/extreme/210872-extremetech-explains-what-is-moores-law.

[2] M. Naik, Keeping up power and performance with cobalt, Semicond. Eng. (2019). Jan, https://semiengineering.com/keeping-up-power-and-performance-with-cobalt/.

[3] S. Davis, IBM Announces 2nm GAA-FET technology—the Sum of "Aha!" moments, Semicond. Dig. (2021). June, https://www.semiconductor-digest.com/ibm-announces-2nm-gaa-fet-technology-the-sum-of-aha-moments/.

[4] J.H. Lau, Recent advances and trends in advanced packaging, in: IEEE Transactions on Components, Packaging and Manufacturing Technology, vol 12, Feb. 2022, pp. 228–252, https://doi.org/10.1109/TCPMT.2022.3144461, no. 2.

[5] S.-W. Kim, et al., Ultra-fine pitch 3D integration using face-to-face hybrid wafer bonding combined with a via-middle through-silicon-via process, in: 2016 IEEE 66th Electronic Components and Technology Conference (ECTC), Las Vegas, NV, USA, 2016, pp. 1179–1185, https://doi.org/10.1109/ECTC.2016.205.

[6] M. DeGroot, CMPUG meeting. https://nccavs-usergroups.avs.org/wp-content/uploads/CMPUG2017/CMP-717-4-DeGroot-Dow.pdf, 2017.

[7] D. Guo, et al., FINFET technology featuring high mobility SiGe channel for 10nm and beyond, in: IEEE Symposium on VLSI Technology, Honolulu, HI, USA, 2016, pp. 1–2, https://doi.org/10.1109/VLSIT.2016.7573360.

[8] H. Kim *et al.*, "Novel flowable CVD process technology for sub-20nm interlayer dielectrics", 2012 IEEE International Interconnect Technology Conference, DOI:10.1109/IITC.2012.6251590

[9] Y. Yan, et al., Flowable CVD process application for gap fill at advanced technology, ECS Trans. 60 (2014) 503, https://doi.org/10.1149/06001.0503ecst.

[10] J. Javanifard, et al., A 45nm self-aligned-contact process 1Gb NOR flash with 5MB/s Program speed, in: 2008 IEEE International Solid-State Circuits Conference—Digest of Technical Papers, San Francisco, CA, USA, 2008, pp. 424–624, https://doi.org/10.1109/ISSCC.2008.4523238.

[11] Process Capability and Performance, in: https://www.infinityqs.com/statistical-process-control/spc-chart-guide/process-capability-index-cp.

[12] Rana, et al., Machine learning and predictive data analytics enabling metrology and process control in IC fabrication, Proc. SPIE 9424 (94241I-1) (2015), https://doi.org/10.1117/12.2087406.

[13] W. Lu, et al., Total measurement uncertainty and total process precision evaluation of a structural metrology approach to monitoring post-CMP processes, in: Proc. SPIE 5375, Metrology, Inspection, and Process Control for Microlithography XVIII, May 24, 2004, https://doi.org/10.1117/12.538051.

[14] W.-T. Tseng, et al., A microreplicated pad for tungsten chemical-mechanical planarization, ECS J. Solid State Sci. Technol. 5 (2016) 546, https://doi.org/10.1149/2.0391609JSS.
[15] W.-T. Tseng, et al., Modulation of within-wafer and within-die topography for damascene copper in advanced technology, in: 2018 IEEE International Interconnect Technology Conference (IITC), Santa Clara, CA, USA, 2018, pp. 82–84, https://doi.org/10.1109/IITC.2018.8430416.
[16] Y.M. Sub, et al., The study on the effect of pattern density distribution on the STI CMP process, AIP Conf. Proc. 1875 (2017) 030023, https://doi.org/10.1063/1.4998394.
[17] A.J. Khanna, et al., Impact of pad material properties on CMP performance for sub-10nm technologies, ECS J. Solid State Sci. Technol. 8 (2019) P3063, https://doi.org/10.1149/2.0121905jss.
[18] C. Gui, et al., The effect of surface roughness on direct wafer bonding, J. Appl. Phys. 85 (1999) 7448, https://doi.org/10.1063/1.369377.
[19] A. Krueger, et al., CMP process for wafer backside planarization, in: ICPT 2017; International Conference on Planarization/CMP Technology, Leuven, Belgium, 2017, pp. 1–5.
[20] A. Elias, et al., Accurate prediction of kill ratios based on KLA defect inspection and critical area analysis, in: Proc. SPIE 2874, Microelectronic Manufacturing Yield, Reliability, and Failure Analysis II, September 12, 1996, https://doi.org/10.1117/12.250854.
[21] W.-T. Tseng, et al., CMP defect reduction and mitigation: practices and future trends, in: 2021 32nd Annual SEMI Advanced Semiconductor Manufacturing Conference (ASMC), Milpitas, CA, USA, 2021, pp. 1–6, https://doi.org/10.1109/ASMC51741.2021.9435652.
[22] J. Zhu, et al., Optical wafer defect inspection at the 10 nm technology node and beyond, Int. J. Extrem. Manuf. 4 (2022) 032001, https://doi.org/10.1088/2631-7990/ac64d7.
[23] S. Kondo, et al., Electrochemical study on metal corrosion in chemical mechanical planarization process, Jpn. J. Appl. Phys. 56 (2017) 07KA01, https://doi.org/10.7567/JJAP.56.07KA01.
[24] J. Jang, et al., Galvanic corrosion effect of Co liner on ALD TaN barrier, in: 2022 IEEE International Interconnect Technology Conference (IITC), San Jose, CA, USA, 2022, pp. 51–53, https://doi.org/10.1109/IITC52079.2022.9881307.
[25] R. Patllola, et al., Copper metal loss in nanometer fine features during chemical-mechanical planarization, in: 2022 International CMP and Planarization Technology Conference (ICPT), Portland, OR, USA, 2022. https://nccavs-usergroups.avs.org/wp-content/uploads/2023/01/ICPT22-Session-B2-Patlolla.pdf.
[26] D. Zhao, et al., Effect of zone pressure on wafer bending and fluid lubrication behavior during multi-zone CMP process, Microelectron. Eng. 108 (2013) 33–38, https://doi.org/10.1016/j.mee.2013.03.042.
[27] C. Lin, et al., Advanced Process Control Approach for Cu Interconnect Wiring Sheet Resistance Control, 2005. US patent # US20050112997A1.
[28] C. Bozdog, I. Turovets, Optical metrology for advanced process control: full module metrology solutions, in: Proc. SPIE 9782, Advanced Etch Technology for Nanopatterning, V 97820E, March 23, 2016, https://doi.org/10.1117/12.2219919.
[29] K.B. Irani, et al., Applying machine learning to semiconductor manufacturing, IEEE Expert 8 (1) (Feb. 1993) 41–47, https://doi.org/10.1109/64.193054.
[30] J. Yu, P. Guo, Run-to-Run control of chemical mechanical polishing process based on deep reinforcement learning, IEEE Trans. Semicond. Manuf. 33 (3) (Aug. 2020) 454–465, https://doi.org/10.1109/TSM.2020.3002896.

[31] M.-H. Hsu, et al., Advanced CMP process control by using machine learning image analysis, in: 2021 IEEE International Interconnect Technology Conference (IITC), Kyoto, Japan, 2021, pp. 1—4, https://doi.org/10.1109/IITC51362.2021.9537421.
[32] T. Yu, et al., Predictive modeling of material removal rate in chemical mechanical planarization with physics-informed machine learning, Wear 426~427 (2019) 1430—1438, https://doi.org/10.1016/j.wear.2019.02.012.
[33] E. Kesters, et al., Cobalt pre-metallization clean and functional water rinse in BEOL interconnects, in: SPCC 2019, Portland, April 1—3rd, 2019. https://www.linx-consulting.com/wp-content/uploads/2019/04/01-02-E_Kesters-imec-Co_pre-metallization_clean_and_functional_water_rinse_in_BEOL_interconnect-1.pdf.
[34] L.M. Cook, Chemical processes in glass polishing, J. Non-Cryst. Solids 120 (1990) 152—171.
[35] T. Tanaka, et al., Nano size cerium hydroxide slurry for scratch-free CMP process, in: Proceedings of International Conference on Planarization/CMP Technology 2014, Kobe, Japan, 2014, pp. 22—24, https://doi.org/10.1109/ICPT.2014.7017236.
[36] W.G. America, S.V. Babu, Slurry additive effects on the suppression of silicon nitride removal during CMP, Electrochem. Solid State Lett. 7 (2004) G327, https://doi.org/10.1149/1.1817870.
[37] B.V.S. Praveen, et al., Abrasive and additive interactions in high selectivity STI CMP slurries, Microelectron. Eng. 114 (2014) 98—104, https://doi.org/10.1016/j.mee.2013.10.004.
[38] Z. Liu, Chemical Mechanical Planarization Compositions for Reducing Erosion in Semiconductor Wafers, 2007. US patent 7,300,603 B2.
[39] M. Stender, et al., Tungsten Chemical Mechanical Planarization (CMP) with Low Dishing and Low Erosion Topography, 2020. European Patent Application EP 3 604 468 A1.
[40] J.A. Siddiqui, et al., Method and Slurry for Tuning Low-$k$ versus Copper Removal Rates during Chemical Mechanical Polishing, 2008. US Patent Application, US 20080149884 A1.
[41] K.L. Stewart, et al., Relationship between molecular structure and removal rates during chemical mechanical planarization: comparison of Benzotriazole and 1,2,4-triazole, J. Electrochem. Soc. 155 (2008) D625.
[42] H.-Y. Ryu, et al., Selection and optimization of corrosion inhibitors for improved Cu CMP and post-Cu CMP cleaning, ECS J. Solid State Sci. Technol. 8 (2019) P3058.
[43] M. Aoki, et al., Wafer-level hybrid bonding technology with copper/polymer co-planarization, in: 2010 IEEE International 3D Systems Integration Conference (3DIC), Munich, Germany, 2010, pp. 1—4, https://doi.org/10.1109/3DIC.2010.5751471.
[44] A. Gupta, et al., Barrierless ALD Molybdenum for buried power Rail and via-to-buried power Rail metallization, in: 2022 IEEE International Interconnect Technology Conference (IITC), San Jose, CA, USA, 2022, pp. 58—60, https://doi.org/10.1109/IITC52079.2022.9881304.
[45] Z. Liu, T. Buley, Advanced CMP pad surface texture characterization and its impact on polishing, CMPUG Spring 2016 (Apr. 7, 2016).
[46] J. McGrath, C. Davis, Polishing pad surface characterization in chemical mechanical planarization, J. Mater. Process. Technol. 153—154 (2004) 666—673, https://doi.org/10.1016/j.jmatprotec.2004.04.094.
[47] W.-T. Tseng, unpublished results (2021).
[48] D. Rosales-Yeomans, et al., Design and evaluation of pad grooves for copper CMP, J. Electrochem. Soc. 155 (2008) H797, https://doi.org/10.1149/1.2963268.
[49] W.-T. Tseng, et al., Optimization of within-die planarity and defectivity for chemical-mechanical planarization, in: 2022 International CMP and Planarization Technology

Conference (ICPT), Portland, OR, USA, 2022. https://nccavs-usergroups.avs.org/wp-content/uploads/2023/01/P1-WTTseng.pdf.

[50] A.J. Khanna, et al., Engineering surface texture of pads for improving CMP performance of sub-10 nm nodes, ECS J. Solid State Sci. Technol. 9 (2020) 104003, https://doi.org/10.1149/2162-8777/abbcb5.

[51] S. Lee, et al., Pad designs—to navigate the fundamentals of CMP, in: 2022 International CMP and Planarization Technology Conference (ICPT), Portland, OR, USA, 2022. https://nccavs-usergroups.avs.org/wp-content/uploads/2023/01/ICPT22-Session-E4-Lee.pdf.

[52] A. Sunamaya et al., "Polyurethane for Polishing Layers, Polishing Layer, Polishing Pad and Method for Modifying Polishing Layer", European Patent Application # EP 3 878 897 A1; see also: https://www.kuraray.com/rd/topics.

[53] R. Bajaj, et al., CMP Pad Construction with Composite Material Properties Using Additive Manufacturing Processes, 2018. US patent # 9,873,180 B2.

[54] N.B. Kenchappa, et al., Soft chemical mechanical polishing pad for oxide CMP applications, ECS J. Solid State Sci. Technol. 10 (2021) 014008, https://doi.org/10.1149/2162-8777/abdc40.

[55] A. Philipossian, S. Olsen, Fundamental tribological and removal rate studies of interlayer dielectric chemical mechanical planarization, Jpn. J. Appl. Phys. 42 (2003) 6371−6379.

[56] M.Y. Tsai, et al., Development and analysis of double-faced radial and cluster-arranged CMP diamond disk, Math. Probl Eng. 2014 (2014) 9, https://doi.org/10.1155/2014/913812. Article ID 913812.

[57] C. Gould et al., "Novel Method to Measure the Sharpness of CMP Pad Conditioner Abrasive Tips", 2016 CMP User Group Meeting (CMPUG). https://nccavs-usergroups.avs.org/wp-content/uploads/CMPUG2016/CMP2016_7gould.pdf.

[58] Y.-C. Kim, S.-J.L. Kang, Novel CVD diamond-coated conditioner for improved performance in CMP processes, Int. J. Mach. Tool Manufact. 51 (6) (2011) 565−568, https://doi.org/10.1016/j.ijmachtools.2011.02.008.

[59] S. Lee, "Recent trend of CMP equipment platform and its requirement of process and consumables: BEOL CMP", 2018 CAMP CMP symposium, https://nccavs-usergroups.avs.org/wp-content/uploads/CMPUG2018/CMP418-2-Lee.pdf.

[60] W.-T. Tseng, et al., Microreplicated conditioners for Cu barrier chemical-mechanical planarization (CMP), ECS J. Solid State Sci. Technol. 4 (2015) P5001, https://doi.org/10.1149/2.0011511jss.

[61] A. S. Lawing, "Pad Conditioning Effects in Chemical Mechanical Polishing", 2004 CMP User Group Meeting (CMPUG), https://nccavs-usergroups.avs.org/wp-content/uploads/CMPUG2004/CMPUG_05_2004_Lawing.pdf.

[62] W.-T. Tseng, et al., Post Cu CMP cleaning process evaluation for 32nm and 22nm technology nodes, in: SEMI Adv. Semicond. Manuf. Conf., Saratoga Springs, NY, USA, 2012, pp. 57−62, https://doi.org/10.1109/ASMC.2012.6212868.

[63] W.-T. Tseng, Chap. 17: approaches to defect characterization mitigation and reduction, in: S. Babu (Ed.), Woodhead Publishing Series in Electronic and Optical Materials, Advances in Chemical Mechanical Planarization (CMP), second ed., Woodhead Publishing, 2022, pp. 591−627, https://doi.org/10.1016/B978-0-12-821791-7.00004-6.

[64] C.F. Graverson, et al., Development of "soft" cleaning chemistries for enhanced STI post-CMP cleaning, ECS Trans. 92 (2019) 165, https://doi.org/10.1149/09202.0165ecst.

[65] L. Yang, et al., Composite complex agent based on organic amine alkali for BTA removal in post CMP cleaning of copper interconnection, J. Electroanal. Chem. 910 (2022) 116187, https://doi.org/10.1016/j.jelechem.2022.116187.

[66] J. Seo, et al., Post-CMP cleaning solutions for the removal of organic Contaminants with reduced galvanic corrosion at copper/cobalt interface for advanced Cu interconnect applications, ECS J. Solid State Sci. Technol. 8 (2019) P379, https://doi.org/10.1149/2.0011908jss.

[67] J.-H. Lee, et al., Comparative evaluation of organic contamination sources from roller and pencil type PVA brushes during the Post-CMP cleaning process, Polym. Test. 90 (2020) 106669, https://doi.org/10.1016/j.polymertesting.2020.106669.

[68] Y. Sampurno, et al., Understanding the Reasons behind defect levels in post-copper-CMP cleaning processes with different chemistries and PVA brushes, ECS J. Solid State Sci. Technol. 10 (2021) 064011, https://doi.org/10.1149/2162-8777/ac0b8d.

[69] J.-H. Lee, et al., A Breakthrough method for the effective conditioning of PVA brush used for post-CMP process, ECS J. Solid State Sci. Technol. 8 (2019) P307, https://doi.org/10.1149/2.0111906jss.

[70] See, for examples C.-L. Chu, et al., Semicond. Sci. Technol. 35 (2020) 045001, https://doi.org/10.1088/1361-6641/ab675d. B. N. Sahoo, et al, Ultrasonics Sonochemistry, 82, 105859 (2022). https://doi.org/10.1016/j.ultsonch.2021.105859; A. A. Busnaina and N. Moumen, Materials Research Society symposia proceedings. Materials Research Society 566(1) (2011). https://doi.org/10.1557/PROC-566-247.

[71] R.K. Singh, et al., Post-CMP clean PVA brush design advancements and characterization in Cu/low-k applications, in: International Conference on Planarization/CMP Technology ICPT, Dresden, Germany, 2007, pp. 1—5.

# CHAPTER 7

# Limits of gate dielectrics scaling

Shahab Siddiqui[1], Takashi Ando[2], Rajan Kumar Pandey[3] and Dominic J. Schepis[4,5]

[1]IBM Research Albany, Albany, NY, United States; [2]IBM T. J. Watson Research Center, Yorktown Heights, NY, United States; [3]Vellore Institute of Technology, Vellore, Tamil Nadu, India; [4]IBM Microelectronics, East Fishkill, NY, United States; [5]GLOBALFOUNDRIES, Austin, TX, United States

## 1. Introduction

Silicon dioxide as a gate oxide reached its scaling limit somewhere around 90–65 nm technology node due to excessive leakage below 1 nm physical oxide thickness. To overcome silicon dioxide scaling limits, hafnium oxide as a high k dielectric material was introduced in the 45 nm node. Introduction of high k dielectric as a gate oxide enabled gate length and EOT scaling for improved device performance and area scaling. However, further physical scaling of hafnium dioxide ($HfO_2$) looks difficult due to increased leakage below 1.5 nm physical thickness. For continued scaling other device innovations such as FINFET and FDSOI technologies have been introduced. Introduction of FinFET allowed us to relax EOT criteria due to excellent short channel control of FinFET devices, however, future gate length scaling of FinFET devices still require EOT to be scaled, especially for ultra-low threshold devices. In this chapter, we will discuss current status of hafnium oxide as a replacement of $SiO_2$, and how EOT can be scaled practically by modifying $HfO_2$, for example by doping hafnium oxide to form HfOxN, HfLaOxN, and potentially some bilayer dielectrics such as $HfO_2/TiO_2$ to increase K value of the dielectric stack without changing the entire material system in gate stack. These modified $HfO_2$ systems are practical approaches and have potential to implement at future highly scaled FinFET, FDSOI devices and beyond FinFETs such as gate-all-around (GAA), and alternative channel (SiGe and Ge) based FINFET and nanowire devices.

We will discuss innovation in high-voltage input/output (I/O) devices gate oxide due to dual channel planar high-performance devices, and later into highly scaled FinFET to gate-all-around architecture. Rise of system on chip (SoC) and Hik/metal gate requires careful integration of thin film gate oxide processes for I/O oxide. It is evident that I/O oxide will play a critical role in FinFET and GAA schemes for overall transistor strategy as we push forward on transistor scaling beyond 5 nm FinFET and new device architectures such as GAA and alternate channel substrates.

Silicon germanium channels as a mobility enhancement, and gate oxide scaling knob results will be reviewed. In this chapter, we will also review how ab-initio calculations (density functional theory) are playing a role in developing gate oxides and overall stack by providing guidance to technologists and engineers, and how gate dielectric can be optimized by utilizing ab-initio calculations as a tool available to technologist, material chemists, and semiconductor process research and development community.

In this edition, we review state-of-the-art nano-sheet transistor technology and transistor integration details. For gate oxides, we briefly touch upon current state of NS logic gate oxide and technology requirements. We will review detailed results and options for I/O transistor gate oxide for nano-sheet devices with various options and their results evaluated at IBM Research. We review and provide results on gate oxide formation for high mobility SiGe (25–80 atomic % Ge on SiGe) substrate for future high-performance devices for further CMOS scaling and drive current enhancements.

## 2. Dennard scaling theory

In early 1970s Robert Dennard at IBM Research introduced the principle of MOS transistor scaling called "Dennard's scaling theory" [1]. It will not be an overreach to claim that Dennard scaling theory is the foundation which drove significant innovation in CMOS miniaturization for over 30 years, which covers device design, device integration, lithography processes, equipment innovation, silicon processes and materials breakthroughs. Dennard scaling theory will be discussed in detail in next section.

### 2.1 Constant electric-field scaling

The basic idea of scaling, shown in Fig. 7.1, is to reduce the dimensions of the MOS transistors and the wires connecting them in integrated circuits. Thus, the right side of Fig. 7.1 is scaled down in size from that on the left by reducing all dimensions by a factor of $\alpha$. The MOS transistor works on the

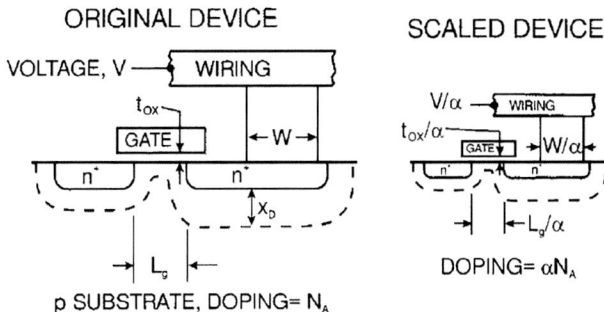

**Figure 7.1** Schematic illustration of the scaling of Si technology by factor alpha. *(Adopted from D.J. Frank, et al., Proc. IEEE 89 (3) (2001) 259–288.)*

principle of modifying the electric field in the silicon substrate underneath the gate in such a way as to control the flow of current between the sources and drain electrodes. Scaling achieves the same electric-field patterns in the smaller transistor by reducing the applied voltage along with all the key dimensions, including the thickness $t_{ox}$ of the insulating oxide layer between the gate and the silicon substrate [2].

In order to keep the same electric field pattern as the original transistor in the substrate of the scaled device, original transistor impurity doping concentration is increased for the smaller device.

Taken along with the reduced applied voltage, this reduces the size of the depletion regions, identified by $x_d$ in Fig. 7.1, underneath all three transistor electrodes (gate, source, and drain) [2]. In general, these depletion regions must be kept separated so that the transistor can be turned off properly by the control gate [3]. The scaled down depletion regions in the transistor on the right of Fig. 7.1 allows the separation L between source and drain to be reduced along with the other physical dimensions. In this simple constant-electric field transformation, the dimension, voltage and doping are all modified by common factor $\alpha$, as noted in Fig. 7.1 [2].

This constant-electric-field scaling gives three important results. First, the density improves by a factor $\alpha^2$ due to the smaller wiring and device dimensions. Next, the speed, which is related to $g_m/C$, improves by a factor $\alpha$ because the capacitance (C) of the shorter wires and smaller devices is reduced by d while the transconductance ($g_m$) of the devices (scaled in both length and width) remains about the same. Finally, the power dissipation per circuit is reduced by factor $\alpha^2$ because of the reduced voltage and current in each device, with the important result that power density is constant [2].

## 2.2 Generalized scaling

In constant electric field scaling theory, the supply voltage reduction proportion to device dimension was not adopted due to reservations to depart from standardized voltage levels. For generalized scaling, the electric field patterns within a scaled device are still preserved, but the intensity of the electric field can be changed everywhere within the device by a multiplicative factor $\epsilon$. Thus, the applied voltage, which is given by $\epsilon/\alpha$, can be scaled less rapidly by allowing $\epsilon$ to increase. The electric field patterns with the device are maintained by increasing the doping impurity concentration by a factor $\epsilon$, which preserves the size of the depletion region $x_d$ defined in Fig. 7.1 [2].

There were fundamental limitations to generalized scaling (increased electric-field), first being as $\epsilon$ increases the long-term reliability of device degradation occurred, such as hot carrier mechanisms, and gate dielectric breakdown. The second limitation is increase in power dissipation, which increases by $\epsilon^2$ when speed is constant, where power is given by $CV^2f$. Summary of constant and generalized scaling is given in Table 7.1 [4].

## 3. Gate oxide and EOT scaling

The continued performance improvement and cost reduction of complementary metal-oxide-semiconductor (CMOS) integrated circuits have been accomplished by a calculated reduction of all dimensions of a transistor: a practice termed "scaling [1]". The keystone for this scaling is the ability to reduce the thickness of gate dielectric, a core part of a field-effect-transistor (FET). Fig. 7.2 shows IBM's historical $SiO_2$ and SiON (silicon oxynitride) gate dielectric thicknesses over the past several decades (after [6]). The trend shows a steady oxide thickness ($T_{ox}$) scaling over many generations starting from several hundred Å, which had been a key driver of CMOS device miniaturization. However, the oxide thickness reached a saturation beyond CMOS10S (the 90 nm node) when the gate dielectric became thin enough to cause quantum mechanical tunneling through the oxide, resulting in unmanageable gate leakage currents.

At this point, replacement of SiON gate dielectric with high dielectric constant (high-k) oxides became mandatory for continued device scaling. Scalability of high-k materials is often evaluated with a metrics called equivalent oxide thickness (EOT), which represents the theoretical thickness of $SiO_2$ that would be required to achieve the same capacitance density as described in the following equation.

Table 7.1 Technology scaling rules for three cases (planar transistor) [4].

| Physical parameter | Constant-electrical field scaling factor | Generalized scaling factor | Generalized selective scaling factor |
|---|---|---|---|
| Channel length, insulator thickness | $1/\alpha$ | $1/\alpha$ | $1/\alpha_d$ |
| Wiring width, channel width | $1/\alpha$ | $1/\alpha$ | $1/\alpha_w$ |
| Electric field in device | 1 | $\epsilon$ | $\epsilon$ |
| Voltage | $1/\alpha$ | $\epsilon/\alpha$ | $\epsilon/\alpha_d$ |
| On-current per device | $1/\alpha$ | $\epsilon/\alpha$ | $\epsilon/\alpha_w$ |
| Doping | A | $\epsilon\alpha$ | $\epsilon\alpha_d$ |
| Area | $1/\alpha^2$ | $1/\alpha^2$ | $1/\alpha_w^2$ |
| Capacitance | $1/\alpha$ | $1/\alpha$ | $1/\alpha_w$ |
| Gate delay | $1/\alpha$ | $1/\alpha$ | $1/\alpha_d$ |
| Power dissipation | $1/\alpha^2$ | $\epsilon^2/\alpha^2$ | $\epsilon^2/\alpha_w\alpha_d$ |
| Power density | 1 | $\epsilon^2$ | $\epsilon^2/\alpha_w\alpha_d$ |

$\alpha$ is the dimensional parameter, $\epsilon$ is the electric field scaling parameter, and $\alpha_D$ and $\alpha_W$ are separate dimensional scaling parameters for the selective scaling case. A $\alpha_D$ is applied to the device vertical dimensions and gate length, while $\alpha_W$ applies to the device width and the wiring

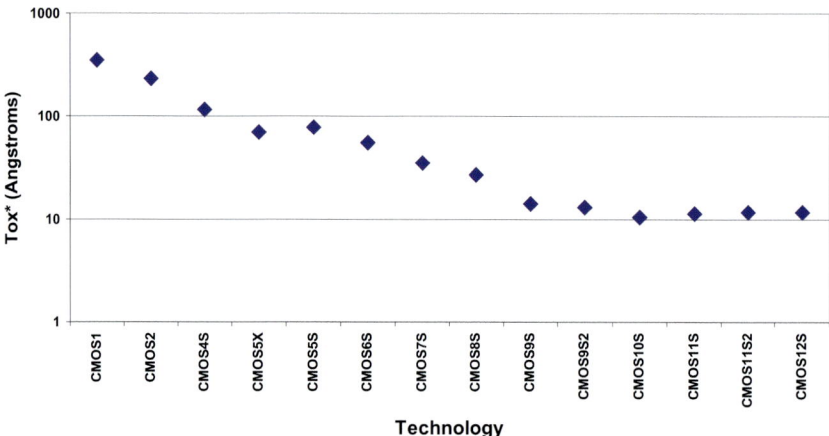

**Figure 7.2** Historical trend of gate dielectric thickness ($T_{ox}$) in IBM technologies over several decades in time. *(Adopted from M. Chudzik, et al., Wiley-VCH Verlag GmbH & Co. KGaA, Chapter 17 (2012).)*

$$\text{EOT} = t_{high-k} \frac{k_{SiO2}}{k_{high-k}} = \frac{k_{SiO2}\varepsilon_0 A}{C_{ox}} \quad (7.1)$$

where $t_{high-k}$ is the thickness of the high-k film, $k_{SiO2}$ is the dielectric constant of $SiO_2$, and $k_{high-k}$ is the dielectric constant of the high-k film, $\varepsilon_0$ is the permittivity of free space (=8.8531023 fF/μm), A is the area of the MOS structure, $C_{ox}$ is the capacitance of gate dielectric. Capacitance-equivalent thickness (CET) is defined by using total capacitance of the MOS stack instead of $C_{ox}$ in Eq. (7.1). CET is used when the capacitance from the semiconductor substrate via quantum mechanical effects is difficult to estimate. CET in inversion state of MOSFET is termed $CET_{inv}$ or $T_{inv}$. As seen in Eq. (7.1), if $k_{high-k}$ is greater than $k_{SiO2}$, the EOT value can be reduced without scaling the physical thickness. After a decade-long search for the appropriate high-κ materials, the semiconductor industry has converged on $HfO_2$ based oxides as will be reviewed in the following section. Fig. 7.3 summarizes the trend of $T_{inv}$ values as a function of CMOS technology node for high-k metal gate stacks reported in literature from various groups [7–13].

The trend for the last generations of SiON based technologies [6] are shown for comparison. The disruptive $T_{inv}$ scaling is clearly seen at the 45–32 nm node when $HfO_2$ gate dielectrics and metal electrodes were introduced to CMOS technology for the first time. This broke the stagnation of gate dielectric scaling in the last couple of generations and re-

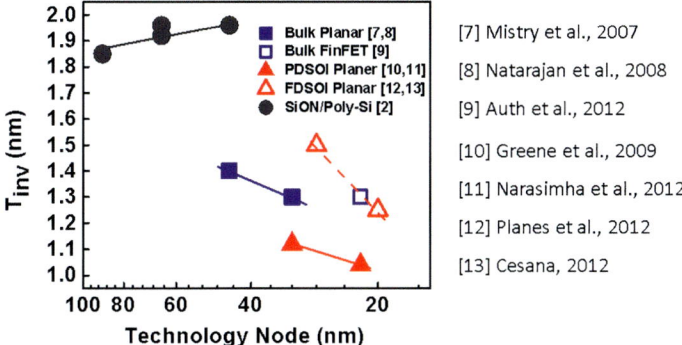

**Figure 7.3** Inversion oxide thickness ($T_{inv}$) as a function of CMOS technology node for high-k metal gate stacks reported in literature from various groups. The trend for the last generations of SiON based technologies are shown for comparison.

enabled Dennard scaling [1]. After the introduction of the first generation high-k products, the gate oxide thickness scaling did not follow exactly the lateral device scaling as in the ideal case (×0.7 scaling per generation), mainly due to a trade-off between EOT scaling and device reliability, which we will review later. Instead, it became a standard practice to scale EOT by 0.1 nm per generation to continually improve electrostatics control, as seen for bulk Si planar technology [7,8] and partially depleted Si on insulator (PDSOI) planar technology [10,11]. When fully depleted device architectures, such as FinFET or fully depleted SOI (FDSOI), were introduced, the gate control of the channel was improved by the device geometry and the requirement for EOT scaling was relaxed, resulting in higher $T_{inv}$ values for [9,12,13] compared to planar technologies at a given technology node. However, even for those new device architectures, a mild EOT scaling (approximately 0.1 nm at a time) was still observed and the trend is expected to continue for the 14 nm and beyond. Predicted EOT from ITRS along with supply voltage trends are shown in Fig. 7.4, which requires continuous EOT scaling. However, we will discuss in depth that EOT scaling with fixed gate leakage is a significant challenge and would be very difficult to meet ITRS projections.

Hafnium Dioxide (High K Dielectrics) for Continued Scaling.

Over last decade hafnium-based dielectrics have emerged as a replacement of $SiO_2$ and SiON gate insulator, and a choice for a future transistor's structures. The Initial evaluation for the gate oxides focused on dielectric constant which are attributed to the polarizability due to electronic and ionic dipoles in the GHz frequency window that is needed in CMOS [14]

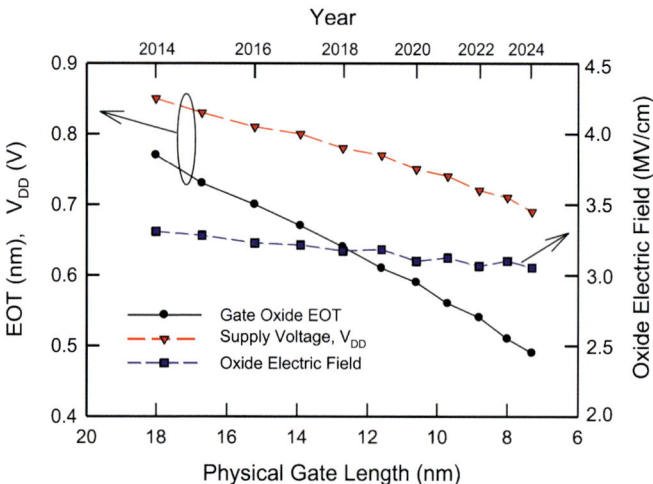

**Figure 7.4** Predicted trends of gate oxide EOT and supply voltage, $V_{DD}$, scaling for the technology node in the coming decade. Oxide electric fields are calculated based on the physical thickness of gate oxide and $V_{DD}$ as predicted by ITRS 2013. *(Adopted from H. Wong, et al., Microelectron. Eng. 138 (2015), 57–76.)*

as shown in Fig. 7.5 [15]. The fact that the dielectric response is mainly based on ionic and electronic polarization at high frequencies dictates a metal element that forms an ionic bond with oxygen and has a large atomic number [14]. Transition metal oxides thus emerged as promising candidates, especially those of heavy metal elements, to maximize the dielectric constants [14]. Besides the dielectric constant value (k), the dielectric bandgap (Eg) and sufficient bandgap offset with silicon conduction and valence band is necessary.

To replace $SiO_2$ and SiON to allow continued scaling, following key dielectric material properties must be met.

(a) Dielectric constant (k) value should be between the ranges of $10 - 30$ (compared to 3.9 for $SiO_2$) [16]. A too large of a k vale (e.g., $TiO_2$ or $SrTiO_3$) is not a good candidate as it will result in a two-dimensional electric fringing field from the drain through the physically thick gate dielectric of the MOS transistors. This fringing field can lower the source-to-channel potential barrier and hence the threshold voltage [14]. From device design perspective, Fig. 7.6 shows a window of dielectric constant k value and insulator thickness for 15 nm depletion depth. This provides good window for engineers and scientist working on finding ways to further scale gate dielectric. It is obvious from

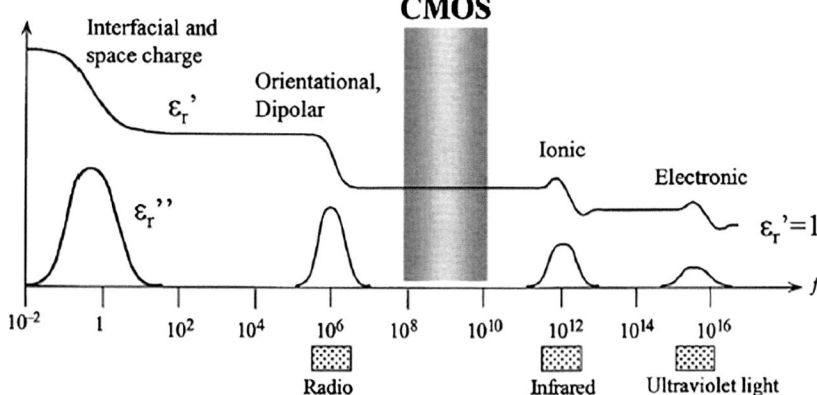

**Figure 7.5** The frequence dependence of the real ($\epsilon_r'$) and imaginary ($\epsilon_r''$) parts of the dielectric permittivity. In CMOS devices, ionic and electronic contributions are present [15].

(Fig. 7.6) plot that innovation beyond hafnium oxide dielectric for EOT scaling is quite challenging. However, some potential modification can be made to HfOx to further increase k value within the window defined in Fig. 7.6 leading to Tinv scaling hence performance improvements.

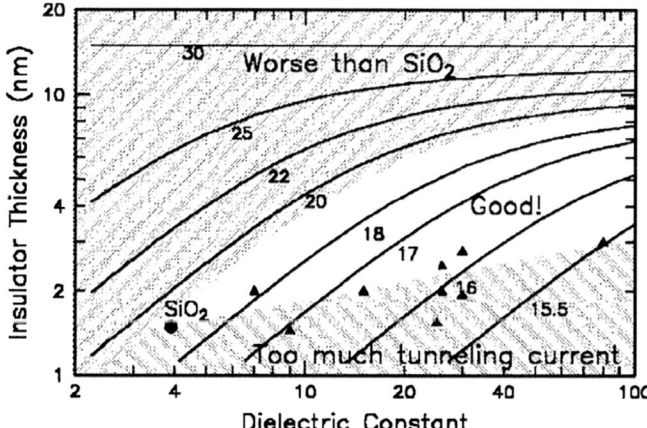

**Figure 7.6** Contour of constant scale length versus dielectric constant and insulator thickness, showing the useful design space for high-k gate dielectrics. Data points are rough estimates of the tunneling constraints for various high-k insulators. Depletion depth is 15 nm here. Useful design space will shrink with decreasing depletion depth [4].

(b) It should have a large band gap (Eg > 5 eV) and large enough band offset (>1 eV) with silicon energy bands (Ec, conduction band, and Ev, valence band). This is to minimize carrier injection into its bands [16]. Band gap vs. k values for different gate oxides and band offsets to silicon conduction and valence bands are shown in Fig. 7.7 for different high k dielectrics.
(c) The dielectric should display low density of defects within its bulk region and as well as its interface with bottom Si channel and with top metal gates [16].
(d) Ability to handle high temperature thermal budget, although replacement metal gate technology relaxes the high temperature requirements, as high temperature processes are performed before gate stack formation.

With careful balance of all the above requirements for gate dielectric material, hafnium dioxide is emerged to replace $SiO_2$ and SiON and successfully implemented at 45 nm and beyond and continue to be a choice of gate dielectric for FinFETs and future architectures such as gate-all-around (GAA) and alternate channel devices.

Hafnium dioxide has a high permittivity, dielectric constant value of 20 and relatively large bandgap of 5.7 eV, large barrier height at interface with silicon, and provides several orders of lower leakage current than $SiO_2$ at same EOT or $T_{inv}$ value. However, introduction of Hik dielectrics required introduction of metal gates instead of heavily doped polysilicon gate electrode due to thermodynamic interface instability of polysilicon and Hi k dielectric. Summary of key dielectric properties between $SiO_2$ and $HfO_2$ is given in Table 7.2.

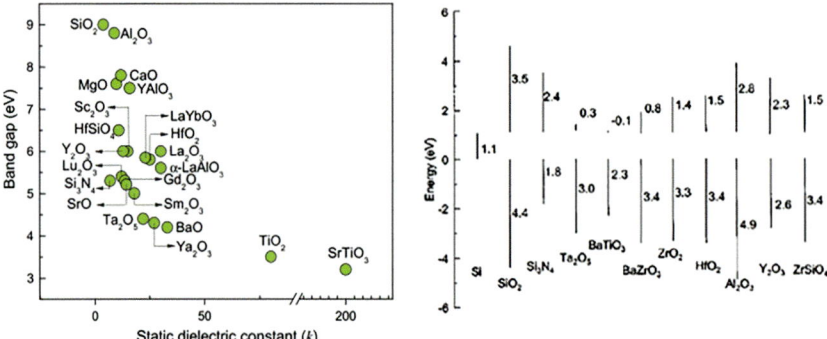

**Figure 7.7** Band gap ($E_g$) versus static dielectric constant (k) for representative high-K material and band offset to silicon conduction and valence bands [15,16].

**Table 7.2** Comparison of different dielectric properties of silicon dioxide and hafnium dioxide critical for gate oxide applications

| Parameter | $SiO_2$ | $HfO_2$ |
|---|---|---|
| Dielectric constant | 3.9 | 20 |
| Bandgap (eV) | 9 | 5.7 |
| Conduction band offset $\Delta E_c$ (eV) | 3.5 | 1.5 |
| Valence band offset $\Delta E_v$ (eV) | 4.4 | 3.4 |
| Direct tunneling thickness (nm) | 3.0 | 3.5 |
| Chemical bonding | Covalent | Ionic |

Hafnium dioxide is well known for its intrinsic bulk defects than $SiO_2$ and would be useful to review fundamental differences with $SiO_2$. The $SiO_2$ possesses such a low concentration of defects for following reasons:
(1) $SiO_2$ has high heat of formation (-218 kcal/mol), as a result, it is difficult to form defects in $SiO_2$ matrix
(2) $SiO_2$ has a covalent bonding with a low coordination. This makes $SiO_2$ an excellent glass former, as a result $SiO_2$ is in amorphous form [17].
(3) Bonding in the a-$SiO_2$ can relax locally to minimize the defect concentration. The defects are dangling bonds, and these can be removed by having these defects react with network, leading to re-bonding. This particularly occurs at the Si:$SiO_2$ interface [17].

For hafnium dioxide, the following are the potential fundamental dielectric properties responsible for high bulk defects [17]
(1) $HfO_2$ also has large heat of formation (-271 kcal/mol), higher than $SiO_2$. This means the equilibrium concentration of non-stoichiometric defects should be low. However, the non-equilibrium concentration of defect is high, because the oxide network is not able to relax and rebound to remove defects [17].
(2) The nature of intrinsic defects in ionic oxides differs from those in $SiO_2$ (covalent). They are oxygen vacancies, oxygen interstitials or oxygen deficiency defects due to multiple valence of the metals [17]. These oxygen vacancies are the most problematic to manage long term gate oxide reliability and proven to be the biggest challenge as CMOS technology transitioned to high k dielectrics.
(3) The high k oxides are ionic bonding in nature, and they have higher coordination number [17]. The greater ionic character of the bonding and the higher coordination mean that the high K oxides are poor glass formers [17]. This is well known as it is difficult to maintain these oxides as amorphous during high temperature processing. However,

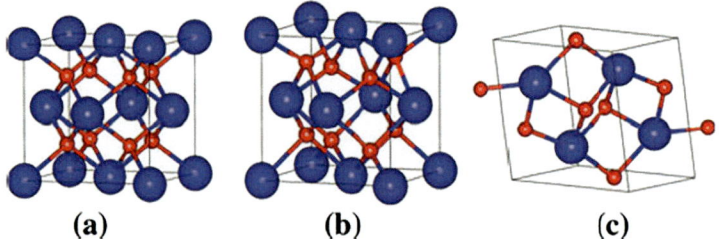

**Figure 7.8** (a) cubic, (b) tetragonal, (c) monoclinic HfO$_2$ [16].

crystallization temperature can be increased by dopants, such as nitrogen, lanthanum, and aluminum incorporation in HfO$_2$ matrix.

### 3.1 Physical structure of hafnium dioxide

HfO$_2$ can exist in several polymorphic phases. However, for advanced CMOS gate dielectric applications atomic layer deposition (ALD) is a choice of HfO$_2$ growth method. For as deposited HfO$_2$ using ALD technique HfO$_2$ is typically amorphous (a-HfO$_2$). This ALD amorphous HfO$_2$ starts to show some crystallinity above 900°C—60 s rapid thermal anneals, but it can have negative effects for dielectric leakage. Stable amorphous phase of HfO$_2$ is current state of structure on most advanced commercially available 14 nm FinFET technology.

The other three low pressure crystalline polymorphs of HfO$_2$ are the monoclinic, tetragonal, and cubic phases as illustrated in Fig. 7.8 [16]. Among the three phases monoclinic phase is stable at low temperature, which has the lowest free energy of formation and the largest volume [14]. Monoclinic phase can be phase transitioned to the tetragonal phase (t-HfO$_2$) at ∼ 2000 K and subsequently to the cubic phase (c-HfO$_2$) at ∼ 2870 K [16].

**m-HfO$_2$:** has four HfO$_2$ units in primitive cell; in each unit the Hf site is sevenfold coordinated, and the two O sites are threefold and fourfold coordinated, respectively [16]

**c-HfO$_2$:** Hf atoms at face centered cubic lattice sites and O atoms occupying all tetrahedral interstitial sites (Fluorite structure).

**t- HfO$_2$:** tetragonal phase can be obtained by deforming cubic phase.

## 4. Hafnium based ternary, quaternary and bilayer oxides for EOT scaling

Hafnium dioxide is well established choice of gate dielectric from 45 nm and beyond. To further scale EOT by decreasing HfO$_2$ thickness or introduction of higher K dielectric lead to unacceptable increase in leakage current. As a result, in this section we will cover few hafnium based ternary and quaternary systems which can be important for further EOT scaling without a major increase in gate leakage. Dielectric systems we will review are HfOxN, HfLaO$_x$N, HfAlO$_x$N and HfO$_2$/TiO$_2$ bilayer systems, which are under research from various research groups and show promising results.

## 4.1 Hafnium oxynitride (HfOxN)

Plasma nitridation of high k dielectrics such as HfSiO$_2$, HfO$_2$, HfLaO, and HfAlO to form nitrided ternary and quaternary dielectrics is one of the practical and effective ways to obtain lower EOT at reasonable gate leakage. EOT scaling due to nitridation of as deposited hafnium-based dielectric is due to reduction in phase separation, and higher crystallization time [18]. In addition, combination of hafnium-based dielectrics and interfacial layer (IL) nitridation leads to nitridation of interfacial layer, which results in higher k IL (SiON) leading to smaller EOT. Fig. 7.9 shows some of the early works in EOT scaling by plasma nitridation of HfSiO oxide. This result clearly shows that increasing nitrogen content % (N%) reduces EOT and gate leakage.

However, there is a small N% window to scale EOT and needs to be balanced carefully with other device parameters, such as pFET threshold voltage (net positive charge from N increase pFET Vt), bias temperature instabilities increase due to trapped charge formation in the oxide and at the interface due to nitridation, carrier mobility also degrades with increasing N%, as increase in nitrogen can place electron scattering 'sites' at the

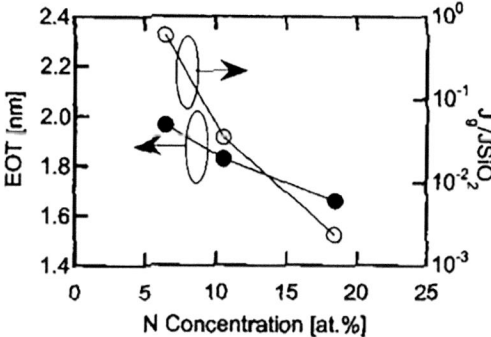

**Figure 7.9** EOT and gate leakage reduction versus nitrogen concentration relationship by plasma nitridation method for hafnium silicate films (40% Si, 4.5 nm thickness) [18].

substrate/IL interface leading to reduction in carrier mobility in the channel.

Min et al. have studied HfO$_2$ nitridation and annealing window to scale EOT at manageable leakage [19]. In this work, we not only studied the HfO$_2$ nitridation mechanism in bulk HfO$_2$ but also characterized nitrogen distribution in interfacial layer (IL) and HfO$_2$ as holistic system to increase dielectric constant value leading to EOT scaling. Fig. 7.10 shows the film leakage current (N-MOSFET) at 1 V and the corresponding EOT for various plasma nitridation/post Nitridation anneal (PNA) conditions. The as-deposited HfO$_2$ film is denoted by A, B denotes the HfO$_2$/IL stack with plasma nitridation only, and D denotes the annealed HfO$_2$/IL stack (750°C annealing without nitridation). It can be seen that by combining the low power plasma nitridation with the 750°C PNA, the film leakage can be reduced at the scaled EOT as denoted by C, compared to A (as-deposited HfO$_2$ film) and D (as-deposited HfO$_2$ film after annealing). When the plasma nitridation power is further increased (and combined with 750°C PNA), the EOT of the film is scaled more aggressively but with a significant degradation in leakage current (denoted by X). The EOT scaling benefit from PNA suggests the k value of the HfO$_2$/IL stack is more likely improved by driving the nitrogen atoms to the lower k IL, and pure nitridation of the HfO$_2$ layer shows less contribution to the k value change as indicated by B at the thicker EOT in Fig. 7.10.

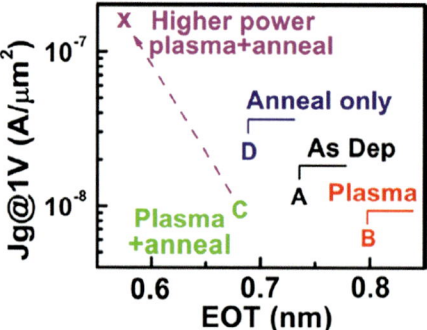

**Figure 7.10** Leakage current and EOT of the HfO$_2$/IL film (N-MOSFET) with various combinations of plasma nitridation conditions and PNA (at 750°C for 10 s): (a)-as-deposited HfO$_2$, (b)-low power plasma nitridation, (c)-low power plasma nitridation and PNA, (d)-as-deposited HfO$_2$ with anneal only, X-more aggressive higher power plasma nitridation and PNA [19].

To confirm nitridation of $HfO_2$ and IL, we used X-ray photoelectron spectroscopy (XPS). Fig. 7.11 first shows that nitrogen indeed is present in $HfO_2$/IL system.

To study the stability of the film after nitridation, PNA was performed in an $N_2$ ambient by RTA. Fig. 7.11 shows the N1s spectra of the $HfO_2$/IL film after 40 s of low power plasma nitridation and then annealed at different PNA temperatures. Compared to the unannealed film, the films annealed at elevated temperatures show less metastable nitrogen as indicated by the reduction in the corresponding peak intensity at 405 eV. In addition, the increasing intensity of the N–Hf peak at 397 eV suggests that part of the metastable nitrogen further reacts with the $HfO_2$ to form N–Hf bonds. Part of the metastable nitrogen also diffuses to the IL, implied by the intensity increasing of the N–Si peak and the Si2p peak. From the magnitude of the intensity change of the metastable nitrogen peak during the annealing, it seems that nitrogen atoms start to move at temperatures between 600 and 650°C. At temperatures of 750°C, $HfO_2$ crystallization may occur and, therefore, the change of nitrogen becomes more substantial.

In summary, plasma nitridation of $HfO_2$ and $HfSiO_2$ is a practical way to scale EOT, with careful balance with leakage and NBTI. Additionally, nitrogen incorporation leads to $HfO_2$ phase stability and crystalline temperature to as high as 1000°C. With advancement in plasma nitridation technology such as low energy radio frequency generators, it is possible to

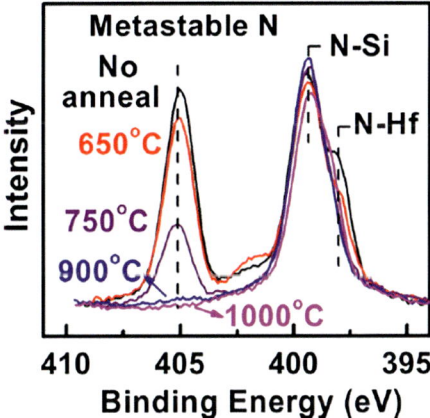

**Figure 7.11** XPS spectra of 10 s of PNA for the $HfO_2$/IL films after the 40 s low power plasma nitridation process. The PNA temperatures are varied from 650 to 1000°C. Metastable nitrogen is suppressed after the annealing (partly diffuses to the IL or is converted to N–Hf) [19].

incorporate nitrogen in sub 2 nm HfO$_2$ and IL without degrading reliability and mobility.

## 4.2 Hafnium lanthanum oxynitride (HfLaOxN)

HfLaOx is another technologically important hafnium-based oxide which shows EOT scaling upon plasma nitridation, higher crystalline temperature (amorphous dielectric up to 1000°C), and potential to implement in advanced nodes without significant disruption. HfLaOx has a higher k dielectric constant compared to HfO$_2$, and lanthanum (La) incorporation allows to tune wide range of threshold voltages due to lanthanum's nature to form dipole at IL and HiK interface. Plasma nitridation of HfLaOx allows to improve thermal stability [20]. Qiuxia et al. and Ariyoshi et al. [20,21] both show EOT scaling down to 0.62—0.65 nm (Fig. 7.12). Nitridation of HfLaOx leads to HfLaO and IL nitridation, IL nitridation not only nitrides the IL, but also scales the IL by forming nitrided silicate layer. Nitridation of IL play a bigger role in EOT scaling as it nitrides the IL, making SiON type interface, and also reacting with bulk dielectric forming silicate type material. XPS results from Ariyoshi et al. in Fig. 7.13 show Si—N formation at the IL. Fig. 7.14 shows TEM image of IL reduction due to plasma nitridation and low temperature anneal, which confirms that IL being scaled due to nitridation and interfacial reaction

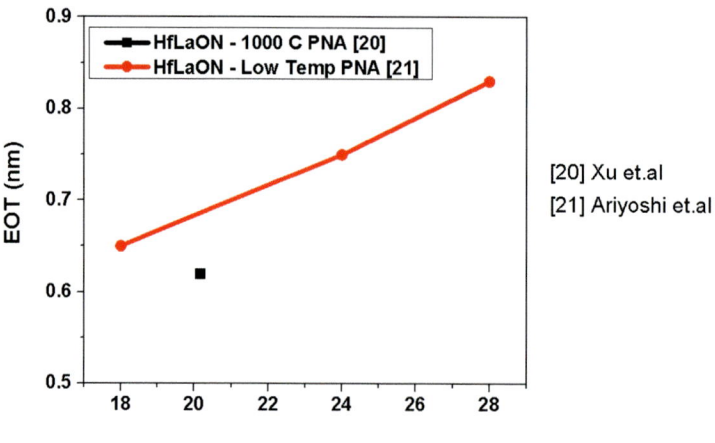

**Figure 7.12** Physical thickness of HfLaOxN and measurement EOT for two different post nitridation anneal temperatures. 0.62—0.65 nm EOT scaling was achieved by nitriding the bulk of HfLOx and IL.

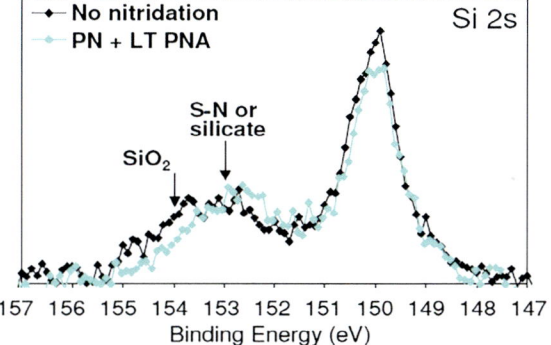

**Figure 7.13** Plasma nitridation and low temperature annealing of HfLaOx and IL: IL nitridation show shift to Si 2s peak indicating nitridation and possible formation of silicate formation [21].

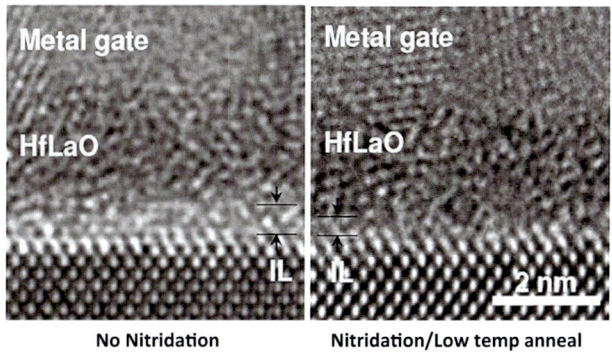

**Figure 7.14** TEM cross section image of HfLaO (a) without nitridation, showing significantly thicker IL, (b) with plasma nitridation where IL is scaled due to interfacial reaction forming SiON or Hf/La based silicate [21].

between hafnium lanthanum oxide and $SiO_2$ IL leading to thinner IL, which in turn leads to thinner EOT.

### 4.3 Bilayer gate dielectrics: HfO$_2$/TiO$_2$ higher 'K" for EOT scaling

It is well known after significant research that $TiO_2$ as a single gate dielectric layer for silicon CMOS transistor has fundamental challenges. Currently it appears that these challenges are difficult to overcome by materials or semiconductor process and integration innovations.

To briefly summarize, following are the key challenges to implement $TiO_2$ as a gate dielectric:

(1) $TiO_2$ has quite small bandgap (3.5 eV) and small $\Delta E_c$ (eV) to Si (1.2 eV). This $TiO_2$/Silicon electrical properties lead to increase in gate leakage [22].
(2) $TiO_2$/Si is thermodynamically unstable, as $TiO_2$ upon contact to Si and with additional thermal annealing tends to phase separate into $SiO_2$ and metal oxide ($M_xO_y$, M = metal). This instability resulting into formation of additional $SiO_2$ at Si/$TiO_2$ which leads to significant EOT increase, resulting in a performance degrade
(3) Oxygen up-diffusion (from $TiO_2$ to metal gates), can grow unnecessary metal oxide complexes [23,24] leading to threshold voltage (Vt) instability to significant Vt shift.

In order to overcome these challenges following strategies can be adopted, and some promising results from several groups have been demonstrated. Most promising approach to inhibit interfacial regrowth is to use $HfO_2$/$TiO_2$ bilayer stack, where there is significant reduction in interfacial layer regrowth in comparison to $TiO_2$/Si [23]. However, there is still significant regrowth of the interface due to $HfO_2$ not being such a good oxygen barrier. Oxygen diffusion can be inhibited further by carefully improving the quality of $HfO_2$, e.g., (a) using high quality amorphous ALD $HfO_2$ (b) further densifying the $HfO_2$ by annealing, (c) low N% incorporation and annealing of $HfO_2$ [25]. These strategies clearly show no interfacial regrowth at Si/HfOx interface, while presence of ultra-thin $TiO_2$ cap show significant gate leakage improvement [25,26]. This improvement in leakage can allow to scale $HfO_2$ in bilayer $HfO_2$/$TiO_2$ system to scale EOT while keeping the leakage at acceptable values. This is a significant milestone in $HfO_2$/$TiO_2$ system, however, $TiO_2$ cap up-diffusion is still a challenge and require careful optimization of the film and overall gate stack. Some strategies for $TiO_2$ can be as simple as TiOxN formation using advanced ultra-low-energy plasma nitridation processes as discussed by Chudzik et al. in U.S.Patent 9,478,425 to stabilize the $TiO_2$ cap layer. However, in advanced replacement metal gate FinFET devices, nFET and pFET metal stacks are different from each other, and it will be critical for any higher K capping layer to be fully compatible and stable with both nFET and pFET metal gate interface. It is possible, with advanced surface modification and atomic layer deposition growth technologies, $TiO_2$ to metal gate interface can be stabilized leading to potential implementation of 'higher k dielectric' for highly scaled FinFET technology.

## 5. EOT scaling through interfacial layer
### 5.1 Nitrided interfacial layer (SiON)

Introduction of $HfO_2$ brought many challenges in gate stack, one of the challenges was to form good dielectric/substrate (silicon) interface. Earlier in HiK/Metal gate development, it was realized that direct contact of hafnium oxide on silicon channel leads to an unstable interface, which results in degrade in carrier mobility. In order to overcome this problem, thin $SiO_2$ interfacial layer (0.7—0.9 nm) has been used as a standard interfacial layer generally formed by good quality wet chemical based thin $SiO_2$.

This bottom interfacial layer must be designed to produce high quality interface by having low interface trap densities (dangling bonds) and also minimize carrier scattering (low mobility) to produce reliable and good performance transistors. Since interfacial layer ($SiO_2$) capacitance is in series with Hi-K oxide capacitance, the lower K ($SiO_2$) becomes a limiting factor in the gate stack to achieve maximum gate capacitance. As a result, further EOT scaling of gate oxide to increase overall gate capacitance for improved short channel effect (SCE) and gate length scaling can be achieved by scaling the IL layer. One potential option can be to use nitrided $SiO_2$ (SiON) to increase dielectric constant of interfacial layer. Scaling Tinv by using nitrided IL will lead to device performance benefits due to reduction in SCE effects at scaled gate length. However, trade-off between Tinv scaling (introduction of N) and other device characteristics, such as carrier mobility and reliability (NBTI) have to be carefully designed. Increase in N% in ultra-thin $SiO_2$ can lead to scattering centers at IL/channel interface which can degrade carrier mobility [6]. As a result, carrier mobility versus Tinv tradeoff has to be carefully optimized as a function of N% in SiON IL for optimum performance. Introduction of nitrogen in IL can lead to NBTI degrade especially in low thermal budget regime, as discussed by Chudzik et al. enhanced nitrogen content tends to reduce NBTI lifetime [6]. Ramp voltage NBTI test shows that heavily nitrided interface lead to reduced NBTI lifetime as illustrated in Fig. 7.15 [6].

In summary, nitridation of IL layer either post IL growth or through Hi-K during Hik nitridation is a potential knob for overall EOT scaling. With careful IL process engineering, and evaluating a trade-off between Tinv, mobility and NBTI, nitrided ILs are an attractive option for further FinFET EOT scaling.

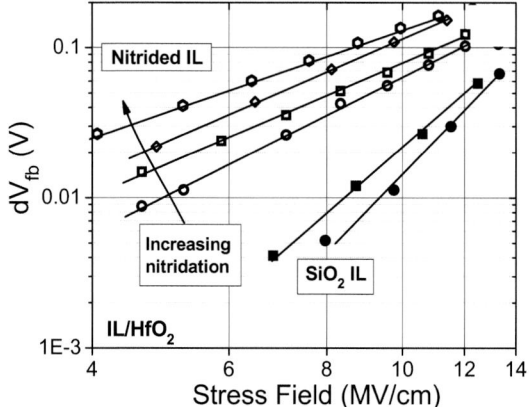

**Figure 7.15** Comparison of NBTI response (ramp voltage test) for HKMG transistor with thermal SiO₂ interlayers and with nitrided interlayers with various degrees of nitridation [6].

### 5.2 Interfacial layer scavenging

As reviewed in the previous section, nitridation of $SiO_2$ IL can provide EOT scaling of up to 2Å with the optimized process. This is a viable option for the short-term device scaling, however, it cannot satisfy the scaling need in the long run since the k-value of SiON (approximately 4–7, depending on the nitrogen concentration) is not significantly larger than that of $SiO_2$. On the other hand, scaling of the physical thickness of the $SiO_2$ IL has more significant impact on EOT. Several techniques going in this direction have been developed, such as IL scaling via scavenging reaction [26–32], cycle-by-cycle atomic layer deposition and annealing of $HfO_2$ [33], and advanced post deposition anneal for epitaxial $HfO_2$ growth [34]. In this section, we mainly review the IL scavenging technique since systematic understanding on the impacts on carrier mobility, EWF control, device reliability has been obtained with this approach [35].

IL scavenging has become a popular approach in recent years to realize aggressive EOT scaling down to 0.5 nm. Choice of scavenging element is one of the most important factors for IL scavenging reaction. The IBM group found that the Gibbs free energy change at 1000 K ($\Delta G°_{1000}$) of the following reaction (1) serves as a guiding principle for the choice of scavenging element [26].

$$Si + \frac{2}{y} M_x O_y \rightarrow \frac{2x}{y} M + SiO_2 \quad (7.2)$$

where M is the scavenging element in the gate stack.

The EOT trend for metal-inserted poly-Si stack (MIPS) with $SiO_2$/$HfO_2$ dual-layer gate dielectrics in the literature is summarized in Fig. 7.16 as a function of $\Delta G°_{1000}$ per oxygen atom ($\Delta G°_{1000}/O$) for the scavenging element. As shown in Fig. 7.16, EOT and $\Delta G°_{1000}/O$ values show a very strong correlation. The IBM group proposed doping of scavenging metals with high $\Delta G°_{1000}/O$ values into a thermally stable TiN electrode. This technique enables highly controllable IL scavenging for both gate-first [26] and gate-last [36] integrations.

Fig. 7.17 shows high field electron mobility as a function of EOT from literature data. It should be noted that most data points reported by various groups show a universal relationship between high field electron mobility and EOT in the sub-nm EOT regime. The experimental data are compared with the estimated mobility-EOT relationship providing the same drive current ($I_{on}$) at a given $L_g$ in Ref. [37]. The mobility-EOT slope for IL scavenging from literature ($\sim 20$ $cm^2$/Vs per 0.1 nm) is shallower than the estimated breakeven relationship for $L_g \leq 30$ nm ($\sim 40$ $cm^2$/Vs per 0.1 nm). This indicates that it is possible to improve the short-channel device performance by employing IL scavenging in conjunction with aggressive $L_g$ scaling in future nodes.

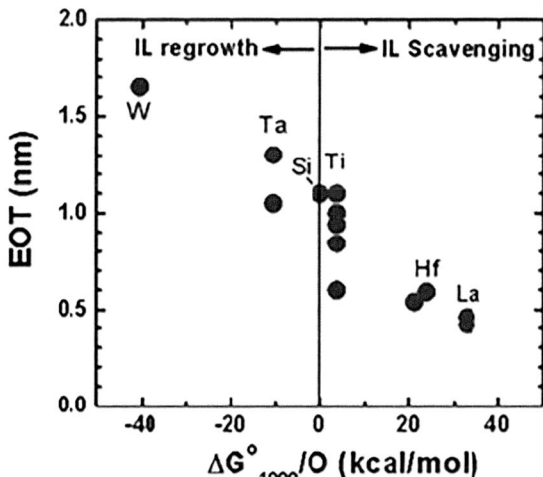

**Figure 7.16** EOT of $SiO_2$/$HfO_2$ MIPS structure as a function of $\Delta G°1000$ per oxygen atom for scavenging element from literature data. *(Adopted from Greene et al., Symp. VLSI Tech. Dig. Tech. Pap (2009), 140; S. Migita, et al., In Proceedings of IEEE International Electron Devices Meeting, San Francisco, CA, USA (2010), 269–272.)*

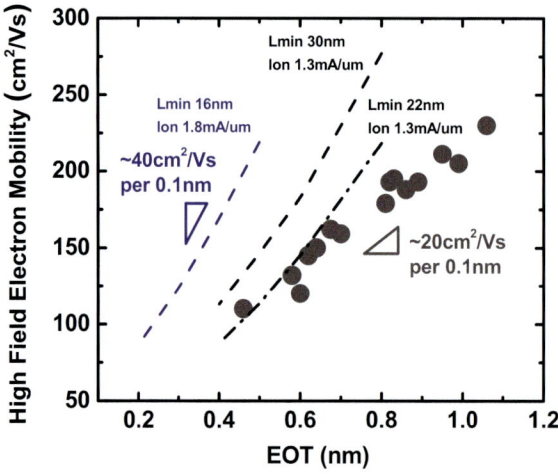

**Figure 7.17** High field electron mobility as a function of EOT from literature data. Simulated contour lines providing the same $I_{on}$ at $L_{min}$ 16, 22, and 30 nm [37] are shown for comparison. *(Adopted from S. Migita, et al., In Proceedings of IEEE International Electron Devices Meeting, San Francisco, CA, USA (2010), 269–272.)*

The impact of EOT scaling on the EWF for gate-last process was investigated by using IL scavenging reaction in Ref. [36]. Fig. 7.18 compares the EWF-EOT trends for the n-type and p-type WF-setting metals.

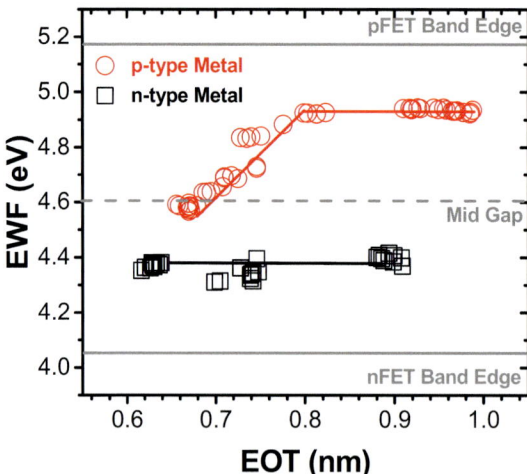

**Figure 7.18** EWF-EOT trend for n-type and p-type WF-setting metals with gate-last process. The EOT was changed via IL scavenging technique. *(Adopted from T. Ando, Materials 5 (3) (2012), 478–500.)*

The n-type WF metal provides a completely flat EWF-EOT trend down to 0.6 nm. On the other hand, the p-type WF metal shows a flat trend down to EOT 0.8 nm and then exhibits a linear trade-off trend toward the midgap with further scaling. Degradation in film quality in the sub-monolayer IL regime may facilitate oxygen vacancy generation in the $HfO_2$ layer and/or the $SiO_2$ IL, resulting in the unfavorable EWF shift for pFET. Thus, leaving an ultra-thin and robust $SiO_2$ IL after scavenging is indispensable for EWF control.

Other device parameters requiring close attention with aggressive EOT scaling are reliability. The IBM group investigated impacts of IL scavenging on reliability by using $SiO_2/HfO_2$ dual-layer stacks with varying IL thicknesses. The change in the device lifetimes, including positive bias temperature instability (PBTI), negative bias temperature instability (NBTI), and time dependent dielectric breakdown (TDDB), are estimated in Ref. [38]. The BTI lifetimes are predicted to decrease by $50-100\times$ for every 0.1 nm of IL scaling. Drastic lifetime reductions also occur for TDDB. Note that the estimated lifetime trends for IL scaling are similar for gate-first and gate-last processes, indicating that these trends arise from the fundamental materials properties of the $SiO_2/HfO_2$ dual-layer stacks and do not depend much on the fabrication method.

As reviewed above, the mobility-EOT trend in literature suggests that short-channel performance improvement is attainable with aggressive EOT scaling via IL scavenging. However, extreme IL scaling is accompanied with loss of EWF control and with a severe penalty in reliability. Therefore, highly precise IL thickness control in an ultra-thin IL regime (<0.5 nm) will be the key technology to satisfy both performance and reliability requirements for future CMOS devices.

## 6. Ab-initio modeling
### 6.1 Tool to evaluate higher K dielectric

Hafnia ($HfO_2$) is technologically important because of its high bulk modulus, high melting point, and high chemical stability, besides having a high dielectric constant compared to Silica. Dictated by the demand of scaling the semiconductor devices, $HfO_2$ has been found to replace $SiO_2$ in order to continue scaling at the cost of much smaller gate leakage. It has been shown that the dielectric response of $HfO_2$ vary with the crystal phase [39]. Among the three phases, namely, monoclinic, cubic, and tetragonal, the monoclinic phase is stable at room temperature. The monoclinic phase has a strongly anisotropic lattice dielectric tensor and a smaller dielectric constant (orientationally averaged) compared to the cubic and tetragonal phases, yet is high enough to replace $SiO_2$ for next generations of semiconductor devices.

In this work, we shall discuss about computing the dielectric properties of insulators through ab-initio techniques. The static dielectric constant can be computed via

$$\varepsilon_s = \varepsilon_\infty + \varepsilon_l$$

where $\varepsilon_s$ is the static dielectric permittivity, the first term is due to the electronic contribution, and the second term is due to the lattice contribution of the dielectric permittivity. The calculation of lattice contribution to the dielectric tensor requires computation of Born Effective charges, and infrared-active phonon modes.

The Born Effective Charges tensor ($Z^*$) is related polarization via

$$\Delta_p = \frac{e}{V} \sum_i Z_i^* \Delta u_i$$

and computed by finite differences of polarizations $\Delta P$ as various sub-lattice displacements ($u_i$).

$$\varepsilon_{\alpha\beta}^s = \varepsilon_{\alpha\beta}^\infty + \frac{4\pi e^2}{M_0 V} \sum_\lambda \frac{z_{\lambda a}^* z_{\lambda b}^*}{\omega_\lambda^2}$$

$$Z^* = \frac{dF}{d\xi}$$

where $\omega_\lambda$ is the frequency of the $\lambda^{th}$ infra-red-active phonon mode, V is the volume of the unit cell, e is the electronic charge, $M_0$ is a reference mass that we take for convenience to be 1 amu, F is the force, $\xi$ is the electric field. This methodology requires three calculations for each model system, including computation of zone center optical phonon modes. It becomes prohibitively difficult and computationally expensive for systems with point defects, as we need to include a large number of atoms in a supercell. This would mean computing a large number of optical phonon modes. A similar situation arises when computing dielectric properties of thin films, and interfaces. Using this method, Fischer et al. [40] reported the dielectric constant of defected $HfO_2$ with relatively small number of atoms (24) using this method. This amounts to a very high doping concentration. For smaller doping (a few %) may require including a few hundred to several hundred atoms in a model supercell, thus, making it computationally very expensive.

We follow a method of computing dielectric response as developed by Umari and Pasquarello [41]. This method requires four calculations for each system, but avoids computing zone center optical phonon modes, which are not straight forward for complex systems, such as those involving defects, surfaces/interfaces.

This method provides a much more practical way of investigating dielectric properties. As widely known, the difficulty of treating finite electric fields is related to the intrinsic non-periodic nature of the position operator. By calculating derivatives with respect to atomic positions, Umari et al. [41] demonstrated that their functional and method was suitable for application in ab initio molecular dynamics in finite homogeneous electric fields. Using bulk MgO as a test case, they calculated the high-frequency dielectric constant and the Born effective charges through finite difference and found excellent agreement with those of perturbative or linear response methods. Following a similar procedure, the static dielectric constant can be obtained by performing a molecular dynamics relaxation, thereby completely avoiding the calculation of phonon modes at the zone center. This method can be used for systems with large sizes, thereby a practically feasible way to study and predict dielectric response of systems with low doping of less than a percent.

For a system obeying periodic boundary conditions, we describe its metastable state induced by the presence of a finite electric field $\xi$ (taken along $x$) by the vibrational energy functional,

$$E^\xi[\{\psi_i\}] = E^0[\{\psi_i\}] - \xi \cdot P[\{\psi_i\}]$$

where

$$P[\{\psi_i\}] = -\frac{L}{\pi} \text{Im}(\ln \det S[\{\psi_i\}])$$

and

$$S_{ij} = \langle \psi_i | e^{2\pi i x/L} | \psi_j \rangle$$

Here $E^{(0)}$ is the Energy functional in the absence of an electric field, P is the polarization along the direction of $\xi$, L is the periodicity of the cell and a matrix calculated for the set of doubly occupied wave functions $\psi_i$. The electronic contribution to the dielectric permittivity is computed through

$$\varepsilon_\infty = \frac{4\pi}{L^3}\frac{\Delta P^\xi}{\xi} + 1$$

$$\Delta P^\xi = P^\xi - P^0$$

Here $\Delta P^\xi$ is the difference in the polarization with and without applied electric field $\xi$.

$$\Delta\varepsilon = \frac{4\pi}{L^3}\frac{P^\xi_{relaxed} - P^\xi_{unrelaxed}}{\xi}$$

Here $P^\xi_{relaxed}$ and $P^\xi_{unrelaxed}$ are the polarization with and without damped dynamics respectively in a finite electric field. The static dielectric constant can be given as sum of electronic, and ionic or lattice contributions,

$$\varepsilon_s = \varepsilon_\infty + \Delta\varepsilon$$

Using above calculations method, we computed electronic, ionic and total dielectric constant of undoped and undoped silicon dioxide and hafnium dioxide. Our results are summarized in Table 7.3. As can be seen from Table 7.3 results, HfOxN, HfLaON, and HfAlON are technological important higher K contenders if can be implemented in scaled FinFET technologies.

We performed the density functional theory (DFT) simulations in the local density approximation (LDA) with Perdew–Zunger parameterization of the exchange-correlation functional to model the electron–electron interaction. The interaction of valence electrons to that of the core has been approximated through the pseudopotentials for all the atoms using Vanderbilt type ultra-soft pseudopotentials. The use of Vanderbilt ultra-soft pseudopotentials allow highly accurate calculations to be achieved with a low energy cutoff. A wave function cutoff of 25 Rydberg, and charge density cutoff of 200 Rydberg were used throughout. About 3% substitutional doping of Al, or La replacing random Hf sites (for HfAlO and HfLaO case), and 6% N (for SiON, and HfON) replacing O site were used. We used conjugate gradient algorithm to compute total energies and forces. The geometry of the supercell was optimized until the forces between the atoms were less than 0.05 eV/Å. This was followed by molecular dynamics simulation with and without the finite electric field to compute electronic contribution of dielectric permittivity, and a damped dynamics calculation to compute the ionic response in a finite electric field. All the calculations were carried out using plane wave basis code Quantum ESPRESSO [42,43].

Table 7.3 Computed electronic, ionic, and total dielectric constant of undoped/doped silica, and hafnia.

| System | $\varepsilon_\infty$ | $\Delta\varepsilon$ | $\varepsilon_s$ | Experiment |
|---|---|---|---|---|
| SiO$_2$ (trigonal) | 2.42 | 1.89 | 4.31 | 3.9–4.5 |
| SiON (trigonal) | 2.48 | 4.97 | 7.45 | 6.5–7.5 |
| HfO$_2$ (monoclinic) | 4.44 | 11.04 | 15.48 | 16.0–18.0 |
| HfAlO (monoclinic) | 4.45 | 13.92 | 18.37 | 15.0–18.0 |
| HfLaO (monoclinic) | 4.63 | 19.52 | 24.15 | 24.0 |
| HfON (monoclinic) | 11.68 | 42.00 | 53.68 | 35.0 |

Table 7.4 shows that by introducing a small amount (few percentage) of Si or transition metal atoms in $HfO_2$, the dielectric constant can be enhanced substantially. These results clearly demonstrate a way to engineer the desired dielectric permittivity for scaled semiconductor devices in the future.

## 6.2 Effective work function engineering

The threshold voltage of a MOSFET is related to the effective work function (EWF) and that, in turn, is related to the amount of charge transfer and dipole strength at the high-k/metal interface. The presence of point defects can modulate this dipole and the EWF [43], thus affecting the threshold voltage. Among the $HfO_2$ native defects, O vacancy is a dominant intrinsic electronic defect. There have been some studies of the EWF computation, and the effects of O vacancies on the Schottky barrier height and EWF modulation, using the phenomenological theory of Metal Induced Gap States (MIGS) model developed by Robertson [44,45]. This method describes the charge transfer at the metal-insulator interface in terms of the alignment of a charge neutrality level with the metal Fermi level. However, this method cannot correctly describe the effect of specific interface termination or the interface stoichiometry. A better approach to computing the EWF is through the band offset method [46,47]. In this method, the EWF is obtained by subtracting the valence band offset between $HfO_2$ and the metal, from the experimental band gap of the $HfO_2$, and then by adding the experimental electron affinity of the $HfO_2$. However, this method suffers from the well-known problem of the DFT, namely, the errors in the band structure (extracted from the bulk calculations on the materials forming the interface) as well as in the band line up (obtained from the interface calculation). This is due to the self-interaction

**Table 7.4** Computed electronic, ionic, and total dielectric constant of doped $HfO_2$.

| System (3% doping in $HfO_2$) | $\varepsilon_\infty$ | $\Delta\varepsilon$ | $\varepsilon_s$ | Experiment |
|---|---|---|---|---|
| HfSiO | 4.56 | 16.09 | 20.65 | 20–26 |
| HfSrO | 9.21 | 27.00 | 36.21 | |
| HfBaO | 6.90 | 30.44 | 37.34 | 38 |
| HfLaO | 4.63 | 19.52 | 24.15 | 24 |
| HfAlO | 4.45 | 13.92 | 18.37 | 15–18 |
| HfYO | 4.46 | 15.46 | 19.92 | 25–32 |
| HfErO | 5.11 | 21.21 | 26.32 | 28 |

error in the electron-electron interaction term of the LDA-DFT. Thus, it has limited predictive capability. A very recent work by Prodhomme et al. [47] uses the band offset method to compute the EWF going beyond standard LDA-DFT. They considered the entire stack of a MOSFET involving $Si/SiO_2/HfO_2/TiN$. However, their EWF values from LDA-DFT simulations differ from experimental values by more than 1.0 eV. They obtained their EWF values by employing computationally expensive GW approximation.

We have investigated the impact of oxygen vacancy, oxygen interstitial or substitutional defects at the $HfO_2/TiN$ interface, and bulk part of TiN through ab initio atomistic simulations [43], employing DFT code as implemented in Quantum ESPRESSO [42]. All the simulations were performed within the LDA, with Perdew-Zunger parametrization of the exchange-correlation functional. The core and valence electron interactions were treated through pseudopotentials for all the atoms using Vanderbilt type ultra-soft pseudopotentials. A wave function cutoff of 50 Ry and the charge density cutoff of 500 Ry were sufficient to converge the results. For the EWF computation, the model interface was created by placing a (100) TiN slab formed with six-unit cells of cubic TiN, on the top of an orthorhombic $HfO_2$ slab (formed with three unit cells of strained monoclinic $HfO_2$). Thus, an orthorhombic supercell (slab) was created, with over 6 Å thick vacuum on both sides of the slab perpendicular to the interface. The vacuum was introduced to avoid interaction of interface with its periodic images in the perpendicular direction. We then equilibrated the interface through Nose thermostat at a temperature of 1000°C for about 5 ps, with a time step of $\sim 1$ fs. After this, we carried out molecular dynamics (MD) for about 15 ps, thereby reducing the temperature from 1000 to 0°C. The MD simulations were carried out with much smaller wave function (25 Ry), charge density (250 Ry) cutoffs, and using a single k-point (gamma) only. Further, the geometry of the stack was optimized, such that the forces between the atoms were less than 0.05 eV/A°. A $5 \times 5 \times 1$ Monkhorst-Pack mesh was sufficient to converge the calculations.

We compute the EWF using a method, proposed by Zhu and Ramprasad [48]. This method allows us to compute the EWF from the interface dipole and the vacuum work function of the metal and eliminates the need to compute the band offsets. Thus, it avoids the errors introduced in the band structure and the valence band offset calculations. The Figure shows the band diagram and the methodology to compute the EWF. The expression to compute the EWF is given as

$$\Phi_{eff} = \Phi + \frac{4\pi D_x}{A},$$

where A is the area of the interface and $D_x = D_n - D_o$ is the $HfO_2$-induced interfacial dipole, $D_n$ is the net dipole of the interface, and $D_o$ is the dipole of the $HfO_2$ free surface. $\Phi$ is the vacuum work function of the TiN and $\Phi_{eff}$ is the EWF of the TiN on top of the $HfO_2$. We compute $D_n$ and $D_o$ by introducing an electric double layer (dipole correction) in the vacuum region [49] of the $HfO_2$/TiN interface and $HfO_2$ free surface slab supercells, respectively as shown in Fig. 7.19.

In Table 7.5 we show the EWF values for several possible O point defects in $HfO_2$/TiN interface, along with defect-free reference interface. The presence of these defects has dramatic impact on interface dipole and the EWF. The EWF of defect-free $HfO_2$/TiN interface to be 4.56 eV. From the data shown in Table 7.5, we find that the presence of oxygen

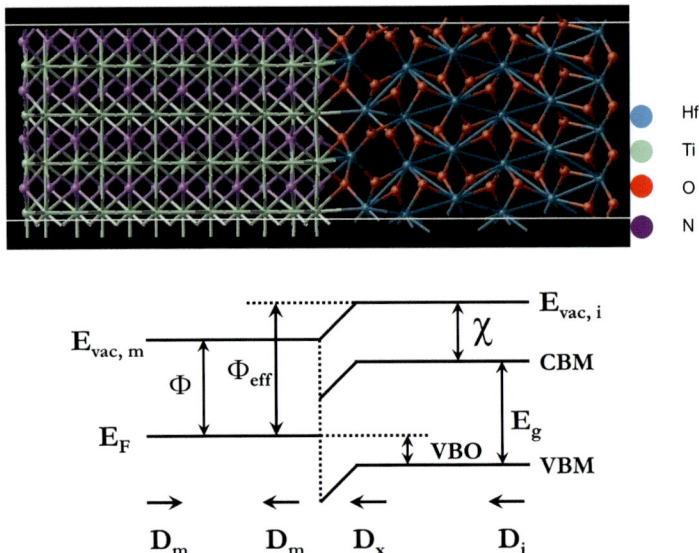

**Figure 7.19** Schematic band alignment of $HfO_2$/(100) TiN interface. $E_{vac, m}$ is the vacuum reference and $E_F$ is the Fermi level of the TiN. $E_{vac, i}$ is vacuum level, CBM - conduction band minimum, VBM - valence band maximum, $\chi$ is the electron affinity, and $E_g$ is the band gap of the insulator ($HfO_2$). The valence band offset (VBO) is the energy difference between the TiN Fermi level and the $HfO_2$ VBM. $D_m$, is the surface dipole moments of TiN free surfaces, and $D_i$ is the surface dipole moment of the $HfO_2$ free surface. $D_x$ is the $HfO_2$-induced interfacial dipole (net interface dipole moment minus $HfO_2$ free surface dipole moment). $\Phi$ is the vacuum work function TiN free surface, and $\Phi_{eff}$ is the effective work function of TiN, deposited on top of the $HfO_2$.s.

**Table 7.5** Effective work function of the HfO$_2$/TiN interface with O defects, and interface engineering. The vacuum work function of TiN ~3.70 eV was used.

| Model system | $\Phi_{eff}$ (eV) |
|---|---|
| Pristine (stoichiometric) interface (HfO$_2$/(100)TiN) | 4.56 |
| O vacancy at HfO$_2$/TiN interface | 4.20 |
| O interstitial at HfO$_2$/TiN interface | 4.66 |
| O interstitial in the bulk (interior)TiN | 4.96 |
| O vacancy at HfO$_2$ side and O interstitial at TiN side of the interface | 4.29 |
| O vacancy at HfO$_2$ side of the interface and O interstitial in bulk TiN | 4.78 |
| O vacancy at HfO$_2$ side of the interface, O interstitial substitutes Ti in bulk TiN | 4.30 |
| O vacancy at HfO$_2$ side of the interface, O interstitial substitutes N in bulk TiN | 4.18 |
| HfO$_2$/(111)TiN with Ti-terminated interface | 4.07 |
| HfO$_2$/(111)TiN with N-terminated interface | 4.91 |

vacancy at the interface decreases the interface dipole and the EWF, compared to the defect-free interface. On the other hand, an oxygen interstitial at the interface enhances the interface dipole and the EWF, compared to the defect-free interface. A similar behavior is seen when an oxygen interstitial is introduced in the bulk TiN, resulting in higher EWF. This implies that one can engineer the EWF for pFET by incorporating certain percentage of oxygen in TiN. In the case when oxygen atom comes from HfO$_2$ side into TiN bulk, leaving behind a vacancy in HfO$_2$, the EWF is smaller by a 180 meV compared to oxygen interstitial in the TiN coming from ambience. Thus, O vacancies at the HfO$_2$/TiN interface may be good for nFET EWF engineering, whereas O in TiN bulk may be good for pFET EWF engineering. Another possibility would be to grow TiN preferentially along (111) orientation. This way we may selectively get N-layer or Ti-layer interfacing with HfO$_2$ slab. For a comparison, we carried out simulations involving a (111) TiN on top of HfO$_2$. We find that a Ti rich interface gives nFET-like EWF, whereas an N-rich interface gives pFET-like EWF. These results clearly demonstrate the possibility of EWF engineering in scaled MOSFET.

In summary, we have demonstrated how ab-initio materials modeling could help guide new experiments and thereby could help achieve device performance targets. We have taken a few representative examples, namely computation of dielectric permittivity, and the EWF. These results could be a guideline to engineer dielectric properties, and the EWF in the scaled devices.

# 7. Gate oxides in FinFET era

Continuous transistor scaling at 20 nm planar bulk transistor node reached the point where depletion width of source and drain region are very close to each other. This lead to increase in Drain-Induced-Barrier-Lowering (DIBL). Increase in DIBL causes loss of gate control over channel. Additionally, source and drain depletion width being close to each other (due to gate length reduction) leads to increase in off-state leakage paths from source to drain, and off-state current starts to dominate. Traditional knobs to improve DIBL such as increase in channel doping to reduce channel depletion width, halo implant doping to control source and drain depletion regions and reduce punch through leakage, and EOT scaling for improved gate control, all these knobs start to have detrimental effect on device variability and reliability problems. In order to overcome scaling challenges, and still increase transistor density, new device architecture called Fin Shaped Field Effect Transistor (FinFET) are introduced first at 22 nm node by Intel, and at 14/16 nm widely accepted as a standard architecture for high performance devices.

FinFETs in general is a structural engineering of a channel and a gate in a field effect transistor, where carriers are confined in a vertical channel, while gate is wrapped around the channel from two sides and top as shown in Fig. 7.20 [50] allowing better short channel control. This structure allows superior gate control and smaller depletion widths.

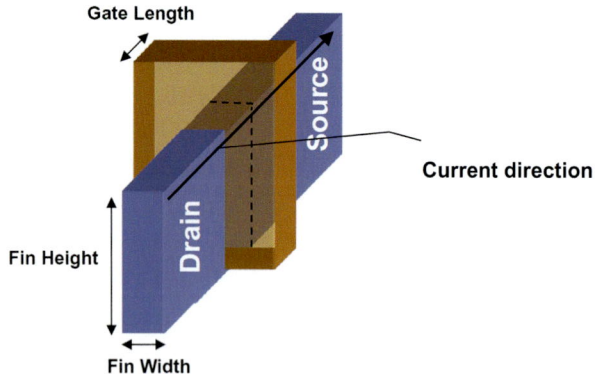

**Figure 7.20** Simple FinFET structure schematic [50].

Substrate doping in FinFET devices is eliminated, leading to reduced device variability due to elimination of dopant fluctuation.

Due to 3D nature of gate, FinFET offers excellent gate control, leading to improve short channel effects. As a result, Tinv scaling criteria for FinFET is relaxed compared to highly scaled planar transistor. As discussed earlier, state of the art bulk 14 nm FinFET has Tinv of 1.3 nm (EOT = 0.9 nm). Future gate length (Lg below 20 nm) scaling below 7 nm node will require EOT scaling for better gate control and some of the strategies we discussed for reducing EOT scaling by plasma nitridation of HiK/IL stack and potential bilayer higher "k" dielectric system may come useful.

Introduction of FinFET brought new challenges to gate oxide due to 3D nature of the transistor structure. We will discuss those challenges in terms of (1) intrinsic (2) extrinsic issues. Intrinsic challenges are mostly driven by requirement of forming conformal gate oxide around Fin structure, as a result, depositing conformal $HfO_2$ thin film (1.5–2 nm) is very critical and fundamental requirement for FinFET devices. Atomic layer deposition (ALD) technology to deposit $HfO_2$ using $HfCl_4/H_2O$ precursor system allows to deposit conformal oxide around the Fin and produces high quality amorphous $HfO_2$. ALD $HfO_2$ is the industry standard to deposit gate oxide. With taller Fin and scaled Fin pitch for 7 nm and beyond it would be critical to validate $HfO_2$ conformality and continuity around the Fin for 1.5–2 nm scaled $HfO_2$. If nitridation is used for HiK/IL nitridation conformality is another process which needs to be carefully developed to make sure oxide is uniformly nitrided.

Introduction of 3D FinFET structure, brought new Fin and gate design dependent leakage and gate dielectric reliability failure modes which were only identified and reported recently [45]. In time dependent dielectric breakdown (TDDB) measurements, breakdown time depends on effective gate area and device design layout such as Fin and gate numbers are not considered a TDDB limiting factor [51]. However, Liu et al. has observed that increased Fin and gate numbers per unit cell in a fixed area for 14 nm SOI RMG FinFET can have increased gate leakage and reduced time to breakdown for gate dielectric TDDB. Results are shown in Fig. 7.21 where initial gate leakage increase, and TDDB time reduction trend is observed with increased number of Fins design.

These dependence were eliminated by optimizing overall FinFET formation. This gate oxide failure mode which highly depends on the device design and how robust FinFET structure is formed is a new

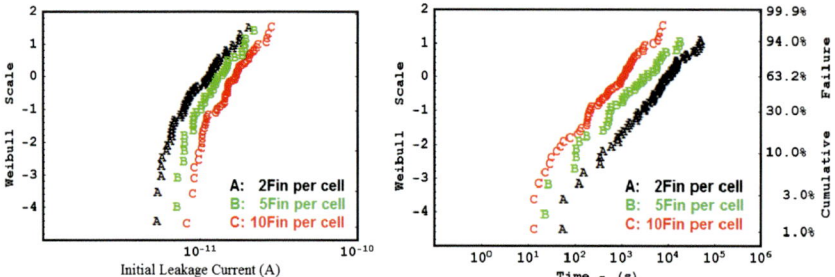

**Figure 7.21** (a) Initial leakage current (Isense0) distributions of devices having 2 fins/5 fins/10 fins per unit cell, (b) $T_{bd}$ distributions of devices having 2 fins, 5 fins, 10 fins per unit cell but equal total gate area [51].

paradigm in CMOS integration in FinFET architecture. As Fin pitch scaling continues from 14 to 10 to 7 and 5 nm, these design to process integration challenges and their impact on gate oxide leakage and reliability will need to be carefully managed. These interactions will become very important in gate-all-around and nano-wire channel device architectures which are even more complex than FinFET.

In summary, higher k material are still desirable at scale gate length for FinFET devices, however, fully working higher k solutions are still a big challenge. $HfO_2$/IL nitridation is one of the practical approaches to scale Tinv by 0.1 nm (1 Å) with reasonable gate leakage, additionally $HfO_2$/$TiO_2$ bilayer system starts to show some encouraging publications and patents and with continuous research and development it is possible to introduce a higher 'K' solution. It is fair to say that $HfO_2$ physical thickness in the range of 1.5–2 nm has reached the physical limit where physical thickness scaling will lead to undesirable gate leakage, and reliability challenges, as a result, it is not expected to have any physical thickness scaling of $HfO_2$ till the end of FinFET roadmap. However, higher K solutions such as bilayer $HfO_2$/$TiO_2$ will be needed beyond 7 nm node with scaled gate length which can provide additional EOT scaling at better gate leakage.

## 8. High voltage (HV) input/output (I/O) gate oxides with HiK/MG for advanced SOC (FinFET and FDSOI)

Rise of system on chip (SoC) and their ubiquitous applications in electronic devices require complex suite of transistor devices with different performance and power requirements. As a result, at 0.18 um technology, thick gate oxides are offered along with low voltage high performance "thin gate oxide" transistors to meet different circuit requirements [46]. Integration of thick gate oxide certainly brings complexity into standard CMOS flow, especially in HiK/Metal gate, and now in FinFET and FDSOI transistor technologies. Following are some features which make complex integration of thick oxide very valuable, and absolute need for modern multipurpose microprocessors.

(1) Thick gate oxide transistor can be used as the sleep transistor to disable the core circuit. This provides a technique to effectively control the circuit leakage current and reduction in standby power [52].
(2) Dual gate oxide thickness process allows that the high-performance digital logic circuits use more advanced transistors to operate at higher frequencies with lower operating voltages, while I/O interface blocks use the thick gate oxide transistor which bridges the higher performance transistor with those implemented in the older technology nodes, in addition peripheral circuits which require high voltage and lower speed.
(3) Thick gate oxide is beneficial for RF and analog circuit designs. Thick gate oxide allows handling of large signals, and the ability to handle higher power for amplifiers. Thick gate oxide is also more suitable to reduce phase noise levels for local oscillators in analog circuits.

In summary, all the modern electronic devices such as laptops, mobile devices, wearables, internet of thing devices, servers, and sensors have thick oxide high voltage devices, and their integration with high performance digital devices are central for microprocessor technology.

For current state-of-the-art 14 nm SOC FinFET technology, it is believed for HV I/O devices that gate oxides are thermal processes to grow 3.5−4.0 nm oxides with maximum operating voltage (Vmax) of 1.98 V. Traditional thermal oxides (wet or dry) are not a viable option for FinFET I/O oxide growth due to differential growth rate between 100 and 110 plane for top and sidewalls of the FinFET respectively. As a result, it is likely that chip makers are using some innovative rapid thermal processing-based oxide growth process at 14 nm (48 and 42 nm Fin pitch), such as oxygen/hydrogen (in-situ steam generation) or nitrous oxides/hydrogen combined (NO or $N_2O$) to grow I/O gate oxide, these RTP processes allows controlled oxide growth on the sidewall better than pure $O_2$ based dry or wet thermal oxide [53]. These innovative thermal processes still consume silicon as their primary mechanism to grow oxide. However, these processes can be significantly challenged at scaled FIN pitch and taller FIN. Regardless of the good quality of thermally grown oxides, all thermal methods consume silicon to grow $SiO_2$. As a result, at highly scaled FinFET technologies (e.g., 7 and 5 nm nodes) where project Fin width is 7—8 nm, Fin height 48—54 nm tall, and Fin pitch is sub 35 nm, silicon consumption, conformal coverage of oxide in high aspect ratio, and retaining "tall" Fin shape are critical requirements. These demanding needs would be difficult with silicon consuming thermal oxide processes. As a result, along with continued aggressive physical scaling of devices and potential introduction of alternate channel materials such as silicon germanium (SiGe), alternate non-consuming methods (ALD $SiO_2$) to grow oxide will play a central role as an enablement element for scaled FinFET technology. For advanced FDSOI technology, thin silicon channel (nFET), and strained silicon germanium channel for pFET with ability to offer 1.8 V I/O oxide will also require no silicon and silicon germanium consuming high quality oxide solutions.

Following new elements in SoC transistor technology brings new challenges to I/O gate oxides and require new innovations

(1) Introduction of FinFET and FDSOI beyond 14 nm technologies require as minimum as possible silicon channel consumption and few angstroms of oxide thickness control. In addition, at scaled FIN pitch, and tall Fin shape conformality of oxide growth to control Tinv and breakdown voltage (Vbd) will require precision like oxide control. Additionally, core devices are built by removing I/O oxides in digital design areas, as a result, impact of I/O oxide processing (FIN geometry control) now are directly linked to core device performance. I/O areas processing can be de-coupled with core areas, but it can bring tremendous cost and additional complexity in already complex and scaled technology.

(2) Introduction of SiGe channel in future SiGe channel FinFET or nano wire devices will require low thermal budget I/O oxide to keep SiGe stoichiometry, minimum Ge out-diffusion into I/O oxides, while also passing 1.8 V reliability requirements.

(3) Scaled FDSOI technology (12 and 5 nm) with ultra-thin silicon channel for nFET devices, and strained SiGe channel for pFET devices, along with strong desire to provide 1.8 V I/O devices for internet-of-things, and automotive technology will require excellent control of I/O oxide along with thermal oxide equivalent quality.

To circumvent the above challenges, introduction of ALD $SiO_2$ with post treatments, such as plasma nitridation, oxidation, and thermal treatments will be critical to meet above requirements.

It has been demonstrated in the marketplace for 32/22 nm SOI based technologies that deposited atomic layer deposition (ALD) $SiO_2$ with innovative post treatments can be equivalent to thermal oxide, as shown in Fig. 7.20 [54,55].

In Figs. 7.22 and 7.45 nm thermal SiON/Poly gate system is compared to 32 nm HiK/MG with ALD $SiO_2$ with post treatments, and at 1.98 V both thermal oxide and ALD $SiO_2$ at same Tinv value have equivalent time to failure. This work puts the foundation that low temperature ALD $SiO_2$ with post treatments and with proper interface engineering can lead to thermal oxide quality results without consuming excessive silicon channel, while providing benefits of conformal deposition of ALD oxide. Future technologies such as GAA where silicon is literally a wire will need to avoid any silicon consumption, and ALD based $SiO_2$, and post treatments can be a potential solution for I/O devices.

For FDSOI technology, ALD $SiO_2$ solution is equivalently important as silicon channel requires minimum to no consumption. FDSOI technology is getting widely accepted for IoT and RF technologies such as 28 nm FDSOI, and 22 nm FDSOI where I/O devices with 1.8 V Vmax offerings are critical for integration of RF, and embedded nonvolatile memory (eNVM). This kind of integration on chip will require 1.8 V I/O offerings with very reliable, nonsilicon consuming gate oxide, and ALD solutions will be able to potentially meet the requirements. In addition, advancement

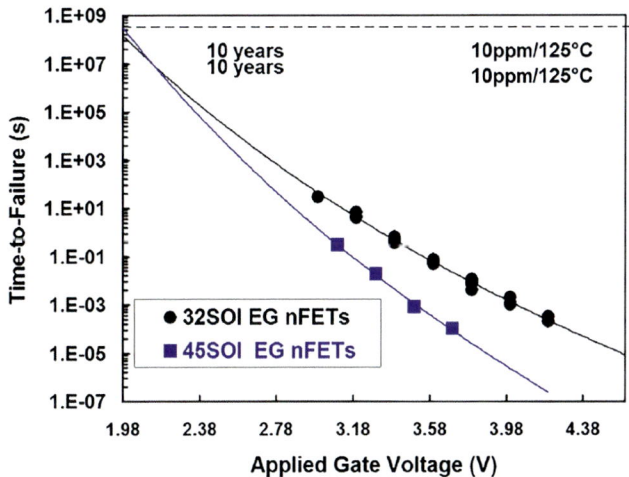

**Figure 7.22** TDDB results ALDSiO$_2$/HKMG Vs SiON/poly [54].

in automotive electronics as a branch of IoT will also require very highly reliable I/O gate oxides while still providing performance.

Integration of ALD $SiO_2$ with SiGe planar channel was also demonstrated and for I/O devices. This integration brought significant surface chemistry and engineering challenges which needed to be overcome for successful enablement. Although the ALD $SiO_2$ process provides superior thickness uniformity, better process control and better dielectric reliability, there is a significant differential growth rate difference between cSi and cSiGe. We used electrical Tinv (inversion layer thickness) difference between pFET (cSiGe) and nFET (cSi) to quantify the differential growth rate difference. As shown below in Fig. 7.23, ALD $SiO_2$ leads to pFET to nFET Tinv delta of 8 Å (Å). This Tinv difference clearly confirms differential growth rate difference between nFET (cSi channel) and pFET (cSiGe channel). However, if we review Fig. 7.22 where Ge3d XPS spectrum after ALD oxide deposition is shown, we find that Ge is almost completely in its elemental form. Therefore, Ge acts as a catalyst during ALD $SiO_2$ deposition for oxidation of Si in SiGe layer, where neither Ge is oxidized nor incorporated into the deposited ALD $SiO_2$ film. This demonstrates that Ge increases the reaction rate of silicon oxidation in SiGe while Ge itself remains completely unchanged by the reaction. This catalytic effect of Ge on the oxidation of Si in SiGe film had been observed and discussed for thermal oxide in depth before by LeGoues et al. [56]. This Tinv difference of 8 Å between nFET and pFET (pFET being 8 A higher) is not acceptable for I/O circuits.

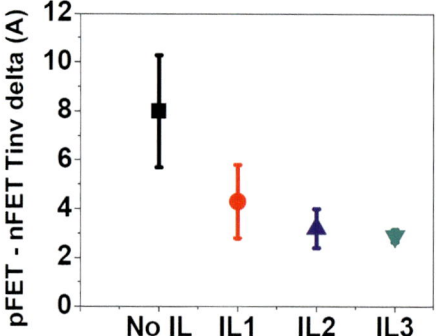

**Figure 7.23** Tinv difference between nFET and pFET I/O devices. Introducing novel passivation layer through pre-ALD $SiO_2$ deposition reduces growth rate difference between Si and SiGe channel [54].

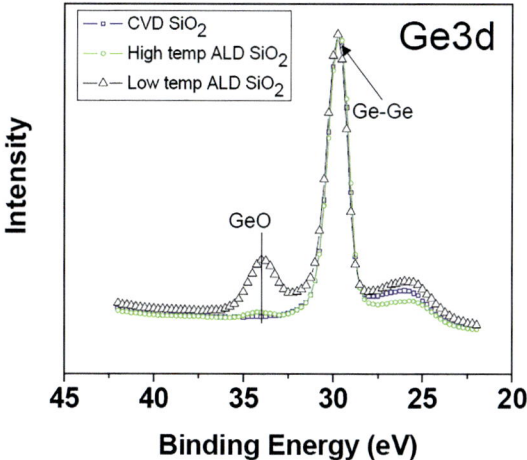

**Figure 7.24** XPS Ge3d spectra showing GeO peak with different deposited oxide processes [54].

In order to reduce this differential growth rate, we developed novel surface passivation layer which resulted in 2.9 Å pFET to nFET Tinv difference without any threshold voltage (Vt) shift, mobility and reliability degrade. Results for pFET Tinv reduction using novel interfacial layer (IL) are shown in Fig. 7.24, where IL3 achieves 2.9 Å of pFET to nFET Tinv delta [54].

## 9. SiGe as a pFET channel (cSiGe) to enable gate oxide scaling

High-k metal gate stacks tend to show effective work function (EWF) corresponding to near Si mid-gap after high temperature processing (>600°C), resulting in unacceptably high pMOSFET $V_t$s. This phenomenon is attributed to formation of electrically charged oxygen vacancies in Hf-based gate dielectrics [57,58]. In order to attain appropriate $V_t$ with the presence of oxygen vacancies, a SiGe epitaxial channel (cSiGe) was introduced selectively only on the pMOSFET channel area [59,60]. The valence band offset between SiGe and Si increases the EWF by more than 300 mV and thereby reduce the $V_t$ of the pMOSFET to appropriate values for high performance and low power CMOS [61].

Thus, the cSiGe was originally introduced for the purpose of EWF control of pMOSFET. However, a unique reliability improvement due to cSiGe has been identified as the technology became mature [62]. Franco et al. found that the valence band offset between SiGe and Si makes the defect band in the gate dielectric inaccessible from the inversion layer as schematically shown in Fig. 7.25. This results in significant improvement in $V_t$ stability of pMOSFET (i.e., negative bias temperature instability or NBTI). In addition to this mechanism, the EWF shift from the cSiGe allows the use of much lower (closer to Si mid-gap) metal gate WF as compared to Si channel pMOSFETs. This fundamentally relaxes the reliability requirements due to lower oxide fields [63–66]. This effect was systematically studied and summarized in Ref. [67]. The normalized bias voltage to cause a $V_t$ shift of 50 mV under an NBTI bias is plotted as a function of metal work function in Fig. 7.26. It was found that the reliability improvement from metal work function shifts toward the Si mid-gap is beyond just the change of oxide fields (i.e., the 1:1 line). Both of these effects for gate stacks on cSiGe are additive, resulting in significant improvement in NBTI.

The reliability improvement from cSiGe can be traded for further EOT scaling since the EOT scaling in the state-of-the-art CMOS has been limited by device reliability to some extent. The semiconductor industry has indeed used these unique advantages of cSiGe for gate oxide scaling. Fig. 7.27 shows pMOSFET $T_{inv}$ as a function of CMOS technology node for high-k metal gate stacks reported in literature, comparing Si and SiGe

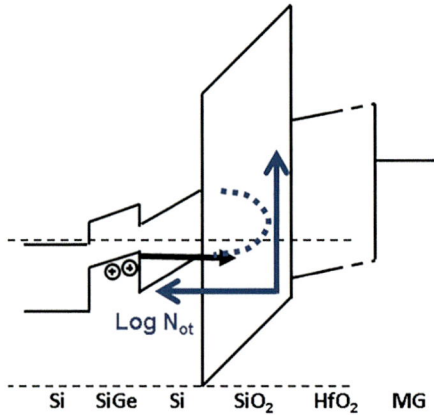

**Figure 7.25** Band diagram for cSiGe/Si-cap/SiO2/HfO2/MG stack, showing U-shaped defect band inaccessible from the inversion changes due to the valence band offset between SiGe and Si. *(Adopted from J. Franco et al., Tech. Dig. Int. Electron Devices Meet (2010), 70.)*

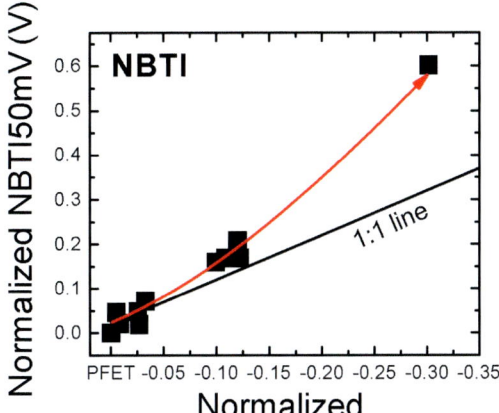

**Figure 7.26** Normalized biased voltage to cause 50 mV threshold voltage shift as a function of metal work function. *(Adopted from P. Barry, Linder, et al., IRPS (2016), 4B.)*

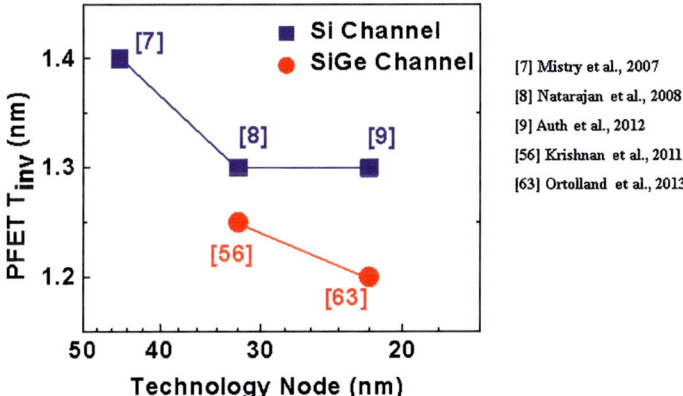

**Figure 7.27** PFET inversion oxide thickness ($T_{inv}$) as a function of CMOS technology node for high-k metal gate stacks reported in literature, comparing Si and SiGe channel devices.

channel devices [60–63]. As demonstrated in Fig. 7.27, the cSiGe devices show more scaled $T_{inv}$ at a given technology node and allow node-to-node $T_{inv}$ scaling.

## 10. Nano-sheet (NS) gate-all-around (GAA) transistor technology and its implication on gate-oxide for logic and I/O transistors

FinFET transistor has reached physical scaling limit on commercially available 3 nm FinFET technology. At 3 nm node, it is projected that Fin pitch is sub 28 nm, and Fin height is greater than 51–55 nm range. Further reduction in Fin pitch and increase in Fin height (high aspect ratio) is not possible due to structural stability to increase transistor density and drive current. As a result, a new transistor architecture named gate-all-around (GAA) nano-sheet (NS) is considered to advance physical scaling and drive more current in a same area as FinFET. Nano-sheet technology is in intense research and development at large industrial labs such as IBM Research, IMEC, and leading semiconductor manufacturers such as Intel, TSMC, Samsung and Rapidus as a next generation transistor manufacturing technology.

## 10.1 FinFET versus nano-sheet gate-all-around structure comparison

Fig. 7.28 provides simple cartoon of FinFET and nano-sheet gate-all-around transistor structure. Highly scaled most advanced FinFET has following challenges which limits further scaling
(1) Fin width ($D_{fin}$) variability
(2) Reduced electrostatic control at scaled $L_g$
(3) Fin pitch scaling and, Fin height increase limitations (structural stability)

Nano-sheet transistor technology offers following fundamental advantages as FinFET successor

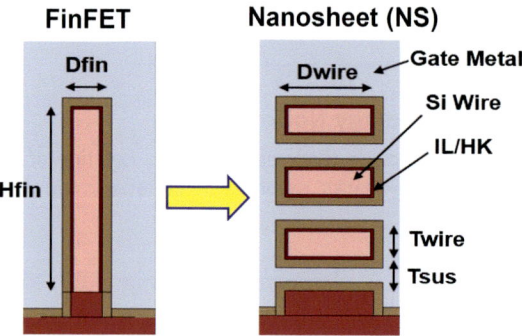

**Figure 7.28** Scaled FinFET versus nano-sheet structure differences where FinFET is a thin body (Si) device and Fin is wrapped with gate oxide and metal gate to control carrier transport. Nano-sheet structure provides three stacked separated silicon sheets separate from each other, while gate dielectric, and metal gate is wrapped around each nano-sheet to operate a nano-sheet transistor. *(Adopted from N. Loubet, VLSI 2020 Virtual Conference, Short Course SC1.1.)*

(1) Improved electrostatics (Due to a wrap-around nature of gate)
(2) Higher performance (Effective channel width increase, allowing higher transport current)
(3) Extend CMOS scaling (Better electrostatic control leads to superior short-channel control, which enables shorter channel length and lower operating voltage)

Fig. 7.29 shows how NS structure at fundamental conceptual level evolves from FinFET by rotating FinFET by 90 degrees, and ability to stack silicon-sheets on top of each other.

NS device in principal doesn't change any transport, and fundamentals of field effect transistor physics, while allowing performance and scaling benefits. However, new nano-sheet transistor structure requires fundamental and creative process integration and process innovations, which have never been integrated together in high volume CMOS microprocessor manufacturing. New innovations in nano-sheet transistor structure will be briefly covered in next section.

## 10.2 Brief review of nano-sheet transistor structure process integration

Loubet et al. provided first comprehensive NS-GAA transistor structure process integration comprehensive details of NS transistor technology in pre-manufacturing environment with complex test-site (test-site, contained various complex circuitry including SRAM). Fig. 7.30 shows step-by-step integration flow of NS transistor [69].

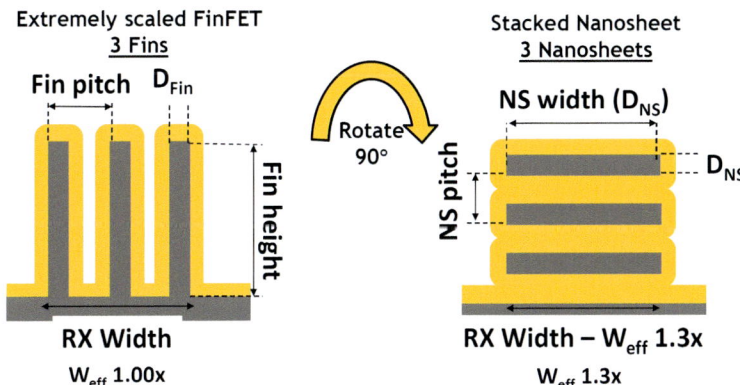

**Figure 7.29** Conceptual evolution of nano-sheet transistor structure by 90 degree rotation of FinFET. *(Adopted from N. Loubet, VLSI 2020 Virtual Conference, Short Course SC1.1.)*

- NS stack epitaxy (a)
- NS "Fin" patterning & STI (b)
- NS "Fin" reveal (c)
- Dummy Gate patterning (d)
- Spacer & Inner Spacer (e)
- Dual SD Epitaxy (f)
- Channel Release (g)
- RMG (h)
- Air Spacer
- Wrap-around contact
- MOL/BEOL (i)

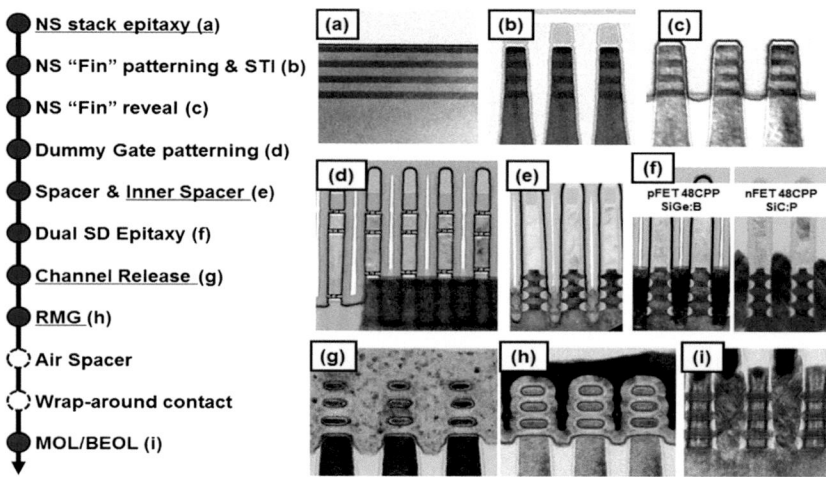

**Figure 7.30** Stacked nano-sheet process sequence and TEM gallery. *(Adopted from Nicolas et al., Stacked nano-sheet gate-all-around transistor to enable scaling beyond FinFET 2017 Symposium on VLSI Technology, Kyoto, Japan, 2017 (Copyright publication year 2017, The Japan Society of Applied Physics).)*

The integration of GAA nanosheet transistor introduces several novel steps which requires significant process and integration innovations. Summary of critical integration modules to build nano-sheet transistor are listed below [70].

1. **Stacked nano-sheet formation**: Superlattice Si/SiGe epitaxial layers grown on the silicon (Si) substrate. Each epitaxial layer are controlled with high precision. Epitaxial channel formation and its control is not lithographically defined, as compare to FinFET.
2. **FIN reveal and shallow trench isolation (STI)**: Active device is lithographically defined (like FINFET), and shallow trench isolation is performed to isolate neighboring devices
3. **Dummy gate formation**: A poly silicon (sacrificial) gate is formed to enable high temperature FEOL processing in junction module.
4. **Inner spacer formation**: Inner spacer formation module is fundamentally new different module in NS transistor. Simply speaking in Fig. 7.30e, before source/drain (S/D) epitaxy formation, there is an indent formation in SiGe portion of nano-sheet. This is a new process which requires lateral SiGe etch with high precision. Additionally, SiGe indent has to be selective to Si portion of nano-sheet (actual silicon channel, where carrier transport will occur). Once indent is formed in

SiGe portion of super-lattice stack, indent is filled with atomic layer deposited dielectric (preferably low K material to reduce capacitance), to inhibit S/D epitaxy formation in SiGe portion of the nano-sheet. Additionally, inner spacer region serves as isolation between the gate and S/D regions [71]. This is a fundamental innovation on how S/D junctions epitaxy is formed in CMOS transistor (planar or FinFET).
5. **Junction formation:** Junction formation using Si:P for nFET and SiGe:B epitaxy for pFET in NS architecture is also novel and require precision engineering, as epitaxy has to grow from silicon nano-sheets which is 5–6 nm in width. Epitaxy over-burden above top nano-sheet needs to be kept at minimum for device external resistance to be kept lowest.
6. **Replacement metal gate formation**
   a. The dummy gate is removed out to form HiK/Metal gate, (like FINFET).
   b. Sacrificial SiGe channel release: The SiGe portion of nano-sheet stack is removed leaving Si nano-sheets. This is a novel step in NS CMOS flow, which was never practiced in CMOS manufacturing.
   c. HiK/Metal Gate (HKMG) formation: An interfacial oxide, a High-K dielectric gate oxide, and the n-type or p-type work functions are selectively deposited on nFET and pFET.

## 10.3 Logic transistor interfacial layer and gate oxide challenges for nano-sheet devices

Nano-sheet GAA transistor requires transmission electron microscopy and transistor electrical characterization for $T_{inv}$ and gate leakage test to understand IL ($SiO_2$) and gate dielectric ($HfO_2$) quality for GAA devices. NS-GAA devices have a unique requirements for gate oxide processing, where $HfO_2$ gate dielectric not only needs to provide conformal coverage of channel on all four sides, but also provide conformal coverage to all three sheets where aspect ratio is quiet high for ALD $HfO_2$ growth. Fig. 7.31 illustrates excellent $HfO_2$ coverage for stacked nano-sheet coverage. This data suggests that atomic layer deposition (ALD) and hafnium oxide will continue to be the choice of technology and material.

Interfacial layer formation is likely will be based on $O_3$ based wet chemistry IL like previously used in FinFET technologies. However, for nano-sheet transistors, silicon consuming IL needs to be carefully evaluated as nano-sheet consumption beyond 6 nm width can lead to quantum confinement effects, which in turn can degrade mobility [71]. Perhaps, ALD based high quality $SiO_2$ can potentially offer better alternate due to

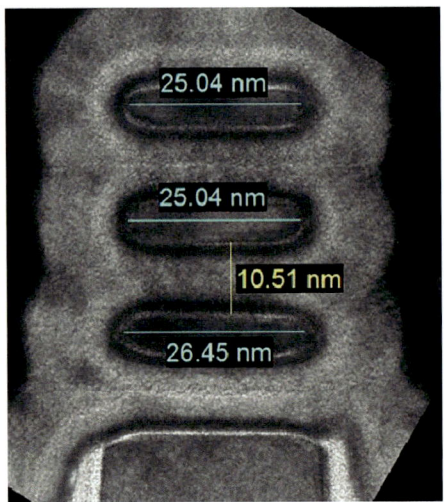

**Figure 7.31** xFIN view of the $W_{sheet}$ design of 20 nm device where $T_{sus}$ is ~10.5 nm. It is important to note that $HfO_2$ dielectric is conformal and wrapped around the silicon nano-sheet within few atomic layer variation. *(Adopted from Zhang J., 2017. IEEE International Electron Devices Meeting (IEDM), San Francisco, CA, USA (2017), 22.1.1–22.1.4.)*

non-consuming silicon nature. In summary, logic gate dielectric and interfacial layer can be extended to nano-sheet GAA transistor technology with careful consideration of minimum silicon channel loss beyond 5–6 nm of channel thickness.

## 10.4 Nano-sheet high voltage I/O transistor gate dielectrics and their Co-integration with logic transistor

I/O transistors and their gate oxide formation for nano-sheet technology is limited by spacing between nano-sheets. Generally, I/O transistors operate at 1.8 V (V), and require 3.5 nm of gate oxide thickness. To form true GAA I/O device, maximum I/O gate oxide thickness can only be limited to 2 nm of $SiO_2$, since 1.6 nm of $HfO_2$ and 6 nm of N and P work function metals are needed. 2 nm of $SiO_2$ as an I/O gate oxides limits maximum allowed voltage to 1.2 V. However, some alternate options where gate oxide is "pinched off" between nano-sheets, and HiK/Metal gate is wrapped around the "pinched-off" I/O gate oxide can break the gate oxide thickness limitation for NS architecture. NS GAA transistor will require I/O last flow where I/O gate oxide is formed after poly pull and SiGe channel release. This is a major innovation, as I/O last flow is never been used in FinFET technology [72]. To evaluate deposited oxide for GAA application, we deposited 3 nm of ALD oxide and densified using commerically available plasma nitridation and rapid thermal anneal (DPN3/PNA) process. Fig. 7.32 shows TEM cross section of nano-sheet device with ALD oxide after nitridation and anneal, where excellent conformality of gate dielectric without consuming silicon sheet is achieved.

Maruf et al. [72] published comprehensive results comparing deposited oxide and selective growth (thermal based) oxide for I/O nano-sheet devices. Maruf et al. results demonstrate that thermal based oxide consumes nano-sheet silicon channel and are limited to 2.1 nm of oxide thickness; on the other hand, deposited oxide does not have thickness limitation due to non-consuming nature of ALD oxide. Since deposited oxides >2.5 nm starts to pinch off the spacing between nano-sheets, metal gate is only deposited on the outside of nano-sheet, while in between sheets, there is no metal gate (due to oxide pinch off). This "pinched-off" oxide devices lead to reduction in current conduction in pinched off areas, hence losing some I/O device performance as a trade-off, while providing increased Vmax operating capability. Maruf et al. data also confirms excellent dielectric deposited oxide with post treatments, which is equivalent to thermal based I/O dielectric. In summary deposited oxides used for FinFET as an I/O oxide is a viable option for NS architecture, where offering unique capabilities for wide range of low voltage/higher performance to high voltage/lower performance device design capabilities.

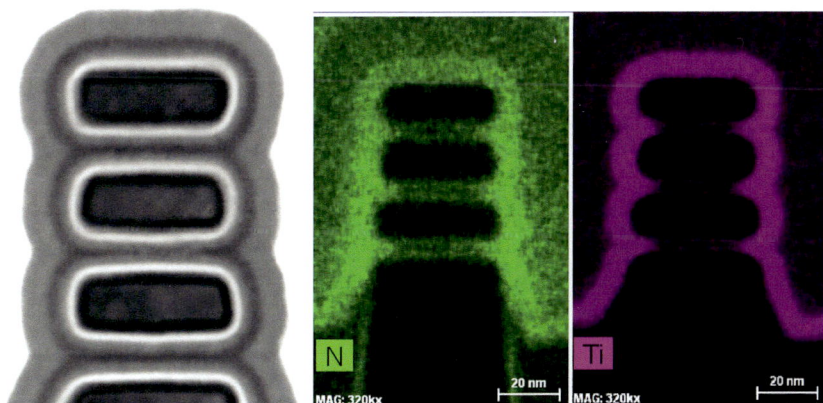

**Figure 7.32** Nano-sheet TEM cross section showing conformal deposited oxide around nano-sheets. EDX maps show conformal nitridation of ALD oxide, while TiN gate is pinched off between nano-sheet.

## 11. Si/SiGe heterostructure-based I/O devices with low temperature ALD oxide and densification

### 11.1 Integrate with nano-sheet gate-all-around device

In this section we propose and demonstrate a device architecture having heterostructure Si/Si$_{1-x}$Ge$_x$ channel for I/O applications and its compatibility with the GAA nanosheet logic technology. High-quality deposited gate oxide is developed as one of the key components for the heterostructure I/O device offering. The novel I/O device architecture shows excellent characteristics including threshold voltage tunability, low gate leakage, low interface state density, significant mobility enhancement, high breakdown voltage, and superior negative bias temperature instability. The unique heterostructure superlattice maintains sharp boundaries at the end-of-process with minimal interdiffusion between Si and Si$_{1-x}$Ge$_x$ layers.

Gate-All-Around (GAA) nanosheet transistor has been introduced as the device architecture to continue logic technology scaling beyond FinFET structure, with benefits in density, power-performance, and design flexibility [69,73]. Previously, it was demonstrated extremely scaled gate length [69], and multi-Vt solution schemes [70] for GAA Nanosheet Technology. Enablement of I/O devices is another key element for any technology. Due to the requirements for handling large voltages, thick gate oxide is the signature for such devices. One of the essential pursuits is to integrate I/O devices with logic devices in the same process integration flow. I/O device voltage offerings can be limited by the required narrow spacing between nanosheets (Tsus in Fig. 7.33a) as higher operating voltage require thicker gate oxide. *We propose and demonstrate Si/Si$_{1-x}$Ge$_x$ heterostructure fins as an*

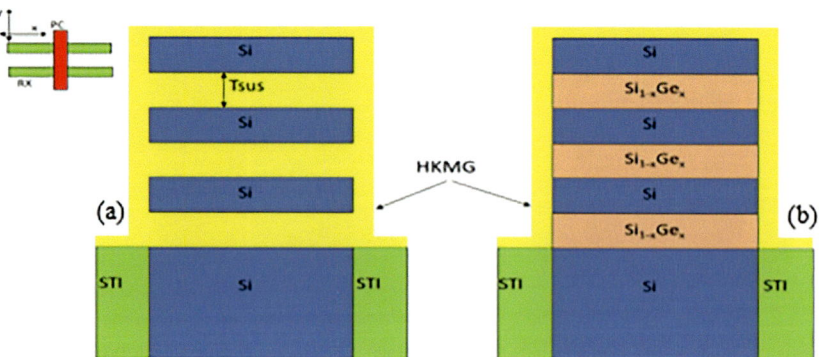

**Figure 7.33** Cross-fin (y-cut) schematic of (a) logic device and (b) proposed Si/Si1-xGex heterostructure-based I/O device for nanosheet technology.

*I/O device offering for the GAA NS technology* (Fig. 7.33b). The major difference in the channel between I/O devices and logic devices is the sacrificial $Si_{1-x}Ge_x$ layers are kept for I/O devices; and thus, the limitation on gate oxide thickness is eliminated toward enabling high voltage offering.

For I/O heterostructure device, we focus on $Si/Si_{1-x}Ge_x$ heterostructure channel with Ge atomic concentration at 25% and 35%. Furthermore, we use plasma oxidation at 400°C densification technique for the thick oxide processing. In this chapter, we present systematic study of the process flow, device structural integrity, and electrical properties for the $Si/Si_{1-x}Ge_x$ heterostructure I/O devices.

## 11.2 Heterostructure device structure results

**Process Flow:** The process flow to form heterostructure $Si/Si_{1-x}Ge_x$ stack is similar to the flow described by Loubet et al. [69] as shown in Fig. 7.34. Thick oxide deposition and densification is carried out right after fin reveal and before dummy gate material disposition. To achieve high quality gate dielectric and robust interface between dielectrics and the heterostructure channel region, both channel epitaxial growth and gate dielectric processes are critical. Regarding the multilayer channel epitaxy, we have verified that all Si and $Si_{1-x}Ge_x$ layers were pseudomorphic with respect to Si and defect-free thanks to optimized surface preparation and well controlled epitaxial growth [69]. For the gate dielectric densification process, plasma nitridation/anneal (***Process A as discussed in*** Chapters 2 and 3 **with 700°C**

**Figure 7.34** Process flow for Si/Si1-xGex heterostructure device.

***rapid thermal anneal)* and Process B *(Plasma oxidation at 400°C, at 2 mT pressure, for 120 s)*** is performed to improve the dielectric quality where thermal budget of Process B > A. For rest of the chapter, we will use process A and B for ease of explanation unless specified.

**Gate Dielectric and Heterostructure Channel** To have high quality thick oxide, generally high thermal budget is preferred. However, high thermal processes can impact heterostructures such as $Si/Si_{1-x}Ge_x$ layers interdiffusion leading to non-crystalline channel material with poor transport properties. The developed novel processes of gate oxide deposition (high temperature plasma enhanced ALD $SiO_2$) and densification (microwave plasma nitridation and novel plasma oxidation) [54] were able to form highly reliable gate oxide while keep the integrity of the $Si/Si_{1-x}Ge_x$ heterostructure. Figs. 7.35 and 7.36 show intact crystalline heterostructure layers after the gate oxide deposition step and after the end of the whole fabrication process, respectively. Figs. 7.35 and 7.36 are technologically important as it demonstrates low temperature thick gate I/O oxide formation without Si/SiGe interdiffusion.

Elemental analysis in Fig. 7.37 further confirms robustness and non-detectable Ge interdiffusion in the heterostructure channel.

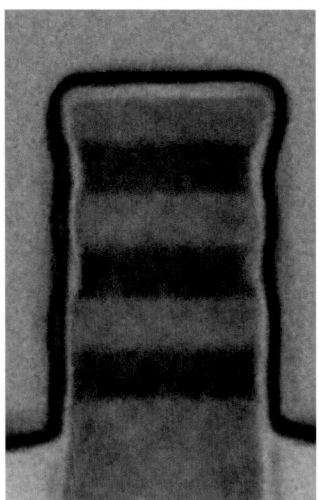

**Figure 7.35** TEM image of $Si/Si_{0.65}Ge_{0.35}$ heterostructure channel after ALD oxide deposition with novel densification (low pressure plasma oxidation at 400°C) demonstrating non-interfused channel.

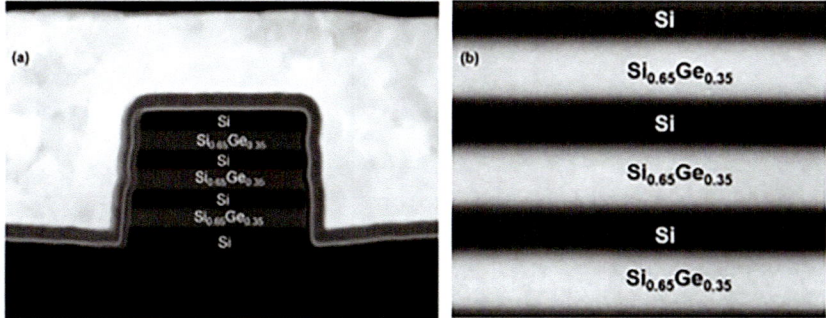

**Figure 7.36** (a) TEM cross section, and (b) zoomed-in image of a fin showing the intact crystalline structure of $Si_{0.65}Ge_{0.35}$ and Si after the full process flow.

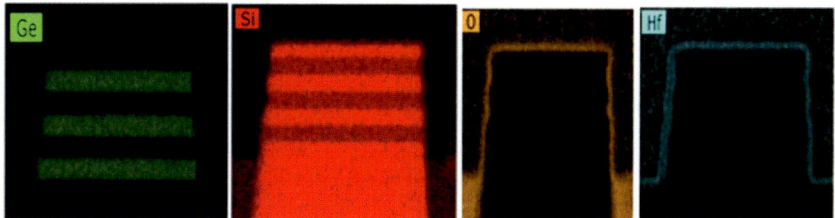

**Figure 7.37** Elemental composition map of (a) Ge, (b) Si, (c) O, (d) Hf in the heterostructure fin showing distinct layers having Ge.

## 11.3 Electrical results and discussions

**I−V Characteristics:** Current-Voltage characteristics of the devices are obtained at two different drain to source voltages ($V_{ds}$) of −50 mV and −1.5 V, with gate voltage sweeping from +1V to −2.6 V, as shown in Fig. 7.38. High gate voltage is applied to ensure proper characterization at the voltage regime desired for I/O applications. For $V_{ds} = -50$ mV, ~8 orders of magnitude $I_{on}-I_{off}$ ratio is demonstrated for both source and drain current, indicating excellent gate modulation of the heterostructure channel.

Devices with $Si/Si_{0.75}Ge_{0.25}$ heterostructure demonstrates higher threshold voltage (more negative) than those with $Si/Si_{0.65}Ge_{0.35}$ heterostructure, as shown in Fig. 7.39. This suggests that the $Si_{1-x}Ge_x$ layers contribute effectively to the on-off states. The higher Ge concentration, the more valence band edge offset is expected, hence lower threshold voltage.

At Vds = −1.5 V, drain current exhibits ~7 orders of $I_{on}$-$I_{off}$ ratio with increasing drain leakage observed at higher off-state gate voltages

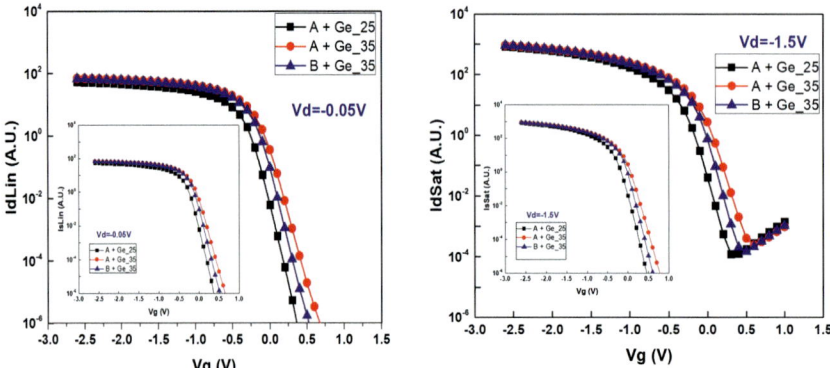

**Figure 7.38** Id versus Vg (inset: Is vs. Vg) curves at (a) Vd = −50 mV and (b) Vd = −1.5 V for the heterostructure transistors, showing well behaved characteristics with on/off ratio in the range of ∼7–9 orders.

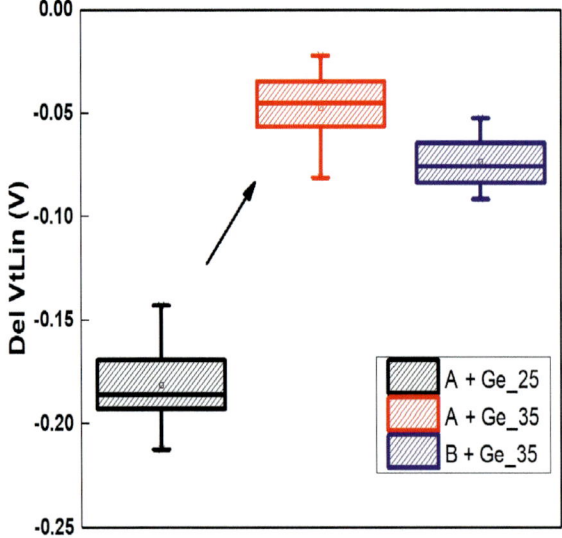

**Figure 7.39** Vt modulation due to variation in Ge composition.

(Fig. 7.38b). At such voltage ranges, drain is leaking to the substrate indicating GIDL as the primary leakage mechanism [74]. In Fig. 7.40, drain leakage is plotted for the devices at electrode voltages of Vg = Vt + 0.8 V, Vd = −1.5 V, and versus = Vsub = 0 V. Increasing Ge composition worsens the drain leakage by ∼3x, whereas Process B is effective in reducing drain leakage by ∼ 0.8x.

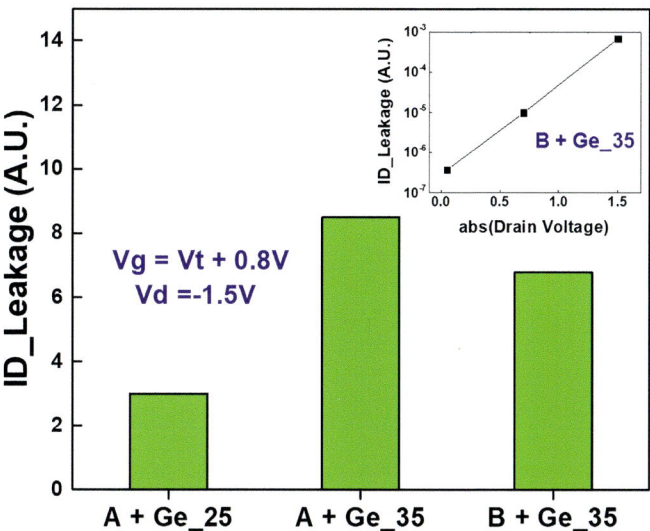

**Figure 7.40** Off state drain leakage at Vd = −1.5 V, showing ∼3x greater value at higher Ge composition (inset drain leakage at various Vd for process B).

Fig. 7.41 shows the off-state source current at Vg = +0.5 V, which is primarily influenced by the Vt difference arising due to Ge composition. Process B is also found to be effective in DIBL reduction (inset Fig. 7.49).

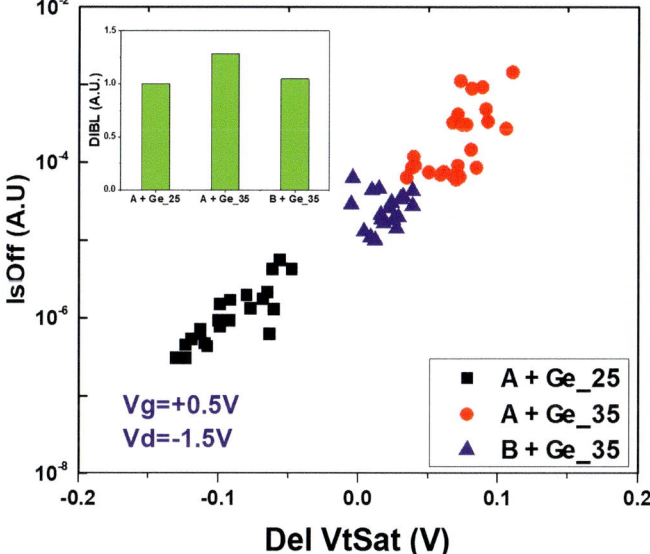

**Figure 7.41** Off state source leakage primarily influenced by Ge composition induced Vt shift in the channel (inset: DIBL for various splits).

## 11.4 Gate stack characteristics

Critical to I/O devices is the gate stack quality. Fig. 7.42 shows the gate leakage versus gate voltage. Excellent gate leakage characteristic is observed with gate current ~8 orders lower than the drain current even at 2.5 V gate voltage, attesting to the excellent quality of the gate oxide making it appropriate for high voltage I/O applications. Fig. 7.42 results are

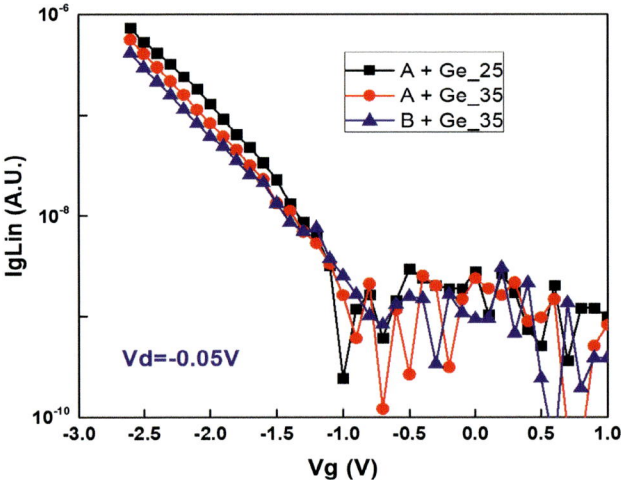

**Figure 7.42** Ig versus Vg characteristics with leakage ~8 orders lower than drive current @Vg = −2.5 V.

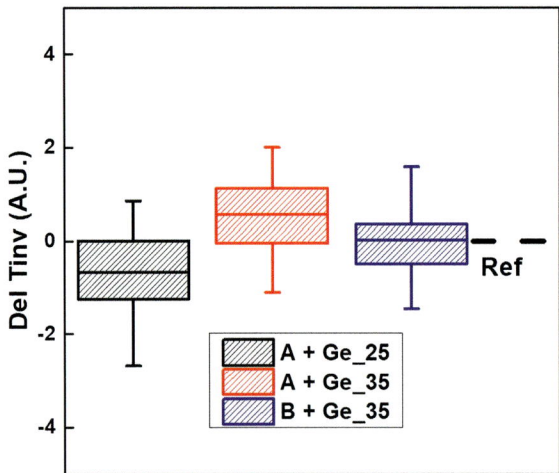

**Figure 7.43** Delta Tinv plot with process B leading to Tinv reduction.

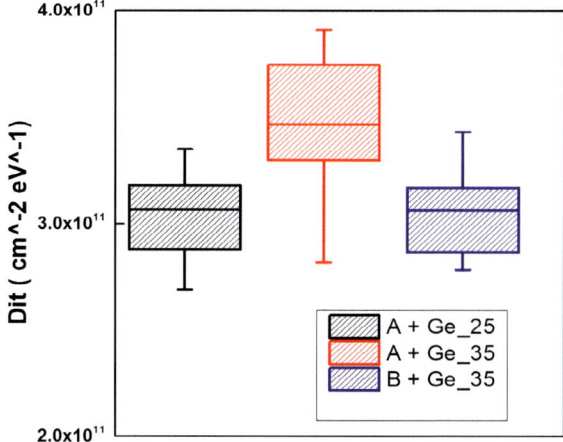

**Figure 7.44** Dit plots with process B showing benefits over process A. However, it is remarkable to achieve Dit values with such a high Ge atomic % (25% and 35%) channel devices.

remarkable showing ALD oxide with densification can lead to thermal oxide equivalent oxide for complex heterostructure silicon/silicon germanium devices.

In Fig. 7.43, Tinv of the different splits are compared. All the splits are within ∼2 A variation, with process B showing reduction in Tinv (∼0.5 A).

With the introduction of $Si_{1-x}Ge_x$ into the channel region of a device, interface state density ($D_{it}$) is always a critical aspect that could impact device behavior. With the optimized heterostructure channel process and gate dielectric formation, we have achieved low $10^{11}$ cm$^{-2}$ level of $D_{it}$, as shown in Fig. 7.44, similar to $D_{it}$ demonstrated for SiGe FinFET technology [75].

Subthreshold swing values (SS) for the different splits are depicted in Fig. 7.45. Higher Ge concentration causes SS degradation which is consistent with $D_{it}$ observed in Fig. 7.45. Process B is effective in mitigating the interface trap states as it leads to both Dit and SS improvement.

**Carrier Mobility and Device Resistances:** It is expected that introduction of $Si_{1-x}Ge_x$ into the channel region will bring benefit for carrier mobility. Fig. 7.46 shows hole mobility versus $N_{inv}$ curves for a variety of experiments. External resistance arising from the source/drain region is corrected for accurate mobility extraction [76]. In general, I/O FETs with $Si/Si_{1-x}Ge_x$ heterostructure channel demonstrate higher

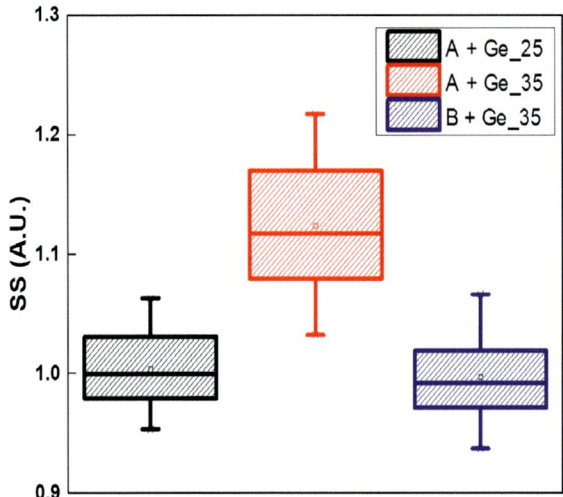

**Figure 7.45** Subthreshold swing plot with process B showing benefits over process A.

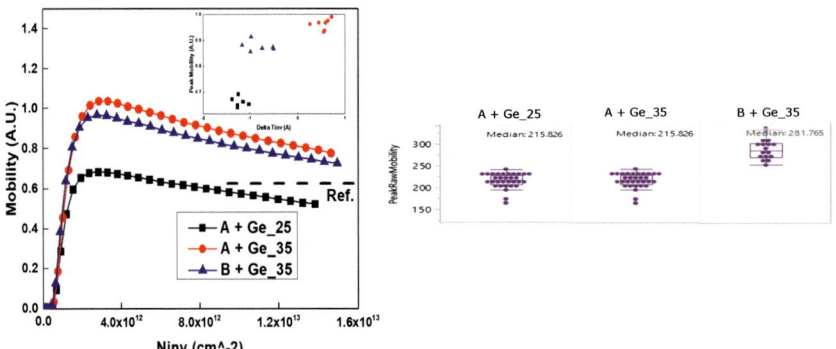

**Figure 7.46** Mobility versus Ninv characteristics with Ge conc. 35% generating ∼1.5x greater peak mobility than Ge 25%. Raw actual peak mobility results for Ge atomic % and gate oxide experiments. With hetero-structure Si/SiGe our pFET mobility values are significantly higher with viability to make thick gate oxide I/O devices is best known reported results.

mobility than I/O FETs with pure silicon channel as shown in Fig. 7.46. It is observed that Ge concentration is effective toward improving hole mobility. From heterostructure with Si/Si$_{0.75}$Ge$_{0.25}$ layers to Si/Si$_{0.65}$Ge$_{0.35}$, peak hole mobility is improved by ∼1.5x.

Due to higher D$_{it}$, coulombic scattering for Ge 35% can be expected to be worse than the Ge 25% atomic concentration. As a result, the mobility

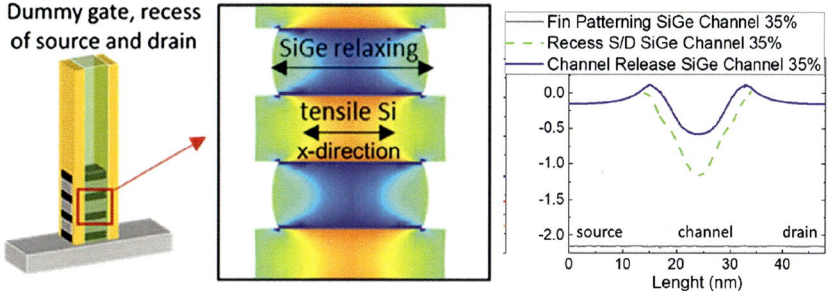

**Figure 7.47** Strain profile in GAA nano-sheet structure after S/D EPI formation showing compressively strained SiGe layer. SiGe layer is a transport layer for SiGe/Si I/O device discussed in this chapter. *(Adopted from S. Reboh et al., IEDM (2019), p246−249.)*

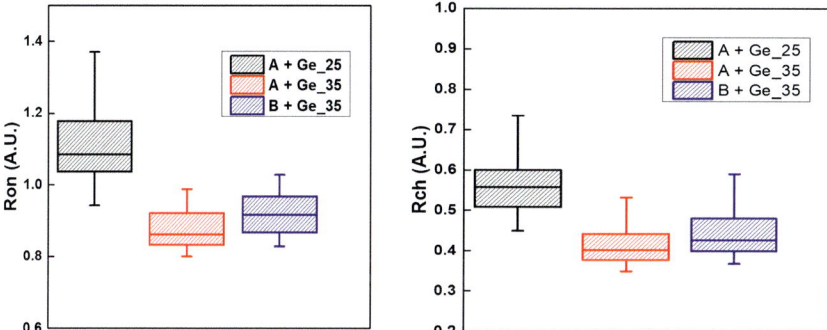

**Figure 7.48** (a) Ron and (b) Rchannel plots showing reduction in values for higher Ge composition.

improvement can be attributed to ameliorated phonon scattering in Ge 35% channel. This is further corroborated when the peak mobility versus Tinv characteristics (inset of Fig. 7.54) of Ge 35% with process B is compared to Ge 25% with process A (A > B, thermal budget) at similar Tinv. Although both splits have similar Tinv and Dit, the higher peak mobility for Ge 35% might be purely arising from phonon scattering related effects.

Leitz et al. [77] has reported compressively strained $Si_{1-y}Ge_y$ alloy channels, despite the y. (1−y) dependence on alloy scattering. Additionally, they reported the strained Si surface layer should have even less influence for higher Ge content channels, where the large band offset between the buried compressive $Si_{1-y}Ge_y$ and surface Si effectively confines holes within the buried layer [77]. In our proposed device, holes are confined in SiGe layer since hole mobility results show strong dependence on Ge atomic % as shown in Fig. 7.54. Murray et al. reported systematic studies on strain measurements in nano-sheet structure we used in our research for heterostructure I/O device, where it is confirmed SiGe layer is compressively strained due to lattice mismatch with underlying Si [78]. Additionally, narrow nano-sheets have more uniform compressive strain across the channel in comparison to wider sheets was reported. Reboh et al. [79] reported systematic strain measurements in GAA - nanosheet transistors as a function of process flow. Strain measurements after S/D Epi formation confirms compressive SiGe layer, however, SiGe sheets starts to relax on the edges. Fig. 7.47 shows strain results in nano-sheet structure from Reboh et al. where they confirm SiGe is compressively strained, however edge of the sheets starts to show some relaxation.

Higher Ge composition leads to reduction of on-resistance (Fig. 7.48a) primarily arising from channel resistance (Rch) reduction of ~30% (Fig. 7.48b). Meanwhile, peak channel transconductance improvement of ~30% is observed, as shown in Fig. 7.49. Process B shows slightly higher Rch over process A, consistent with the mobility trend.

**I/O Gate Oxide Reliability:** Reliability characteristics of the heterostructure devices are evaluated as shown in Fig. 7.50. Experiments with a variety of robust dielectric deposition and densification processes show excellent breakdown voltage of above 5 V, showing thermal oxide like dielectric quality [54]. NBTI characterization show above 4 V $V_g$ to 50 mV

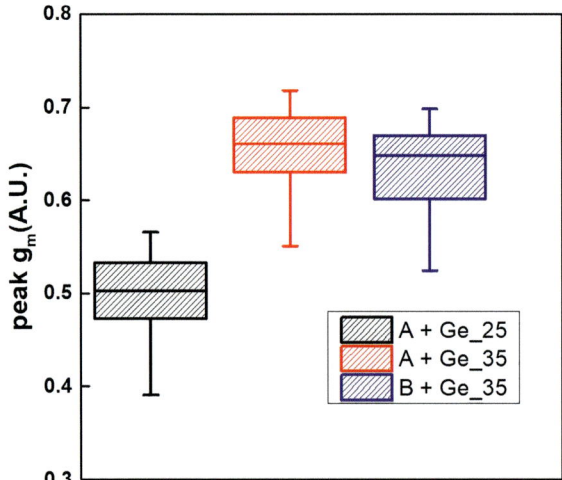

**Figure 7.49** Gm plots showing improvement for higher Ge conc., consistent with Rch trend.

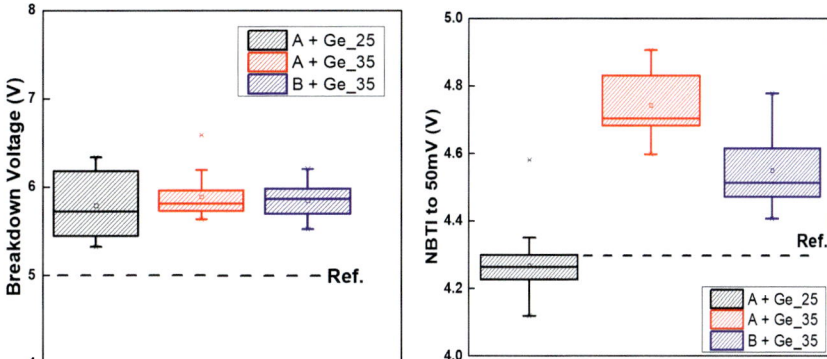

**Figure 7.50** (a) Breakdown voltage exceeding 5 V at target Tinv and (b) NBTI, Vg to 50 mV is above 4 V for all the devices.

shift, comparable to pure silicon fin I/O devices. This is attributed to lower electric field across gate dielectric from thick gate oxide and enhanced valence band edge offset with the introduction of $Si_{1-x}Ge_x$ layer. Increased Ge composition improves NBTI characteristics, thanks to further enhancement of valence band offsets.

In summary, the $Si/Si_{1-x}Ge_x$ heterostructure channel is a suitable solution for I/O devices offering, compatible with the base integration scheme for GAA Nanosheet Technology. Proposed I/O device heterostructure eliminates the thickness limitation of I/O gate oxide imposed by narrow space between nanosheets required for parasitic capacitance reduction for GAA nanosheet technology. We demonstrate high quality deposited oxide and low thermal budget densification showing high $V_{bd}$ (>5 V), and excellent gate oxide/$Si/Si_{1-x}Ge_x$ interface quality leading to $10^{11}$ cm$^{-2}$ level of $D_{it}$. Additional benefits demonstrated of this device structure for I/O applications includes channel mobility improvement by Ge composition tuning, and excellent NBTI characteristics.

## 12. High mobility (high atomic % Ge, SiGe) channel: Logic IL and I/O gate oxide research results

Beyond silicon channel, it will be imperative to evaluate and potentially implement SiGe as pFET channel material for nano-sheet technology, as SiGe pFET channel material was implemented for IBM's 32SOI, 22SOI planar, Global Foundries and S.T Microelectronics' FDSOI planar, and TSMC's most advanced FinFET technologies.

We studied In-situ steam generated (ISSG) thermal oxide deposited at 800°C for $Si_{1-x}Ge_x$ [x = 0 to 0.8], where we targeted 1.5 nm of oxide thickness on silicon surface as a control. Thermal oxidation behavior for Ge atomic % range from 0% (Si) to 80% is a first of a kind of studies to understand gate oxide growth for logic transistor IL and I/O gate oxide. Fig. 7.51 shows oxide thickness (regrowth) as a function of Ge atomic %, where oxidation clearly has three different regimes for ISSG oxide. From 10 to 50 atomic % Ge (regime I), there is a very small regrowth of 0.5 nm, where silicon is preferentially being oxidized. From 50 to 65 atomic % Ge, there is a significant oxide regrowth (regime II), Ge—Ge bonds breaks first, then Si—Ge bonds break leading to excessive regrowth of 14 nm at 65 atomic % Ge, while 65 to 80 atomic % there is decline in regrowth, and eventually leading to expected thickness on silicon substrate, in this regime Ge—Ge bonds dominates, and SiGe surface starts to sublime. For regime-II, higher atomic Ge% in SiGe lattice plays a dominant role as a catalyst leading to enhanced oxidation. In regime-III, beyond 65 atomic % Ge, Ge—Ge bonds dominates, and SiGe is completely sublimated.

This result show that it is possible to form thin thermal IL for SiGe with low atomic % Ge. However, beyond 50 atomic % Ge, high temperature thermal oxide IL or dielectric deposition is not a viable option due to excessive oxidation, and damage to SiGe crystalline structure.

Fig. 7.52 shows high resolution transmission electron microscopy (HRTEM), electron energy loss (EELS) and X-ray photoelectron microscopy (XPS) results of 25, 55 and 80 atomic % Ge SiGe substrate results where 1.5 nm of ISSG (14 s growth time) was used to grow thermal oxide.

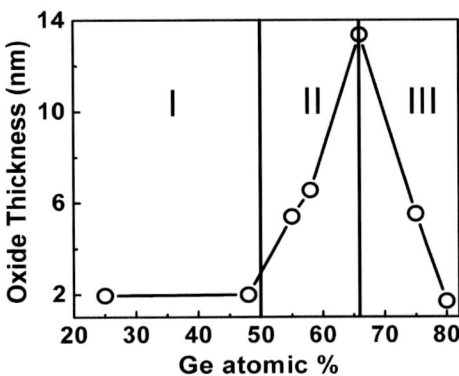

**Figure 7.51** Ge atomic % versus oxidation regrowth for ISSG [80].

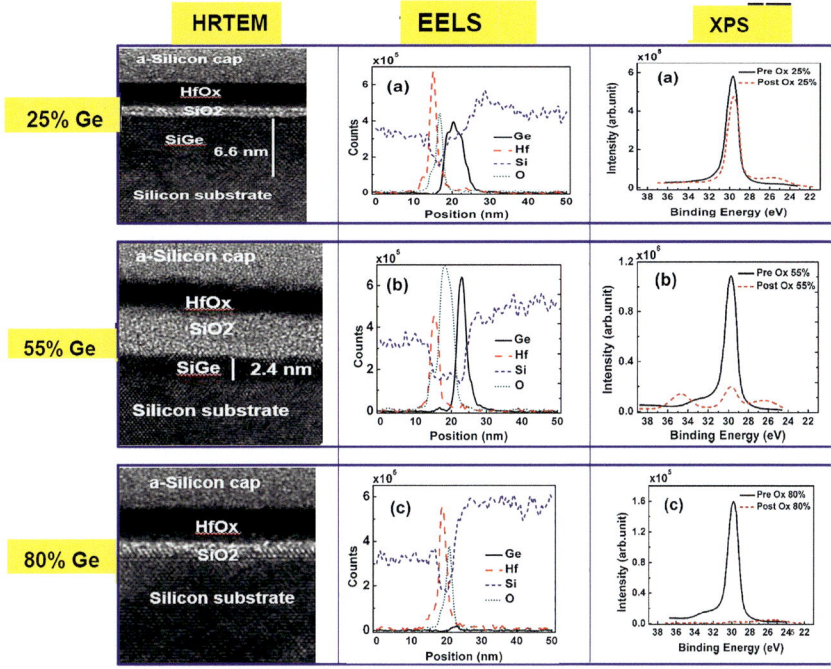

**Figure 7.52** HRTEM, EELS, and XPS results for ISSG oxide (fixed time at 14 s to target 1.5 nm on silicon substrate) on 25, 55, and 80 atomic % Ge [80].

For HRTEM results, we observe 0.5 nm of oxide growth for 25 atomic % Ge, 6 nm of oxide for 55 atomic % Ge (4.5 nm of regrowth), and at 80 atomic % Ge we get 1.5 nm of oxide with complete sublimation (consumption) of SiGe substrate. XPS results for Ge-3d spectra confirms consumption of SiGe channel as increase in Ge atomic % for ISSG growth.

Next, we studied ALD oxide deposition at room temperature (RT), 75°C, and 300°C on 25 and 55 atomic % Ge SiGe substrate. We again used HRTEM and XPS to understand regrowth amount on SiGe substrates and changes in Ge3d spectra to understand Ge—Ge bond evolutions. At room temperature ALD oxide depositions, there is no regrowth observed, and small GeO formation was observed as shown in Fig. 7.53. This may not be the best approach to form gate oxide, but this kind of approach with densification through plasma oxidation can lead to high Ge% SiGe and oxide interface in nano-sheet structure, e.g., inner spacer processing etc.

For 75°C deposition, we observed very uniform ALD oxide formation with very small GeO formation as shown in Fig. 7.54. This provides

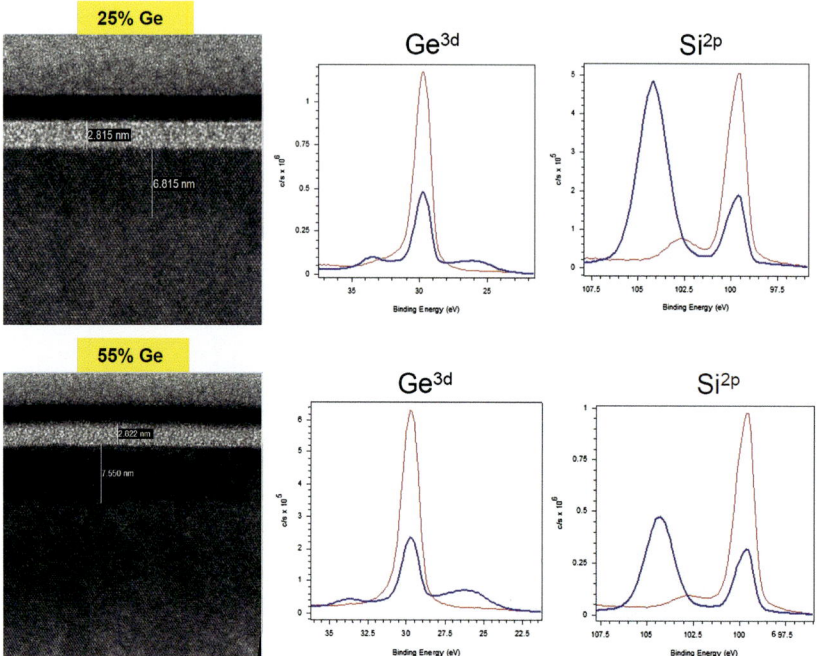

**Figure 7.53** HRTEM, and XPS Ge3d and Si2p spectra of 25 and 55 atomic % Ge in SiGe for room temperature (RT) ALD oxide deposition [80].

additional degree of freedom to form either sacrificial hard mask or insulating uniform film in nano-sheet structures.

Fig. 7.55 shows HRTEM and XPS results for 300°C, where Ge3d spectra shows significant increase in GeO formation at 34.5 eV for 25 and 55 atomic % Ge. 0.5 nm of increased oxide growth is observed for 25% and 55% atomic Ge.

Table 7.6 provides summary of ISSG and PEALD oxide deposition, $SiO_2$ regrowth, and GeO formation during oxide growth. In our results, ISSG oxide grown at 800°C is not a viable option to grow oxide at high Ge atomic % SiGe due to excessive regrowth and complete sublimation of 80 atomic % Ge substrate.

In summary, growing low temperature and high quality ALD oxides dielectric on SiGe with high Ge atomic % is critical to enable gate dielectrics (IL and I/O gate oxides) for future high mobility pFET channel materials. Additionally in NS technology understanding of oxides and nitrides dielectrics deposition on high atomic % Ge SiGe substrate is a critical need to develop NS and future stacked FET devices.

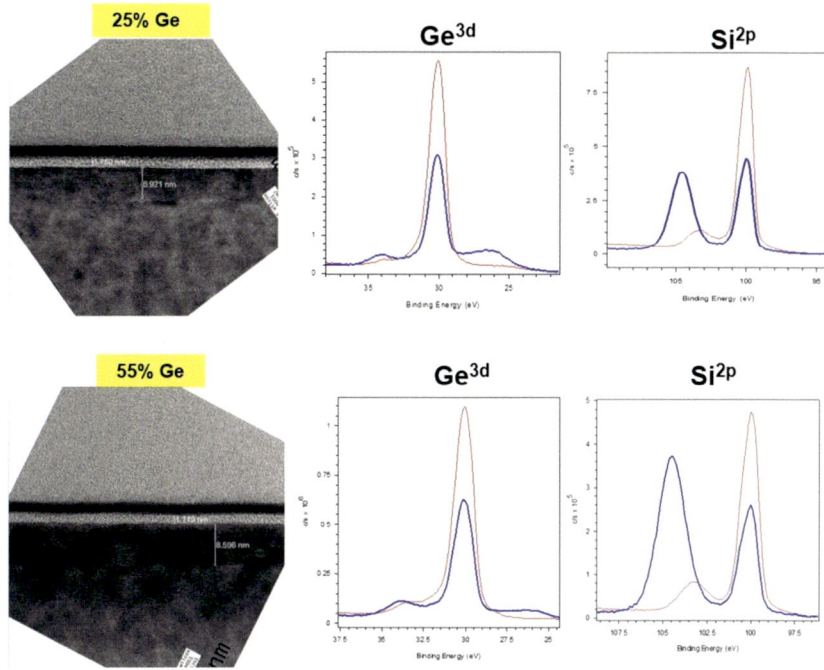

**Figure 7.54** HRTEM, and XPS Ge3d and Si2p spectra of 25 and 55 atomic % Ge in SiGe for 75°C ALD oxide deposition [80].

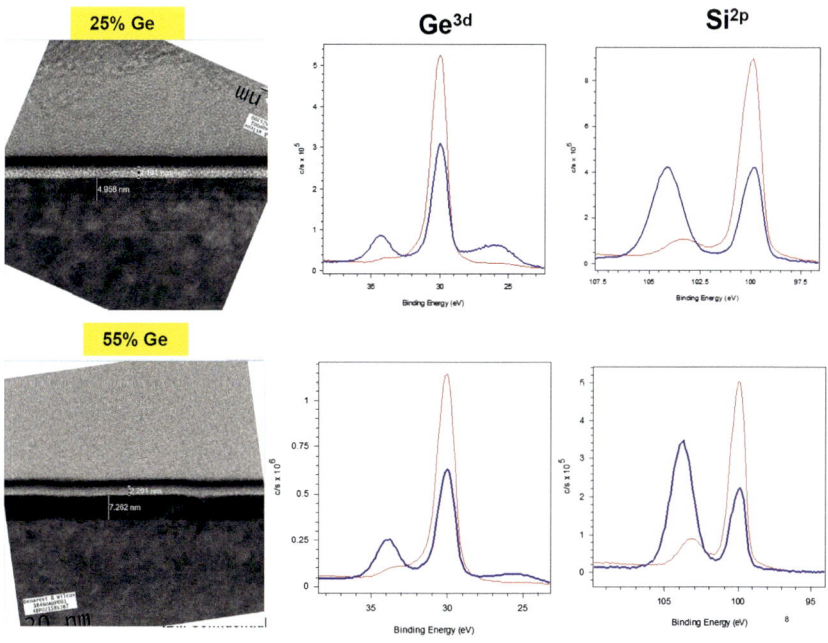

**Figure 7.55** HRTEM, and XPS Ge3d and Si2p spectra of 25 and 55 atomic % Ge in SiGe for 300°C ALD oxide deposition [80].

**Table 7.6** Summary of ISSG and ALD oxide growth on 25, 55, and 80 atomic % SiGe [80].

| Oxides | Si Sub | SiGe, (Ge 25%) | SiGe (Ge 55 %) | SiGe (Ge 80%) | Key Observation/Comments | |
|---|---|---|---|---|---|---|
| | | SiO2/ GeO Thickness | | | SiO regrowth | GeO formation |
| ISSG 800 °C | 1.4 nm | 1.9 nm | 3.5 nm | 1.4 – 1.5 nm | 0.5 nm of SiO growth for 25% SiGe<br><br>2.5 nm regrowth for SiGe 55% | No GeO formation for 25% SiGe<br><br>Some GeO formation for 55% |
| PEALD 75 °C | 1.8 nm | 1.8 nm | 1.7 nm | | No regrowth | Very small GeO formation for 25% and 55% SiGe |
| PEALD 300 °C | 1.5 nm | 2.1 nm | 2.1 nm | | 0.6 nm regrowth | Highest GeO formation for 25% and 55% |
| RT PEALD | 2.8 nm | 2.815 | 2.822 | | No regrowth | Small GeO formation even at RT |

## 13. Conclusion

It is author's view that nano-sheet transistor can potentially enable two to three technology nodes (beyond 3 nm FinFET), and perhaps be a foundational building block of complementary FET (CFET), where NMOS and PMOS devices are stacked vertically. If successful, basic nano-sheet structure and device can enable 15 years of CMOS manufacturing without changing fundamental field effect transistor physics principals or significant material changes. Vertical FET is another potential disruptive contender which can enable dimensional scaling, however, significant architectural and process integration, and process development innovations are required. Three additional critical technologies, back side power rail (BSPR), hybrid bonding, and chiplet technologies (with photonics) will continue unprecedented integration and system performance.

For gate oxide scaling there is no major breakthrough beyond $HfO_2$, HfZrO, and HfON with $SiO_2$ as an IL is expected. However, high mobility SiGe channel for pFET NS may require evaluating low temperature ALD oxide with post treatments as a potential IL material for HiK dielectrics. I/O devices with deposited gate oxides and post densification allows multiple innovative options to co-integrate I/O devices with NS logic transistors, such as Si/SiGe heterostructure I/O devices, NS I/O devices with 'pinched off' dielectrics, and thin dielectrics with fully GAA devices. Thermal oxide based I/O dielectrics will be severely challenged due to thin body nature of GAA or any future CMOS devices.

In author's view, beyond silicon technologies there is not a single material system which can replace entire silicon CMOS transistor, its components, and FET architecture in near term for high volume manufacturing. Perhaps carbon-based transistor in far future will be a potential contender to replace silicon. In author's view, it is also possible to develop heterostructure Si/SiGe channel devices on nano-sheet platform as a major disruptive future transistor technology, allowing significant extension of nano-sheet technology. Introduction of extreme ultraviolet lithography (EUV) with high NA EUV will allow continued physical scaling, but transistors will be limited by self-heating, leakage, and structure feasibility. Hybrid bonding and chiplet technologies in CMOS will become a future of microprocessor innovations, and perhaps complex multi-functional, and non-silicon computation units integration on chip (III−V semiconductors based) allowing un-imagined computing capabilities.

## References

[1] H.R. Dennard, et al., IEE J. Solid State Circ. SC-9 (1974) 256−268.
[2] B. Davari, et al., Proc. IEEE 83 (4) (1995) 595−606.
[3] B. Hoeneisen, et al., Solid State Electron. 15 (1972) 819−829.
[4] D.J. Frank, et al., Proc. IEEE 89 (3) (2001) 259−288.
[5] H. Wong, et al., Microelectron. Eng. 138 (2015) 57−76.
[6] M. Chudzik, et al., Wiley-VCH Verlag GmbH & Co. KGaA, Chapter 17, 2012.
[7] K. Mistry, et al., in: Technical Digest International Electron Devices Meet, 2007, p. 247.
[8] S. Natarajan, et al., in: Technical Digest International Electron Devices Meet, 2008, p. 941.
[9] C. Auth, in: Symposium VLSI Technical Digest of Technical Papers, 2008, p. 128.
[10] Greene, et al., in: Symposium VLSI Technical Digest Technical Papers, 2009, p. 140.
[11] S. Narasimha, et al., in: Technical Digest International Electron Devices Meet, 2012, p. 52.
[12] N. Planes, et al., in: Symposium VLSI Technical Digest of Technical Papers, 2012, p. 133.
[13] G. Cesana, Workshop Fully Depleted SOI, 2012. http://www.soiconsortium.org/fully-depleted-soi/presentations/february-2012/1%20-%20Giorgio%20Cesana%20%20 28%20&%2020nm%20FDSOI%20Technology%20Platforms.pdf.
[14] H.J. Choi, et al., Mater. Sci. Eng., AR 72 (2011) 97−136.
[15] D.G. Wilk, et al., J. Appl. Phys. 89 (2001) 5243.
[16] H. Zhu, et al., J. Mater. Sci. 47 (2012) 7399.
[17] J. Robertson, Eur. Phys. J. Appl. Phys. 28 (2004) 265.
[18] S. Inumiyal, et al., in: Symposium of VLSI Technical Digest of Technical Papers, 2003, p. 17.
[19] M. Dai, et al., J. Appl. Phys. 113 (2013) 044103.
[20] Q. Xu, et al., Appl. Phys. Lett. 93 (2008) 252903.
[21] K. Ariyoshi, et al., in: 40th IEEE Semiconductor Interface Specialists Conference, 2009, p. 1.

[22] R.H. Huff, et al., Springer-Verlag Berlin Heidelberg, 2005.
[23] M.M. Frank, et al., Microelectron. Eng. 86 (2009) p1603.
[24] S. Kim, et al., J. Appl. Phys. 107 (2010) 054102.
[25] Chudzik, M., et al., Fabrication of higher-k dielectrics. U.S. Patent 9,478,425.
[26] T. Ando, et al., in: Proceedings of IEEE International Electron Devices Meeting, Washington, DC, USA, 7–9 December 2009, pp. 423–426.
[27] H. Kim, et al., J. Appl. Phys. 96 (2004) 3467–3472.
[28] Choi, C et al., In Proceedings of VLSI Technology Symposium, Kyoto, Japan, 14–18 June 2005; pp. 208–209.
[29] Huang, J et al., In Proceedings of VLSI Technology Symposium, Kyoto, Japan, 15–18 June 2009; pp. 34–35.
[30] Choi, K et al., In Proceedings of VLSI Technology Symposium, Kyoto, Japan, 15–18 June 2009; pp. 138–139.
[31] T. Ando, et al., Appl. Phys. Lett. 96 (2010), 132904:1–132904:3.
[32] L.-Å. Ragnarsson, et al., in: 2009 IEEE International Electron Devices Meeting (IEDM), Baltimore, MD, USA, 2009, pp. 1–4, https://doi.org/10.1109/IEDM.2009.5424254.
[33] Y. Okuno, et al., in: Proceedings of IEEE International Electron Devices Meeting, Baltimore, MA, USA, 7–9 December 2009, pp. 663–666.
[34] M. Takahashi, et al., in: Proceedings of IEEE International Electron Devices Meeting, Washington, DC, USA, 2007, pp. 523–526.
[35] S. Migita, et al., in: Proceedings of IEEE International Electron Devices Meeting, San Francisco, CA, USA, 6–8 December 2010, pp. 269–272.
[36] T. Ando, Materials 5 ([3]) (2012) 478–500.
[37] T. Ando, et al., IEEE Electron. Device Lett. 34 (6) (2013) 729.
[38] K. Tatsumura, et al., in: Proceedings of IEEE International Electron Devices Meeting, Washington, DC, USA, 7–9 December 2007, pp. 1–4.
[39] X. Zhao, et al., Phys. Rev. B 65 (2002) 233106.
[40] D. Fischer, et al., Appl. Phys. Lett. 92 (2008) 012908.
[41] P. Umari, et al., Phys. Rev. Lett. 89 (15) (2002) 157602.
[42] P. Giannozzi, et al., J. Phys. Condens. Matter 21 (39) (2009) 395502.
[43] http://www.quantum-espresso.org/.
[44] R.K. Pandey, et al., J. Appl. Phys. 114 (2013) 034505.
[45] J. Robertson, et al., Appl. Phys. Lett. 91 (2007) 132912.
[46] K. Tse, J. Robertson, Microelectron. Eng. 85 (2008) 9.
[47] P.-Y. Prodhomme, F. Fontaine-Vive, A. Van Der Geest, P. Blaise, J. Even, Appl. Phys. Lett. 99 (2011) 022101.
[48] H. Zhu, R. Ramprasad, Phys. Rev. B 83 (2011), 081416(R).
[49] L. Bengtsson, Phys. Rev. B 59 (1999) 12301.
[50] T. Ando, et al., World Scientific Publishing Co. Pte. L.
[51] W. Liu, et al., IRPS, IEEE, 2016, p. 7A-3-2.
[52] H. Xu, Thick-gate-oxide MOS Structures with Sub-design-rule Channel Lengths for Digital and Radio Frequency Circuit Applications, PhD Thesis, University of Florida, 2007.
[53] S.C. Sun, et al., MRS Proc. 342 (1994) 181–186.
[54] S. Siddiqui, et al., ECS Trans. 53 (3) (2013) 137–146.
[55] S. Mittl, et al., IRPS, IEEE, 2012, p. 6A.5.1.
[56] K.F. LeGoues, et al., J. Appl. Phys. 65 (4) (1989) 1724.
[57] K. Shiraishi, et al., Jpn. J. Appl. Phys. 43 (2004) L1413.
[58] E. Cartier, et al., in: Symposium of VLSI Technical Digest of Technical Papers, 2005, p. 230.
[59] B. Winstead, et al., IEEE Electron. Device Lett. 28 (8) (2007) 719.

[60] H.R. Harris, et al., in: Symposium of VLSI Technical Digest of Technical Papers, 2007, p. 154.
[61] S. Krishnan, et al., in: Technical Digest of International Electron Devices Meet, 2011, p. 634.
[62] J. Franco, et al., in: Technical Digest of International Electron Devices Meet, 2010, p. 70.
[63] M. Wang, et al., in: Electron Device Letters vol. 34, 2013, pp. 837–839, 7.
[64] S. Ramey, et al., IRPS, 2015, p. 3B-2.
[65] H. Arimura, et al., IRPS, IEEE, 2014, p. 3C-4.
[66] W. McMahon, et al., IRPS, 2013, p. 4C-4.
[67] P. Barry, Linder, et al., IRPS, 2016, p. 4B-1.
[68] N. Loubet, VLSI 2020 Virtual Conference, Short Course SC1.1.
[69] N. Loubet, et al., in: 2017 Symposium on VLSI Technology, Kyoto, Japan, 2017, pp. T230–T231.
[70] S. Mukesh, et al., Electronics 11 (21) (2022) 3589.
[71] J. Zhang, et al., in: 2017 IEEE International Electron Devices Meeting (IEDM), San Francisco, CA, USA, 2017, pp. 22.1.1–22.1.4.
[72] M. Bhuiyan, et al., in: 2021 IEEE International Electron Devices Meeting (IEDM), San Francisco, CA, USA, 2021, pp. 1–4.
[73] G. Bae, et al., IEDM, 2018, pp. 28.7.1–28.7.4.
[74] T. Chan, et al., IEDM, 1987, pp. 31–33.
[75] G. Tsutsui, et al., in: Symposium of VLSI Technology, 2018, p. T87.
[76] C. Yeung, et al., IEDM, 2018, pp. 28.6.1–28.6.4.
[77] C. Leitz, et al., Materials Research Society Symposium Proceedings, Vol. 686, Materials Research Society, 2002.
[78] C. Murray, et al., Commun. Eng. 1 (2022) 11.
[79] S. Reboh, et al., IEDM, 2019, pp. 11.5.1–11.5.4.
[80] Y. Song, et al., AVS, 2019.

## Further reading

[1] E. Cartier, et al., in: Proceedings of IEEE International Electron Devices Meeting, Washington, DC, USA, 5–7 December 2011, pp. 18.4.1–18.4.4.
[2] C. Ortolland, et al., in: Technical Digest of International Electron Devices Meet, 2013, p. 236.
[3] G. Gaddemane, et al., in: ESSDERC 2022 - IEEE 52nd European Solid-State Device Research Conference (ESSDERC), Milan, Italy, 2022, pp. 265–268.

**PART III**

# Applications and limitations

# CHAPTER 8
# Semiconductor reliability overview

Fernando Guarin[1,2] and Ed Hostetter, Jr.[1,2]
[1]GlobalFoundries, East Fishkill, NY, United States; [2]IBM, Reliability Engineering, East Fishkill, NY, United States

## 1. Introduction

In today's highly interconnected world, semiconductor devices have become the backbone of modern technology, impacting everyday life, powering everything from smartphones and laptops to medical equipment, clean energy generation through photovoltaics, and providing sensors and increased electronic content in automobiles. As these devices continue to shrink in size while increasing in complexity, the reliability of semiconductors becomes paramount. Semiconductor reliability refers to the ability of these devices to perform consistently and accurately over their intended lifespan. This chapter explores the challenges, methodologies, and strategies employed to ensure the reliability of semiconductor devices.

Modern semiconductor devices originated with the realization of the first bipolar point-contact transistor at Bell Labs on December 23, 1947 [1] and subsequent invention of the integrated circuit by Jack Kilby [2] and Robert Noyce [3]. While Germanium transistors dominated industry through 1950s and early 1960s, Silicon has been the dominant material ever since. Silicon's material quality with very low defectivity [4], ability to grow stable oxides down to around 10 Å in thickness, fabrication infrastructure (ion Implantation, lithography, oxidation, metalization, design tools, cost, etc.) have made Silicon the workhorse for the semiconductor industry over the past 7 decades with no viable replacement in the foreseeable future. While our discussion will concentrate on silicon reliability, it should be noted that many of the mechanisms will be relevant for other material systems (III–V, SiGe). SiGe integrates bipolar and CMOS technologies in BiCMOS. The difference between bipolar junction transistors (BJTs) and heterojunction bipolar transistors (HBTs) is negligible from the reliability standpoint. SiGe uses the same oxide and back end of line (BEOL) materials. In the case of III–V semiconductors there are significant deltas in the reliability mostly driven by the large crystalline defectivity, as well as the material systems used for oxidation as well as metalization and interconnection that lead to less reliable circuits than their silicon counterparts. In the rest of this chapter, we will concentrate on the reliability of CMOS silicon devices and circuits.

## 2. Definition of reliability

Reliability is defined as the probability that a given circuit made of semiconductor devices and associated materials will perform its intended function for a given period of time, lifetime, under a given set of conditions. The probability is the likelihood that some given event will occur and as a probability of successful operation, a value between 0 and 1 is assigned. The intended function of the circuit and its use condition are specifications, which need to be defined and stated. The period of time is often referred to as lifetime and depends on the circuit application. In real practice, reliability is more complicated, some mechanisms hot carrier (HC) and bias temperature instability (BTI) exhibit gradual degradation so the failure criteria are arbitrary and may not yield actual circuit fails, but values need to be assigned in order to assess reliability during the technology qualification phase.

## 3. Semiconductor reliability balancing act

Achieving reliability in modern semiconductor technology requires a carefully crafted balance between performance, yield, and reliability. During the development stage, many decisions must be made. While the highest performance is desired, it must be achieved in a manner that will not sacrifice yields and reliability. Device designers can achieve higher performance by increasing electric fields and reducing channel length, but there is a point of diminishing returns that must be considered. Reducing feature size will increase performance but will pay a price in yield and reliability. As we scale technologies, this balancing act is illustrated by Fig. 8.1. Semiconductor technology development requires tradeoffs during the development phase to find the optimum balance between, reliability, performance, and yield. All three elements must be taken into account for the successful development and qualification of advanced technology nodes.

**Figure 8.1** Development of semiconductor technologies require trade-offs between reliability, performance and yield in carefully balanced approach.

## 4. Challenges and principal degradation mechanisms in semiconductor reliability

Semiconductor reliability faces a range of challenges due to the intricacies of the materials and processes involved in their manufacturing. Some of the key challenges include:

1. **Process variability:** The inherently stochastic nature of semiconductor manufacturing processes can lead to variations in device parameters, potentially affecting device performance and reliability.
2. **Electromigration:** The movement of atoms within the conductor material over time due to the passage of current can cause open circuits and degradation in semiconductor devices.
3. **Thermal stress:** Rapid temperature changes during operation or thermal cycling can lead to mechanical stress, causing material fatigue and potential failure.
4. **Bias temperature instability:** Over time, the characteristics of a transistor can change due to biasing and temperature fluctuations, affecting device reliability.
5. **Gate oxide breakdown:** It is also known as time-dependent dielectric breakdown (TDDB). Electric fields at the gate oxide interface can lead to dielectric breakdown, resulting in short circuits and potential device failure.
6. **Hot carriers:** Over time, the characteristics of a transistor can change due to high electric fields induced interface state trapping/detrapping, affecting device reliability.
7. **PID:** Plasma-induced damage will produce charging in the metalization layers, affecting device reliability.

8. **Soft error rate:** Is a mechanism that affects the data state of memories and sequential elements and are caused by random radiation events that occur naturally in the terrestrial environment. For space applications other high energy radiation mechanisms that must be taken into account for space applications.

## 5. Front end of line reliability, back end of line reliability, and middle of the line reliability

To better understand the major wear out mechanisms and their location, we will use Fig. 8.2 to clearly outline the BEOL, MOL, and FEOL regions, as well as the locations where the degradation mechanisms take place.

During a typical semiconductor technology qualification, a set of reliability specifications will be given for FEOL and BEOL mechanisms.

Table 8.1 provides a typical example for the FEOL reliability specifications in a thin oxide device with a nominal 1 V operating voltage. In this example, we have used a 5% supply margin, yielding a maximum voltage of 1.05 V and a maximum temperature of 125 C. Note that for the thick

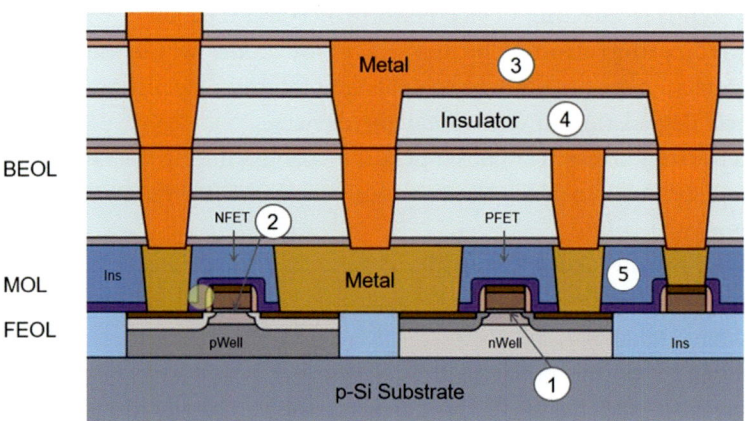

1. Bias Temperature Instability and Hot Carrier at Si-SiO2 Interface
2. Gate Dielectric Breakdown
3. Electro- and Stress-migration at Metal Interconnect
4. BEOL Dielectric Breakdown
5. MOL

**Figure 8.2** Cross sectional illustration of Chip layers and their separation into three major groups: Front End Of Line (FEOL), Middle Of Line (MOL) and Back End Of Line (BEOL). This figure shows the relevant locations for the various reliability degradation mechanisms.

Semiconductor reliability overview 363

Table 8.1 Semiconductor manufacturers provide reliability specifications for each degradation mechanism indicating the worst case use condition and the criteria for failure. A typical illustration for a given technology is provided in this table.

| Device | Mech | Usage (C) | Deliverable |
|---|---|---|---|
| Thin oxide | TDDB | 1.05 V, 125 | 0.1% failure @10 years DC, 10 mm2 |
|  | HCI | 1.05 V, 125 | Degradation of 10% Idsat shift @0.05 years DC |
|  | BTI | 1.05 V, 125 | Degradation of 50 mV Vt shift @5 years DC |
|  | PID | 2 × VDD, 25 | Max 20% Ig increase <5% tailing in lognormal CDF plot |
| Thick oxide | TDDB | 1.98 V, 125 | 0.1% failure @10 years DC, 1.5 mm2 |
|  | HCI | 1.98 V, 125 | Degradation of 10% Idsat shift @0.2 years DC |
|  | BTI | 1.98 V, 125 | Degradation of 50 mV Vt shift @5 years DC |
|  | PID | 2 × VDD, 25 | Max 20% Ig increase <5% tailing in lognormal CDF plot |

oxide device with a nominal voltage of 1.8 V, we have used a 10% supply margin, yielding a maximum voltage of 1.98 V.

## 6. Reliability assessment methodologies

Ensuring semiconductor reliability involves a combination of simulation, testing, and modeling to identify potential failure mechanisms and develop strategies for mitigation. Some key methodologies include:

1. **Accelerated life testing:** Semiconductor devices are subjected to conditions that simulate years of use in a short amount of time. By observing failures under these accelerated conditions, manufacturers can estimate the device's lifespan and identify potential weak points.
2. **Failure mode and effects analysis (FMEA):** FMEA is a systematic approach to identifying potential failure modes, their causes, and their effects. This method helps prioritize design improvements and mitigation strategies.
3. **Simulation and modeling:** Advanced computer simulations and modeling allow engineers to predict the behavior of semiconductor devices under various conditions, aiding in identifying potential reliability concerns and optimizing designs.
4. **Reliability physics analysis:** This involves studying the physical mechanisms underlying device degradation, such as electromigration, oxide breakdown, and thermal stress, to better understand and predict failure mechanisms.

## 7. Mitigation strategies

To enhance semiconductor reliability, manufacturers employ a range of strategies:

1. **Redundancy:** Implementing redundant circuits or components can improve device reliability. If one part fails, the redundant component can take over, reducing the risk of complete system failure.
2. **Error correction codes (ECC):** ECC techniques can detect and correct errors in memory or data storage, ensuring accurate data retention even in the presence of hardware faults.
3. **Temperature management:** Careful thermal design and management can mitigate thermal stress, ensuring that devices operate within safe temperature ranges.
4. **Process control:** Tight control over manufacturing processes reduces variability and ensures consistent device performance.

5. **Design for reliability:** Engineers design devices with reliability in mind, considering factors such as layout, material selection, and stress distribution to minimize failure risks.

## 8. Test structures and methodologies for reliability assessment

One of the major challenges to reliability engineers is the accurate validation of the reliability measurements, predictions, and models employed for reliability evaluations. There is also the need to separate mechanisms and they are often intermingled; a particularly challenging case is the separation of hot carriers amd bias temperature instabilities, which has the added complication of separating self-heating contributions. To overcome these challenges' reliability engineers, use specially designed test structures. To illustrate this point, we provide the ring oscillator (RO) designs [5], one with 13 stages that will be dominated by HC contributions and a second one with 101 stages that will be dominated by BTI. By using RO's, the self-heating component is mostly eliminated as the GHz frequencies of the oscillators will be much faster than the thermal response. Fig. 8.3 clearly shows the difference between BTI and HC contributions, as well as the validation of the models for each of these mechanisms against the actual

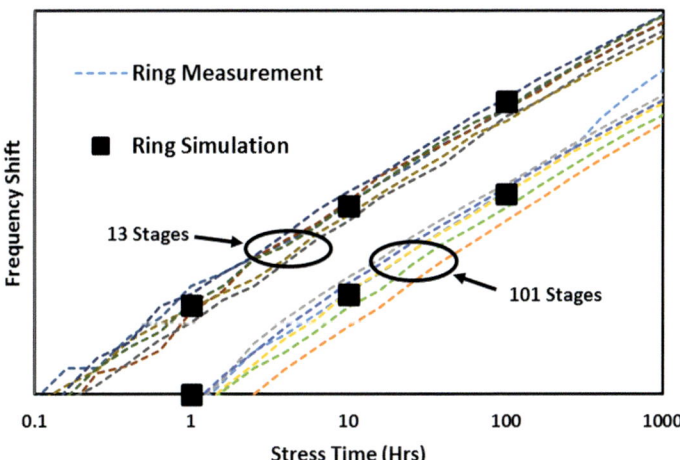

**Figure 8.3** Validation of the measured degradation in reliability Ring Oscillators (RO). One ring with 13 stages to better highlight the Hot Carrier (HC) degradation, the second RO with 101 stages, running at a lower frequency (7.77 lower ratio) which minimizes the number of transitions and consequently decreasing the HC component while making the BTI more noticeable.

RO measurements. Note that the BTI (101 stage) measurements exhibit more variability across the various devices tested as there is still a remaining small thermal component.

## 9. Conclusion

Semiconductor reliability is a critical aspect of modern technology, influencing the performance and longevity of devices we rely on daily. By understanding the challenges inherent in semiconductor manufacturing, employing assessment methodologies, and implementing effective mitigation strategies, manufacturers can create devices that meet the demands of today's fast-paced technological landscape. As semiconductor technology continues to evolve, the pursuit of reliability remains central to unlocking the full potential of scaled devices. Over the last 5 decades, reliability has kept pace with Moore's law. Accomplishing an impressive eight orders of magnitude improvement in the failure rate for transistors [6] while continuing the relentless drive to scale; making the oxides thinner, reducing the dimensions of devices and interconnects, increasing the number of metal layers in the BEOL, all while significantly increasing the transistor count in complex designs. Semiconductor reliability has been able to evolve from an empirical fitting of failure data to generate failure models to a more comprehensive understanding of the physics and material science driving forces of the major reliability degradation mechanisms. Microelectronic reliability has been able to keep the relentless pace required to meet the challenges of scaling up to the 3 nm node. It will undoubtedly continue to play a major role as we continue beyond gate all around and heterogeneous integration.

## References

[1] W. Shockley, The path to the conception of the junction transistor, IEEE Trans. Electron Devices 23 (1976) 597—620.
[2] C. Lécuyer, Making Silicon Valley: Innovation and the Growth of High Tech, 1930—1970, MIT Press, 2006.
[3] A.N. Saxena, Invention of integrated circuits: untold important facts, World Sci. (2009) 140.
[4] W. Heywang, K.H. Zaininger, Silicon: the semiconductor material, in: P. Siffert, E.F. Krimmel (Eds.), Silicon: Evolution and Future of a Technology, Springer Verlag, 2004.
[5] S. Mittl, F. Guarín, Self-heating and its implications on hot carrier reliability evaluations, in: Proc. IEEE International Reliability Physics Symposium (IRPS), 2015, pp. 4A.4.1—4A.4.6.

[6] A. Kelleher, Intel, on the advance of Moore's law and resulting trends in reliability, in: IEEE International Reliability Physics Symposium (IRPS) Keynote, 2023.

## Further reading

[1] J.D. Cressler, Re-engineering silicon: Si—Ge heterojunction bipolar technology, IEEE Spectr. (1995) 49—55.
[2] Hot carrier degradation in semiconductor devices, in: The Energy Driven Hot Carrier Model, Springer International Publishing, 2015, pp. 29—56.
[3] Considerations for reliability, in: Extreme Environment Electronics 10, CRC, 2012, p. 451.

CHAPTER 9

# Thin film development for LED technologies

J. Lee[1,2], Y.C. Chiu[1,2], J.-P. Leburton[1,2,3] and C. Bayram[1,2,a]

[1]Department of Electrical and Computer Engineering, University of Illinois at Urbana-Champaign, Champaign, IL, United States; [2]Micro and Nanotechnology Laboratory, University of Illinois at Urbana-Champaign, Champaign, IL, United States; [3]Department of Physics, University of Illinois at Urbana-Champaign, Champaign, IL, United States

## 1. Introduction

Since Professor Nick Holonyak Jr. invented the world's very first visible light-emitting diodes (LEDs) in 1962 [1], the LED market has been rapidly growing until it reached the market size of 83.8 billion dollars in 2021 [2] and it is expected to grow further to 124.7 billion dollars in 2027 [3]. Interesting applications include solid-state lighting (SSL) [4], display (i.e., televisions, micro-displays, VR/AR) [5], and free-space communication (i.e., Li-Fi, Lidar, underwater communication) [6]. Especially, LED applications in SSL achieved great success, rapidly being adopted with a 14.4% annual market growth rate and contributing to energy savings of ∼25% already in terms of light consumption per household. However, targeting even higher performance is desired due to the expected population growth and the lagging adoption of the current SSL roadmap. The performance of a commercial light source is evaluated by its luminous efficacy, which is a measure of how efficiently the light source produces visible light. Although SSL achieved 200 lm/W of luminous efficacy in laboratory settings [7] based on the blue InGaN LEDs and yellow-converting phosphors, this approach is theoretically limited to 250 lm/W of luminous efficacy due to the significant loss in secondary lenses and phosphors. A new and accelerated SSL roadmap is needed in order to sustain the success of SSL, which can only be enabled by directly mixing blue, green, and red LEDs. This

---

[a] Innovative COmpound semiconductoR (ICOR) Laboratory, Intel Alumni Endowed Faculty Scholar

approach can ultimately achieve 414 lm/W of luminous efficacy if we have highly efficient LEDs for all three colors. The department of energy (DOE) in the United States targets 80%, 55%, and 60% of wall plug efficiency at 100 A/cm$^2$ of current density for blue, green, and red, respectively, by 2035. Table 9.1 shows the state-of-the-art efficiency of LEDs for each color and the 2035 goal set by DOE [8].

In LEDs, electrons and holes recombine and create photons as an output. The efficiency of a single-color LED is measured by various metrics (e.g., wall plug efficiency, external quantum efficiency, internal quantum efficiency, etc.) that are related to each other. Wall plug efficiency (WPE) refers to the ratio of the output optical power to the input electrical power, external quantum efficiency (EQE) is the ratio of the output photons to the injected electrons, and internal quantum efficiency (IQE) is defined as the ratio of the internally generated photons to the injected electrons in the active region. In LEDs, carriers recombine both nonradiatively (i.e., generating heat) and radiatively (i.e., generating light), and the IQE can be simply expressed as:

$$\text{IQE} = \frac{Bn^2}{An + Bn^2 + Cn^3} \quad (9.1)$$

where $A$ is the nonradiative Shockley–Read–Hall (SRH) recombination coefficient, $B$ is the radiative recombination coefficient, $C$ is the nonradiative Auger recombination coefficient, and $n$ is the injected carrier density, assuming electron and hole densities are the same. Therefore, IQE dictates how much radiative recombination we get from an LED compared to the total recombination of carriers. Although many aspects determine the final performance of an LED (i.e., WPE), IQE sets the upper limit of the WPE, thereby being a significant metric.

Theoretically, IQE is determined by the design of the thin film structure ("stack") such as quantum-well (QW) width or doping density of the p–n junction. Experimentally, defects in the crystal structure such as atom vacancies, interstitial impurities (0D), screw and edge type threading dislocations (1D), stacking faults (2D), or voids (3D) all degrade the crystal quality, deteriorating the IQE either by acting as nonradiative centers or increasing the radiative carrier lifetime. Therefore, it is pivotal to understand the design and the growth of thin-film structure to eventually improve the WPE.

LED technology heavily relies on the growth of thin-film structures where III–V materials are grown and fabricated. The typical structure of

**Table 9.1** Current state-of-the-art efficiency of LEDs reported by the department of energy, United States. Peak external quantum efficiency (EQE) and wall plug efficiency at a certain current density and a certain junction temperature ($T_j$) are shown.

| Efficiency metric | 2021 status | | | 2025 DOE target | | | | 2035 DOE target | | |
|---|---|---|---|---|---|---|---|---|---|---|
| | Blue (%) | Green (%) | Red (%) | Blue (%) | Green (%) | Red (%) | | Blue (%) | Green (%) | Red (%) |
| Peak EQE | 80 | 62 | 63 | 88 | 60 | 69 | | 93 | 78 | 70 |
| WPE at 35 A/cm², $T_j \leq 25°C$ | 71 | 27 | 51 | 80 | 50 | 60 | | 86 | 70 | 75 |
| WPE at 100 A/cm², $T_j \leq 85°C$ | 54 | 19 | 30 | 65 | 30 | 43 | | 80 | 55 | 60 |

the LED is a p-i-n junction with a single-quantum-well (SQW) or multi-quantum-well (MQW) structure as i-layer. Here, injected electrons from the n-layer and injected holes from the p-layer recombine inside the QW, creating photons whose energy is the same as the energy difference between the two carriers. Therefore, the wavelength (color) of the emitted light is dictated by the bandgap of the material used. InGaN/GaN QW with a wurtzite crystal structure is typically used for blue LEDs, while AlGaInP/AlGaInP QW with a zincblende crystal structure is typically used for red LEDs. Fig. 9.1 shows the difference between the wurtzite and zincblende crystal structure. Fortunately, due to the amazing bandgap adjustability of III–Vs, thin film semiconductors can theoretically cover a large range of spectrum by controlling the composition of alloys such as InGaN or AlGaInP and. For instance, adding more Indium in InGaN lowers the effective bandgap, thereby enabling longer wavelength light. Nevertheless, highly efficient green emission is challenging in both nitride and phosphide material systems.

The green gap refers to the lack of high WPE green LEDs especially at high current densities such as 100 A/cm$^2$. The state-of-the-art green LED shows 19% of WPE, while the targets set by the DOE are 30% in 2025%, and 55% in 2035 [8]. This value of 19% is about one-third of the WPE in the state-of-the-art blue LED (54%) at high current densities. Indeed, IQE roughly decreases with longer emission wavelengths in InGaN/GaN LEDs. Fig. 9.2 shows the IQE of the InGaN/GaN LED extracted by

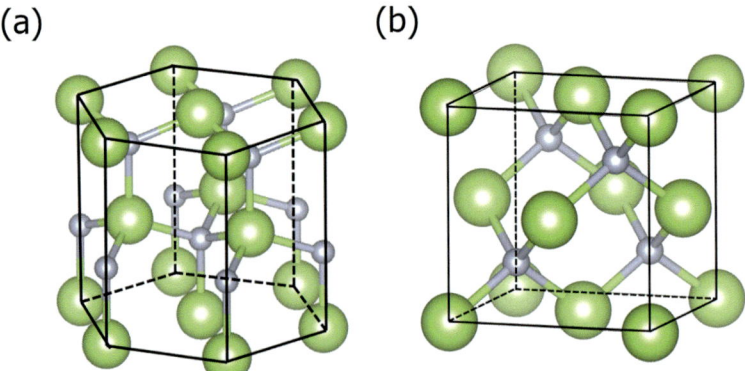

**Figure 9.1** Crystal structures of the (a) InGaN (Wurtzite structure), and (b) AlGaInP (zincblende structure) are shown, which are the typical materials used for LED application. Here, small gray spheres indicate III-element, while large green spheres indicate V-element.

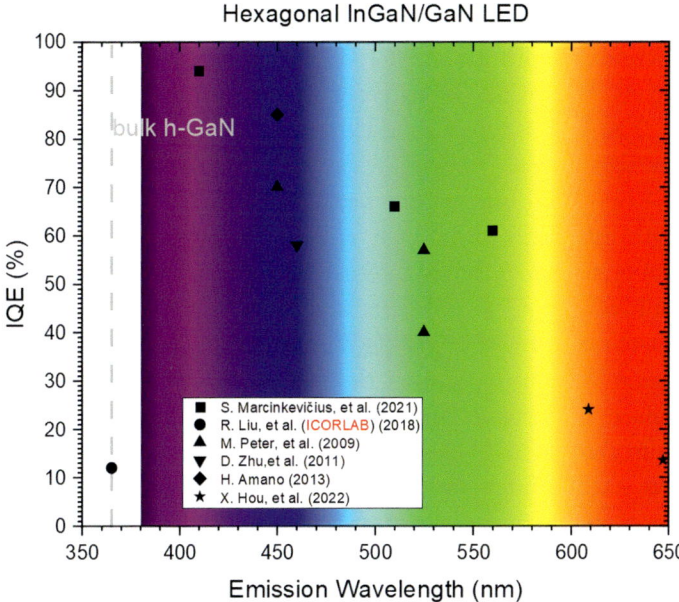

**Figure 9.2** The IQE values of InGaN/GaN LEDs in the literature are shown as a function of wavelength. Decreasing IQE for longer emission wavelength is displayed.

photoluminescence (PL) measurement as a function of wavelength reported in the literature [9–14]. It can be observed that the IQE significantly increases when moving from a bulk GaN material to the InGaN/GaN QW and then decreases to longer emission, illustrating the green gap issue from the GaN side. Researchers have strived to solve the green gap by exploring the cause of low efficiency in green InGaN/GaN LEDs.

Here, we introduce the issue with the state-of-the-art green-emitting hexagonal (i.e., wurtzite) InGaN/GaN LED development and explore one of the potential solutions: the growth of cubic InGaN/GaN LEDs. The development of cubic InGaN/GaN LEDs will be detailed from both theoretical and experimental perspectives in order to clarify the true potential of those cubic materials.

## 2. Development of green-emitting hexagonal InGaN/GaN LEDs

The typical structure for the InGaN/GaN LED is a p-i-n junction with a layer called electron blocking layer (EBL): [Si-doped n-GaN/active layer/

Mg-doped p-AlGaN (EBL)/Mg-doped p-GaN/heavily Mg-doped $p^+$ GaN for contact], with the active layer consisting of unintentionally doped i-GaN layers and InGaN layers creating a single- or multi-quantum-well structure. The EBL is typically added in order to suppress electron leakage current at high current densities, which significantly lowers the IQE of the LED. The two main issues of the hexagonal InGaN/GaN green LEDs are the growth difficulties from high indium incorporation and the limited theoretical performance due to internal piezoelectric polarization, caused by the strain inside the QW and their hexagonal symmetry.

Green-emitting InGaN/GaN QWs require a high (23% [9] ~30% [15]) percentage of indium to achieve the desired wavelength. Higher indium incorporation leads to a larger lattice mismatch between the resultant InGaN film and the GaN barrier, leading to the composition pulling effect [16] where some indium atoms are excluded from the growth to reduce the deformation energy due to lattice mismatch. Not only it makes the growth of a high indium concentration layer difficult, but also it causes indium segregation or indium clustering at the dislocation [17]. In addition, in contrast to the low temperature growth of InGaN, the following high temperature growth of GaN barriers leads to the indium out-diffusion, further deteriorating the quality of the InGaN film [18]. Furthermore, the growth of InGaN includes the generation of pits called V-defects [19], which are inverted hexagonal pyramids with (101) sidewalls exposed. All the growth challenge mentioned above lowers the IQE of the LED compared to the theory, and several works were done to improve the quality of the InGaN film in order to improve the efficiency.

Historically, GaN films have been grown on Sapphire [4] due to the good quality of growth and relatively cheap cost. Other alternatives include GaN on Silicon (111) [20], SiC [21], freestanding GaN [22], or recent Qromis substrate technology (QST) [23]. Fig. 9.3 shows the topographical characterization of the blue-emitting InGaN/GaN LED structure grown on Sapphire, Silicon, and QST [24]. GaN on Sapphire is under compressive strain, GaN on Silicon is under tensile strain, and QST is nearly lattice-matched to GaN as an engineered substrate. From the figures, one can observe the effect of strain on the growth. For instance, the images reveal the v-defects as a main type of defects in LED on Sapphire, while revealing hexagonal spiral hillocks in LED on Silicon. The top surface is the smoothest in LED on QST, which reveals the impact of strain on the surface roughness and the number of defects. It is also reported that residual tensile strain in LED on Silicon is indeed beneficial toward high and

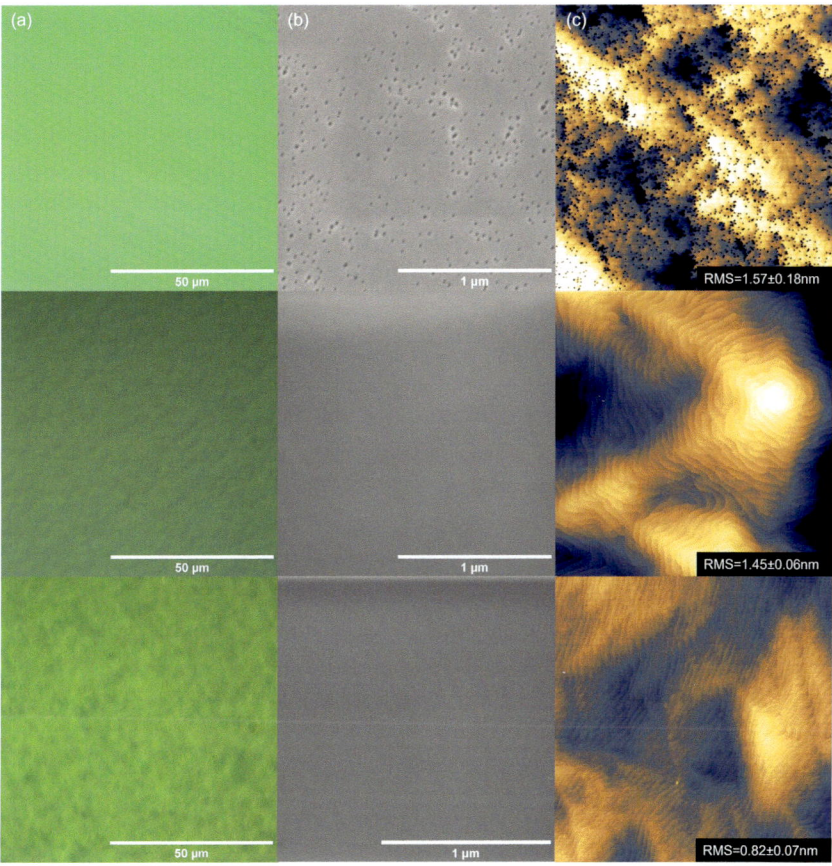

**Figure 9.3** (a) Nomarski, (b) top-view scanning electron, and (c) atomic force microscopy images of InGaN/GaN LED on Sapphire (top row), silicon (middle row), and QST (bottom row) are shown. *(The data for LED on Sapphire and Silicon are reprinted from Y.C. Chiu, C. Bayram, "Low temperature absolute internal quantum efficiency of InGaN-based light-emitting diodes," Appl. Phys. Lett. 122 (9) (2023) 091101, with the permission of AIP Publishing.)*

uniform indium incorporation [25]. Indeed, the IQE of the three samples measured from PL shows 27% for LED on Sapphire, 58% for LED on Silicon, and 57% for LED on QST. However, it is worth noting that the LED on Sapphire in Fig. 9.3 is worse than commercial LEDs on Sapphire. Typically, commercial LEDs on Sapphire film are of better quality than LEDs on Silicon, as GaN growth on Silicon poses difficulties such as substrate cracking or melt-back etching [26]. Substrate cracking occurs due to the thermal expansion coefficient mismatch between GaN and Si, while the melt-back etching occurs due to the alloying and the chemical reaction between Si and Ga atoms. Furthermore, larger lattice mismatch leads to a higher density of threading dislocations if buffer layers are not engineered carefully. Table 9.2 summarizes the structural characterization results for these three LEDs, revealing relatively stress-free bulk GaN but higher defectivity in silicon. Here, defect density from the atomic force microscopy (AFM) image shows the top surface defect density, while XRD shows the underlying n-GaN defect density, revealing the good-quality bulk GaN on Sapphire.

In addition to growth challenges, piezoelectric polarization in the wurtzite InGaN/GaN LED structure poses significant design challenges to overcome, which deteriorates its theoretical performance. Polarization plays a significant role in InGaN/GaN LEDs. Not only do InGaN and GaN show inherent spontaneous polarization due to the hexagonal symmetry in the wurtzite crystal, but also piezoelectric polarization in the growth direction exists when the wurtzite crystal is strained, which is the case when a thin InGaN layer is grown on top of the GaN layer. This polarization difference between the interface manifests itself as a polarization charge at the interface, creating an internal electric field inside the LED. As a result, three issues can be identified: efficiency droop, spectral blueshift, and the voltage penalty.

Efficiency droop refers to the phenomenon where the efficiency of the LED decreases as the current density increases, which is an issue because high output optical power (i.e., high input current and high efficiency at that current) is needed in SSL. The efficiency droop in InGaN/GaN LEDs is exceptionally strong compared to other material systems, and various explanations were suggested to explain this strong efficiency droop. One explanation is that the Auger recombination is especially strong in InGaN due to the hole localization caused by the polarization along with the large hole effective mass of InGaN [15,27]. This localization also decreases the spatial overlap between the electron and hole wave function, decreasing radiative recombination and eventually reinforcing the efficiency droop. Recently, it was suggested that the efficiency droop in InGaN/GaN LEDs occurs mainly due to the interplay between electron leakage and the Auger recombination, both enhanced by the piezoelectric polarization [28]. Electron leakage occurs because the polarization lowers the conduction band minimum of p-GaN, making electrons not confined inside the QW, especially for high current densities. This leakage can be suppressed by adding a heavily p-doped AlGaN layer, which is called EBL. Fig. 9.4a

Table 9.2 Tabulated structural characterization results from AFM, SEM, XRD and Raman spectroscopy for LED on Sapphire, Silicon, and QST are shown.

| LED substrate Material | AFM surface roughness (nm) | Defect density | | | Stress/strain | | |
|---|---|---|---|---|---|---|---|
| | | AFM (cm$^{-2}$) | XRD Screw (cm$^{-2}$) | XRD Edge (cm$^{-2}$) | In-plane strain | XRD-RSM In-plane stress (GPa) | Raman In-plane stress (GPa) |
| Sapphire | 1.57 ± 0.18 | 6.3 ± 0.5 × 10$^9$ | 7.62 × 10$^7$ | 8.49 × 10$^7$ | −0.0111 | −5.3122 | −0.91 |
| Si(111) | 1.45 ± 0.06 | 5.4 ± 0.8 × 10$^8$ | 2.97 × 10$^8$ | 1.03 × 10$^9$ | −0.0030 | −1.4567 | ≈ 0 |
| QST | 0.82 ± 0.07 | 2.6 ± 0.8 × 10$^9$ | 2.05 × 10$^8$ | 1.44 × 10$^8$ | −0.0034 | −1.6181 | −0.23 |

The data for LED on Sapphire and Silicon are reprinted from Y.C. Chiu, and C. Bayram, "Low temperature absolute internal quantum efficiency of InGaN-based light-emitting diodes," Appl. Phys. Lett. 122(9), 091101 (2023), with the permission of AIP Publishing.

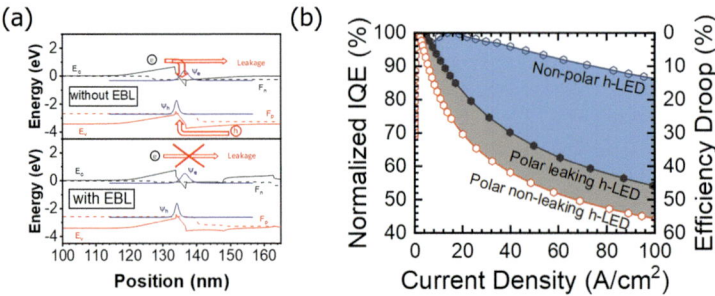

**Figure 9.4** (a) Band diagram of the InGaN/GaN LED without EBL (*top*) and with EBL (*bottom*), which reveals how EBL suppresses the electron leakage by band engineering. (b) Efficiency droop of the InGaN/GaN LED with electron leakage (*black*), without electron leakage (*red*), and without polarization (*blue*). It is shown that suppressing the leakage current will increase the efficiency droop, while removal of polarization quenches the efficiency droop by ~70%. (Figure (b) is adapted from Y.-C. Tsai, C. Bayram, J.-P. Leburton, "Interplay between Auger recombination, carrier leakage, and polarization in InGaAlN multiple-quantum-well light-emitting diodes," J. Appl. Phys. 131 (19) (2022) 193102, with the permission of AIP Publishing.)

shows the band diagram of the InGaN/GaN LED without and with EBL, pointing to the way the EBL suppresses the leakage current. However, suppressing the leakage current does not necessarily improve the efficiency droop. It is suggested that excess carriers mostly contribute to the Auger recombination due to the abovementioned carrier localization, thereby increasing the efficiency droop. Fig. 9.4b shows three InGaN/GaN LEDs with electron leakage (black), without electron leakage (red), and without polarization (blue) [28]. It is shown that removing the electron leakage indeed increases the efficiency droop. It is also shown that removing the polarization quenches the efficiency droop to a large extent, pointing to the role of polarization in this interplay between carrier leakage and the Auger recombination.

Large spectral blueshift in InGaN/GaN LEDs is also caused by polarization. Quantum confined stark effect (QCSE) refers to the change in the quantized energy levels due to the external electric field [29]. As a result of the internal polarization field in InGaN/GaN LED, the energy level inside the QW is lowered, inducing a longer wavelength emission than the case without polarization. However, increased injection of carriers screen the polarization field, partially recovering the original emission wavelength. Therefore, the spectrum blueshifts (i.e., heads toward the shorter wavelength) as the current density increases. Similar to the polarization field inside the QW layer including spectral blueshift, the polarization field inside the barrier layer causes the voltage penalty (i.e., excess forward voltage needed to turn-on the device compared to what is expected from the bandgap) [30]. That is, the polarization field inside the barrier impedes the vertical carrier transfer, acting as a local potential barrier. This voltage penalty stacks up as the number of QWs increases, which decreases the WPE compared to IQE. Fig. 9.5 illustrates the spectral blueshift and the voltage penalty shown in the commercial LED.

In order to solve the abovementioned issue, various solutions are put forward: the use of n-ZnO instead of n-GaN [18], injection through V-defects [31] to solve the voltage penalty, use of multiple junctions [32], N-polar GaN [33], inverted p−n configuration [34], staggered InGaN quantum-well [35], non-polar GaN, and recently the use of cubic (i.e., zincblende) GaN instead of traditional hexagonal (i.e., wurtzite) one [28], which will be highlighted here.

## 3. State-of-the art of bulk cubic GaN and InGaN/GaN LEDs

Cubic GaN crystal is in a zincblende structure compared to the traditional hexagonal GaN. Cubic GaN has numerous benefits over hexagonal GaN in the application of LEDs and lasers: absence of polarization in the <100> growth direction, smaller bandgap [36], smaller electron-heavy hole effective masses [36], smaller Auger loss [37], shorter radiative recombination lifetime [38], lower p-doping activation energy [39], higher hole mobility [39], higher drift velocity [40], and larger optical gain [41]. However, almost all research and development in the GaN system have focused on its hexagonal

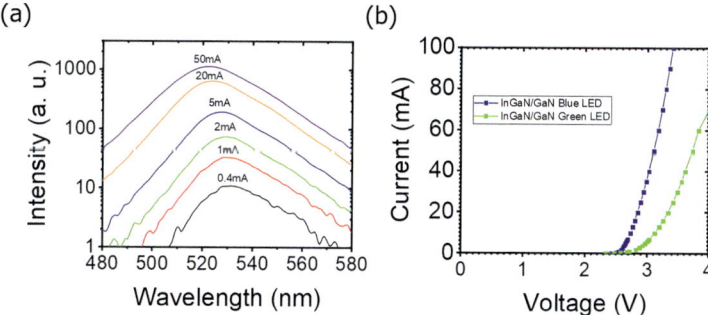

**Figure 9.5** Issues in green-emitting InGaN/GaN LEDs are shown from the commercial device. (a) Spectral blueshift with increasing current densities, and (b) the increase in forward voltage due to polarization are shown.

phase because cubic GaN is metastable: That is, cubic phase GaN is dominant only when certain conditions are met. Growth of cubic GaN can be done via either direct deposition on substrates with cubic symmetry or via the phase transition method. Currently, three major methods are available in the literature, which achieved decent film quality of cubic GaN: direct deposition on 3C−SiC (100) [42], phase transition on V-grooved Si (100) [43], and phase transition on U-grooved Si (100) [44].

Direct deposition on cubic substrates such as MgO [45], GaAs [46], 3C−SiC [47], and Si [48] has been demonstrated in the early studies either by metal organic chemical vapor deposition (MOCVD) [46,47] or molecular beam epitaxy (MBE) [45,48], being a relatively straightforward method. This direct deposition often results in a highly defective and a mixed-phased (i.e., mixture of hexagonal and cubic phase) film. 3C−SiC (100) is a common choice for direct deposition of cubic GaN due to low lattice mismatch (∼3.5%) between GaN and 3C−SiC. Still, such direct deposition suffers from large-area hexagonal GaN inclusion [42] and high stacking fault density (i.e., typically $10^5$ cm$^{-1}$) [49].

The phase transition from hexagonal to cubic GaN was first reported in 2004 by cubic GaN on V-grooved Si (100) [43], and cubic GaN on U-grooved Si (100) emerged as a different technique in 2014 [44]. Fig. 9.6 shows how the phase transition takes place [10,44]. Here, Silicon (100) substrates are nano-patterned so that {111} facets are exposed. Then, selective growth of GaN on Si {111} planes takes place, forming a hexagonal GaN growth line with an angle of 54.74 degrees with the Si (100) plane. When two growth fronts from each {111} plane meet each other, the phase transition from hexagonal GaN to cubic GaN occurs, not only because the apex angle ($\theta$) is identical to the bonding angle of the cubic GaN (∼109.47 degrees), but also because the hexagonal <0001> and cubic <111> crystal directions are equivalent with the only difference in the stacking sequence (i.e., ABABAB … in hexagonal <0001> while ABCABCABC. in cubic <111>). Although this phase transition method enables pure cubic GaN on top surface, there are always hexagonal GaN shoulders propagating along with the triangular cubic GaN, which impedes the scaling up of the technology. Cubic GaN on U-grooved Si (100) intended to solve this issue by designing oxide sidewalls that block the propagation of hexagonal GaN shoulders physically.

Fig. 9.7 shows the PL efficiency of cubic InGaN/GaN LEDs reported in the literature [10,50−53]. Although green-emitting cubic InGaN/GaN LEDs are predicted to show better theoretical performance, the current

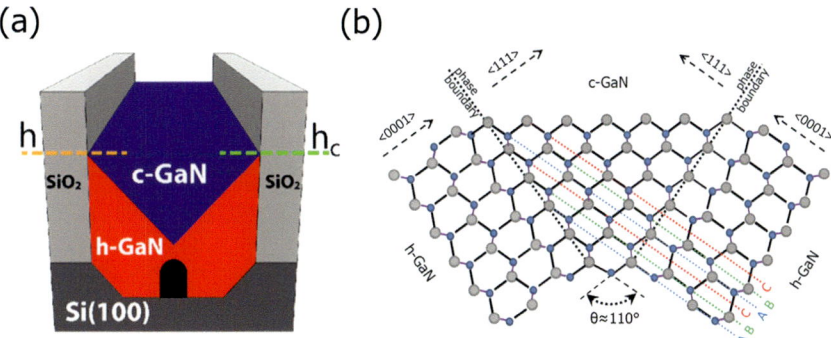

**Figure 9.6** Figures showing the hexagonal to cubic phase transition method. (a) Schematic of cubic GaN on U-grooved Si (100) is shown, where cubic GaN surface without hexagonal GaN is achievable. (b) Atomic arrangement when the phase transition occurs is shown. One can observe the crystallographic truths that enable the phase transition. *(a) is reprinted with permission from R. Liu, R. Schaller, C.Q. Chen, C. Bayram, High internal quantum efficiency ultraviolet emission from phase-transition cubic GaN integrated on nanopatterned Si(100), ACS Photon. 5 (3) (2018) 955–963. https://doi.org/10.1021/acsphotonics.7b01231, Copyright 2018 American Chemical Society. (b) is reproduced with permission from C. Bayram, J.A. Ott, K.-T. Shiu, C.-W. Cheng, Y. Zhu, J. Kim, M. Razeghi, D.K. Sadana, Adv. Funct. Mater. 24 (2014) 4492, Copyright 2014 WILEY-VCH Verlag GmbH & Co. KGaA, Weinheim.)*

state-of-the-art cannot compete against hexagonal InGaN/GaN LEDs. Aside from <0.1% efficiency of bulk GaN on GaAs [50], there are four reported PL IQE values for cubic InGaN/GaN LEDs, including bulk GaN material. Here, IQE is calculated from the following equation:

$$\eta_{\mathrm{IQE}} = \frac{I(300\mathrm{K})}{I_{\mathrm{low}}} \tag{9.2}$$

where $I(300\ \mathrm{K})$ is the integrated intensity of the PL peak at 300 K, and $I_{\mathrm{low}}$ is the integrated intensity of the PL peak at the lowest temperature (i.e., typically 4 K or 10 K). This equation is valid under the assumption that low temperature IQE is 100% due to the freezing-out of nonradiative centers, which is not necessarily the case. Interested readers could read *Y. C. Chiu and C. Bayram* [24]. Cubic InGaN/GaN on 3C–SiC (100) reported only 5% of IQE near 450 nm emission, possibly due to the defects inside the crystal. Cubic InGaN/GaN on V-grooved Si (100) reported a higher, 36% of IQE near 498 nm emission. However, it is asserted that this is the upper limit of the IQE, as the authors used 77 K as a low temperature, which is not low enough for the nonradiative centers to completely freeze out, overestimating the IQE.

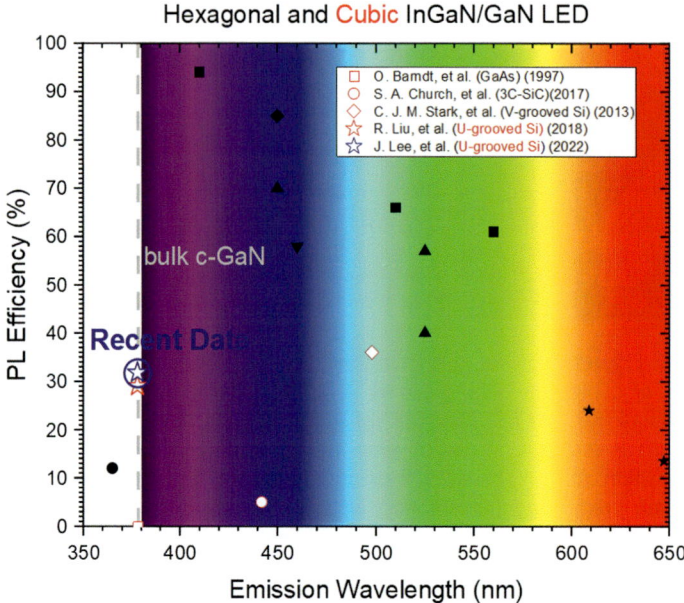

**Figure 9.7** PL efficiency of cubic InGaN/GaN LEDs reported in the literature are shown. Circled data is the recent data for cubic GaN on U-grooved Si (100).

The relatively low efficiency of cubic InGaN/GaN LEDs reported in the literature can be understood from two perspectives: theoretical (i.e., the lack of optimized structure for cubic InGaN/GaN LEDs) and experimental (i.e., the lack of high-quality bulk cubic GaN). In this regard, cubic GaN on U-grooved Si (100) is explored in two different directions. First, the theoretical design that is specific to cubic InGaN/GaN LEDs has been verified to achieve high IQE. Second, the growth of a high-quality GaN layer is demonstrated as seen from the IQE value of 29%–32%, which shows promise as a bulk material. It is expected that the IQE will be even higher once the QW structure is made, similar to the increase from 12% to 94% in hexagonal InGaN/GaN LEDs. The following sessions will report the details of the abovementioned work regarding cubic GaN on U-grooved Si (100).

## 4. Computation-based design of cubic InGaN/GaN LED

Compared to the experimental work reported regarding cubic GaN and its alloys, theoretical studies on cubic InGaN/GaN LEDs (c-LEDs) are scarce

[15,28,54]. Specifically, although c-LEDs require different design rules from hexagonal InGaN/GaN LEDs (h-LEDs), there is no or few studies on methodologies to configure c-LEDs (i.e., design rules of c-LED). As a result, scientific literature on c-LEDs frequently shows non-optimal design originating from "intuitive" h-LED designs, which obscures the true potential of c-LEDs [15,28,51,52,54,55] by limiting their theoretical performance. Indeed, not only the absence of polarization enables alternative design rules to h-LED design, but also the difference in physical parameters or growth challenges in InGaN imply different design rules from the same zincblende GaAs or AlGaInP LED design. In this section, three crucial design rules for green-emitting c-LEDs are reported from our recent work [56].

Theoretical c-LEDs design is carried out by the Open Boundary Quantum LED Simulator (OBQ-LEDsim) [57]. OBQ-LEDsim is a one-dimensional quantum-corrected drift-diffusion solver implemented by Dr. Yi-Chia Tsai (email: yichiat2@illinois.edu) for his Ph.D. dissertation entitled "Theoretical exploration of efficiency droop mechanisms in III-nitride visible light-emitting diodes," supervised by Professor Can Bayram (email: cbayram@illinois.edu) and Professor Jean-Pierre Leburton (email: jleburto@illinois.edu) at the University of Illinois Urbana—Champaign. OBQ-LEDsim was also employed in several prior works [15,27,28,58,59].

Common approaches to simulate LEDs use the quantum-corrected drift-diffusion model [60], where the device is divided into classical and quantum regions, dealing with the quantum and classical carrier densities separately. Here, the Schrödinger equation is solved numerically only at the quantum region and the boundary condition that the wave function vanishes to 0 is applied at a certain artificial boundary. This, along with the common wave function truncation outside the QW, may lead to nonphysical discontinuities or uncertainty in modeling the recombination outside the QW. OBQ-LEDsim, in contrast, makes use of the variational principles with the following analytic form of trial wave function to describe the ground-state electron and heavy hole wave functions inside the QW:

$$\psi(x) = \sqrt{\frac{\beta^2}{2\pi\alpha} \sin\left(\frac{\alpha\pi}{\beta}\right)} \frac{e^{\alpha(x+\gamma)}}{\cosh[\beta(x+\gamma)]} \quad (9.3)$$

Here, $\alpha$, $\beta$, and $\gamma$ are variational parameters controlling the wave function symmetry, width, and position, respectively. The trial function with the condition $\alpha < \beta$ ensures that the boundary condition where the wave function vanishes to 0 at $x = \pm\infty$ is automatically satisfied. This solution is then integrated into the classical drift-diffusion solver in a way that correctly captures the carrier density from the wave function outside the QW. In this way, it couples quantum and classical physical quantities without enforcing artificial boundaries between the QW and the classical continuum. The software details and the material parameters are described in earlier publications [15,27,28,36,61,62]. Although ambiguity exists in determining the recombination parameters, the Shockley—Read—Hall (SRH) recombination lifetime of 50 ns, ambipolar Auger recombination coefficient of $3 \times 10^{-30}$ cm$^6$ s$^{-1}$, and Auger electron-hole asymmetry of 0.4 are assumed for both c-LED and h-LED in this work, which agrees very well with experimental data for h-LED [28]. In h-LED, polarization charge is modeled without the polarization screening factor (i.e., using the theoretical value without down-scaling), similar to the recent literature [30,63]. In c-LED, an isotropic effective mass approximation is used, ignoring the heavy hole transverse effective mass reduction of the strained cubic crystal [64].

Fig. 9.8 shows the peak emission wavelengths of both c-LED and h-LED structures used for the optimization (described in Fig. 9.8 inset, which both show ~520 nm (green) emission at 20 A/cm$^2$ injection). A lower indium content due to QCSE with a large spectral blueshift is shown in h-LED, whereas c-LED has a stable green emission without noticeable blueshift. Fig. 9.8 inset shows the c- and h-LED QW structures considered in this optimization.

■ c-LED: [120 nm n-GaN: Si (doping: $5 \times 10^{18}$ cm$^{-3}$)/11 nm i-GaN/3 nm i-In$_{0.33}$Ga$_{0.67}$N/11 nm i-GaN/100 nm p-GaN: Mg (doping: $10^{19}$ cm$^{-3}$)/15 nm p$^+$ GaN: Mg (doping: $10^{20}$ cm$^{-3}$)]

■ h-LED: [120 nm n-GaN: Si (doping: $5 \times 10^{18}$ cm$^{-3}$)/11 nm i-GaN/3 nm i-In$_{0.22}$Ga$_{0.78}$N/11 nm i-GaN/15 nm p$^+$-Al$_{0.2}$Ga$_{0.8}$N: Mg (doping: $10^{20}$ cm$^{-3}$)/100 nm p-GaN: Mg (doping: $10^{19}$ cm$^{-3}$)/15 nm p$^+$ GaN: Mg (doping: $10^{20}$ cm$^{-3}$)]

The two LED structures are identical except: (1) the QW indium content is 33% in the c-LED and 22% in the h-LED for similar peak emission wavelength at 20 A/cm$^2$, and (2) the EBL is only present in the h-LED to prevent electron leakage current. The indium content of 33% in c-LED is due to the uncertainty in the current material parameters for cubic

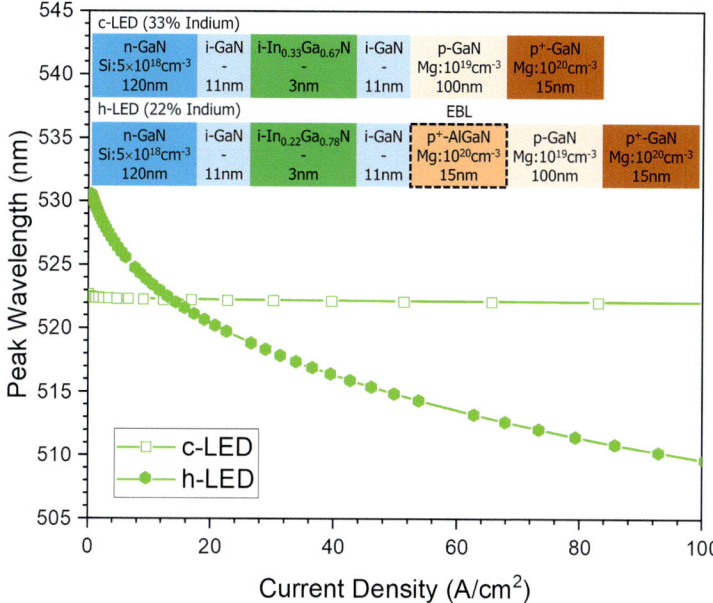

**Figure 9.8** Basic designs used for this work. The peak emission wavelengths for the given cubic and hexagonal InGaN/GaN LED (c-LED and h-LED) structures as a function of current density are shown. Large spectral blueshift is shown in h-LED, whereas stable green emission is shown in c-LED. *(Reproduced with permission from J. Lee, J.-P. Leburton, C. Bayram, "Design tradeoffs between traditional hexagonal and emerging cubic In X Ga (1−X) N/GaN-based green light-emitting diodes," J. Opt. Soc. Am. B 40 (2023) 1017. Copyright 2023 Optica Publishing Group.)*

InGaN, and it is recently shown that green emission in c-LED is possible with only 16% of indium content in contrast to this choice [71]. However, the design rules presented here are primarily bandgap-dependent instead of indium-content-dependent, which deems the results still valid.

The first design rule is that EBL is not needed for c-LED, pointing to the availability of the Aluminum-free design. Fig. 9.9a displays the leakage current density with respect to the total current density for the c-LED and h-LED, without and with EBL. In contrast to the h-LED showing significant leakage current without EBL, the c-LED does not experience such leakage current even without EBL, which is attributed to the absence of piezoelectric polarization. Fig. 9.9b displays the band structures of c-LED and h-LED without EBL at 100 A/cm$^2$ that show the influence of polarization on the leakage current. From the band structure of the h-LED, one can observe that the conduction band of the p-side barrier is lower than the

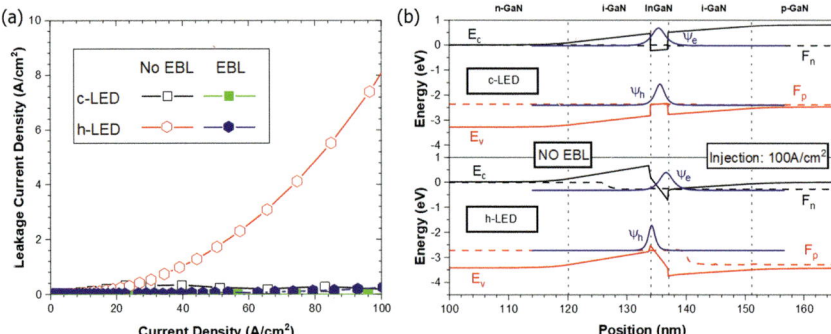

**Figure 9.9** (a) Leakage current as a function of current density in c-LED and h-LED, with and without EBL. It turns out that h-LED shows significant leakage current density without EBL, while the leakage current density of c-LED is negligible even without EBL. (b) Band structures of both c-LED and h-LED without EBL are shown. Large leakage current occurs in h-LED without EBL, whereas p-GaN successfully suppresses leakage current in c-LED. *(Reproduced with permission from J. Lee, J.-P. Leburton, C. Bayram, "Design tradeoffs between traditional hexagonal and emerging cubic In X Ga (1−X) N/GaN-based green light-emitting diodes," J. Opt. Soc. Am. B 40 (2023) 1017. Copyright 2023 Optica Publishing Group.)*

electron quasi-Fermi level, making electrons easily escape the QW. In c-LED, in contrast, the p-GaN layer suppresses electron leakage with the effective potential barrier height of 510 meV due to the absence of conduction band edge lowering. This removal of EBL is mainly beneficial from the growth perspective, due to the reported reliability issue in AlGaN growth [65].

The second design rule is that a wider QW is desirable in c-LEDs not only for higher IQE, but also for lower indium concentration. Fig. 9.10a shows the effect of QW thickness on the SQW c-LED performance (solid lines) and h-LED performance (dotted lines). It is seen that increasing the QW thickness increases the IQE at all biases for c-LED, which is attributed to the increased active volume. This increased radiative recombination rate also drives the onset of efficiency droop to higher current densities and decreases the efficiency droop because more Auger recombination is needed to saturate the radiative recombination. On the opposite, increasing the QW thickness in h-LED decreases the IQE at all biases due to the decreasing overlap integral between electron and hole wave function with thicker QW under polarization. It is worth noting that only the electron and hole ground states in the 2−5 nm QW are considered to ensure that the effect of excited energy levels is negligible. Therefore, extremely wide

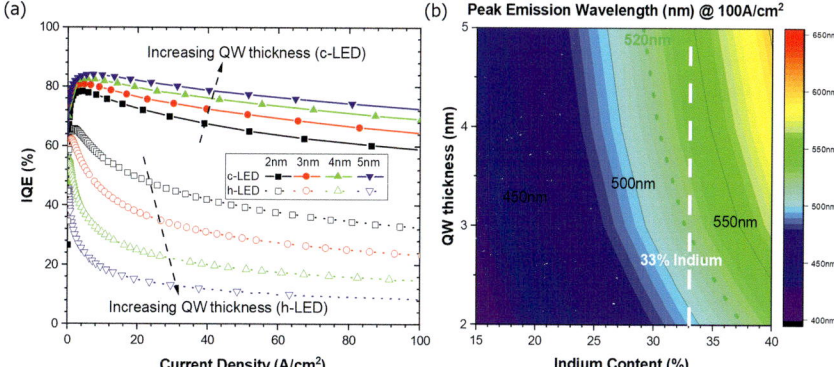

**Figure 9.10** (a) The effect of the quantum well thickness on the IQE is shown for both c-LED and h-LED. Increasing quantum well thickness increases IQE in c-LED, in contrary to decreasing IQE in h-LED. (b) The effect of quantum well thickness and the indium content on the peak emission wavelength is shown in c-LED. The dashed white line shows the achievable peak emission wavelength with the indium content of 33%, while the dotted green line shows the 520 nm emission line. By increasing the quantum well thickness from 2 to 5 nm, indium content can be reduced by ∼7% to achieve 520 nm emission. *(Reproduced with permission from J. Lee, J.-P. Leburton, C. Bayram, "Design tradeoffs between traditional hexagonal and emerging cubic In X Ga (1−X) N/GaN-based green light-emitting diodes," J. Opt. Soc. Am. B 40 (2023) 1017. Copyright 2023 Optica Publishing Group.)*

(i.e., over 10 nm) QW [66] is not of consideration here. Another benefit of wide QWs can be found in the wavelength tuning. Fig. 9.10b shows the peak emission wavelength of c-LED at 100 A/cm$^2$ injection as we change QW thickness and indium content. One can observe that increasing the QW thickness increases the peak emission wavelength, making green emission achievable with less indium content. For instance, a 2 nm QW requires 37% indium content for ∼520 nm emission, while it is only ∼30% indium for a 5 nm QW. Therefore, thick QW in c-LED is desirable in that it enables both lower indium content and higher IQE. In contrast, thick QW in h LED enables indium "savings" in the same way but compromises IQE to a large extent, which makes thick QW in h-LED unattractive.

The third design rule is related to the design cautions for MQW c-LED. Fig. 9.11 illustrates the effect of barrier thickness in MQW c-LED, where the barrier thickness ranges from 5 to 14 nm and the number of QWs varies from 1 to 8. Fig. 9.11a displays the IQE curves for the 14 nm barrier MQW

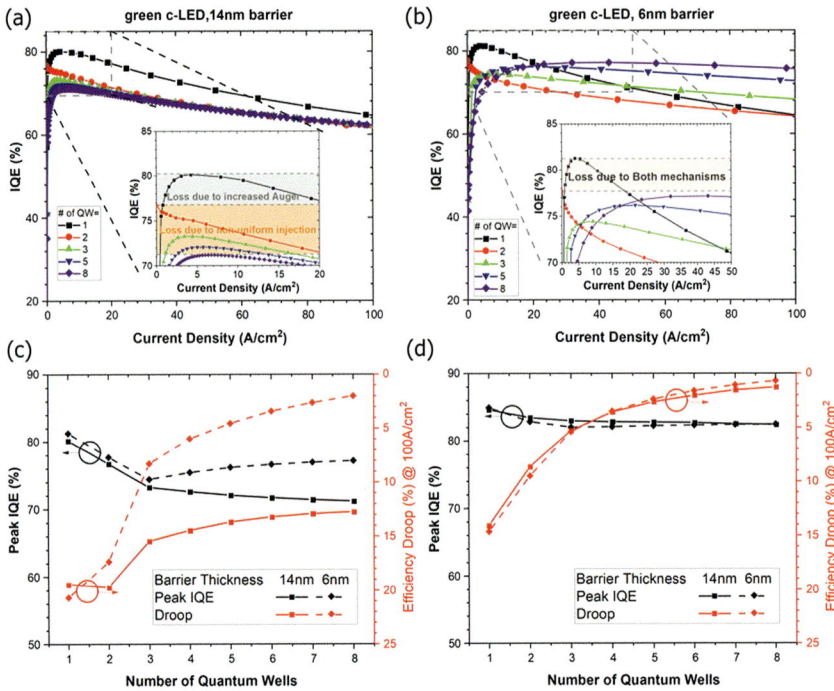

**Figure 9.11** The IQE curves of green c-LEDs with the different number of quantum wells are presented when the barrier thickness is (a) 14 nm (thick) and (b) 6 nm (thin). (c) The calculated peak IQE and the efficiency droop as a function of the number of quantum wells are presented for both the 6 nm barrier and 14 nm barrier structures. The trade-off between the peak IQE and the efficiency droop is clear in the 14 nm barrier case, while it is quenched in 6 nm barrier case. (d) The same analysis as (c) is done for blue-emitting c-LEDs. It is shown that the barrier thickness does not significantly affect the performance for lower indium content c-LEDs. *(Reproduced with permission from J. Lee, J.-P. Leburton, and C. Bayram, "Design tradeoffs between traditional hexagonal and emerging cubic In X Ga (1−X) N/GaN-based green light-emitting diodes," J. Opt. Soc. Am. B 40 (2023) 1017. Copyright 2023 Optica Publishing Group.)*

structures with varying numbers of QWs, as a function of current densities. Interestingly, the trade-off between the peak IQE value and the efficiency droop is clear (i.e., increasing the number of QWs decreases the peak IQE value, while decreasing the efficiency droop). Fig. 9.11b shows different IQE trends for the 6 nm barrier MQW structures. The reduction in the peak IQE value persists from SQW to 2QW device, but high injection performance improves due to the decrease in carrier density. Fig. 9.11c summarizes the results. Here, the efficiency droop is defined as.

$$\text{Droop} = \frac{\eta_{\text{peak}} - \eta_{100}}{\eta_{\text{peak}}} \tag{9.4}$$

where $\eta_{\text{peak}}$ is the peak IQE and $\eta_{100}$ is the IQE at the current density of 100 A/cm$^2$. For thick barrier (i.e., 14 nm) structures, the trade-off between peak IQE value and efficiency droop is shown clearly. In contrast, for thin barrier (i.e., 6 nm) structures, the peak IQE value bounces back after reaching the minimum at the three QW structure, although it is still lower than SQW peak IQE. Also, the efficiency droop significantly improves in thin barrier structures, truly benefiting from employing MQW. Therefore, barrier thicknesses directly affect the high current density performance. Fig. 9.11d shows the trade-off between peak IQE and the efficiency droop in blue-emitting c-LED by reducing the indium content in the active region. Here, although the trade-off exists, the peak IQE decrease is not pronounced, as well as thin barriers do not affect the result. From the comparison between Fig. 9.11c and d, it is suggested that the depth of the well (determined by the indium content) is the main cause of the decrease in peak IQE, so MQW c-LEDs need to be carefully designed when targeting green emission.

Fig. 9.12 suggests that the trade-off between peak IQE and the efficiency droop occurs due to the inter-QW carrier transfer. In Fig. 9.12a, we show the sheet electron density, sheet hole density, and recombination current for each QW in the eight QW structure, for both the 6 and 14 nm barrier structures at the IQE peak point. We labeled the QWs as first QW (closest to n-GaN) to eighth QW (closest to p-GaN). $J^{(i)}/J_{\text{total}}$ is the ratio of the recombination current in i-th QW to the total current $J_{\text{total}}$. It is shown that electron and hole density separation (i.e., highest electron density in the first QW and highest hole density in the eighth QW) occurs in the high-indium-content MQW structure for both 6 and 14 nm barrier structures. As a result, two mechanisms dominate the drop of the peak IQE value, i.e., enhancement of Auger recombination in the last QW and low injection level in middle QWs (i.e., second−seventh QWs). The former is caused by the carrier density asymmetry in the last QW. That is, compared to the SQW structure, the electron density is significantly lowered in the last QW, which leads to the decrease in all three recombination processes (SRH, radiative, and Auger recombination) considered. Therefore, the IQE now peaks at higher carrier densities, thereby enhancing the effect of Auger recombination. In addition, the hole density is now higher than the electron density, further enhancing the *hhe* Auger recombination due to the

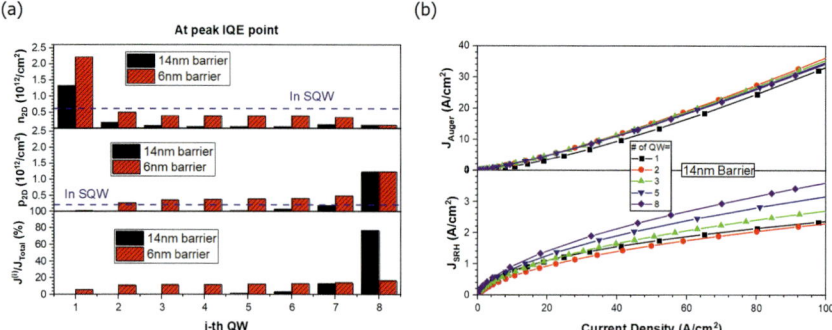

**Figure 9.12** The explanation for the trade-off between peak IQE and the efficiency droop is suggested. (a) The electron ($n_{2D}$) and hole ($p_{2D}$) sheet charge densities and recombination current ($J^{(i)}$) in each quantum well are presented for both 6 nm barrier and 14 nm barrier cases when eight quantum wells exist. Blue dotted line indicates the sheet charge density inside the quantum-well in the SQW device. Thinner barriers enable much more uniform carrier injection and thus much more uniform recombination in the middle quantum wells. (b) The Auger and the SRH recombination current densities are shown for a different number of quantum wells. Increased Auger recombination from 1 × QW to 2 × QW while increased SRH recombination from 2 × QW to 8 × QW are observed. (Reproduced with permission from J. Lee, J.-P. Leburton, and C. Bayram, "Design tradeoffs between traditional hexagonal and emerging cubic In X Ga (1−X) N/GaN-based green light-emitting diodes," J. Opt. Soc. Am. B 40 (2023) 1017. Copyright 2023 Optica Publishing Group.)

imbalanced carrier densities. As for the low injection level in middle QWs, the latter is caused by the low carrier densities in the middle QWs due to barrier-quenching inter-QW transfer of carriers that results in increased SRH recombination in MQW structure compared to the SQW structure. In the 14 nm barrier structure, the two mechanisms are clearly separated, as shown in Fig. 9.12b. For instance, Auger recombination increases from SQW to 2 × QW structure, whereas SRH recombination significantly increases when the number of QWs is added from 2 × QW to 8 × QW. Using the 6 nm barrier structure mitigates both effects, improving the IQE in the MQW structure. The same effect is expected for InGaN barriers (i.e., lower barriers) [63].

Finally, Fig. 9.13 illustrates how the discussed design rules improve the IQE of the 520 nm-emitting c-LED when combined together. Starting with an arbitrary and non-optimized structure [120 nm n-GaN: Si (doping: $5 \times 10^{18}$ cm$^{-3}$)/11 nm i-GaN/3 nm i-In$_{0.33}$Ga$_{0.67}$N/11 nm i-GaN/15 nm p$^+$-Al$_{0.2}$Ga$_{0.8}$N: Mg (doping: $10^{20}$ cm$^{-3}$)/100 nm p-GaN: Mg (doping: $10^{19}$ cm$^{-3}$)/15 nm p$^+$ GaN: Mg (doping: $10^{20}$ cm$^{-3}$)] that is

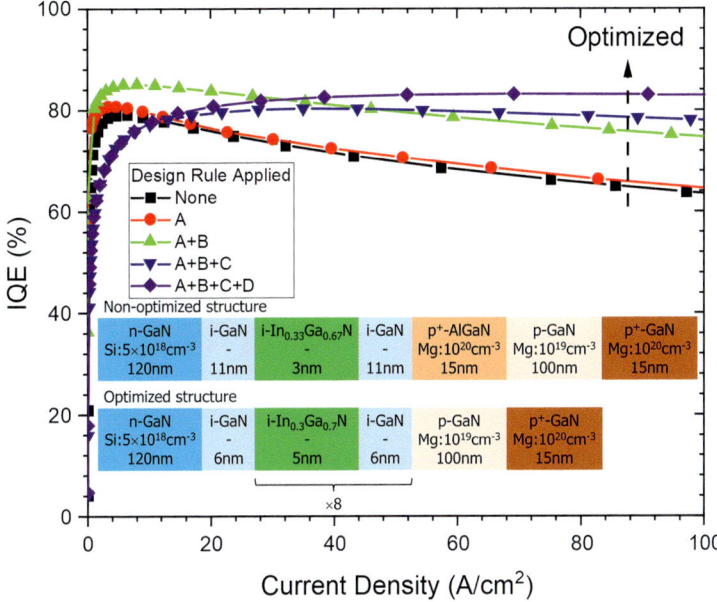

**Figure 9.13** The application of the above mentioned design rules are shown. Starting from a non-optimized structure, the IQE improvement of ~30.5% is observed, applying the following optimization steps: (A) Removal of the EBL, (B) the use of 5 nm $In_{0.3}Ga_{0.7}N$ QW instead of 3 nm $In_{0.33}Ga_{0.67}N$ QW, (C) changing from SQW to eight QW structure, and (D) the use of 6 nm barriers. *(Reproduced with permission from J. Lee, J.-P. Leburton, and C. Bayram, "Design tradeoffs between traditional hexagonal and emerging cubic $In_XGa_{(1-X)}N$/GaN-based green light-emitting diodes," J. Opt. Soc. Am. B 40 (2023) 1017. Copyright 2023 Optica Publishing Group.)*

equivalent to an EBL-added c-LED, four design rules are sequentially applied: (A) removal of the EBL, (B) change of the 3 nm $In_{0.33}Ga_{0.67}N$ QW to a 5 nm $In_{0.3}Ga_{0.7}N$ QW, (C) addition of seven additional QWs to the SQW structure, and (D) decrease of the inter-QW barrier thicknesses from 11 to 6 nm. It is seen that the IQE at 100 A/cm² progressively increases from 63.52% to 82.89%, which is equivalent to ~30.5% of IQE improvement.

In Summary, we detailed three crucial design rules for green-emitting (520 nm ≤ λ ≤ 550 nm) c-LEDs. First, we show that an electron blocking layer (EBL) is redundant in c-LEDs, removing the necessity of AlGaN growth. Second, we suggest that wider QWs are beneficial for c-LEDs, which enables both higher IQE and the reduction of the indium content in the active layer. Third, we indicate that thinner barrier layers are suitable for achieving high IQE in green-emitting MQW c-LEDs. We expect a higher theoretical performance of c-LEDs when applying these critical design rules.

Finally, it is worth reiterating that our analysis relies on a value of $3 \times 10^{-30}$ cm$^6$ s$^{-1}$ for the Auger recombination coefficient of c-LED, which is assumed to be the same as h-LEDs due to the lack of data in literature but predicted to be smaller than h-LEDs [37]. In this context, further studies may determine the Auger coefficient of c-LED small enough to "erase" the efficiency droop in thick SQWs at high current densities. In this case, SQW c-LEDs would outperform MQW structures as the MQW shows a lower peak IQE than the SQW structure.

## 5. Experimental growth of cubic GaN on U-grooved Si (100) for green LEDs

As mentioned above, the synthesis of cubic GaN is not straightforward due to its metastability. However, high quality bulk cubic GaN is critical for the development of cubic InGaN/GaN LEDs, as the growth of uniform InGaN layer becomes harder if we have propagating defects from the bulk layer [55]. In this section, we introduce the cubic GaN growth technique based on U-grooved Si (100) substrate, based on our recent publication [53]. We enabled single-phase single-crystalline pure cubic phase GaN for the very first time on large area ($\sim$ cm$^2$) dies based on MOCVD and studied the structural and optical properties. Through temperature-dependent and time-resolved photoluminescence measurements, it revealed a high IQE of $\sim$26%. We further identify the optical defect levels in phase transition c-GaN and report a selective etching technique, where we can remove hexagonal GaN selectively, increasing the efficiency to $\sim$32%. This high bulk GaN efficiency promises even higher efficiency for green-emitting cubic InGaN/GaN LEDs, pointing to the path of high-efficiency, droop-free green LEDs.

Fig. 9.14 (top) shows the U-grooved Si (100), prepared through (1) RCA cleaning (2) Oxide (i.e., SiO$_2$), growth, (3) Lithography and reactive ion etching (RIE) of the oxide, and (4) Selective Si wet etching (i.e., exposing {111} planes) via KOH solution. The pattern is generated uniformly all over the is $1 \times 1$ cm die, and the pattern dimensions are $p = 677 \pm 3$ nm, $t_d = 94 \pm 3$ nm, $\alpha = 2.9 \pm 0.3$ degrees, and $t_{ox} = 415 \pm 4$ nm. Fig. 9.14 (bottom) shows cubic GaN growth on the U-grooved Si (100) die. First a low-temperature AlN (LT-AlN) buffer layer

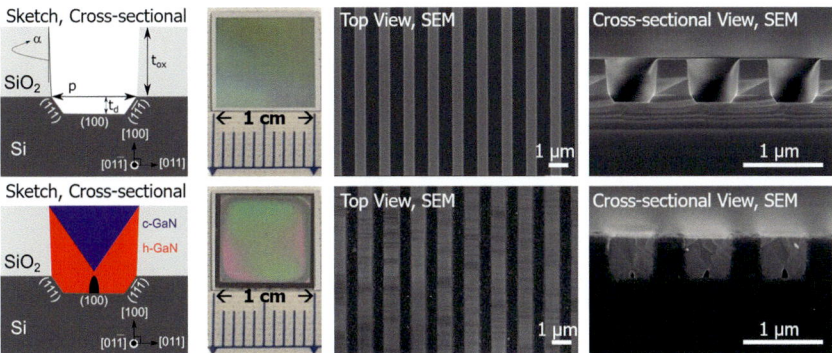

**Figure 9.14** Cubic GaN epitaxy on large-area (~1 cm²) U-grooved silicon (100) dies is demonstrated. (rom left to right) cross-sectional sketches, top-view photographs, top-view SEM images, and cross-sectional view SEM images of (top row) U-grooved silicon (100) and (bottom row) cubic GaN atop. *(Reprinted from J. Lee, Y.C. Chiu, M.A. Johar, C. Bayram, "Structural and optical properties of cubic GaN on U-grooved Si (100)," Appl. Phys. Lett. 121 (3) (2022) 032101, with the permission of AIP Publishing.)*

deposition, and then selective GaN deposition only on Silicon {111} planes takes place under selective MOCVD growth condition, experiencing the phase-transition from hexagonal GaN to cubic GaN when two growth facets meet, as explained earlier. It is proved that such epitaxy enables the growth of pure cubic GaN with a clear phase boundary between the hexagonal and cubic GaN [67], and the epitaxial quality is exceptional due to the aspect ratio trapping (ART) reported earlier [44]. That is, the threading dislocations that are formed from the lower GaN layer will no longer propagate to the top surface as they will be terminated by the oxide wall due to the aspect ratio of these nano-patterns.

Epitaxial quality is examined by microscopy imaging. Fig. 9.15 shows the AFM and scanning transmission electron microscopy (STEM) images of the as-grown cubic GaN on U-grooved Si (100). The AFM image visualizes ~600 nm wide cubic GaN straps with narrow hexagonal GaN exposed on the side, which we named hexagonal GaN shoulders. The plan-view STEM reveals exceptional top surface quality with no threading dislocations or other types of dislocations on top, thanks to the ART. Stacking faults with a density of $3.27 \pm 0.18 \times 10^4 \text{ cm}^{-1}$ are observed that are mostly perpendicular to the grooves. This density is an order of magnitude lower than what is reported in cubic GaN grown on 3C—SiC (i.e., $10^5 \text{ cm}^{-1}$) [49], showing the promises of cubic GaN on U-grooved Si.

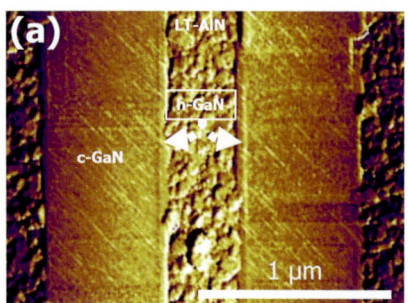 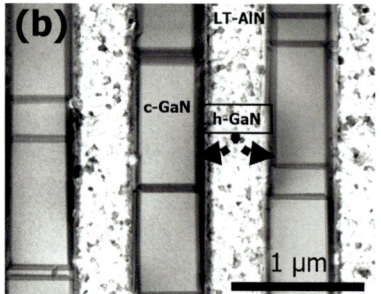

**Figure 9.15** Cubic GaN structural characterization. (a) Tapping mode AFM and (b) Bright field top-view STEM images are shown. The plan-view STEM reveals no threading dislocations or other types of dislocations on the cubic GaN surface. Stacking faults with a density of $3.27 \pm 0.18 \times 10^4 \mathrm{cm}^{-1}$ are observed. Cubic (c-) and hexagonal (h-) GaN, and LT-AlN buffer are labeled. *(Reprinted from J. Lee, Y.C. Chiu, M.A. Johar, C. Bayram, "Structural and optical properties of cubic GaN on U-grooved Si (100)," Appl. Phys. Lett. 121(3) (2022) 032101, with the permission of AIP Publishing.)*

In addition to the topographical studies, optical studies are done. Fig. 9.16a shows the PL spectra of as-grown cubic GaN at different temperatures ranging from 20 to 300 K. A 2 mW continuous-wave diode-pumped solid-state laser ($\lambda = 266$ nm) with the spot size of nearly $400 \times 500$ μm is used to excite cubic GaN inside the helium-bath cryostat, collecting the emission from hundreds of stripes. The PL spectra reveal two major peaks: the cubic GaN main peak and the hexagonal GaN near band emission (NBE) peak. The low-temperature cubic GaN main peak resolves into two Gaussian peaks at 3.19 eV (Donor–Acceptor Pair ($DAP_1$)) and 3.22 eV (Bound Exciton (BX)), which are identified based on our previous cathodoluminescence study [68] (Fig. 9.16a inset). Fig. 9.16b shows the Arrhenius plot of the integrated intensity of the $DAP_1$ and BX cubic GaN peaks. Here, we plot the integrated PL intensity as a function of the inverse of temperature to find out the activation energy of each luminescence center. The integrated intensity ($I$) obeys [69]:

$$I(T) = \frac{I_0}{1 + C \exp(-E_{\mathrm{act}}/k_\mathrm{B} T)} \quad (9.5)$$

where $I_0$ and $C$ are fitting coefficients, $E_{\mathrm{act}}$ is the activation energy of the luminescence center, $k_\mathrm{B}$ is the Boltzmann's constant, and $T$ is temperature. The fitting in Fig. 9.16b reveals the activation energy of $17.7 \pm 1.1$ meV for $DAP_1$ and $24.0 \pm 2.1$ meV for BX. Using the activation energy and the peak position of $DAP_1$ here, and the bandgap of cubic GaN as $3.31 \pm 0.01$ eV [68], the acceptor activation energy is calculated as

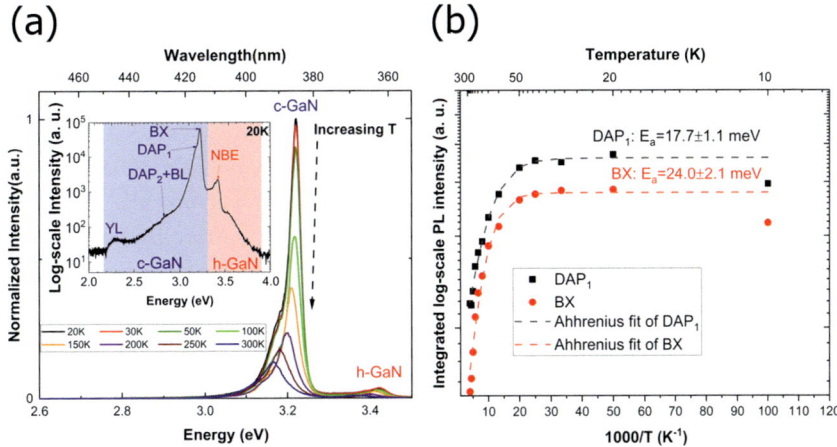

**Figure 9.16** The optical characterization of cubic GaN is shown. (a) The temperature-dependent PL spectra of the as-grown cubic GaN are shown. The data are normalized to cubic GaN peak intensity at 20 K. Inset shows the low-temperature(20 K) log-scale spectrum and the corresponding peaks. Cubic (c-) and hexagonal (h-) GaN are labeled. **(b)** Arrhenius plots and the calculated activation energies for the Donor-acceptor Pair (DAP$_1$) and the Bound Exciton (BX) peak are shown. 10 K data are excluded from the fitting as suspected outliers. *(Reprinted from J. Lee, Y.C. Chiu, M.A. Johar, C. Bayram, "Structural and optical properties of cubic GaN on U-grooved Si (100)," Appl. Phys. Lett. 121 (3) (2022) 032101, with the permission of AIP Publishing.)*

100.9 ± 11.1 meV. The IQE of the band-to-band transition of cubic GaN ($\eta_{IQE}$) is calculated from the abovementioned "relative" IQE method as 25.6 ± 0.9%, which is slightly lower than our earlier report (i.e., 29% [10]) but is achieved on much larger areal measurement.

Although structural and optical studies reveal that cubic GaN on U-grooved Si (100) is of high quality, achieving a top cubic GaN surface free of hexagonal phase requires precise control over the Si U-groove parameters as well as the epitaxy conditions. It has been previously demonstrated that U-groove parameters defined in Fig. 9.14 affect the cubic GaN surface coverage atop the U-grooves. Cubic GaN surface without hexagonal GaN shoulders occurs when the thickness ($h$) of GaN deposited above the Si(100) substrate surface equals the critical thickness ($h_c$) [67]:

$$h_c = \frac{1.06p - 0.75t_d}{1 - \dfrac{\tan \alpha}{0.71}} \tag{9.6}$$

For a large ($\sim$cm$^2$) area demonstration as in this work, it is challenging to achieve the exact conditions in an academic facility. One way to circumvent these stringent conditions and achieve a top GaN surface free of hexagonal phase is to use selective etching. In addition to the control sample labeled as (A) as-grown, three different wet-etch experiments are conducted and labeled as (B) LT-AlN removed, (C) LT-AlN and SiO$_2$ removed, and (D) LT-AlN, SiO$_2$, and h-GaN removed. In addition to temperature-dependent PL, time-resolved PL (TRPL) measurements are carried out to extract effective carrier lifetimes ($\tau_{\text{eff}}$) with a frequency-tripled Ti:Al$_2$O$_3$ laser ($\lambda = 266$ nm) under a pulse width of 100 fs and a repetition rate of 80 MHz. Fig. 9.18 shows the low temperature PL spectra and the TRPL intensity decay curve for sample A and sample D as representative samples. One can clearly observe the decrease in hexagonal GaN peak and increase in effective carrier lifetime in sample D, caused by the removal of hexagonal GaN shoulders. The intensity decay curves of all four samples show biexponential decay, and fit the following equation:

$$I(t) = C_1 \exp(-t/\tau_1) + C_2 \exp(-t/\tau_2) + y_0 \tag{9.7}$$

where $\tau_1$ and $\tau_2$ represent the fast and slow decay lifetimes, and $C_1$ and $C_2$ are the fast and slow decay amplitude. Under this condition, $\tau_{\text{eff}}$ is calculated as [70]:

$$\tau_{\text{eff}} = \frac{C_1 \tau_1^2 + C_2 \tau_2^2}{C_1 \tau_1 + C_2 \tau_2} \tag{9.8}$$

From the effective lifetime from TRPL and the IQE values from temperature-dependent PL, radiative ($\tau_r$), and nonradiative ($\tau_{nr}$) recombination lifetimes are calculated per:

$$\eta_{\text{IQE}} = \frac{\tau_{\text{eff}}}{\tau_r} \tag{9.9}$$

$$\frac{1}{\tau_{\text{eff}}} = \frac{1}{\tau_r} + \frac{1}{\tau_{nr}} \tag{9.10}$$

Fig. 9.17 shows the results. The IQE of sample D increased from $25.6 \pm 0.9\%$ to $31.6 \pm 0.8\%$ compared to sample A, which is mostly attributed to the decreased radiative recombination lifetime. It has been previously shown that hexagonal GaN shoulders are defective due to the stacking faults created near the cubic GaN/hexagonal GaN interface and the crystal/oxide interface [10]. Therefore, the removal of hexagonal GaN shoulders (i.e., removal of defective layers) accounts for the decrease in the

| Label | Information | Sample SEM Image | IQE (%) | $\tau_{eff}$(ps) | $\tau_r$(ps) | $\tau_{nr}$(ps) |
|---|---|---|---|---|---|---|
| A | As-grown | | 25.6±0.9 | 518±28 | 2025±133 | 697±59 |
| B | LT-AlN removed | | 23.9±0.7 | 564±39 | 2358±179 | 741±76 |
| C | LT-AlN & SiO$_2$ removed | | 24.4±0.8 | 508±33 | 2078±149 | 672±65 |
| D | LT-AlN & SiO$_2$ & h-GaN removed | | 31.6±0.8 | 551±30 | 1742±106 | 806±66 |

**Figure 9.17** Selective etching of LT-AlN, SiO$_2$ sidewall, and the defective hexagonal GaN in the cubic GaN on U-grooved Si (100) sample is demonstrated. Sample information, cross-sectional SEM images, IQE, effective carrier lifetime, and radiative and nonradiative recombination lifetimes from PL and TRPL experiments are shown. Cubic (c-) and hexagonal (h-) GaN are labeled. *(Reprinted from J. Lee, Y.C. Chiu, M.A. Johar, and C. Bayram, "Structural and optical properties of cubic GaN on U-grooved Si (100)," Appl. Phys. Lett. 121 (3) (2022) 032101, with the permission of AIP Publishing.)*

radiative recombination lifetime. The recombination lifetimes of sample C are within the error range compared to sample A, whereas the radiative recombination lifetime of sample B increased due to the damage caused by the etching method used in sample B. Overall, the high IQE of sample D implies that a cubic GaN surface without hexagonal GaN shoulders can be achieved via selective etching without compromising the quality of the cubic GaN to a large extent. Overgrowth of GaN on sample D will enable

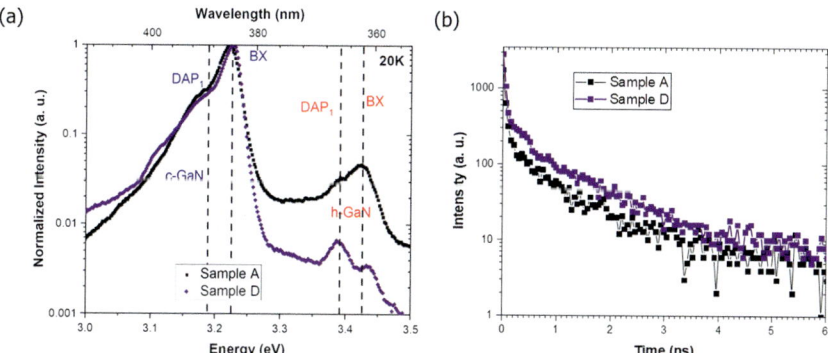

**Figure 9.18** (a) Low temperature PL and (b) room temperature TRPL comparison between sample A (as-grown) and sample D (LT-AlN, SiO$_2$, and h-GaN removed) are shown.

a potential path to coalescence and achieving a continuous cubic GaN film, which will accelerate this cubic GaN technology to be adopted faster.

To sum up, the large-area ($\sim \text{cm}^2$) epitaxy of cubic GaN over U-grooved Silicon (100) dies by MOCVD, utilizing the phase-transition from hexagonal GaN to cubic GaN, is introduced in this session with structural and optical characterization. Cubic GaN growth on U-grooved Si (100) shows exceptional quality with the stacking fault density of $3.27 \pm 0.18 \times 10^4 \text{ cm}^{-1}$ and the bulk IQE of $25.6 \pm 0.9\%$. Furthermore, selective etching including the removal of hexagonal GaN shoulders is demonstrated, which lightens the U-groove design requirements. In this process, a record room temperature cubic GaN band-edge emission IQE of $31.6 \pm 0.8\%$ is achieved. Overall, cubic GaN on U-grooved Si (100) promises to accelerate the progress of cubic GaN materials and devices thereof, by providing a high-quality bulk material for the growth of active layers.

## 6. Future work

So far, we have shown the theoretical design of InGaN/GaN LEDs that can achieve high IQE combined with experimental growth of the high-quality bulk cubic GaN whose top surface is free of hexagonal GaN. Based on the cubic GaN "template" generated in this work, two main directions toward high-efficiency green LED can be identified as illustrated in Fig. 9.19. First, cubic InGaN/GaN QWs can be grown on top of these bulk cubic GaN stripes, thereby enabling pure cubic InGaN/GaN QW

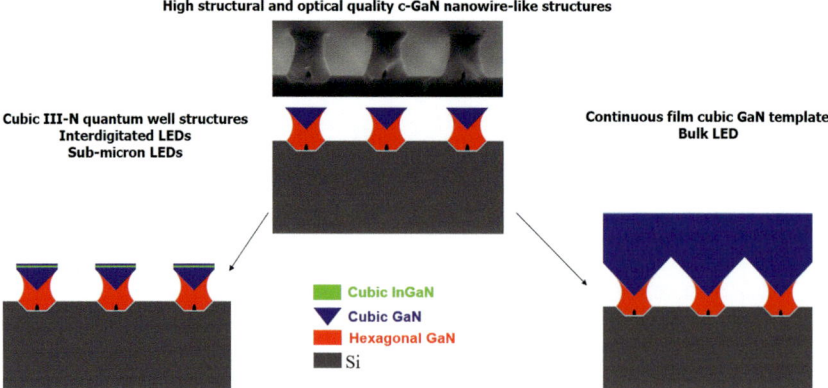

**Figure 9.19** Figure visualizing the two future works toward high-efficiency green-emitting cubic InGaN/GaN LEDs.

study. Furthermore, it can be processed to create an interdigitated LED or sub-micron LED. Second, continued MOCVD growth will enable the triangular cubic GaN layer to expand further, eventually coalescing and creating a continuous-film bulk cubic GaN template. This will be a real breakthrough for cubic GaN society, as it can be utilized not only for LEDs but also for various applications such as power electronics or integrated photonics.

Regarding scalability, this technology bares two clear benefits in that it is based on CMOS compatible process as well as MOCVD growth, making it relatively easy to scale up and commercialize. Three major milestones toward the scaling up are: (1) scaling of the patterning process on Silicon wafer, (2) development of continuous film cubic GaN by the coalescence of stripes, and (3) scaling of the epitaxy process.

## Acknowledgments

The information, data, or work presented herein was funded in part by the Advanced Research Projects Agency-Energy (ARPA-E), U.S. Department of Energy, under Breakers Program, Award Number DE−AR0001109 and the OPEN Program, Award Number DE−AR0001558, in part by the National Science Foundation Faculty Early Career Development (CAREER) Program under Award NSF-ECCS-16−52871, in part by the Office of Naval Research (ONR), under Award Number N00014-23-1-2423, and in part by the Computational resources Allocated by the Extreme Science and Engineering Discovery Environment (XSEDE) under Grant TG-DMR180050 and TG-DMR180075. The views and opinions of the authors expressed herein do not necessarily state or reflect those of the United States Government or any agency thereof. The authors also acknowledge valuable grant support from Coherent / II-VI Foundation, valuable characterization support from the EUROFINS EAG MATERIALS SCIENCE, LLC and valuable material support from the QROMIS INC. Also, J. Lee and Y. C. Chiu acknowledge support from the Promise of Excellence Fellowship and the Nick and Katherine Holonyak, Jr. Graduate Student Award from the Department of Electrical and Computer Engineering, University of Illinois at Urbana−Champaign, IL, USA. J. Lee also acknowledges support from SPIE with Optics and Photonics Education Scholarship. This work was carried out in the Micro and Nanotechnology Laboratory and Frederick Seitz Materials Research Laboratory Central Research Facilities, University of Illinois at Urbana−Champaign, IL, USA.

## References

[1] N. Holonyak, S.F. Bevacqua, Coherent (visible) light emission from $Ga(As_{1-x}P_x)$ junctions, Appl. Phys. Lett. 1 (4) (1962) 82−83.
[2] Statista, Worldwide − LED Lighting Market Size 2021, 2023. Available from: https://www.statista.com/statistics/753939/global-led-luminaire-market-size/.
[3] R. Markets, GlobeNewswire News Room, The Worldwide LED Lighting Industry Is Projected to Reach $124.7 Billion by 2027, 2022. Available from: https://www.

globenewswire.com/en/news-release/2022/07/05/2473791/28124/en/The-Worldwide-LED-Lighting-Industry-is-Projected-to-Reach-124-7-Billion-by-2027.html.
[4] S. Nakamura, T. Mukai, M. Senoh, Candela-class high-brightness InGaN/AlGaN double-heterostructure blue-light-emitting diodes, Appl. Phys. Lett. 64 (13) (1994) 1687−1689.
[5] T. Wu, C.W. Sher, Y. Lin, C.F. Lee, S. Liang, Y. Lu, et al., Mini-LED and micro-LED: promising candidates for the next generation display technology, Appl. Sci. 8 (9) (2018) 1557.
[6] S. Dimitrov, H. Haas, Principles of LED Light Communications: Towards Networked Li-Fi, first ed., Cambridge University Press, 2015. Available from: https://www.cambridge.org/core/product/identifier/9781107278929/type/book.
[7] P. Morgan Pattison, M. Hansen, J.Y. Tsao, LED lighting efficacy: status and directions, Compt. Rendus Phys. 19 (3) (2018) 134−145.
[8] K. Lee, 2022 Solid-State Lighting R&D Opportunities, 2022 [cited 2022 Jun 21] p. DOE/EE-2542, 1862626, 8851. Report No.: DOE/EE-2542, 1862626, 8851. Available from: https://www.osti.gov/servlets/purl/1862626/.
[9] S. Marcinkevičius, R. Yapparov, Y.C. Chow, C. Lynsky, S. Nakamura, S.P. DenBaars, et al., High internal quantum efficiency of long wavelength InGaN quantum wells, Appl. Phys. Lett. 119 (7) (2021) 071102.
[10] R. Liu, R. Schaller, C.Q. Chen, C. Bayram, High internal quantum efficiency ultraviolet emission from phase-transition cubic GaN integrated on nanopatterned Si(100), ACS Photon. 5 (3) (2018) 955−963.
[11] M. Peter, A. Laubsch, W. Bergbauer, T. Meyer, M. Sabathil, J. Baur, et al., New developments in green LEDs: new developments in green LEDs, Phys. Stat. Sol. (A). 206 (6) (2009) 1125−1129.
[12] D. Zhu, C. McAleese, M. Häberlen, C. Salcianu, T. Thrush, M. Kappers, et al., Efficiency measurement of GaN-based quantum well and light-emitting diode structures grown on silicon substrates, J. Appl. Phys. 109 (1) (2011) 014502.
[13] H. Amano, Progress and prospect of the growth of wide-band-gap group III nitrides: development of the growth method for single-crystal bulk GaN, Jpn. J. Appl. Phys. 52 (5R) (2013) 050001.
[14] X. Hou, S.S. Fan, H. Xu, D. Iida, Y.J. Liu, Y. Mei, et al., Optical properties of InGaN-based red multiple quantum wells, Appl. Phys. Lett. 120 (26) (2022) 261102.
[15] Y.C. Tsai, J.P. Leburton, C. Bayram, Quenching of the efficiency droop in cubic phase InGaAlN light-emitting diodes, IEEE Trans Electron Dev. 69 (6) (2022) 3240−3245.
[16] K. Hiramatsu, Y. Kawaguchi, M. Shimizu, N. Sawaki, T. Zheleva, R.F. Davis, et al., The composition pulling effect in MOVPE grown InGaN on GaN and AlGaN and its TEM characterization, MRS Internet J. Nitride Semicond. Res. 2 (1997) e6.
[17] N. Duxbury, U. Bangert, P. Dawson, E.J. Thrush, W. Van der Stricht, K. Jacobs, et al., Indium segregation in InGaN quantum-well structures, Appl. Phys. Lett. 76 (12) (2000) 1600−1602.
[18] C. Bayram, F.H. Teherani, D.J. Rogers, M. Razeghi, A hybrid green light-emitting diode comprised of n-ZnO/(InGaN/GaN) multi-quantum-wells/p-GaN, Appl. Phys. Lett. 93 (8) (2008) 081111.
[19] S.M. Ting, J.C. Ramer, D.I. Florescu, V.N. Merai, B.E. Albert, A. Parekh, et al., Morphological evolution of InGaN/GaN quantum-well heterostructures grown by metalorganic chemical vapor deposition, J. Appl. Phys. 94 (3) (2003) 1461−1467.
[20] A. Krost, A. Dadgar, GaN-based optoelectronics on silicon substrates, Mater. Sci. Eng.: B. 93 (1−3) (2002) 77−84.
[21] J. Edmond, A. Abare, M. Bergman, J. Bharathan, K. Lee Bunker, D. Emerson, et al., High efficiency GaN-based LEDs and lasers on SiC, J. Cryst. Growth 272 (1−4) (2004) 242−250.

[22] C.R. Miskys, M.K. Kelly, O. Ambacher, M. Stutzmann, Freestanding GaN-substrates and devices, Phys. Stat. Sol. (C). (6) (2003) 1627–1650.
[23] Qromis Substrate Technology, 2017. Available from: http://www.qromis.com.
[24] Y.C. Chiu, C. Bayram, Low temperature absolute internal quantum efficiency of InGaN-based light-emitting diodes, Appl. Phys. Lett. 122 (9) (2023) 091101.
[25] R. Liu, C. McCormick, C. Bayram, Comparison of structural and optical properties of blue emitting $In_{0.15}Ga_{0.85}N/GaN$ multi-quantum-well layers grown on sapphire and silicon substrates, AIP Adv. 9 (2) (2019) 025306.
[26] G. Li, W. Wang, W. Yang, Y. Lin, H. Wang, Z. Lin, et al., GaN-based light-emitting diodes on various substrates: a critical review, Rep. Prog. Phys. 79 (5) (2016) 056501.
[27] Y.C. Tsai, C. Bayram, J.P. Leburton, Effect of Auger electron–hole asymmetry on the efficiency droop in InGaN quantum well light-emitting diodes, IEEE J. Quantum Electron 58 (1) (2022) 1–9.
[28] Y.C. Tsai, C. Bayram, J.P. Leburton, Interplay between Auger recombination, carrier leakage, and polarization in InGaAlN multiple-quantum-well light-emitting diodes, J. Appl. Phys. 131 (19) (2022) 193102.
[29] M. Leroux, N. Grandjean, M. Laügt, J. Massies, B. Gil, P. Lefebvre, et al., Quantum confined Stark effect due to built-in internal polarization fields in (Al,Ga)N/GaN quantum wells, Phys. Rev. B 58 (20) (1998) R13371–R13374.
[30] C. Lynsky, A.I. Alhassan, G. Lheureux, B. Bonef, S.P. DenBaars, S. Nakamura, et al., Barriers to carrier transport in multiple quantum well nitride-based c-plane green light emitting diodes, Phys. Rev. Mater. 4 (5) (2020) 054604.
[31] C.H. Ho, J.S. Speck, C. Weisbuch, Y.R. Wu, Efficiency and forward voltage of blue and green lateral LEDs with V-shaped defects and random alloy fluctuation in quantum wells, Phys. Rev. Appl. 17 (1) (2022) 014033.
[32] Z. Jamal-Eddine, B.P. Gunning, A.A. Armstrong, S. Rajan, Improved forward voltage and external quantum efficiency scaling in multi-active region III-nitride LEDs, Appl. Phys. Express 14 (9) (2021) 092003.
[33] F. Akyol, D.N. Nath, S. Krishnamoorthy, P.S. Park, S. Rajan, Suppression of electron overflow and efficiency droop in N-polar GaN green light emitting diodes, Appl. Phys. Lett. 100 (11) (2012) 111118.
[34] S.I. Rahman, Z. Jamal-Eddine, A.M. Dominic Merwin Xavier, R. Armitage, S. Rajan, III-Nitride p-down green (520 nm) light emitting diodes with near-ideal voltage drop, Appl. Phys. Lett. 121 (2) (2022) 021102.
[35] X. Zhao, B. Tang, L. Gong, J. Bai, J. Ping, S. Zhou, Rational construction of staggered InGaN quantum wells for efficient yellow light-emitting diodes, Appl. Phys. Lett. 118 (18) (2021) 182102.
[36] I. Vurgaftman, J.R. Meyer, Band parameters for nitrogen-containing semiconductors, J. Appl. Phys. 94 (6) (2003) 3675–3696.
[37] K.T. Delaney, P. Rinke, C.G. Van de Walle, Auger recombination rates in nitrides from first principles, Appl. Phys. Lett. 94 (19) (2009) 191109.
[38] J. Simon, N.T. Pelekanos, C. Adelmann, E. Martinez-Guerrero, R. André, B. Daudin, et al., Direct comparison of recombination dynamics in cubic and hexagonal GaN/AlN quantum dots, Phys. Rev. B 68 (3) (2003) 035312.
[39] C.A. Hernández-Gutiérrez, Y.L. Casallas-Moreno, V.T. Rangel-Kuoppa, D. Cardona, Y. Hu, Y. Kudriatsev, et al., Study of the heavily p-type doping of cubic GaN with Mg, Sci. Rep. 10 (1) (2020) 16858.
[40] E.W.S. Caetano, R.N. Costa Filho, V.N. Freire, J.A.P. da Costa, Velocity overshoot in zincblende and wurtzite GaN, Solid State Commun. 110 (9) (1999) 469–472.
[41] D. Ahn, S. Park, Optical gain of strained hexagonal and cubic GaN quantum-well lasers, Appl. Phys. Lett. 69 (22) (1996) 3303–3305.

[42] A. Gundimeda, M. Frentrup, S.M. Fairclough, M.J. Kappers, D.J. Wallis, R.A. Oliver, Investigation of wurtzite formation in MOVPE-grown zincblende GaN epilayers on Al$_x$Ga$_{1-x}$N nucleation layers, J. Appl. Phys. 131 (11) (2022) 115703.
[43] S.C. Lee, X.Y. Sun, S.D. Hersee, J. Lee, Y.B. Ziang, H. Xu, et al., Growth of GaN on a nanoscale periodic faceted Si substrate by metal organic vapor phase epitaxy, in: 2003 International Symposium on Compound Semiconductors: Post-Conference Proceedings (IEEE Cat No03TH8767), IEEE, San Diego, CA, USA, 2004, pp. 15–21. Available from: http://ieeexplore.ieee.org/document/1354425/.
[44] C. Bayram, J.A. Ott, K.T. Shiu, C.W. Cheng, Y. Zhu, J. Kim, et al., Cubic phase GaN on nano-grooved Si (100) via maskless selective area epitaxy, Adv. Funct. Mater. 24 (28) (2014) 4492–4496.
[45] M. Pérez Caro, A.G. Rodríguez, E. López-Luna, M.A. Vidal, H. Navarro-Contreras, Critical thickness of β-InN/GaN/MgO structures, J. Appl. Phys. 107 (8) (2010) 083510.
[46] J. Wu, H. Yaguchi, K. Onabe, R. Ito, Y. Shiraki, Photoluminescence properties of cubic GaN grown on GaAs(100) substrates by metalorganic vapor phase epitaxy, Appl. Phys. Lett. 71 (15) (1997) 2067–2069.
[47] J. Komiyama, Y. Abe, S. Suzuki, H. Nakanishi, Suppression of crack generation in GaN epitaxy on Si using cubic SiC as intermediate layers, Appl. Phys. Lett. 88 (9) (2006) 091901.
[48] B. Yang, O. Brandt, A. Trampert, B. Jenichen, K.H. Ploog, Growth of cubic GaN on Si(001) by plasma-assisted MBE, Appl. Surf. Sci. 123–124 (1998) 1–6.
[49] L.Y. Lee, M. Frentrup, P. Vacek, M.J. Kappers, D.J. Wallis, R.A. Oliver, Investigation of stacking faults in MOVPE-grown zincblende GaN by XRD and TEM, J. Appl. Phys. 125 (10) (2019) 105303.
[50] O. Brandt, H. Yang, J.R. Müllhäuser, A. Trampert, K.H. Ploog, Properties of cubic GaN grown by MBE, Mater. Sci. Eng.: B. 43 (1–3) (1997) 215–221.
[51] S.A. Church, M. Quinn, K. Cooley-Greene, B. Ding, A. Gundimeda, M.J. Kappers, et al., Photoluminescence efficiency of zincblende InGaN/GaN quantum wells, J. Appl. Phys. 129 (17) (2021) 175702.
[52] C.J.M. Stark, T. Detchprohm, S.C. Lee, Y.B. Jiang, S.R.J. Brueck, C. Wetzel, Green cubic GaInN/GaN light-emitting diode on microstructured silicon (100), Appl. Phys. Lett. 103 (23) (2013) 232107.
[53] J. Lee, Y.C. Chiu, M.A. Johar, C. Bayram, Structural and optical properties of cubic GaN on U-grooved Si (100), Appl. Phys. Lett. 121 (3) (2022) 032101.
[54] D.R. Elsaesser, M.T. Durniak, A.S. Bross, C. Wetzel, Optimizing GaInN/GaN light-emitting diode structures under piezoelectric polarization, J. Appl. Phys. 122 (11) (2017) 115703.
[55] B. Ding, M. Frentrup, S.M. Fairclough, M.J. Kappers, M. Jain, A. Kovács, et al., Alloy segregation at stacking faults in zincblende GaN heterostructures, J. Appl. Phys. 128 (14) (2020) 145703.
[56] J. Lee, J.P. Leburton, C. Bayram, Design tradeoffs between traditional hexagonal and emerging cubic In$_X$Ga$_{(1-X)}$N/GaN-based green light-emitting diodes, J. Opt. Soc. Am. B 40 (5) (2023) 1017.
[57] Y.C. Tsai, C. Bayram, J.P. Leburton, An Open Boundary Quantum LED Simulator (OBQ-LEDsim), 2021. Available from: https://obqledsim.ece.illinois.edu/.
[58] Going cubic halves the efficiency droop in InGaAlN light-emitting diodes, MRS Bull. 47 (8) (2022) 759.
[59] J. Lee, Y.C. Tsai, C. Bayram, Green light emitting diodes for the ultimate solid-state lighting, in: 2022 Compound Semiconductor Week (CSW), IEEE, Ann Arbor, MI, USA, 2022, pp. 1–2. Available from: https://ieeexplore.ieee.org/document/9930416/.

[60] C. de Falco, E. Gatti, A.L. Lacaita, R. Sacco, Quantum-corrected drift-diffusion models for transport in semiconductor devices, J. Comput. Phys. 204 (2) (2005) 533−561.
[61] Y.C. Tsai, C. Bayram, Band alignments of ternary wurtzite and zincblende III-nitrides investigated by hybrid density functional theory, ACS Omega 5 (8) (2020) 3917−3923.
[62] Y.C. Tsai, C. Bayram, Structural and electronic properties of hexagonal and cubic phase AlGaInN alloys investigated using first principles calculations, Sci. Rep. 9 (1) (2019) 6583.
[63] C. Lynsky, G. Lheureux, B. Bonef, K.S. Qwah, R.C. White, S.P. DenBaars, et al., Improved vertical carrier transport for green III-nitride LEDs using (In, Ga) N alloy quantum barriers, Phys. Rev. Appl. 17 (5) (2022) 054048.
[64] S.L. Chuang, Physics of Photonic Devices, second ed., John Wiley & Sons, Hoboken, N.J, 2009, p. 821. Wiley series in pure and applied optics).
[65] L. Tang, B. Tang, H. Zhang, Y. Yuan, Review—Review of research on AlGaN MOCVD growth, ECS J. Solid State Sci. Technol. 9 (2) (2020) 024009.
[66] M. Hajdel, M. Chlipała, M. Siekacz, H. Turski, P. Wolny, K. Nowakowski-Szkudlarek, et al., Dependence of InGaN quantum well thickness on the nature of optical transitions in LEDs, Materials 15 (1) (2021) 237.
[67] R. Liu, C. Bayram, Maximizing cubic phase gallium nitride surface coverage on nano-patterned silicon (100), Appl. Phys. Lett. 109 (4) (2016) 042103.
[68] R. Liu, C. Bayram, Cathodoluminescence study of luminescence centers in hexagonal and cubic phase GaN hetero-integrated on Si(100), J. Appl. Phys. 120 (2) (2016) 025106.
[69] J. Krustok, H. Collan, K. Hjelt, Does the low-temperature Arrhenius plot of the photoluminescence intensity in CdTe point towards an erroneous activation energy? J. Appl. Phys. 81 (3) (1997) 1442−1445.
[70] J.R. Lakowicz, Principles of Fluorescence Spectroscopy, third ed., Springer, New York, 2006, p. 954.
[71] J. Lee, C. Bayram, Green-emitting cubic GaN/In$_{0.16}$Ga$_{0.84}$N/GaN quantum well with 32% internal quantum efficiency at room temperature, Appl. Phys. Lett. 124 (1) (2024) 011101.

# CHAPTER 10

# Emerging ferroelectric thin films: Applications and processing

Santosh K. Kurinec[1], Uwe Schroeder[2], Guru Subramanyam[3] and Roy H. Olsson III[4]

[1]Electrical & Microelectronic Engineering, Rochester Institute of Technology, Rochester, NY, United States; [2]NaMLab gGmbH, Dresden, Germany; [3]Electrical and Computer Engineering, University of Dayton, Dayton, OH, United States; [4]Electrical and Systems Engineering, University of Pennsylvania, Philadelphia, PA, United States

## 1. Introduction

A dielectric is an electrical insulator. A perfect insulator does not allow charge to conduct. However, it can be polarized by an applied electric field, where electric charges shift from their equilibrium positions, resulting in a dielectric polarization. Ferroelectricity is a property of materials that exhibit spontaneous electric polarization that can be reversed by the application of an external electric field. The presence of polarization charges through mechanical strain is called piezoelectricity. Pyroelectricity means that the dielectric has a spontaneous polarization, whose amplitude can be modulated with temperature. Paraelectricity is the ability of materials to become polarized under an applied electric field with polarization returning to zero on removal of the electric field. Ferroelectric materials have plurality of properties that show dependence on electric field, strain, and temperature. Therefore, they are useful in a wide range of applications including capacitors, memory cells, sensors, actuators, energy storage, electro-optic and more as depicted in Fig. 10.1.

## 2. History

Most ferroelectric materials do not contain iron but were so named because of their similarity to iron-based ferromagnetic materials, which exhibit magnetic polarization in a similar way. Ferroelectricity was first reported in 1921 by Joseph Valasek while studying Rochelle salt, a compound ($KNaC_4H_4O_6 \cdot 4H_2O$) extracted from wine, which was known to be

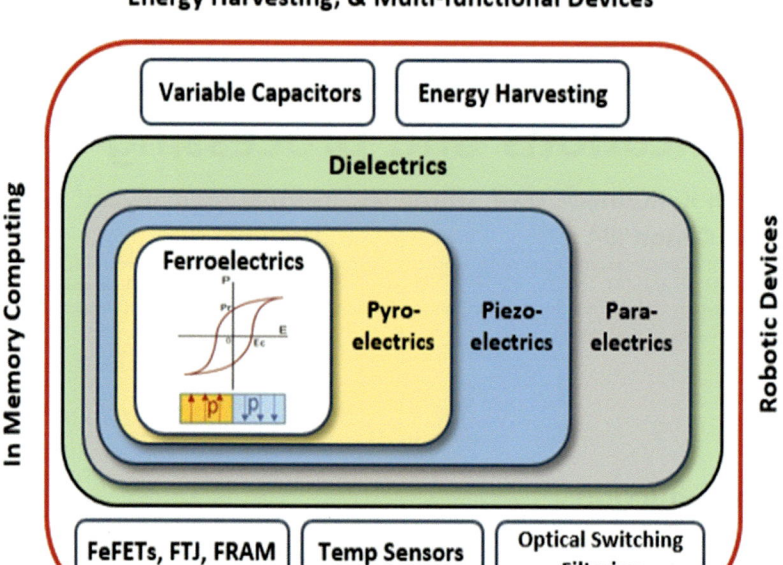

**Figure 10.1** Multiple properties of ferroelectric dielectrics spanning a wide range of applications.

piezoelectric. Valasek measured the electrical polarization of Rochelle salt and observed hysteresis when the external electric field switched direction. Two decades later, ferroelectricity was discovered in stable perovskites like BaTiO$_3$ and PbTiO$_3$ [1,2]. This subsequently led to the discovery of exceptional piezoelectric material, lead-zirconium, titanate (PZT) (PbZrO$_3$–PbTiO$_3$), a ceramic solid solution with composition PbZr$_{0.55}$Ti$_{0.45}$O$_3$ with a Curie temperature of about 350°C. The Curie temperature ($T_C$) is the temperature at which the material undergoes the phase transition from a low-temperature ferroelectric phase to a high-temperature paraelectric phase upon heating. Ferroelectricity and piezoelectricity disappear at $T_C$. PZT materials found widespread use due to their high operational temperatures, high piezoelectric coefficients, and the ability to tailor the properties of PZT ceramics by compositional modifications. Research from ceramics to thin film fabrication, opened a wide spread of new applications. In 1950s, the first patents of using thin film ferroelectrics onto silicon for nonvolatile memory applications were filed. Ferroelectric *materials* in devices such as ceramic capacitors, positive temperature coefficient *(PTC) thermistors* have been used in various commercial electric applications like automotive, smart home electric appliances, the aerospace industry, and other electronic segments as well, with projections to reach *a USD 7.12 billion market share by 2032* [3].

With the rapid growth of microelectronics technology, ferroelectric thin-film materials started to be added as functional components to microelectromechanical systems (MEMs). In 1993, ferroelectric thin films (PZT, $SrBi_2Ta_2O_9$ (SBT)) were successfully integrated with Si complementary metal oxide semiconductor (CMOS), and ferroelectric memory entered the semiconductor industry [2]. In the 1980s, polymer ferroelectrics such as polyvinylidene fluoride (PVDF) and P(VDF-TrFE) were discovered and subsequently investigated [4]. Scaling and CMOS integration of PZT, PVDF, SBTs remained challenging. As a result, product could only be used for niche market applications. In 2011, Böscke et al. reported the discovery of ferroelectricity in $HfO_2$ thin films when doped with silicon and capped with TiN electrodes [5]. Thermal treatment of doped $HfO_2$ in the presence of a capping layer results in its crystallization into a noncentrosymmetric orthorhombic structure, Pbc21, which is believed to be the origin of ferroelectric properties [6]. Since then, several groups have published on FE-$HfO_2$ crystallized using different dopants such as Si, Al, Y, Gd, Zr, Sr, etc. [5,7−12]. $HfO_2$, a high-k dielectric to replace $SiO_2$ in high-performance transistors, which could not be scaled below 15 nm due to high leakage, rose to prominence [13] when Intel announced the deployment as the gate dielectric in 32 nm technologies in 2007 [14]. Subsequently, a number of dopants in $HfO_2$ have been investigated and led to the emergence of ferroelectric hafnium zirconium oxide (HZO) for ferroelectric field effect transistors (FeFETs) and ferroelectric tunnel junction (FTJ) devices [15,16]. Fig. 10.2 provides the time line of developments in the field of ferroelectrics over the last century.

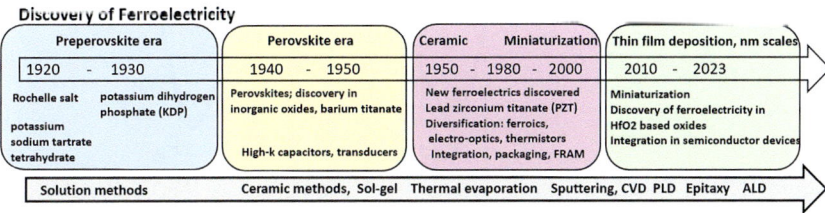

**Figure 10.2** Timeline of innovations in ferroelectrics since its discovery ∼100 years ago.

## 3. Principle

In ferroelectric materials, the polarization possesses two equilibrium states, which can be obtained by the application of a strong electric field [17]. These states are characterized as up or down, are two thermodynamically stable states with energy minima for the system, and have equal and opposite polarizations, called remnant polarization $P_R$, separated by a potential barrier shown in Fig. 10.3. Either one of these configurations are energetically more stable than a nonpolar configuration ($P = 0$). In the presence of an external electric field, the potential barrier is lowered, so that the central ion can change its position, reversing the polarization. A ferroelectric material has a crystalline structure such that charges do not balance to zero around a central point, which gives rise to electrical dipoles. The dipoles originate in small domains within the crystal, making it possible for the bulk material to look more or less ferroelectric. Generally, the polarization in these materials is not uniform. Regions of the crystal with uniformly oriented spontaneous polarization, called domains, exist, caused by the thermal and electrical history of the sample. The more domains are aligned, the stronger the overall net dipole polarization results.

If an electric field is applied to a ferroelectric material and then slowly reversed, the resulting change in polarization of the material follows the hysteresis loop in Fig. 10.4, depicting the important parameters—saturation and remnant polarizations ($P_S$, $P_R$) and coercive field ($E_C$). The polarization vectors in the domains are illustrated at these points in the P-E loop. There

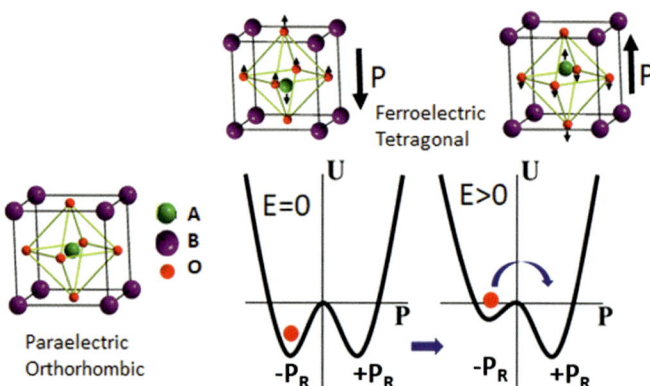

**Figure 10.3** (a) Equilibrium states in a unit cell of a perovskite (ABO$_3$) crystal in the paraelectric and tetragonal ferroelectric phase: A = Ba/Pb, B=Ti/Zr and O = oxygen. (b) Energy as a function of polarization for ferroelectric materials, with and without an external electric field.

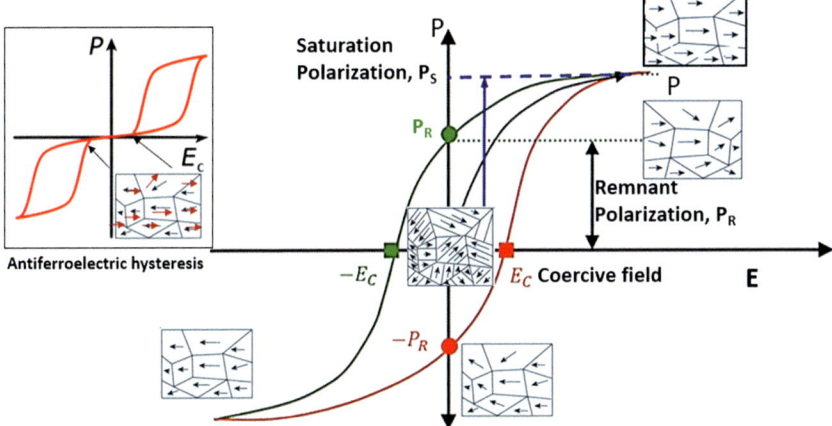

**Figure 10.4** Ferroelectric polarization-electric field (P-E) hysteresis loop with corresponding domain alignments. The inset on top left depicts an antiferroelectric loop showing a double hysteresis loop.

is another type of property known as antiferroelectric (AFE), where the polarization vector inside the material becomes anti-parallel under zero-bias condition. Kittel first discussed AFE crystals, noting that materials with an antipolar lattice structure can transition to a polar phase when an electric field is applied, resulting in characteristic double hysteresis loops in their polarization–electric field curves due to field-induced reversible transitions [18]. Further conditions are discussed for doped $HfO_2$ or $ZrO_2$ thin films. For pure $ZrO_2$ a field induced transition from the nonpolar tetragonal phase to the polar orthorhombic phase was found [19]. In addition, a similar pinched hysteresis loop can be induced by internally biased or pinned FE domains or by ferroelastic switching.

## 3.1 Ferroelectric models

Numerous models have been proposed in an effort to understand ferroelectrics and their behavior in electronic systems. The Preisach model, a basic model adopted from ferromagnetic device studies, serves as a good introduction to ferroelectric devices [20]. The Miller model, proposed by Miller et al. in 1990, presents a mathematical description of the hysteresis curve observed in experimental hysteresis measurements [21]. This mathematical description, given by

$$P_+(E) = P_S \tanh\left(\frac{E - E_C}{2\delta}\right) + \varepsilon_r \varepsilon_0 E \qquad (10.1)$$

where

$$\delta \equiv E_C \left( \ln \left( \frac{1 + \frac{P_R}{P_S}}{1 - \frac{P_R}{P_S}} \right) \right)^{-1} \quad (10.2)$$

is used to calculate half of the hysteresis loop and is then rotated 180 degrees to give the other branch. The linear term of Eq. (10.1) represents the normal dielectric behavior of the ferroelectric while the first term represents the hysteretic component, where + superscript signifies the positive-direction branch of the loop. The negative-direction branch is given by, $P_{S-}(E) = -P_{S+}(E)$. This formalism was extended further by Lue et al. in 2002, which included additional terms to describe the behavior of a ferroelectric material when an electric field is applied that switches some, but not all domains [22]. These minor or nonsaturated loops are useful for describing some ferroelectric behavior but are limited in their ability to describe nonsaturated ferroelectric switching that is not symmetric about the origin.

Since the 1940s, phenomena in ferroelectric materials have been successfully modeled based on the Landau theory of phase transitions, which was first applied to ferroelectrics by Ginzburg and Devonshire. The model based on this Landau–Ginzburg–Devonshire (LGD) approach has been an essential tool in understanding the basic physics of ferroelectricity [23]. In LGD theory, a ferroelectric below the Curie temperature $T_C$, is described by a double-well free energy landscape U as a function of the polarization P as

$$U = \alpha_0 (T - T_C) P^2 + \beta P^4 - \gamma P^6 - EP \quad (10.3)$$

where $\alpha_0$, $\beta$, and $\gamma$ are empirical material constants, TC represents the temperature at which ferroelectric behavior disappears, and $E$ represents the applied electric field. If $T < Tc$ while $\alpha_0$, $\beta$, and $\gamma$ are the material-dependent anisotropy constants, one obtains the ferroelectric energy landscape seen in Fig. 10.3. If $T > Tc$, the ferroelectric behavior disappears and the film acts similar to a normal dielectric. Any crystal is in a thermodynamic equilibrium state that can be completely specified by the values of a number of variables, for example temperature $T$, entropy $S$, electric field $E$, polarization $P$, stress $\sigma$ and strain $\varepsilon$. A small change in one of the variables produces a corresponding change in the other.

The capacitance of a ferroelectric material can be obtained by

$$C_{FE} = \left(\frac{d^2U}{dQ_{FE}^2}\right)^{-1} \quad (10.4)$$

The two degenerate energy minima define two stable spontaneous polarization states in the material, which can be reversed by the application of an electric field. By differentiating $U$ with respect to $P$, one obtains the "S"-shaped P-E curve, where $E$ is the electric field in the ferroelectric (Fig. 10.5). This "S"-shape of the P-E curve implies that in a certain region around $P\sim0$ the ferroelectric possesses a negative differential capacitance (NC), because the capacitance C is proportional to the slope $dP/dE$, which is a direct consequence of the energy barrier in the double-well free energy landscape. This NC region can be modeled by the state-of-the-art [24–26] approach for modeling the dynamics of ferroelectric capacitors relying on the Landau-Khalatnikov (LK) equation:

$$\rho\frac{dP}{dt} + \nabla_p U = 0 \quad (10.5)$$

The negative capacitance (negative permittivity) cannot be achieved in real capacitance measurements due to instability of the ferroelectric in this region. It is said that this region can be stabilized by adding another linear dielectric capacitor in series with the ferroelectric capacitor such that the overall capacitance of the system becomes positive. The negative acceptance is actively being explored for NC-FETs mentioned in Section 8.

There are 32 classes or point groups of crystals, of which 11 have a center of symmetry (centrosymmetric), and 21 do not have a center of

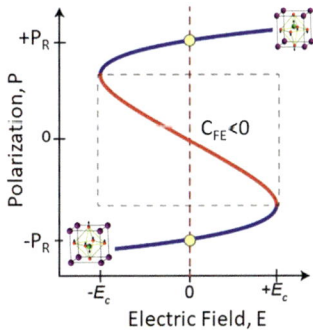

**Figure 10.5** "S"-shaped P-E curve of the ferroelectric material showing the negative capacitance regime between two stable polarization states.

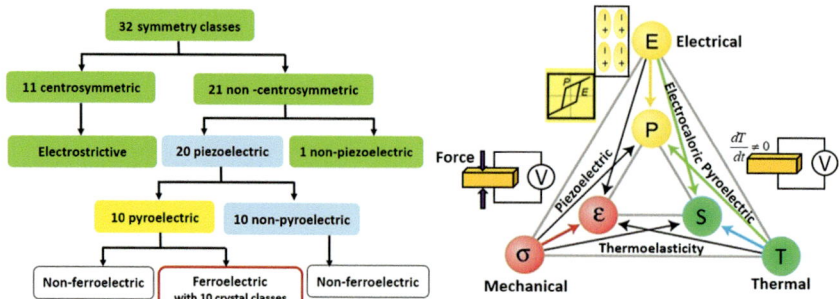

**Figure 10.6** Classification of piezoelectric, pyroelectric and ferroelectric classes of crystal structures (left); each corner represents energy states in the crystal: Mechanical, electrical, and thermal energy. P, S and ε stand for electrical displacement, entropy, and strain, respectively. A small change in one of the variables produces a corresponding change in the other (right) [28].

symmetry (noncentrosymmetric) (Fig. 10.6) [27]. Lack of symmetry results in the net movement of positive and negative ions with respect to each other on application of electric field or stress that will produce electric dipole polarization. Spontaneous polarization is defined to be the magnitude of the dipole moment per unit volume.

## 4. Thin films

Thin films are structural materials in order of 1 μm or less in thickness. Thin films can be realized either by physical or chemical techniques. Research on the development of thin films of a range of ferroelectric materials has emerged in the last decade, owing to the need for device miniaturization and their wide spectrum of appealing applications. Ultra-thin ferroelectric films (<100 nm) with uniform, conformal, and controllable thickness are promising for advancement in technology of future ferroelectric-based devices [29]. The multifunctional properties of ferroelectrics have ignited huge interest in incorporating them onto integrated circuits platforms for nonvolatile memory, MEMS, and sensors. In Fig. 10.2 (bottom row), the timeline of developments in deposition methods is presented. Ferroelectrics have complex oxide crystal structures that need to be stabilized in thin films. The deposition methods, annealing, substrate and thickness play a vital role. Scaling film thickness to nanometer scales can become challenging. In addition, the formation of interfacial layers starts to play a major role in the film characteristics. Table 10.1 lists ferroelectric properties of some of the commonly investigated ferroelectric thin films.

Table 10.1 Ferroelectric properties of promising ferroelectric thin films.

| Ferroelectric properties | $Pb(Zr,Ti)O_3$ | $SrBi_2Ta_2O_9$ | $BiFeO_3$ | $Ba_xSr_{1-x}TiO_3$ | Doped $HfO_2$ $Hf_xZr_{1-x}O_2$ | $Al_xSc_{1-x}N$ |
|---|---|---|---|---|---|---|
| Crystal structure | Perovskite | Perovskite | Perovskite | Perovskite | Fluorite | Wurtzite |
| Remnant polarization $P_R$ ($\mu C/cm^2$) | 10–40 | 5–10 | 90–95 | 15 (BTO)–0.05 ($Ba_{0.5}Sr_{0.5}TiO_3$) | 10–40 | 80–120 |
| Coercive field, $E_c$ (MV/cm) | 0.05–0.08 | 0.03–0.05 | ~0.05 | ~0.4 | 0.8–2 | 2–6.5 |
| Relative permittivity, $\varepsilon_r$ | ~400 | ™ 200 | ~50 | 200–1000 | ~30 | 12.5–25 |
| Curie temperature, (°C) | ~400 | ~400 | ~700 | 124 (BTO)–20 (BST) | 200–500 | >600 |

## 5. Thin film deposition processes

Various processes, classified as vapor-based physical and chemical, and chemical solutions based, have been employed in the deposition of ferroelectric thin films [30,31] as depicted in Fig. 10.7.

The schematics of three types of deposition processes described in this chapter for deposition of $HfO_2$, BST and AlScN are shown in Fig. 10.8.

### 5.1 Chemical solution deposition (CSD)

Chemical solution deposition (CSD) is a highly versatile technique with deposition options ranging from spin-coating, dip-coating, to inkjet printing or aerosol jet printing [32]. CSD is already being employed in the field of electronic thin films due to its cost-effectiveness, ease of use, and high yield. Other advantages of the CSD method include the wide tunability with various dopant systems and the possibility of generating ferroelectricity in thicker films. CSD Sol-gel technique involves dispensing a precursor solution onto a substrate, rotating at a high velocity to produce planar film followed by subsequent annealing and crystallization to produce the film with desired properties. For HfZrO films, the precursor solution was prepared by dissolving the desired amount of hafnium 2,4-pentandionate and zirconium 2,4-pentadionate in propionic acid and propionic acid anhydrite (5:3) at 140°C for 6 h. The prepared solution was spin-coated on a substrate [33].

### 5.2 Sputter deposition

In sputter deposition methods, direct current (DC) or radio frequency (RF) are vastly used for deposition of polycrystalline thin films. High energy ions (e.g., Argon ions) are bombarded on the targets to emit the desired species and deposit on the substrate. It is a line-of-sight method, and thin conformal depositions are hard to achieve. Various forms of sputtering—purely physical using multiple targets, reactive ion or

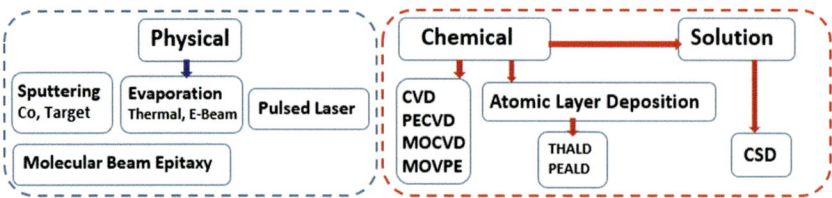

Figure 10.7 Classification of various thin film deposition methods.

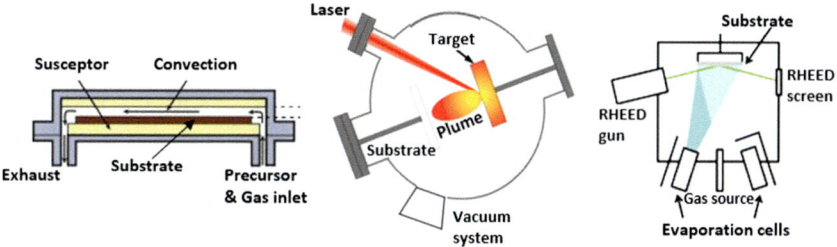

Figure 10.8 Basic schematics of ALD, PLD and MBE.

magnetron are being utilized. Multiple stacks can be deposited, such as bottom electrode (TiN) deposition by DC sputtering followed by RF reactive magnetron sputtering of HZO using $HfO_2$ and $ZrO_2$ targets at a base pressure of $\sim 10^{-4}$–$10^{-5}$ Pa [30,31]. In Ref. [31], the authors have provided detailed comparisons of different deposition methods.

## 5.3 Pulsed laser deposition

Pulsed laser deposition (PLD) is one of the several physical vapor deposition techniques employed to fabricate thin films [34]. A laser is used to ablate the material of the target, creating a plasma containing the energetic species, which are deposited on the substrate and form the film. The most common lasers used in PLD systems are gas (excimers, or excited dimers; electrically pumped) or solid-state lasers (optically pumped), for example the higher harmonics of a yttrium aluminum garnet (YAG) laser due to their high output power. The laser source is chosen depending on the material specifics. The high fluence of the excimer laser pulses generates atomic species with a high degree of ionization and high kinetic energy. One of the most important and enabling characteristics in PLD is the ability to realize stoichiometric transfer of ablated material from targets for many materials [35,36]. The parameters that have to be optimized in a PLD process include the laser power and repetition rate, the pressure and composition of the gas(es) in the PLD chamber, and the deposition temperature. With regard to the formation of the ferroelectric phase in the film, subsequent annealing steps are required (Fig. 10.9).

As an example, in Ref. [37], authors have reported deposition of Gd:$HfO_2$ using an $O_2$ flow of 20 sccm applied to the chamber to reach a deposition pressure of 0.1 mbar, and 2800 pulses of a KrF laser ($\lambda = 248$ nm) having a fluence of 1.5 J/$cm^2$.

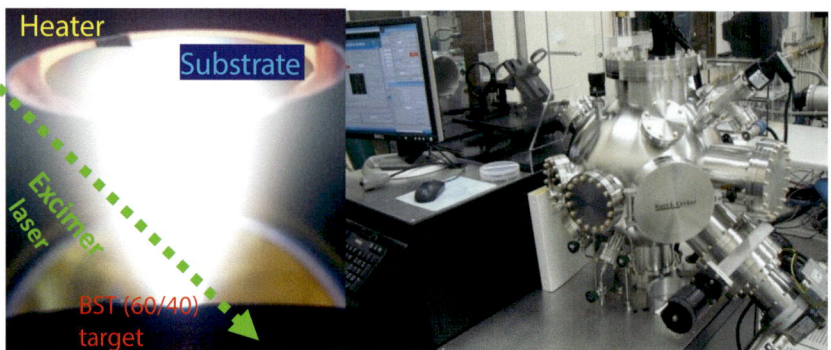

**Figure 10.9** A PLD system employed in deposition of BST thin films [34].

## 5.4 Atomic layer deposition (ALD)

The principle of ALD was discovered twice; in the 1960s under the name "molecular layering" in the Soviet Union, and in the 1970s under the name "atomic layer epitaxy" (ALE) in Finland [38]. ALD is a thin film growth method that belongs to the general class of CVD techniques. ALD chemical process is broken down into steps that isolate different adsorption and reaction steps to have self-limiting reactions. The process employs separate pulses of precursors and reactants that pass sequentially through the process chamber. With the substrate in the process chamber and under high vacuum, an initial precursor is introduced into the chamber. The molecular character of the precursor is such that it will form a chemically-bound monolayer on the substrate surface. Any layers beyond the monolayer are only bound by physisorption forces, which are weak enough to allow any precursor other than that in the monolayer to be pumped away under high vacuum. Once the monolayer is present on the substrate, the chamber is re-evacuated and purged to remove excess precursor. Next, a co-reactant is introduced to the process chamber. It reacts with the monolayer material to form the desired compound on the substrate surface. Byproducts of this reaction are pumped away. A variety of different precursors were tested in literature [39].

The ALD process for hafnium oxide thin films, shown in Fig. 10.10 provides a useful representative example of this process. A pulse of hafnium alkyl compound, in this case tetrakis (dimethylamino) hafnium (TDMAH) is introduced to the process chamber. The substrate is prepared prior to the ALD process so that it has a well-ordered covering of hydroxyls on the surface. TDMAH can be converted into the unsaturated hafnium atom on

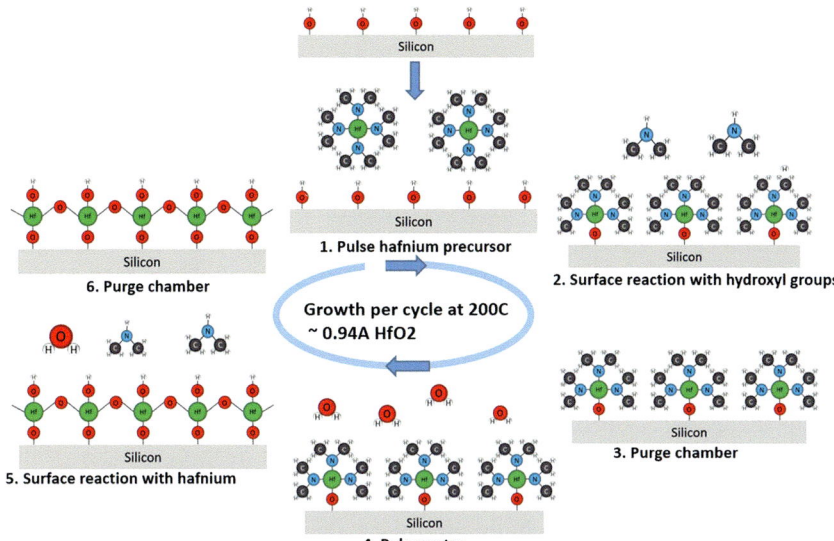

**Figure 10.10** Schematic showing the ALD process for depositing HfO2 on silicon using tetrakis (dimethylamido) hafnium (TDMAH) precursor.

the hydroxylated Si (1 0 0) surface through the ligand exchanged reactions (LERs) between the adsorbed TDMAH and the surface hydroxyl groups losing a dimethylamine (DMA) ligand. Water is now introduced in the chamber as a co-reactant, which enables oxygen bonding with Hf, releasing remaining DMA. The cycle is repeated with precise combination of effective precursor delivery, pumping and purging. The overall reaction is given as;

$$Hf[(CH_3)_2N]_4 + 2H_2O \rightarrow HfO_2 + 4HN(CH_3)_2$$

Dopants can be added in an ALD process by adding dopant precursors with designed periodicity to achieve the composition desired.

Various modifications in ALD method have been reported. These include thermally enhanced ALD and plasma enhanced ALD [40,41]. The inclusion of plasma in atomic layer deposition processes offers the benefit of substantially reduced growth temperatures and greater flexibility in tailoring the gas-phase chemistry to produce specific film characteristics [42], but deposition in 3D structures is critical. In production, typically CpHf[N(CH$_3$)$_2$]$_3$ and the according Zr precursor CpZr[N(CH$_3$)$_2$]$_3$ are used with O$_3$ oxidant when ZrO$_2$ and HfO$_2$ based films need to be deposited in high aspect ratio structures with low carbon content.

## 5.5 Molecular beam epitaxy (MBE)

MBE is a technique for the growth of epitaxial thin film. The term "epitaxial" is applied to a film grown on top of the crystalline substrate in ordered fashion that atomic arrangement of the film accepts crystallographic structure of the substrate. In MBE the desired elements are sublimated in special devices called evaporators (by thermal annealing) or introduced in the chamber in gas phase. In both cases the atoms stick on a substrate creating a solid solution in thin film form. MBE is generally performed only in ultra high vacuum (UHV) condition [43]. The technique is characterized by a very small deposition rate (typically less than 1 nm per minute) that permits the with the best possible thickness precision. RHEED (reflection high-energy electron diffraction) is usually applied in an MBE system to characterize atomic surface structure and morphology information of the sample during deposition. This technique provides thin films with high purity stoichiometry and a high control on their crystal structure. MBE of ferroelectric thin films has been demonstrated [42]. The deposition of AlScN involves fundamental challenges. MBE offers the advantage that high amounts of scandium can be incorporated in the compound.

## 5.6 Metal-organic chemical vapor deposition (MOCVD)

Metal-organic chemical vapor deposition (MOCVD) has been shown to be an outstanding process to prepare thin films of multi-component oxides. Reference [43] describes deposition of oxide perovskite thin films on both single crystal and nonsingle crystal substrates through MOCVD processes. Their study discusses the principles and the basic rules governing conventional, plasma-enhanced, and liquid-assisted MOCVD processes. It is necessary to adjust the performance of ferroelectric thin film materials. These include composition control, crystal structure, interfaces, leakage reduction, fatigue improvements and ferroelectric domain size distribution [44,45]. Shimizu et al. reports the effect of the heat treatment conditions on the constituent phase and electrical properties of MOCVD $(Hf_{0.5}Zr_{0.5})O_2$ films [46].

To summarize, a wide variety of deposition techniques have been employed and explored that can be generally classified into two categories—physical and chemical. Table 10.2 lists the variety of deposition methods employed along with their strengths and weaknesses [30,31].

**Table 10.2** Strengths and weaknesses of different thin film deposition techniques.

| | Deposition techniques | Strengths | Weaknesses |
|---|---|---|---|
| **Physical** | Sputter deposition<br>• Cosputtering<br>• Alloyed target sputtering | • Room temperature deposition possible<br>• Variety of reactant gases, doping possible<br>• Low carbon contamination | • Vacuum (moderate to low) required<br>• Limited conformality |
| | Evaporation techniques<br>• Thermal evaporation<br>• Electron beam evaporation | • Simpler tool required | • Difficult to control stoichiometry<br>• Limited conformality |
| | Pulsed laser deposition (PLD) | • Epitaxial films possible<br>• High deposition rate<br>• Various stoichiometric films achievable<br>• Good thickness control | • Advance mechanism<br>• Angular distribution<br>• Large area uniformity limited<br>• Back sputtering<br>• High cost tool |
| | Molecular beam epitaxy (MBE) | • Epitaxial films possible<br>• Low deposition rate<br>• Good thickness control | |
| **Chemical** | Chemical solution deposition (CSD) | • Thicker films possible<br>• Simpler equipment, no vacuum<br>• High yield | • Precursors availability<br>• Nonuniformity<br>• thin films difficult to obtain |
| | Chemical vapor deposition (CVD)<br>• Plasma-enhanced CVD<br>• Metal-organic chemical vapor deposition (MOCVD)<br>• Metal-organic vapor phase epitaxy (MOVPE) | • Low temperature deposition possible<br>• Good step coverage | • High contamination<br>• Precursors availability |
| | Atomic layer deposition (ALD)<br>• Thermally atomic layer deposited (THALD)<br>• Plasma-enhanced ALD (PEALD) | • High conformality<br>• Self-limiting reaction gives thickness control<br>• Variety of precursors<br>• Thinner films possible | • Carbon contamination from metalorganic precursors<br>• Narrow temperature window |

## 6. Patterning of ferroelectric thin films

Patterning of thin films is commonly achieved using wet etching or dry etching. Wet etching utilizes liquid chemical solutions to dissolve and selectively remove the material based on etchant's reaction with the material [47]. Dry etching, also known as plasma etching, involves the use of reactive gases and plasma, which chemically reacts with the material resulting in its removal from the substrate. The key parameters are etch rates, selectivity with the substrate, masking material (soft or hard mask) and anisotropy. Reactive Ion Etching (RIE) is a dry etching technique widely used in semiconductor manufacturing. The use of reactive gases enables RIE to offer better control and precision over the etching process compared to other methods. The power and bias voltage applied to the electrodes in the RIE system significantly influence the plasma density and the energy of the ions accelerated toward the substrate. The ion bombardment in RIE allows better control of the etching direction compared to isotropic plasma etching. The direction control results in sharp and well-defined features with minimal undercutting. Inductively coupled plasma (ICP) RIE is an advanced RIE technique. In this configuration, an RF coil wrapped around the vacuum chamber generates the plasma instead of parallel electrodes. When the RF power is applied, an oscillating magnetic field is generated that creates a high-density plasma inside the chamber. This configuration enables faster etching rates and improved anisotropic etch. The choice between wet and dry processes depends on the specific material stack, the requirements of the fabrication process, and the desired results. Atomic layer etching (ALE) of ferroelectric HZO in combination with RIE has also been reported [41]. Table 10.3 provides some reported etch recipes for the three exemplary ferroelectric thin films discussed in this chapter under Section 9.

## 7. Characterization of ferroelectric films

As thin films, general scanning and transmission microscopic techniques equipped with elemental analysis are widely used. Reference [30] lists commonly used materials characterization techniques for ultra-thin ferroelectric films and an updated list is provided in Table 10.4. Crystalline structures are determined using grazing incidence X-ray diffraction (GIXRD) and film thicknesses are also measured using X-ray reflectometry (XRR) [58]. This section provides introduction to ferroelectric specific characterizations.

Table 10.3 Various etching techniques for ferroelectric thin films reported.

| Films | Dry etching | Wet etching | References |
|---|---|---|---|
| $HfO_2$, HZO | • Dry etching can be achieved with both chlorine- and fluorine-based chemistries. However, chlorine-based chemistries are usually preferred due to larger volatility of the etching by-products compared to their fluorine counterparts. Some examples of inductively coupled plasma (ICP) etching, are reported below:<br>○ (amorphous HZO) $Ar/Cl_2 = 1:3 — P = 5$ mTorr — $P_{RF} = 50$ W — $P_{ICP} = 700$ W → etching rate ≈ 13 nm/min.<br>○ ($HfO_2$) $SiCl_4/Cl_2 = 1:1 — P = 5$ mTorr — $P_{RF} = 30$ W — $P_{ICP} = 1000$ W → etching rate ≈ 20 nm/min.<br>○ (crystalline $HfO_2$) $BCl_3 — P = 5$ mTorr — $P_{RF} = 20$ W — $P_{ICP} = 800$ W → etching rate ≈ 22 nm/min. | • Hardly etch in diluted hydrofluoric acid (HF) solutions → highly concentrated HF solutions are needed; thus, posing challenges for selectivity toward other materials, such as silicon oxide.<br>• Non–HF–based solutions, such as SC1 or phosphoric acid ($H_3PO_4$), were found unsuccessful in completely removing $HfO_2$ films with a practical etching rate.<br>• Etching rate depends on thermal treatment and crystallographic structure of the film. In general, wet etching is possible if the film is in the amorphous state. | [48–53] |
| BST | • BST thin films are etched with inductively coupled $Cl_2/Ar, BCl_3/(Cl_2+Ar)$, and $CF_4/(Cl_2+Ar)$ plasmas. A chemically assisted physical etch of BST was experimentally confirmed by ICP under various gas mixtures. | • BST thin films deposited by chemical solution deposition are etched with a commercial BHF (65% pure) catalyzed with nitric acid $HNO_3$. A volume ratio of 70/30 for the $BHF/HNO_3$ solution was found to be a selective etching agent for the BST layers. | [54,55] |

*Continued*

Table 10.3 Various etching techniques for ferroelectric thin films reported.—cont'd

| Films | Dry etching | Wet etching | References |
|---|---|---|---|
| $Al_{1-x}Sc_xN$ | • Dry etching achieved with chlorine-based chemistries due to the larger volatility of etching by-products compared to their fluorine counterparts. For instance, inductively coupled plasma (ICP) etching, of epitaxial $Al_{0.84}Sc_{0.16}N$ was performed using $Cl_2:BCl_3:Ar = 2:1:1 - P = 5$ mTorr—$P_{RF} = 50$ W —$P_{ICP} = 200$ W—T $= 38°C$ → etching rate $\approx 12$ nm/min.<br>• The low $ScCl_3$ volatility is the $Al_{1-x}Sc_xN$ etching rate-limiting factor. | • Etching achievable with aqueous solution of potassium hydroxide (KOH), phosphoric acid ($H_3PO_4$) or tetramethylammonium hydroxide (TMAH). For instance:<br>  ○ etching rate $\approx 60$ nm/min for $Al_{0.73}Sc_{0.27}N$ in $H_3PO_4:H_2O = 4:1$ at 80°C.<br>  ○ etching rate $\approx 520$ nm/min for $Al_{0.73}Sc_{0.27}N$ in $TMAH:H_2O = 1:3$ at 80°C. (Etching rates refer to film in N-polarity. M- polar surfaces do barely etch)<br>• Presence of areas with highly anisotropic etch rate (due to defects in the crystal structure and slow etching facets) result in con-like residues → extensive overetching required to remove these residuals. | [56,57] |

**Table 10.4** Commonly used characterization techniques for ultrathin ferroelectric films.

| Characterizations | Techniques |
|---|---|
| Film thickness | Spectroscopic ellipsometer, X-ray reflectivity (XRR), XTEM |
| Structure, surface morphology | SEM, TEM, AFM, reflection high-energy electron diffraction (RHEED), low-energy electron diffraction (LEED) |
| Composition | SIMS, XRD, EDX, XPS, AES, EELS |
| Crystal structure | XRD, grazing angle XRD (GIXRD), electron backscatter diffraction (EBSD) |
| Electron structure | Photoluminescence (PL), cathodoluminescence (CL) |
| Ferroelectric domains | High angle annular dark-field (HAADF)-STEM aberration corrected TEM, PFM, EBSD |
| Ferroelectric properties | PFM, C-V, P-E hysteresis, PUND, FORC, fatigue cycling |

## 7.1 Hysteresis P-E loop

Compared to traditional dielectric films, ferroelectric films are quite complex, with a host of stress, temperature, and field-dependent parameters that modulate their behavior. These include electric field-dependent permittivity—hysteresis loop measurements, temperature and stress-dependent properties measurements and traditional thin film structural measurements—thickness, composition and crystalline structure. Over the past few decades, various methods for obtaining the electrical polarization loops of ferroelectric capacitors have been reported [59]; they include the Sawyer−Tower method (STM) [60], virtual ground method (VGM) [61], shunt method [62], constant current (CC) method [63−65] and triangular current (TC) method. In the dynamic current-voltage hysteresis measurement, a voltage pulse is applied such that the ferroelectric film switches first to one state then to the other, resulting in a complete picture of the hysteresis. For the ferroelectric material, the basic equation of total current I(t) is given by

$$I(t) = A\frac{dD(t)}{dt} + I_C = A\frac{dP(t)}{dt} + C\frac{dV(t)}{dt} + I_C \qquad (10.6)$$

Where $D(t)$ is the dielectric displacement, $P(t)$ is the ferroelectric polarization, $I_C$ is the conduction (leakage) current, $C$ is the parallel plate capacitance, $V(t)$ is the applied voltage, $A$ is the area, and $t$ is the time [10].

With triangular voltage waveform $dV/dt$ is constant and in the case of a negligible conduction current ($I_C$), the total current can be written as

$$I(t) = A\frac{dD(t)}{dt}; \quad D(t) = \frac{1}{A}\int I(t)\,dt \tag{10.7}$$

Using this hysteresis loops can be calculated from the switching current of a ferroelectric capacitor, as shown in Fig. 10.11.

Another measurement that is often used in characterization of ferroelectric materials is the positive-up negative-down, or PUND, measurement [66]. It uses successively the two positive and two negative electrical pulses which allows for capturing both the switching and nonswitching current characteristics of the film, the difference of which essentially cancels out nonferroelectric effects, giving a cleaner view of the ferroelectric properties of the film. First Order Reversal Curves (FORC) are polarization-field dependences described between saturation field $E_{sat}$ (electric field to reach saturation polarization) and a variable reversal field $E_r = (-E_{sat}, E_{sat})$. The FORC diagrams were proposed to describe some characteristics of the switching process in ferroelectrics [67]. The approach is related to the Preisach model which considers the distribution of the elementary switchable units over their coercive and bias fields. For in depth study of ferroelectric properties of thin films, data from polarization, leakage current, and FORC measurements need to be complemented and analyzed [68].

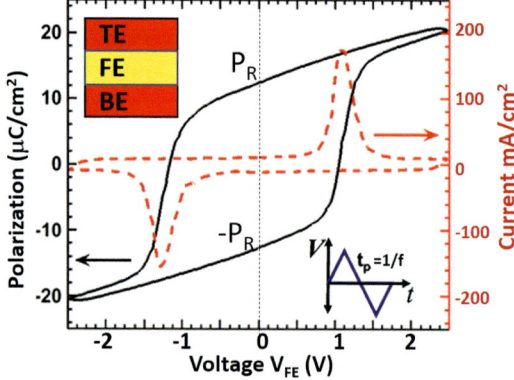

**Figure 10.11** Measured current, capacitance and polarization voltage curves of a metal-ferroelectric-metal (MFM, TiN/10nmHZO/TiN) capacitor with a triangular waveform obtaining the PV hysteresis loop.

To implement ferroelectric films in commercial devices the stability of these ferroelectric parameters over many switching cycles needs to be measured. A so-called fatigue measurement of a film consists of taking a series of hysteresis measurements while cycling the film in between with different pulse heights and widths [69]. The fatigue pulses are meant to simulate application-relevant waveform shapes with a specified rise, fall time and hold time. This allows the user to map the evolution of ferroelectric parameters over the entire cycling lifetime of a film information that can be used to choose an optimal film and design appropriate noise margins into the final application.

## 7.2 Piezoresponse force microscopy (PFM)

PFM is a functional atomic force microscopy (AFM) mode, which probes electromechanical material properties on the nanometer scale in addition to the sample topography [70,71]. As a conductive tip scans the surface in contact, an AC voltage is applied, which results in a surface oscillation with the same frequency as the applied voltage. If the sample domain polarization is oriented parallel to the applied electric field, the piezoelectric sample response will be in-phase with the excitation. The response will be out-of-phase if the domain polarization is oriented anti-parallel to the applied electric field. Thus, the phase-contrast between two oppositely oriented vertical domains is 180 degrees. The amplitude and phase of the sample oscillation are directly detected as deflection signal of the AFM cantilever, which is in contact with the surface, and read out via the lock-in amplifier. PFM has gained increasing recognition for the unique information it can offer on the electromechanical coupling characteristics of various materials, including actuators, sensors, and capacitors for modern communication technology. Common applications of PFM include: local characterization of electromechanical properties of ferroelectric materials, detailed domain mapping and study of domain switching dynamics, visualized in the material's ferroelectric hysteresis. Fig. 10.12 shows a PFM study done on Al doped $HfO_2$ to explore its ferroelectric behavior. PFM measurements for doped $HfO_2$ samples are performed on thin top electrode films to reduce the impact of interface charges [72].

The standard optical beam PFM has been modified to interferometric displacement sensor (IDS) PFM measurement. The IDS PFM technique eliminates long-time electrostatics and cantilever resonance artifacts [73]. Along with IDS, scanning capacitance microscopy (SCM) provides another advanced scanning probe technique to conclusively probe ferroelectricity in ultrathin ferroelectrics [74].

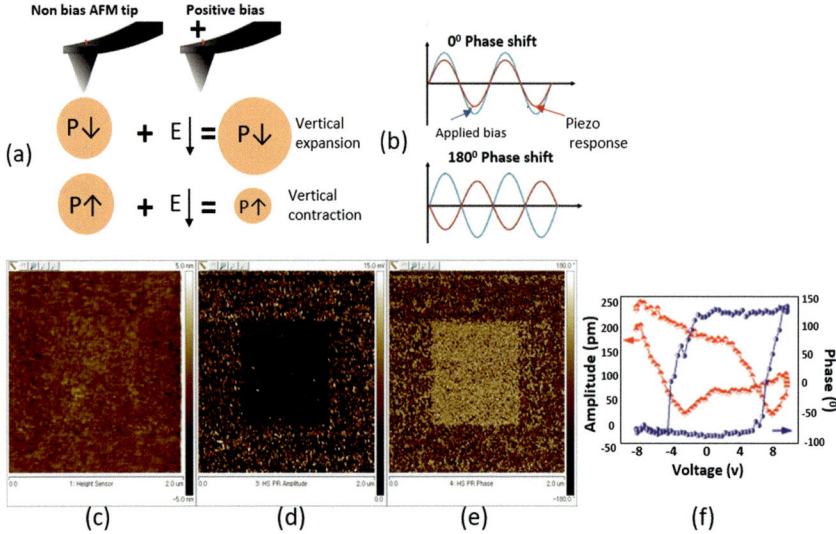

Figure 10.12 PFM working principle: Ferroelectric domains with (a) downward polarization and upward polarization; (b) an AC voltage between the sample and the conductive cantilever introduces a periodic piezoelectric response of the sample depending on the polarization orientation of the domain below the tip, the piezoresponse signal is in phase or 180 degrees out-of-phase with the excitation; (c) AFM topographic image of the Al:HfO$_2$ film deposited by ALD, annealed at $T_g = 800°C$, 60s. (d), (e) out-of-plane PFM amplitude, (e) phase image after applying −8 V voltage; (f) PFM amplitude and phase hysteresis loops.

## 7.3 The corona-Kelvin technique

The conventional characterization of dielectrics requires fabrication of capacitor test structures by patterning electrodes which limits inline rapid feedback characterization of their ferroelectric properties. A noncontact approach is to use corona charge-based characterization that does not require test structure fabrication [75]. Similar corona-Kelvin metrology has been applied to a broad range of high-k gate dielectrics, including HfO$_2$, ZrO$_2$, Al$_2$O$_3$, and SiON on silicon wafers [76,77]. The technique has demonstrated high precision and high repeatability in measuring key dielectric and interfacial parameters for dielectric films with and without interfacial SiO$_2$ layers, and for dielectric stacks [78]. The corona-Kelvin metrology is used for characterization of the ferroelectric behavior of thin films, as seen in Fig. 10.13. The corona charging pulses serve as noncontact poling that induces the ferroelectric property. Measurement on wafers with a free top dielectric surface enables access to any surface site and acquisition of whole wafer maps of the ferroelectric properties.

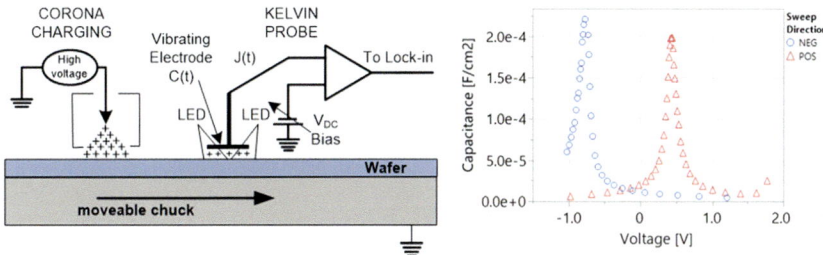

**Figure 10.13** Noncontact corona-Kelvin metrology schematic (left); It consists of a corona charge source, point source electrode and Kelvin probe; A capacitance voltage plot of a PZT ferroelectric thin film with up and down polarization (right) [75].

## 7.4 Other ultrathin advanced techniques

In recent years, scanning transmission electron microscope (STEM) has become an effective way to characterize the grain structure and domain walls (DWs) in ultra-thin ferroelectric films. High angle annular dark field (HAADF) mode detects heavier atoms while integrated differential phase contrast (iDPC) STEM enables imaging lighter atoms and their displacement [79]. Fig. 10.14 shows the HAADF-STEM image of W/5 nm HZO/W stack, showing two adjacent grains and a grain boundary enhanced by Inverse Fast Fourier Transformation (FFT).

In a recent study [79], the authors investigated domain wall evolution in TiN/HZO/TiN sample employing the HAADF-STEM image with the corresponding iDPC-STEM image showing the location of oxygen atoms

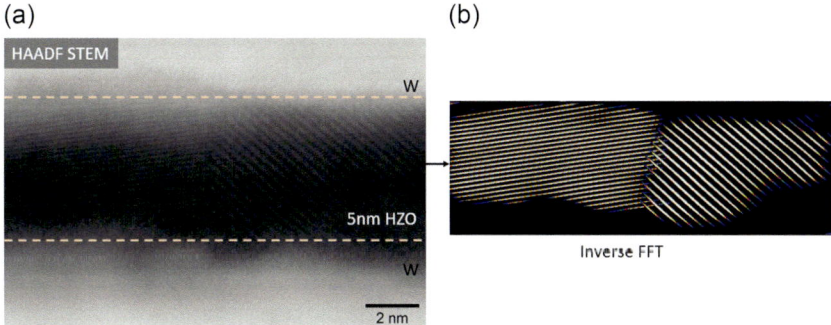

**Figure 10.14** (a) STEM image of W/HZO/W metal-ferroelectric-metal (MFS) structure captured in high angle annular dark field (HAADF) mode. The ferroelectric film shows a grain boundary with two adjacent grains; (b) inverse fast Fourier transformation (FFT) of the ferroelectric film where the orientation of the adjacent grains and their boundary are prominently visible. *(Image provided by Nashrah Afroze, Asif I. Khan, Georgia Tech, USA.)*

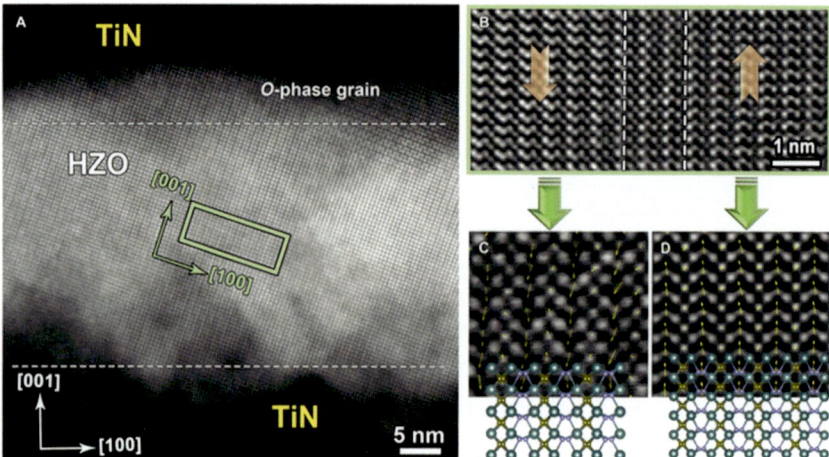

**Figure 10.15** (a) An atomically resolved HAADF-STEM image of the TiN/HZO/TiN stack after waking up. (b) iDPC-STEM image acquired from the green square area in (A), the $O_{II}$ atomic columns of the left image shifted along the [00$\bar{1}$] direction, and the $O_{II}$ atomic columns of the right image shifted along the [001] direction. The polarization mapping of the left image (c) and the right image (d) in (b). *(Reproduced from S. Zhang, Q. Zhang, F. Meng, T. Lin, B. Zeng, L. Gu, M. Liao, Y. Zhou, Domain wall evolution in Hf0. 5Zr0. 5O$_2$ ferroelectrics under field-cycling behavior, Research 6 (2023) 0093 with permission.)*

can be observed clearly (Fig. 10.15). From these images, the polarization mapping of the film with 180 degrees DW could be observed.

The other nonlinear optical technique, second harmonic generation (SHG) is used to investigate piezoelectric and ferroelectric thin films as the photon doubling process is allowed only in materials lacking inversion symmetry [80,81]. In addition, the local crystal phase and orientation of ferroelectric grains e.g., inside TiN/Hf0.5Zr0.5O2/TiN can be studied by electron backscatter diffraction (EBSD), where the local electron beam scattering results in Kikuchi patterns that can be analyzed [82].

## 8. Ferroelectric thin film applications

The multifunctionality of ferroelectric polarization response to an electric field, temperature, and stress offers their numerous applications. Some of the upcoming applications are summarized in this section and depicted in Fig. 10.16.

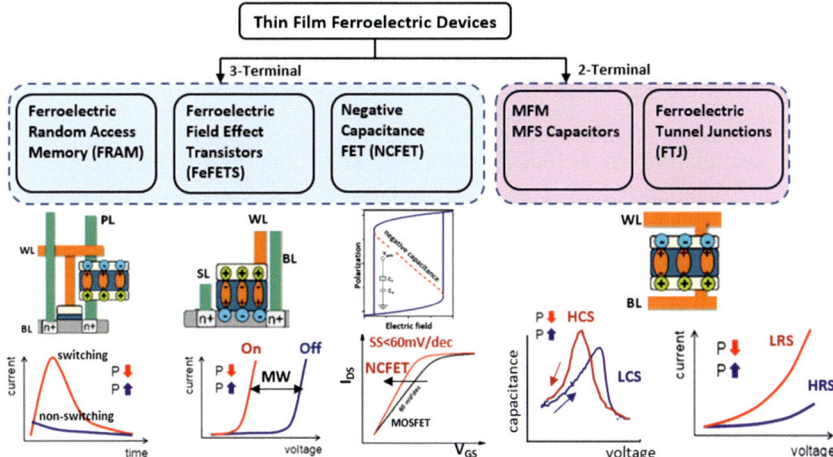

**Figure 10.16** Thin film ferroelectric devices. Three terminal devices include FRAM, FeFET and NC-FET. The polarization is read by switching or not in FRAM; threshold voltage shift in FeFETs; negative capacitance lowering the subthreshold slope below the Boltzmann limit of 60mV/dec in NC-FETs. Two terminal devices include MFM/MFS tunable capacitors and MFM ferroelectric tunnel junctions (FTJs). The capacitance and current flowing through the structure depends on the polarization direction respectively.

## 8.1 Ferroelectric devices

Due to the presence of two stable states, ferroelectric materials are particularly attractive for nonvolatile memory (NVM) data storage. Among various emerging NVM technologies, the simplest ferroelectric memory design, FRAM, integrates a metal/ferroelectric/metal capacitor into a back end of the line (BEOL) process, placing a conventional metal oxide field effect transistor (MOSFET) underneath each cell (1T-1C) [83]. To store data, an electric field switches the capacitor between the two up and down polarization states. FeRAM has been widely developed and commercialized [84] for PZT based ferroelectric capacitors. For doped $HfO_2$ first product type chips are available [85]. The main drawback of this memory is the destructive readout scheme. To sense the stored memory state, the polarization of the cell is reversed. When a voltage pulse is applied to the capacitor during reading, the polarization either changes or remains same, resulting in a different value of the current. Therefore, the state must be rewritten each time after reading, requiring a high endurance of the ferroelectric material [86]. Traditional perovskite ferroelectrics are incompatible with standard BEOL, CMOS-compatible processes [87,88]. As PZT-based materials are scaled to thinner layers, their ferroelectric properties degrade [86,89]. FeRAM reached its scaling limit at 130 nm node.

A ferroelectric FET (FeFET) is a MOSFET having the gate oxide as a ferroelectric material or a stack of dielectrics with one ferroelectric layer (MFMIS or MFIS) [90−92]. The threshold voltage of the FeFET is different for the different polarization states. The "on-state" is obtained by applying a positive pulse, the "off state" is obtained by applying a negative pulse. The ferroelectric polarization acts as a permanent gate bias, either positive or negative, depending on the polarization state. The shift in the $I_D$-$V_G$ characteristics is expressed as a memory window (MW). It is a critical parameter and is considered an important figure of merit of FeFETs as memory devices. The MW should be large enough to ensure a sufficient read margin and thus minimize the read error rate possibly caused by noise and disturb. It is given by $\sim 2E_C t_{Fe}$, where $t_{Fe}$ is the ferroelectric gate dielectric thickness when the remnant polarization is much larger than permittivity times the coercive field [93]. One of the advantages of this device is that the read-out is nondestructive. By sensing the source-drain resistance or current level using a voltage less than the coercive voltage but greater than the threshold voltage, it is possible to know the state of the device. Ferroelectric FETs (FeFETs) are in discussion to achieve gate voltage amplification through the negative capacitance (NC) effect of ferroelectric materials [94−96].

### 8.1.1 Variable capacitors: metal ferroelectric metal (MFM), metal ferroelectric semiconductor (MFS)

An MFM capacitor can be designed to exhibit asymmetric capacitance-voltage (C−V) curve due to which distinct memory states, high capacitance (HCS) and low capacitance (LCS) states at zero bias are observed. The asymmetry is attributed to oxygen vacancies that cause different domain wall pinning during the programming and erasing pulses. Asymmetry can further be enhanced by asymmetric stacks. To increase the HCS/LCS on/off ratio, an MFS stack is desired. A semiconductor (silicon) layer is added between the ferroelectric layer and the bottom electrode. The HCS state corresponds to the formation of an inversion layer and LCS state is dominated by the depletion capacitance. A high on/off ratio of $\sim 125$ is reportedly achieved [97].

## 8.1.2 Ferroelectric tunnel junctions: FTJ

An FTJ is a two-terminal nonvolatile memory device with two metal electrodes sandwiching a thin ferroelectric layer, allowing quantum mechanical tunneling through it [98–100]. The switchable polarization of the ferroelectric layer enables a significant change in tunneling electro resistance (TER) of FTJ devices. Unlike FRAM, FTJ has a nondestructive read-out process since the tunneling current is sensed and the polarization state in the ferroelectric switching layer is retained. In addition, FTJ devices offer a significant advantage with regard to superior scaling capability since scaling in these devices is not limited by the stored amount of polarization charge. It is also necessary for the two ferroelectric/metal interfaces to be asymmetrical to provide different potential barrier heights for different polarization orientations. FTJs based on binary oxides have been extensively explored but have TER values below 100 due to the limitations of the traditional FTJ modulation mechanism and intrinsic material properties. Importantly, the TER in both types of FTJ structure is constrained by the relatively small modulation of the tunneling barrier height [101].

The recent discovery of Sc-doped AlN (referred to as AlScN) as a ferroelectric presents a promising avenue for the realization of practical two-terminal ferroelectric nonvolatile memory devices [92,93]. AlScN exhibits a wurtzite structure similar to AlN and shows N polar or metal polar states. The key to ferroelectric switching is the energy barrier between these two polarization states, which can be lowered by increasing Sc substitutional doping that occupies Al sites or via strain engineering. AlScN shows large coercive fields, $E_C$, of 2–6.5 MV/cm, which enables scaling to thinner ferroelectric layers, while maintaining a large memory window. When combined with high remnant polarizations ($P_R$ of 80–120 μC/cm$^2$) this leads to significant resistive switching due to the strong tunnel barrier modulation, and thus a high on/off ratio [102,103].

## 8.2 In-memory and neuromorphic computing

FeFETs can be used for realizing multibit nonvolatile weight cells, which can store weights) for supporting in-memory computing, TCAM (ternary content addressable memories), FPGAs, in addition to their applications in logic integrated circuits [104]. The separation of computing units and memory in the traditional computer architecture constitutes energy-intensive data transfers creating the von Neumann bottleneck. Non-von-Neumann architectures by implementing deep neural networks (DNNs) or spiking neural networks (SNNs) for data-centric computing with higher energy efficiency and lower latency are being developed. Neural nets consist of multiple neurons and synapses. A neuron could consist of a memory cell with logic gates. The neurons are connected with a link called a synapse. Neural networks consist of three layers—input, hidden, and output, described in Fig. 10.17. In operation, a pattern is first written in a neuron in the input layer. The pattern is broadcast to the other neurons in the hidden layers. The minimum requirement for a synaptic device is that it can act as a programmable weight for Hebbian learning. The programmable weights should have multi-level resistance states [105]. In FeFETs, partial polarization switching can be induced by applying a voltage pulse train with either identical pulses or varying pulses of different amplitudes or pulse widths to the gate of the FeFET, such that a gradual tuning of the threshold voltage and subsequently the channel conductance can be realized. This enables the application of FeFETs as multi-state nonvolatile weight cells or synaptic memory elements—suitable for integration in dense memory cross-point arrays in neuromorphic systems [106].

In FTJs, the direction and degree of polarization affect the tunneling current that flows between two terminals. Since the FTJs are based on domain switching, they have the advantage of enabling nondestructive operation [107,108]. Therefore, based on this advantage, imitating synaptic plasticity through gradual conductance changes is possible. The research on neuromorphic applications with FTJs shows promise.

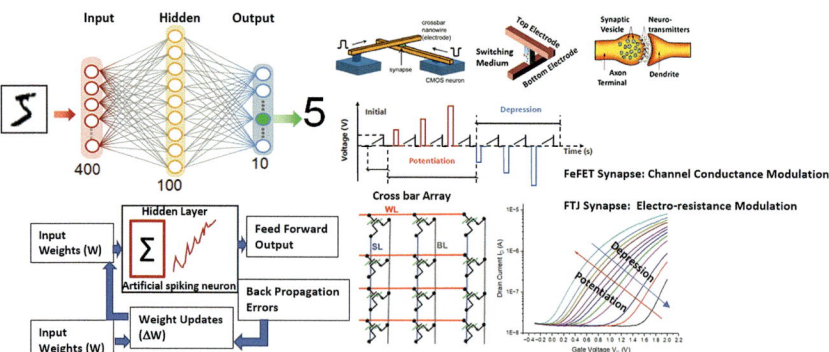

**Figure 10.17** (a) Schematic representation of the neural network architecture used for simulating the performance of FeFET-based synaptic devices. The input is an image of the hand-written digit. To simplify the hardware implementation task, the image is reshaped with a size of 20 × 20 pixels. Therefore, the neural network has 400 input layers, 100 hidden layers, and 10 output layers. (b) Memory array architecture shows the synaptic core used with FeFET/FTJ as synaptic weight devices mimicking the brain synapse (top right).

## 8.3 Pyroelectric, piezoelectric and energy harvesting

The coupled response between electrical energy and thermal energy is a feature present in all ferroelectric materials arising primarily from the temperature dependence of the polarization state. These pyroelectric properties have been identified as the potential conversion mechanisms enabling waste-heat scavenging [109] solid-state cooling, and power-beaming applications (Fig. 10.18). The pyroelectric effect in Si-doped $HfO_2$ thin films has been demonstrated as thermal-electric energy conversion [110]. Antiferroelectric materials are promising candidates for energy storage. A recent summary of the status of such applications has been provided in Ref. [111]. To utilize these properties in thin films is challenging due to lower signal levels. Different 3D architectures, System on Chip (SoC) are being investigated [112].

## 8.4 Tunable microwave devices

The tunable microwave devices such as tunable oscillators, phase shifters and resonators, fabricated by electrically tunable dielectric materials, have wide application prospects in the field of radar and communication [113]. Ferroelectric materials are widely used in RF and mmWave applications as tunable devices. Due to their high dielectric permittivity, they are providing low-loss operation at frequencies above 20 GHz, where they outperform conventional dielectric materials. The most exploited ferroelectrics so far are perovskite-based ferroelectrics, like barium strontium titanate (BST) and lead zirconium titanate (PZT), especially BST thin films. Strong electrically manipulated dielectric permittivity and low dielectric loss have made them excellent alternatives for tunable microwave devices.

Recently, it has been shown that the integration of ferroelectric $HfO_2$ in the back-end-of-line (BEOL) as a high-k ferroelectric capacitor is also

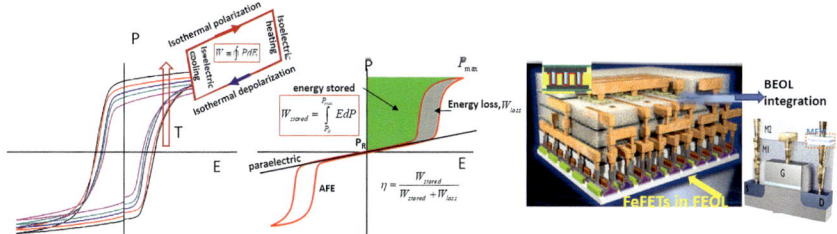

**Figure 10.18** Pyroelectric and energy storing applications. For high density vertical deep trenches will be desirable to integrate at the back end of the line (BEOL), while FeFETs can be implemented at the front end of the line.

feasible in advanced node CMOS technologies [114]. The ferroelectric can change its permittivity upon applied bias and performs as a varactor. Compared to the BST and PZT varactors, the main advantages of ferroelectric $HfO_2$ are its low annealing temperature, good manufacturing properties for etching and deposition, and low tuning voltages. This makes it compatible with advanced CMOS node implementation and hence will make it possible to design much higher frequency varactor-tuned low-power millimeter wave systems such as needed for 6G communications, imaging radar, or THz imaging.

## 9. Three exemplary ferroelectric films

This section presents three promising ferroelectric films—$HfO_2$ based, barium strontium titanate (BST), and AlScN deposited using different deposition techniques: atomic layer deposition (ALD), pulsed laser deposition (PLD) and molecular beam epitaxy (MBE). Fig. 10.19 shows the P-E loops and $P_R$-$E_c$ map of these films.

### 9.1 Hafnium oxide based ferroelectric films

A variety of different applications of doped $HfO_2$ and $ZrO_2$ films are discussed in the literature. Depending on the film composition, annealing and cooling conditions, film thickness, and substrate materials, crystallization of the deposited layer leads to different phases of the material. Multiple nonpolar centrosymmetric and polar noncentrosymmetric hafnia phases are reported. The most common ones are the nonpolar monoclinic (m-phase: space group $P2_1/c$; $\varepsilon_r \sim 20$) and tetragonal (t-phase: space group $P4_2/nmc$;

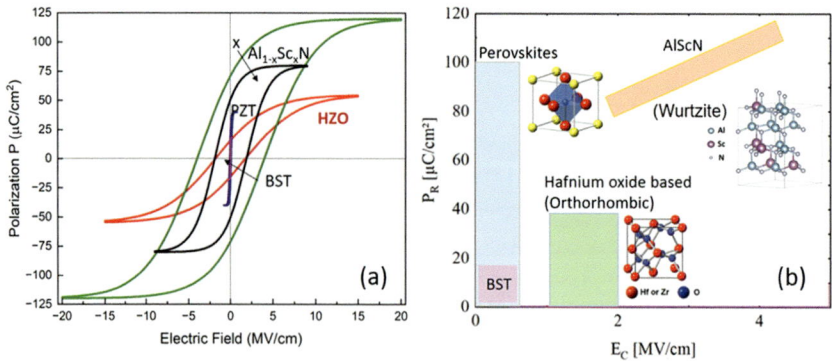

**Figure 10.19** Hysteresis loops (a) and PR-EC map (b) of different ferroelectric films.

$\varepsilon_r \sim 35$) phase if a dielectric property of the material is expected such as for simple capacitor or transistor applications. Doping $HfO_2$ with smaller or larger dopant atoms can lead to the polar orthorhombic (o-phase: space group Pbc2$_1$; $\varepsilon_r \sim 30$) or rhombohedral (r-phase: space group R3m; $\varepsilon_r \sim 30$) with ferroelectric properties as currently discussed for various semiconductor and sensor applications [115−117].

Here, dopant and oxygen vacancy content have counteracting effects. Typically, about 5% dopant content in $HfO_2$ (e.g., Si, Al, Y, La, Sr, …) causes a phase transition from the m-to the o-phase (Pbca), which can be reduced by increasing the oxygen content [118]. A wider process window was found for mixed $Hf_xZr_{1-x}O_2$ films ($0 < x < 1$), where ferroelectric properties are found in a 30%−70% Zr concentration range with the highest polarization values for a 50% $ZrO_2$ content in $HfO_2$. Other factors impacting the ferroelectric properties are the surface energy of grains and stress/strain in the layer [119]. Small grains (<10 nm) with high surface energy tend to cause a t-phase lattice structure in contrast to larger grains (>30 nm) that result in a primary m-phase. In between, the ferroelectric o-phase is preferred. Accordingly, the deposition of films with different thicknesses can lead to various phases of the layer. In addition, stress/strain in the layer due to densification during crystallization anneal, the lattice mismatch between the substrate material and the doped $HfO_2$ layer, and other factors can impact the phase formation.

Different deposition techniques [120] have been used to fabricate thin doped $HfO_2$ and $ZrO_2$ films. Depending on the application and required thickness, atomic layer deposition (ALD), chemical vapor deposition (CVD), physical vapor deposition by sputtering from a target (PVD), pulsed laser deposition (PLD), molecular beam deposition (MBD) or sol-gel deposition are utilized. For semiconductor applications, very uniform deposition with precise dopant control is necessary on 200- or 300-mm size wafers to fabricate capacitor or transistor devices with a low margin between different devices. As a result, the main focus is on ALD- and CVD-deposited films. Out of these two possibilities and different doped $HfO_2$ material options, the most commonly reported $Hf_{0.5}Zr_{0.5}O_2$ ALD is chosen for a detailed process discussion.

An ALD process consists of one ALD cycle that is repeated several times to achieve a thicker $Hf_{0.5}Zr_{0.5}O_2$ layer. In the first cycle, a metal-organic Hf precursor is typically pulsed into a process chamber, leading to a monolayer of the Hf precursor on the substrate surface. In a second pulse, an inert gas like $N_2$ or Ar purges the Hf-precursor out of the chamber. Due to the steric hindrance of the precursor molecules, only a fraction of the surface is covered now with precursor molecules. In the next step, an oxygen precursor (e.g., $H_2O$, $O_3$, $O_2$ plasma) is introduced into the chamber to remove ligands and form $HfO_2$. Multiple repetitions of this ALD cycle form a $HfO_2$ layer. Applying a 1:1 cycle ratio between $HfO_2$ and $ZrO_2$-based cycles, a mixed $Hf_{0.5}Zr_{0.5}O_2$ can be deposited. Typical deposition temperatures below the decomposition temperature of the precursor are in the range between 200–300°C. Different oxygen precursor dose times would lead to different oxygen content in the layers. For shorter dose times, a higher amount of hydrocarbon-containing ligands remain in the layer. Accordingly, a higher t-phase concentration is visible for short oxygen precursor dose times and a higher m-phase content for long dose times [118]. The optimum polar o-phase content was detected for medium dose times. Different metal-organic or halide Hf or Zr precursors can cause other film growth rates, changing carbon, hydrogen, halide contamination levels, film densities, and process windows.

In most cases, films are deposited as amorphous layers and crystalized in a subsequent step. Improved o-phase content was found when the $Hf_{0.5}Zr_{0.5}O_2$ layer was crystallized with a top electrode layer on the dielectric layer. Different crystallization temperatures are necessary depending on the used dopant in $HfO_2$ and layer thickness. If the dopant atomic radius is similar to Hf, a thermal budget of 500°C is required to form the o-phase. For small dopants like Si or large dopants like Sr, the annealing temperature needs to be increased above 800°C to establish the polar phase.

The most common electrode material is TiN, used as a top or bottom electrode. TiN can scavenge oxygen from the doped $HfO_2$ layer during anneal and cause the formation of oxygen vacancies. These vacancies again impact the phase formation process during anneal and, later on, the reliability behavior of the ferroelectric layer of the whole semiconductor device [117].

## 9.2 Barium–strontium–titanate (BST)

Ferroelectrics in the paraelectric phase are useful for realizing voltage tunable variable capacitors (varactors). In standard parallel plate capacitors, the dielectric layer is sandwiched between two electrodes in a metal-insulator-semiconductor (MIM) structure. Unlike conventional dielectrics used in the integrated circuit technology, the relative permittivity ($\varepsilon_r$) of a ferroelectric material is a nonlinear function of the electric field E.

Ferroelectrics in the polar phase are not considered for applications in tunable microwave devices because most ferroelectrics in polar phase are also piezoelectric with higher loss-tangent at microwave frequencies [113]. Domain wall movements may also cause additional losses at microwave frequencies.

In the paraelectric (nonpolar) phase, a ferroelectric thin film is characterized by a high dielectric permittivity, which depends strongly on temperature, applied external electric field and mechanical stress/strain introduced by the substrate or a buffer layer/superstrate. The dependence of the permittivity on the applied electric field in the paraelectric phase is the main characteristic used in frequency and phase agile microwave devices. A variable capacitor (varactor) is the basic building block for voltage tunable microwave circuits such as bandpass/band-reject filters, phase shifters and programmable delay lines. Thin film varactors have advantages over competing technologies in RF/microwave performance, reduced power consumption, sizes, and cost.

Barium–Strontium–Titanate ($Ba_xSr_{1-x}TiO_3$,(BST)), has shown promising material properties and ease of back end of the line integration. By mixing $BaTiO_3$ with $SrTiO_3$ a BST composition is created. Bulk single crystalline $BaTiO_3$ has a Curie temperature $T_C$ of 388 K, is in the ferroelectric phase at room temperature, and has a high tuning ratio. Solid solutions of $BaTiO_3$ and $SrTiO_3$ shift the transition temperature close to room temperature to tailor the tuning ratio and loss. Among the BST compositions, $Ba_{0.6}Sr_{0.4}TiO_3$, BST (60/40), is the most attractive one due to its $T_c \sim 285$ K, high tuning ratio, and low dielectric loss-tangent for microwave applications [121].

Several deposition techniques such as pulsed laser deposition (PLD), metal-organic chemical vapor deposition (MOCVD), sol-gel, sputtering, chemical solution deposition and molecular beam epitaxy (MBE) have been investigated for depositing thin films of BST. Achieving a large area deposition with low cost and high throughput is very significant. In this work, PLD technique is used for the deposition of BST thin films. A study was able to demonstrate deposition of large area BST thin films (up to 4″ in diameter) with uniform film thickness, grain size, crystal structure, orientation, and dielectric properties [121]. Due to its ease of operation, excellent stoichiometry transfer of the material from target to substrate, as well as the rapid deposition time, high quality crystalline films, epitaxial films, multilayer oxides can be achieved using PLD. Due to the advancements in the laser technology, high repetition rate and short pulse durations have made PLD a very feasible option to grow complex oxide films.

BST varactors in the MIM configuration utilize a metal stack of Ti + Au + Pt for the bottom metal and Ti + Pt + Au for the top metal. Metal stack thickness of a minimum of 750 nm is required for lower conductor losses and improving the overall Q. Pt metal provides the best interface for BST thin film varactors in preventing oxygen ion migration. A typical BST thin film varactor using a 250 nm thick BST layer provides a capacitance tuning of roughly 4:1, whereas a thicker film (>500 nm) provides a capacitance tuning >5:1 on sapphire substrates. 500 nm thick BST(60/40) provides the highest unbiased $\varepsilon_r$ of 800–900 at room temperature, and tunable by more than 5:1 with an applied bias of less than 20 V in the MIM configuration. To date, the best performance of BST thin film varactors has been on sapphire substrates.

Recently, BST thin films' dielectric properties have been enhanced by utilizing a samarium scandate ($SmScO_3$ henceforth SSO) substrate [122]. Enhanced domain wall resonances contributed to higher $\varepsilon_r$, high tuning ratio and higher quality factor (Q) for BST (80/20) thin films on SSO. Unfortunately, large area SSO substrates are not currently available. The author Guru Subramanyam's group studied the use of SSO thin film as a buffer layer for growth of BST on large area substrates such as sapphire. 200 nm thick SSO thin films were deposited on sapphire substrate at low temperature (<500°C) and annealed at 1000°C to crystallize the films. BST (60/40) films were deposited using the PLD process developed in our group at 720°C on the SSO buffer layer. 200 nm thick SSO buffered BST thin films showed higher $\varepsilon_r$, lower leakage currents and higher capacitance tuning ratio in varactors. MIM varactors built using the SSO buffered BST(60/40) thin films showed tuning ratio as high as 8:1 on sapphire substrates. The TEM study of the interface between SSO and BST showed excellent interface with a columnar growth of BST thin film on SSO and no interdiffusion between the layers. The enhancement of dielectric properties and tuning are most likely coupled to the strain induced by SSO layer on the BST layer.

## 9.3 Aluminum scandium nitride (AlScN)

The enhanced piezoelectric properties of aluminum scandium nitride ($Al_{1-x}Sc_xN$ or AlScN) were discovered in 2009 by Morito Akiyama's team [123]. Aluminum nitride (AlN) has been widely utilized as a thin film piezoelectric material to create filters since the roll out of third generation

(3G) wireless mobile communications [124]. For many decades AlN was thought to lack ferroelectric properties. Alloying AlN with scandium (Sc) to form scandium aluminum nitride ($Sc_xAl_{1-x}N$), originally pursued to increase the piezoelectric coefficients, gave rise to the discovery of ferroelectricity in $Sc_xAl_{1-x}N$ materials in 2019 [125]. Scandium is alloyed into AlN via a substitution of Sc for Al. The larger Sc atom distorts the lattice structure giving rise to an increase in the a-axis lattice constant and, for x > ~0.3, a reduction in the c-axis lattice constant [123].

Compared to well-established thin film ferroelectric materials, $Sc_xAl_{1-x}N$ is characterized by large remnant polarization, $P_R$, between 80–135 $\mu C/cm^2$, large coercive field, $E_C$, between 2 and 6.5 MV/cm and low dielectric relative permittivity between 9.8 and 22 [125–129]. Each of these properties can be tuned by changing the Sc alloying fraction, x. The reduction of c/a with increased Sc alloying leads to a decrease in $E_c$, while the reduction in c-axis dimension with Sc alloying for x > 0.3 leads to a reduction in $P_R$. AlN is a wurtzite crystal with a spontaneous polarization, while scandium nitride (ScN) is a cubic crystal lacking a spontaneous polarization and thus lacks ferroelectricity. For Sc alloying above x = ~0.43 cubic ScN phases begin to appear in the material and the properties are degraded [123]. Thin film stress also alters the ScAlN lattice parameters and has been shown to be another method for tailoring the ferroelectric properties [125]. Alternative III-N alloys such as Aluminum Boron Nitride ($Al_{1-x}B_xN$) [130] and $Sc_xGa_{1-x}N$ [131] ferroelectric thin films have been demonstrated and are also characterized by their large $P_R$ and $E_c$ similar to that of $Sc_xAl_{1-x}N$.

Ferroelectric $Sc_xAl_{1-x}N$ films have been demonstrated using sputtering [125–127], pulsed laser deposition (PLD) [132], and molecular beam epitaxy (MBE) [133]. Sputtering and MBE have both been utilized to deposit $Sc_xAl_{1-x}N$ materials with thicknesses less than 10 nm [134–137] and greater than 100 nm [125]. The most common approach for synthesis of $Sc_xAl_{1-x}N$ thin films is reactive sputtering in a nitrogen atmosphere. Both co-sputtering from separate Al and Sc targets and sputtering from alloyed AlSc targets have been demonstrated. Both methods can achieve Sc alloying for any composition between x = 0 and x ≥ 0.4. The underlying template is incredibly important for forming highly c-axis oriented $Sc_xAl_{1-x}N$ thin films using sputtering. The underlying template should present hexagonal symmetry with a hexagonal close pack (hcp) spacing similar to the a-axis lattice constant of $Sc_xAl_{1-x}N$. The (111) plane of face-centered cubic (fcc) metals such as aluminum (Al), platinum (Pt) and titanium nitride (TiN) are commonly employed templates as are (110) body-centered cubic (bcc) metals such as molybdenum (Mo) and tungsten (W). Typical sputter deposition temperatures are 300–450°C but ferroelectric $Sc_xAl_{1-x}N$ thin films have been sputter deposited at room temperature [138]. Sputtered $Sc_xAl_{1-x}N$ films have a spontaneous polarization as deposited and do not require subsequent annealing steps to demonstrate stable ferroelectricity. $Sc_xAl_{1-x}N$ films have demonstrated stable polarization retention up to 1100°C and ferroelectric measurements have been demonstrated at temperatures above 500°C [139–143]. The $E_C$ sharply declines with temperature, reaching half its room temperature value at 300°C while the $P_R$ remains relatively constant with temperature. The maximum achievable temperature where $Sc_xAl_{1-x}N$ exhibits stable ferroelectricity remains unknown. $Sc_xAl_{1-x}N$ grown by MBE has been demonstrated on both gallium nitride (GaN) and (002) Mo templates with Sc alloying over the range of x = 0 to x = 0.36 [131,137]. MBE growth is performed under nitrogen rich conditions at temperatures ranging from 400–900°C and does not require subsequent anneals.

## 10. Reliability of ferroelectric films

The main reliability factors in ferroelectric thin film are: (*i*) wake-up effect or rise of the switching polarization with cycling, which is observed at the initial cycling stage, (*ii*) fatigue effect, which is the cycle-by-cycle decrease in the remnant polarization value observed after a certain number of cycles, (*iii*) endurance limitations, (*iv*) imprint or the gradual shift of the hysteresis curve toward positive/negative electric field when upward (P↑)/downward (P↓) polarization state is stored, which is one of the major contributors to the (*v*) retention loss phenomenon.

Table 10.5 and Fig. 10.20 classify these in terms of their effects.

**Table 10.5** Reliability properties of ferroelectrics.

| Favorable | Unfavorable |
|---|---|
| Endurance | Fatigue |
| Retention | Aging, imprint, leakage current, breakdown |

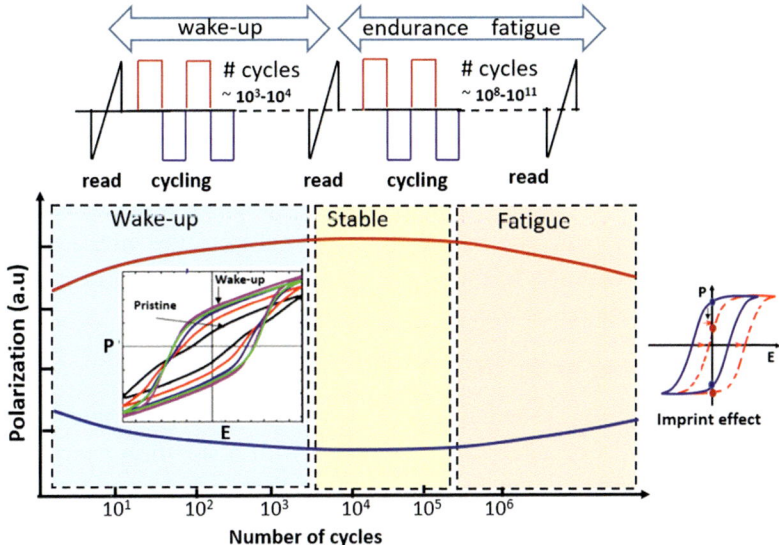

**Figure 10.20** Wake-up, endurance, and fatigue measurements using cycling to determine the endurance and retention. The imprint effect causes horizontal shift in the P-E loop.

## 10.1 Wake-up

The wake-up effect is attributed to an increase in remnant polarization observed in ferroelectrics with repeated initial cycling state of the order of $\sim 10^3$ cycles. This was first studied by Zhou et al. in ferroelectric Si:HfO$_2$ thin films [144]. This effect has also been observed in perovskite-structure ferroelectrics, [145]. The increase in $P_R$ is generally given as

$$P_{R,N} = P_{R,0} + A \ln N \tag{10.8}$$

$P_{R,0}$ and $P_{R,N}$ are the initial and after N cycles remnant polarizations respectively. Slope A represents the "wake-up" rate or the acceleration factor. It depends on the pulse amplitude and frequency. This effect is attributed to a number of mechanisms such as—domains previously perpendicular to the electrodes re-aligning to parallel and thus adding to the net polarization, migration or rearrangement of defect charges within the film which reduce an internal bias and free up previously pinned domains, or a crystal phase transformation in response to the applied field [146,147].

## 10.2 Fatigue

Polarization fatigue is the phenomenon in which the switchable polarization gradually decreases under an increasing number of electrical field cycles. With continued cycling, the polarization drop may be significant enough to affect device performance and is considered a serious issue when determining the lifetime of a particular device. Fatigue is often attributed to injection of charge into the ferroelectric material from the interfaces. This injected charge then acts to pin dipoles, preventing switching or inducing an internal bias that also reduces switching [148–150].

## 10.3 Field cycling endurance

Endurance is defined as the number of switching cycles achievable before a loss of $P_R$ by 50% or dielectric breakdown with a significant increase in leakage current. Endurance decreases with increasing fatigue pulse height, whereas the achievable $P_R$ value increases with increasing pulse height utilized for the P–E measurement. As a result, the $P_R$ and endurance are inversely proportional to each other owing to their different dependence on electric pulse height.

## 10.4 Retention

Retention quantifies the ability of a ferroelectric to retain the stored polarization state and is a critical property for nonvolatile memories. Loss of data can occur due to aging of the written state, or to the progressive development of global or local imprint. Any factor that produces a local change in the stability of the domain state, including injected charges, asymmetric electrodes, strain, grain boundaries, thermodynamically favored domain orientations, or defects can change the retention characteristics [151]. Retention is quantified in terms of both same-state and opposite-state retention; typically, the polarization drops with the wait time, $t_d$ given by

$$P_R = P_0 t_d^{-n} \qquad (10.9)$$

The polarization at 10 years is extrapolated as $P_R$ (10 years) = $P_0$ $(3.15 \times 10^8 \text{ s})^{-n}$. However, different models have been explored for short term, intermediate and long-term retention loss.

## 10.5 Imprint effect

Besides the increase of $P_R$, $E_C$ may also shift in a certain direction during the storage at a certain temperature, which is called an "imprint." This voltage shift in the polarization hysteresis loop causes an increase in operating voltage, degradation of retention characteristics, and write failures in ferroelectric memory devices [152]. Recently, it was reported that the imprint effect in a HfO2-based ferroelectric capacitor occurs during polarization retention, even at room temperature [153]. The time dependent imprint also occurs in ferroelectric-gate field effect transistors.

Several models have been proposed to describe mechanism of the imprint phenomenon; defect-dipole alignment in the internal electric field is the most likely cause in ferroelectric thin films. It is induced by charge trapping, interface layer between electrode and ferroelectrics, thermally and optically stimulated process [154].

Fig. 10.20 summarizes these effects and their measurement strategies.

## 10.6 Scaling

Polarization switching characteristics such as ferroelectric hysteresis loops, remnant polarization ($P_R$) and coercive fields ($E_C$) are adversely affected by thickness scaling [155,156].

Empirically, the coercive field needed to reverse the polarization in a ferroelectric increase with decreasing film thickness $E_C(d) \propto d^{-2/3}$ [157]. This semi-empirical scaling law has been used successfully to describe the thickness dependence of the coercive field in ferroelectric films ranging from 100 μm to 200 nm. The increase in $E_c$ values with decreasing film thickness suggests that the operation voltage for the ferroelectric memory devices should also increase with scaling down of the device. The critical size limit of electric polarization remains a fundamental issue in nanoscale ferroelectrics. Traditional perovskite oxides such as PZT have reached limitations in further scaling and are not suitable for integration in modern semiconductor processes. For applications in nanoscale devices, such as FeFETs and FTJs in high-density memory, films of approximately 5 nm or less are necessary. In very thin films depolarization field effects, leakage currents, the nonferroelectric interface regions at the electrodes start to dominate. Interfacial engineering is needed to be developed to suppress these issues [158–161].

In HfO$_2$ based ferroelectrics, that have ~ 10 times higher coercive field than the conventional perovskites, scalability is superior. 5.7 nm doped HZO layers are used in production type FeRAM memory arrays [85]. A 4-nm-thick epitaxial Hf$_{0.5}$Zr$_{0.5}$O$_2$ film on an epitaxial LSMO bottom electrode showing reasonable ferroelectric properties ($P_R \approx 2\ \mu C\ cm^{-2}$) has been reported [162]. Stabilized HZO films ~1.5 nm (~5 lattice constants) have been achieved by ALD [163].

## 11. Conclusions

This chapter provides a comprehensive overview on the vastly emerging field of ferroelectric thin films. Over the last decade, many new materials with novel ferroelectric properties have been invented and investigated. Being primarily complex oxides/nitrides, the material composition, crystallinity, domain structures, and physical properties are very challenging to optimize and integrate onto manufacturing platforms. Various deposition techniques for nanoscale thin film ferroelectrics have been developed. $HfO_2$-based ferroelectric thin films are promising for nonvolatile memory applications in both logic and memory. These films can be deposited via ALD using a manufacturable process for CMOS integration. Since nonferroelectric high-k $HfO_2$ has already been implemented in CMOS, ferroelectric $HfO_2$ nonvolatile FeFETs were integrated in the front end of the line and first product-type memory arrays are available [85]. Barium strontium titanate (BST) thin films have a pure perovskite structure and exhibit excellent unique ferroelectric characteristics. BST is widely studied and investigated due to its excellent properties including piezoelectric and electro optic properties providing variety of applications starting from memory devices to tunable capacitors and sensors, etc. BST thin films promise a powerful design paradigm toward on-chip integrations with diverse electronics into sensors via CMOS-compatible techniques. Pulsed laser deposition (PLD) technique is one of the most widely employed physical deposition techniques available for the deposition of multicomponent oxide thin films having high stoichiometry. Wurtzite-type materials like Aluminum scandium nitride (AlScN) with a high Sc content have shown ferroelectric properties, which provide a new option for CMOS-process-compatible ferroelectric memory, sensors, and actuators, as well as tunable devices. A coercive field of approximately 3 MV/cm and a large remnant polarization of more than 100 $\mu C/cm^2$ were exhibited on the Pt surface. AlScN-based films are promising candidates for nonvolatile memory applications and high-temperature sensors due to their outstanding functional and thermal stability exceeding most other ferroelectric thin film materials. Various deposition methods have been employed, with sputtering molecular beam epitaxy showing promising results. Here, issues like a breakdown field close to the coercive field need to be overcome.

## Acknowledgments

The authors would like to acknowledge Paul Jacob (RIT, NY), Malia Harvey and Eunsung Shin (Dayton, OH) involved in ferroelectric research. Special thanks to Nashrah Afroze and Asif I. Khan, Georgia Tech, USA for providing HAADF -STEM information. The authors acknowledge the supported by the National Science Foundation, Grant #EEC- 2123863 that fostered the collaboration between their groups. Any opinions, findings, and conclusions or recommendations expressed in this material are those of the author(s) and do not necessarily reflect the views of the National Science Foundation.

## References

[1] L.E. Cross, R.E. Newnham, History of Ferroelectrics, 1986.
[2] J.A. Gonzalo, G. Lifante, F. Jaque, 100 Years of Ferroelectricity: 1921-2021, World Scientific, 2022. ISBN 9811243093.
[3] Precedence Research: Report Code 2933, 2022 Ferroelectric Materials Market.
[4] R.G. Kepler, R.A. Anderson, Ferroelectricity in polyvinylidene fluoride, J. Appl. Phys. 49 (1978) 1232−1235.
[5] T.S. Böscke, J. Müller, D. Bräuhaus, U. Schröder, U. Böttger, Ferroelectricity in hafnium oxide thin films, Appl. Phys. Lett. 99 (2011).
[6] X. Sang, E.D. Grimley, C. Niu, D.L. Irving, J.M. LeBeau, Direct observation of charge mediated lattice distortions in complex oxide solid solutions, Appl. Phys. Lett. 106 (2015).
[7] J. Muller, T.S. Boscke, U. Schroder, S. Mueller, D. Brauhaus, U. Bottger, L. Frey, T. Mikolajick, Ferroelectricity in simple binary ZrO2 and HfO2, Nano Lett. 12 (2012) 4318−4323.
[8] T. Olsen, U. Schröder, S. Müller, A. Krause, D. Martin, A. Sıngh, J. Müller, M. Geidel, T. Mikolajick, Co-sputtering yttrium into hafnium oxide thin films to produce ferroelectric properties, Appl. Phys. Lett. 101 (2012).
[9] S. Mueller, C. Adelmann, A. Singh, S. Van Elshocht, U. Schroeder, T. Mikolajick, Ferroelectricity in Gd-doped HfO2 thin films, ECS Journal of Solid State Science and Technology 1 (2012) N123.
[10] J. Müller, U. Schröder, T.S. Böscke, I. Müller, U. Böttger, L. Wilde, J. Sundqvist, M. Lemberger, P. Kücher, T. Mikolajick, Ferroelectricity in yttrium-doped hafnium oxide, J. Appl. Phys. 110 (2011).
[11] T. Schenk, S. Mueller, U. Schroeder, R. Materlik, A. Kersch, M. Popovici, C. Adelmann, S. Van Elshocht, T. Mikolajick, Strontium doped hafnium oxide thin films: wide process window for ferroelectric memories, in: Proceedings of the 2013 Proceedings of the European Solid-State Device Research Conference (ESSDERC), IEEE, 2013, pp. 260−263.
[12] S. Starschich, D. Griesche, T. Schneller, U. Böttger, Chemical solution deposition of ferroelectric hafnium oxide for future lead free ferroelectric devices, ECS Journal of Solid State Science and Technology 4 (2015) P419.
[13] R. Chau, S. Datta, M. Doczy, B. Doyle, J. Kavalieros, M. Metz, High-/Spl kappa// metal-gate stack and its MOSFET characteristics, IEEE Electron. Device Lett. 25 (2004) 408−410.
[14] M.T. Bohr, R.S. Chau, T. Ghani, K. Mistry, The high-k solution, IEEE Spectr 44 (2007) 29−35.

[15] W. Yang, C. Yu, H. Li, M. Fan, X. Song, H. Ma, Z. Zhou, P. Chang, P. Huang, F. Liu, Ferroelectricity of hafnium oxide based materials: current statuses and future prospects from physical mechanisms to device applications, J. Semiconduct. 44 (2023) 1–45.
[16] H. Qiao, C. Wang, W.S. Choi, M.H. Park, Y. Kim, Ultra-thin ferroelectrics, Mater. Sci. Eng. R Rep. 145 (2021) 100622.
[17] M.I. Khan, T.C. Upadhyay, General introduction to ferroelectrics, Multifunctional Ferroelectric Materials 7 (2021).
[18] C. Kittel, Theory of antiferroelectric crystals, Phys. Rev. 82 (1951) 729.
[19] P.D. Lomenzo, L. Collins, R. Ganser, B. Xu, R. Guido, A. Gruverman, A. Kersch, T. Mikolajick, U. Schroeder, Discovery of nanoscale electric field-induced phase transitions in ZrO2, Adv. Funct. Mater. (2023), 2303636.
[20] F. Preisach, Über die magnetische nachwirkung, Z. Phys. 94 (1935) 277–302.
[21] S.L. Miller, R.D. Nasby, J.R. Schwank, M.S. Rodgers, P.V. Dressendorfer, Device modeling of ferroelectric capacitors, J. Appl. Phys. 68 (1990) 6463–6471.
[22] H.-T. Lue, C.-J. Wu, T.-Y. Tseng, Device modeling of ferroelectric memory field-effect transistor (FeMFET), IEEE Trans. Electron. Dev. 49 (2002) 1790–1798.
[23] A. Tselev, A.V. Ievlev, R. Vasudevan, S.V. Kalinin, P. Maksymovych, A. Morozovska, Landau-ginzburg-devonshire theory for domain wall conduction and observation of microwave conduction of domain walls, in: Domain Walls: From Fundamental Properties to Nanotechnology Concepts, 2020, pp. 271–292.
[24] L.D. Landau, I.M. Khalatnikov, On the anomalous absorption of sound near a second order phase transition point, Proceedings of the Dokl. Akad. Nauk SSSR 96 (1954) 25.
[25] V.C. Lo, Simulation of thickness effect in thin ferroelectric films using Landau–Khalatnikov theory, J. Appl. Phys. 94 (2003) 3353–3359.
[26] W. Zhang, K. Bhattacharya, A computational model of ferroelectric domains. Part I: model formulation and domain switching, Acta Mater. 53 (2005) 185–198.
[27] D.B. Litvin, Ferroelectric space groups, Acta Crystallogr. A 42 (1986) 44–47.
[28] L. Jin, F. Li, S. Zhang, Decoding the fingerprint of ferroelectric loops: comprehension of the material properties and structures, J. Am. Ceram. Soc. 97 (2014) 1–27.
[29] N. Setter, D. Damjanovic, L. Eng, G. Fox, S. Gevorgian, S. Hong, A. Kingon, H. Kohlstedt, N.Y. Park, G.B. Stephenson, Ferroelectric thin films: review of materials, properties, and applications, J. Appl. Phys. 100 (2006).
[30] Y. Wang, W. Chen, B. Wang, Y. Zheng, Ultrathin ferroelectric films: growth, characterization, physics and applications, Materials 7 (2014) 6377–6485.
[31] H.A. Hsain, Y. Lee, M. Materano, T. Mittmann, A. Payne, T. Mikolajick, U. Schroeder, G.N. Parsons, J.L. Jones, Many routes to ferroelectric HfO2: a review of current deposition methods, J. Vac. Sci. Technol. A 40 (2022).
[32] I. Bretos, R. Jiménez, J. Ricote, A.Y. Rivas, M. Echániz-Cintora, R. Sirera, M.L. Calzada, Low-temperature sol–gel methods for the integration of crystalline metal oxide thin films in flexible electronics, J. Sol. Gel Sci. Technol. (2023) 1–9.
[33] S. Starschich, T. Schenk, U. Schroeder, U. Boettger, Ferroelectric and piezoelectric properties of Hf1-XZrxO2 and pure ZrO2 films, Appl. Phys. Lett. 110 (2017).
[34] N.A. Shepelin, Z.P. Tehrani, N. Ohannessian, C.W. Schneider, D. Pergolesi, T. Lippert, A practical guide to pulsed laser deposition, Chem. Soc. Rev. 52 (7) (2023) 2294–2321.
[35] F. Craciun, T. Lippert, M. Dinescu, Pulsed laser deposition: fundamentals, applications, and perspectives, in: Handbook of Laser Micro-and Nano-Engineering, Springer, 2021, pp. 1291–1323.
[36] C.V. Varanasi, K.D. Leedy, D.H. Tomich, G. Subramanyam, Large area Ba1−XSrxTiO3 thin films for microwave applications deposited by pulsed laser ablation, Thin Solid Films 517 (2009) 2878–2881.

[37] M. Cavalieri, E. O'Connor, C. Gastaldi, I. Stolichnov, A.M. Ionescu, Experimental investigation of pulsed laser deposition of ferroelectric Gd: HfO2 in a CMOS BEOL compatible process, ACS Appl. Electron. Mater. 2 (2020) 1752–1758.
[38] R.L.A. Puurunen, Short history of atomic layer deposition: Tuomo Suntola's atomic layer epitaxy, Chem. Vap. Depos. 20 (2014) 332–344.
[39] M. Materano, C. Richter, T. Mikolajick, U. Schroeder, HfxZr1− XO2 thin films for semiconductor applications: an Hf-and Zr-ALD precursor comparison, J. Vac. Sci. Technol. A 38 (2020).
[40] A.G. Chernikova, D.S. Kuzmichev, D.V. Negrov, M.G. Kozodaev, S.N. Polyakov, A.M. Markeev, Ferroelectric properties of full plasma-enhanced ALD TiN/La: HfO2/TiN stacks, Appl. Phys. Lett. 108 (2016).
[41] N.A. Strnad, D.M. Potrepka, B.M. Hanrahan, G.R. Fox, R.G. Polcawich, J.S. Pulskamp, R.R. Knight, R.Q. Rudy, Extending atomic layer deposition for use in next-generation PiezoMEMS: review and perspective, J. Vac. Sci. Technol. A 41 (2023).
[42] R.K. Jha, P. Singh, M. Goswami, B.R. Singh, Plasma enhanced atomic layer deposited HfO 2 ferroelectric films for non-volatile memory applications, J. Electron. Mater. 49 (2020) 1445–1453.
[43] L. Mazet, S.M. Yang, S.V. Kalinin, S. Schamm-Chardon, C. Dubourdieu, A review of molecular beam epitaxy of ferroelectric $BaTiO_3$ films on Si, Ge and GaAs substrates and their applications, Sci. Technol. Adv. Mater. 16 (3) (2015) 036005.
[44] I. Fina, F. Sanchez, Epitaxial ferroelectric $HfO_2$ films: growth, properties, and devices, ACS Appl. Electron. Mater. 3 (2021) 1530–1549.
[45] Y. Otani, S. Okamura, T. Shiosaki, Recent developments on MOCVD of ferroelectric thin films, J. Electroceram. 13 (2004) 15–22.
[46] T. Shimizu, T. Yokouchi, T. Shiraishi, T. Oikawa, P.S.S.R. Krishnan, H. Funakubo, Study on the effect of heat treatment conditions on metalorganic-chemical-vapor-deposited ferroelectric Hf0. 5Zr0. 5O2 thin film on Ir electrode, Jpn. J. Appl. Phys. 53 (2014) 09PA04.
[47] L.A. Dinu, C. Romanitan, M. Aldrigo, C. Parvulescu, F. Nastase, S. Vulpe, R. Gavrila, P. Varasteanu, A.B. Serban, R. Noumi, Investigation of wet etching technique for selective patterning of ferroelectric zirconium-doped hafnium oxide thin films for high-frequency electronic applications, Mater. Des. 233 (2023) 112194.
[48] T. Mauersberger, J. Trommer, S. Sharma, M. Knaut, D. Pohl, B. Rellinghaus, T. Mikolajick, A. Heinzig, Single-step reactive ion etching process for device integration of hafnium-zirconium-oxide (HZO)/Titanium nitride (TiN) stacks, Semicond. Sci. Technol. 36 (2021) 095025.
[49] V. Lowalekar, S. Raghavan, Etching of zirconium oxide, hafnium oxide, and hafnium silicates in dilute hydrofluoric acid solutions, J. Mater. Res. 19 (2004) 1149–1156.
[50] P. Bodart, G. Cunge, O. Joubert, T. Lill, SiCl4/Cl2 plasmas: a new chemistry to etch high-k materials selectively to Si-based materials, J. Vac. Sci. Technol. A 30 (2012).
[51] E. Sungauer, E. Pargon, X. Mellhaoui, R. Ramos, G. Cunge, L. Vallier, O. Joubert, T. Lill, Etching mechanisms of Hf O 2, Si O 2, and poly-Si substrates in B Cl 3 plasmas, J. Vac. Sci. Technol. B 25 (2007) 1640–1646.
[52] J. Chen, W.J. Yoo, Z.Y.L. Tan, Y. Wang, D.S.H. Chan, Investigation of etching properties of HfO based high-K dielectrics using inductively coupled plasma, J. Vac. Sci. Technol. A: Vacuum, Surfaces, and Films 22 (2004) 1552–1558.
[53] D. Shamiryan, M. Baklanov, M. Claes, W. Boullart, V. Paraschiv, Selective removal of high-k gate dielectrics, Chem. Eng. Commun. 196 (2009) 1475–1535.
[54] S. Pavy, C. Borderon, S. Baron, R. Renoud, H.W. Gundel, Study of wet chemical etching of BaSrTiO 3 ferroelectric thin films for intelligent antenna application, J. Sol. Gel Sci. Technol. 74 (2015) 507–512.

[55] G.-H. Kim, K.-T. Kim, C.-I. Kim, Dry etching of (Ba, Sr) TiO3 thin films using an inductively coupled plasma, J. Vac. Sci. Technol. A 23 (2005) 894–897.
[56] S. Fichtner, Development of High Performance Piezoelectric AlScN for Microelectromechanical Systems: Towards a Ferroelectric Wurtzite Structure, BoD–Books on Demand, 2020. ISBN 3750431426.
[57] M.T. Hardy, B.P. Downey, D.J. Meyer, N. Nepal, D.F. Storm, D.S. Katzer, Epitaxial ScAlN etch-stop layers grown by molecular beam epitaxy for selective etching of AlN and GaN, IEEE Trans. Semicond. Manuf. 30 (2017) 475–479.
[58] K. Florent, Ferroelectric $HfO_2$ for Emerging Ferroelectric Semiconductor Devices (thesis), Rochester Institute of Technology, 2015. ISBN 1339323923 Accessed from, https://repository.rit.edu/theses/8889.
[59] D.-Y. Wang, C.-Y. Chang, Triangular current: method for measuring hysteresis loops of ferroelectric capacitors, Jpn. J. Appl. Phys. 43 (2004) 6225.
[60] C.B. Sawyer, C.H. Tower, Rochelle salt as a dielectric, Phys. Rev. 35 (1930) 269.
[61] A.M. Glazer, P. Groves, D.T. Smith, Automatic sampling circuit for ferroelectric hysteresis loops, J. Phys. 17 (1984) 95.
[62] F.L. Schloss, C. McIntyre, P. Polarization, Recovery of fatigued Pb (Zr, Ti) O 3 thin films: switching current studies, J. Appl. Phys. 93 (2003) 1743–1747.
[63] J.A. Giacometti, C. Wisniewski, W.A. Moura, P.A. Ribeiro, Constant current: a method for obtaining hysteresis loops in ferroelectric materials, Rev. Sci. Instrum. 70 (1999) 2699–2702.
[64] J.A. Giacometti, C. Wisniewski, P.A. Ribeiro, W.A. Moura, Electric measurements with constant current: a practical method for characterizing dielectric films, Rev. Sci. Instrum. 72 (2001) 4223–4227.
[65] D.-Y. Wang, C.-Y. Chang, Switching current study: hysteresis measurement of ferroelectric capacitors using current–voltage measurement method, Jpn. J. Appl. Phys. 44 (2005) 1857.
[66] J.J.T. Evans, The Relationship between Hysteresis and PUND Responses, Radiant Technologies, Inc, 2008.
[67] L. Mitoseriu, L. Stoleriu, A. Stancu, C. Galassi, V. Buscaglia, First order reversal curves diagrams for describing ferroelectric switching characteristics, Processing and Application of Ceramics 3 (2009) 3–7.
[68] J.D. Anderson, Measurement of Ferroelectric Films in MFM and MFIS Structures, Rochester Institute of Technology, 2017. ISBN 0355169347.
[69] S. Hong, Nanoscale Phenomena in Ferroelectric Thin Films, 2004.
[70] A. Gruverman, S.V. Kalinin, Piezoresponse force microscopy and recent advances in nanoscale studies of ferroelectrics, J. Mater. Sci. 41 (2006) 107–116.
[71] N. Balke, T. Schenk, I. Stolichnov, A. Gruverman, Piezoresponse force microscopy (PFM). Ferroelectricity in Doped Hafnium Oxide: Materials, Properties and Devices, 2019, pp. 291–316.
[72] N. Balke, P. Maksymovych, S. Jesse, A. Herklotz, A. Tselev, C.-B. Eom, I.I. Kravchenko, P. Yu, S. Kalinin, V differentiating ferroelectric and nonferroelectric electromechanical effects with scanning probe microscopy, ACS Nano 9 (2015) 6484–6492.
[73] L. Collins, Y. Liu, O.S. Ovchinnikova, R. Proksch, Quantitative electromechanical atomic force microscopy, ACS Nano 13 (2019) 8055–8066.
[74] S.S. Cheema, D. Kwon, N. Shanker, R. Dos Reis, S.-L. Hsu, J. Xiao, H. Zhang, R. Wagner, A. Datar, M.R. McCarter, Enhanced ferroelectricity in ultrathin films grown directly on silicon, Nature 580 (2020) 478–482.
[75] D. Marinskiy, P. Polakowski, P. Edelman, M. Wilson, J. Lagowski, J. Metzger, R. Binder, J. Müller, Ferroelectric HfO2 thin film testing and whole wafer mapping with non-contact corona-Kelvin metrology, Physica Status Solidi (A) 214 (2017) 1700249.

[76] D.K. Schroder, Semiconductor Material and Device Characterization, John Wiley and Sons, 2015. ISBN 0471739065.
[77] M. Wilson, D. Marinskiy, A. Byelyayev, J. D'Amico, A. Findlay, L. Jastrzebski, J. Lagowski, The present status and recent advancements in corona-Kelvin non-contact electrical metrology of dielectrics for IC-manufacturing, ECS Trans. 3 (2006) 3.
[78] M. Wilson, J. Lagowski, J. D'Amico, P. Edelman, A. Savtchouk, Investigation of high-K dielectric properties with the non-contact SASS technique, Proceedings of Electrochemical society PV 22 (2004) 425–440.
[79] S. Zhang, Q. Zhang, F. Meng, T. Lin, B. Zeng, L. Gu, M. Liao, Y. Zhou, Domain wall evolution in Hf0. 5Zr0. 5O2 ferroelectrics under field-cycling behavior, Research 6 (2023) 0093.
[80] X. Xu, F.-T. Huang, Y. Qi, S. Singh, K.M. Rabe, D. Obeysekera, J. Yang, M.-W. Chu, S.-W. Cheong, Kinetically stabilized ferroelectricity in bulk single-crystalline HfO2:Y, Nat. Mater. 20 (2021) 826–832, https://doi.org/10.1038/s41563-020-00897-x.
[81] S.A. Denev, T.T.A. Lummen, E. Barnes, A. Kumar, V. Gopalan, Probing ferroelectrics using optical second harmonic generation, J. Am. Ceram. Soc. 94 (2011) 2699–2727.
[82] M. Lederer, T. Kämpfe, R. Olivo, D. Lehninger, C. Mart, S. Kirbach, T. Ali, P. Polakowski, L. Roy, K. Seidel, Local crystallographic phase detection and texture mapping in ferroelectric Zr doped HfO2 films by transmission-EBSD, Appl. Phys. Lett. 115 (2019).
[83] J.S. Meena, S.M. Sze, U. Chand, T.-Y. Tseng, Overview of emerging nonvolatile memory technologies, Nanoscale Res. Lett. 9 (2014) 1–33.
[84] J. Rodriguez, K. Remack, J. Gertas, L. Wang, C. Zhou, K. Boku, J. Rodriguez-Latorre, K.R. Udayakumar, S. Summerfelt, T. Moise, et al., Reliability of ferroelectric random access memory embedded within 130nm CMOS, in: Proceedings of the 2010 IEEE International Reliability Physics Symposium, 2010, pp. 750–758.
[85] EeNews Analog, IEDM: Micron to Present DRAM-like Non Volatile Memory for AI, Available online: https://www.eenewseurope.com/en/iedm-micron-to-present-dram-like-non-volatile-memory-for-ai/. (Accessed 21 January 2024).
[86] P. Gao, Z. Zhang, M. Li, R. Ishikawa, B. Feng, H.-J. Liu, Y.-L. Huang, N. Shibata, X. Ma, S. Chen, Possible absence of critical thickness and size effect in ultrathin perovskite ferroelectric films, Nat. Commun. 8 (2017) 15549.
[87] Y. Arimoto, H. Ishiwara, Current status of ferroelectric random-access memory, MRS Bull. 29 (2004) 823–828.
[88] D.C. Yoo, B.J. Bae, J.E. Lim, D.H. Im, S.O. Park, S.H. Kim, U.-I. Chung, J.T. Moon, B.I. Ryu, Highly reliable 50nm-thick PZT capacitor and low voltage FRAM device using Ir/SrRuO/sub 3//MOCVD PZT capacitor technology, in: Proceedings of the Digest of Technical Papers. 2005 Symposium on VLSI Technology, 2005, IEEE, 2005, pp. 100–101.
[89] Y.S. Kim, D.H. Kim, J.D. Kim, Y.J. Chang, T.W. Noh, J.H. Kong, K. Char, Y.D. Park, S.D. Bu, J.-G. Yoon, Critical thickness of ultrathin ferroelectric BaTiO3 films, Appl. Phys. Lett. 86 (2005).
[90] S.L. Miller, P.J. McWhorter, Physics of the ferroelectric nonvolatile memory field effect transistor, J. Appl. Phys. 72 (1992) 5999–6010.
[91] H. Mulaosmanovic, E.T. Breyer, S. Dünkel, S. Beyer, T. Mikolajick, S. Slesazeck, Ferroelectric field-effect transistors based on HfO2: a review, Nanotechnology 32 (2021) 502002.
[92] H. Ishiwara, Ferroelectric random access memories, J. Nanosci. Nanotechnol. 12 (2012) 7619–7627.

[93] K. Toprasertpong, M. Takenaka, S. Takagi, Memory window in ferroelectric field-effect transistors: analytical approach, IEEE Trans. Electron. Dev. 69 (2022) 7113–7119.
[94] A.I. Khan, K. Chatterjee, B. Wang, S. Drapcho, L. You, C. Serrao, S.R. Bakaul, R. Ramesh, S. Salahuddin, Negative capacitance in a ferroelectric capacitor, Nat. Mater. 14 (2015) 182–186.
[95] D. Kwon, S. Cheema, N. Shanker, K. Chatterjee, Y.-H. Liao, A.J. Tan, C. Hu, S. Salahuddin, Negative capacitance FET with 1.8-nm-thick Zr-doped $HfO2$ oxide, IEEE Electron. Device Lett. 40 (2019) 993–996.
[96] I. Luk'yanchuk, A. Razumnaya, A. Sene, Y. Tikhonov, V.M. Vinokur, The ferroelectric field-effect transistor with negative capacitance, npj Comput. Mater. 8 (2022) 52.
[97] Z. Zhou, J. Leming, J. Zhou, Z. Zheng, Y. Chen, K. Han, Y. Kang, X. Gong, Experimental demonstration of an inversion-type ferroelectric capacitive memory and its 1 Kbit Crossbar array featuring high C HCS/C LCS, Fast speed, and long retention, in: Proceedings of the 2022 IEEE Symposium on VLSI Technology and Circuits (VLSI Technology and Circuits), IEEE, 2022, pp. 357–358.
[98] V. Garcia, M. Bibes, Ferroelectric tunnel junctions for information storage and processing, Nat. Commun. 5 (2014) 4289.
[99] M. Liehr, J. Hazra, K. Beckmann, V. Mukundan, I. Alexandrou, T. Yeow, J. Race, K. Tapily, S. Consiglio, S.K. Kurinec, Implementation of high-performance and high-yield nanoscale hafnium zirconium oxide based ferroelectric tunnel junction devices on 300 Mm wafer platform, J. Vac. Sci. Technol. B 41 (2023).
[100] J. Wu, H.-Y. Chen, N. Yang, J. Cao, X. Yan, F. Liu, Q. Sun, X. Ling, J. Guo, H. Wang, High tunnelling electroresistance in a ferroelectric van Der Waals Heterojunction via giant barrier height modulation, Nat Electron 3 (2020) 466–472.
[101] U. Sharma, G. Kumar, S. Mishra, R. Thomas, Ferroelectric tunnel junctions: current status and future prospect as a universal memory, Front Mater 10 (2023), 1148979.
[102] S. Fichtner, F. Lofink, B. Wagner, G. Schönweger, T.-N. Kreutzer, A. Petraru, H. Kohlstedt, Ferroelectricity in AlScN: switching, imprint and sub-150 Nm films, in: Proceedings of the 2020 Joint Conference of the IEEE International Frequency Control Symposium and International Symposium on Applications of Ferroelectrics (IFCS-ISAF), IEEE, 2020, pp. 1–4.
[103] K.D. Kim, Y.B. Lee, S.H. Lee, I.S. Lee, S.K. Ryoo, S. Byun, J.H. Lee, H. Kim, H.W. Park, C.S. Hwang, Evolution of the ferroelectric properties of AlScN film by electrical cycling with an inhomogeneous field distribution, Adv Electron Mater 9 (2023), 2201142.
[104] J. Ajayan, P. Mohankumar, D. Nirmal, L.M.I.L. Joseph, S. Bhattacharya, S. Sreejith, S. Kollem, S. Rebelli, S. Tayal, B. Mounika, Ferroelectric field effect transistors (FeFETs): advancements, challenges and exciting prospects for next generation non-volatile memory (NVM) applications, Mater. Today Commun. (2023) 105591.
[105] S. Thomann, H.L.G. Nguyen, P.R. Genssler, H. Amrouch, All-in-Memory brain-inspired computing using fefet synapses, Frontiers in Electronics 3 (2022) 833260.
[106] M. Jerry, P.-Y. Chen, J. Zhang, P. Sharma, K. Ni, S. Yu, S. Datta, Ferroelectric FET analog synapse for acceleration of deep neural network training, in: Proceedings of the 2017 IEEE International Electron Devices Meeting (IEDM), IEEE, 2017, pp. 2–6.
[107] T. Moon, H.J. Lee, S. Nam, H. Bae, D.-H. Choe, S. Jo, Y.S. Lee, Y. Park, J.J. Yang, J. Heo, Parallel synaptic design of ferroelectric tunnel junctions for neuromorphic computing, Neuromorph. Comput. Eng. 3 (2023) 024001.
[108] D. Kim, J. Kim, S. Yun, J. Lee, E. Seo, S. Kim, Ferroelectric synaptic devices based on CMOS-compatible HfAlO x for neuromorphic and reservoir computing applications, Nanoscale 15 (2023) 8366–8376.

[109] B. Hanrahan, C. Mart, T. Kämpfe, M. Czernohorsky, W. Weinreich, A. Smith, Pyroelectric energy conversion in doped hafnium oxide (HfO2) thin films on area-enhanced substrates, Energy Technol. 7 (2019), 1900515.

[110] K. Kühnel, M. Czernohorsky, C. Mart, W. Weinreich, High-density energy storage in Si-doped hafnium oxide thin films on area-enhanced substrates, J. Vac. Sci. Technol. B 37 (2019).

[111] A. Chauhan, S. Patel, R. Vaish, C.R. Bowen, Anti-ferroelectric ceramics for high energy density capacitors, Materials 8 (2015) 8009–8031.

[112] G. Wang, Z. Lu, Y. Li, L. Li, H. Ji, A. Feteira, D. Zhou, D. Wang, S. Zhang, I.M. Reaney, Electroceramics for high-energy density capacitors: current status and future perspectives, Chem. Rev. 121 (2021) 6124–6172.

[113] G. Subramanyam, M.W. Cole, N.X. Sun, T.S. Kalkur, N.M. Sbrockey, G.S. Tompa, X. Guo, C. Chen, S.P. Alpay, G.A. Rossetti, Challenges and opportunities for multi-functional oxide thin films for voltage tunable radio frequency/microwave components, J. Appl. Phys. 114 (2013).

[114] S. Abdulazhanov, Q.H. Le, D.K. Huynh, D. Wang, D. Lehninger, T. Kämpfe, G. Gerlach, THz thin film varactor based on integrated ferroelectric HfZrO2, ACS Appl. Electron. Mater. 5 (2022) 189–195.

[115] T. Mikolajick, S. Slesazeck, H. Mulaosmanovic, M.H. Park, S. Fichtner, P.D. Lomenzo, M. Hoffmann, U. Schroeder, Next generation ferroelectric materials for semiconductor process integration and their applications, J. Appl. Phys. 129 (2021).

[116] M.H. Park, Y.H. Lee, H.J. Kim, Y.J. Kim, T. Moon, K.D. Kim, J. Mueller, A. Kersch, U. Schroeder, T. Mikolajick, Ferroelectricity and antiferroelectricity of doped thin HfO2-based films, Adv. Mater. 27 (2015) 1811–1831.

[117] U. Schroeder, C.S. Hwang, H. Funakubo, Ferroelectricity in Doped Hafnium Oxide: Materials, Properties and Devices, Woodhead Publishing, 2019. ISBN 0081024312.

[118] M. Materano, P.D. Lomenzo, A. Kersch, M.H. Park, T. Mikolajick, U. Schroeder, Interplay between oxygen defects and dopants: effect on structure and performance of HfO2-based ferroelectrics, Inorg. Chem. Front. 8 (2021) 2650–2672.

[119] B. Xu, P.D. Lomenzo, A. Kersch, T. Schenk, C. Richter, C.M. Fancher, S. Starschich, F. Berg, P. Reinig, K.M. Holsgrove, Strain as a global factor in stabilizing the ferroelectric properties of ZrO2, Adv. Funct. Mater. (2023), 2311825.

[120] J.P.B. Silva, R. Alcala, U.E. Avci, N. Barrett, L. Bégon-Lours, M. Borg, S. Byun, S.-C. Chang, S.-W. Cheong, D.-H. Choe, Roadmap on ferroelectric hafnia-and zirconia-based materials and devices, Apl. Mater. 11 (2023).

[121] S.K. Dey, S. Kooriyattil, S.P. Pavunny, R.S. Katiyar, G. Subramanyam, Analyses of substrate-dependent broadband microwave (1–40 GHz) dielectric properties of pulsed laser deposited Ba0. 5Sr0. 5TiO3 films, Crystals 11 (2021) 852.

[122] M.C. Harvey, G. Subramanyam, E. Shin, Enhancements of BST thin film varactors with SmScO 3 buffer layer on sapphire substrate, in: Proceedings of the 2022 IEEE International Symposium on Antennas and Propagation and USNC-URSI Radio Science Meeting (AP-S/URSI), IEEE, 2022, pp. 952–953.

[123] M. Akiyama, T. Kamohara, K. Kano, A. Teshigahara, Y. Takeuchi, N. Kawahara, Enhancement of piezoelectric response in scandium aluminum nitride alloy thin films prepared by dual reactive cosputtering, Adv. Mater. 21 (2009) 593–596.

[124] A. Žukauskaitė, Editorial for special issue "piezoelectric aluminium scandium nitride (AlScN) thin films: material development and applications in microdevices." Micromachines 14 (2023) 1067.

[125] S. Fichtner, N. Wolff, F. Lofink, L. Kienle, B. Wagner, AlScN: a III-V semi-conductor based ferroelectric, J. Appl. Phys. 125 (2019).

[126] D. Wang, P. Musavigharavi, J. Zheng, G. Esteves, X. Liu, M.M.A. Fiagbenu, E.A. Stach, D. Jariwala, I.I.I.,R.H. Olsson, Sub-microsecond polarization switching in (Al, Sc) N ferroelectric capacitors grown on complementary metal−oxide−semiconductor-compatible aluminum electrodes, Physica Status Solidi (RRL)−Rapid Research Letters 15 (2021), 2000575.

[127] S. Yasuoka, T. Shimizu, A. Tateyama, M. Uehara, H. Yamada, M. Akiyama, Y. Hiranaga, Y. Cho, H. Funakubo, Effects of deposition conditions on the ferroelectric properties of (Al1− XScx) N thin films, J. Appl. Phys. 128 (2020).

[128] P. Wang, D. Wang, S. Mondal, M. Hu, J. Liu, Z. Mi, Dawn of nitride ferroelectric semiconductors: from materials to devices, Semicond. Sci. Technol. 38 (2023) 043002.

[129] K.-H. Kim, I. Karpov, R.H. Olsson III, D. Jariwala, Wurtzite and fluorite ferroelectric materials for electronic memory, Nat. Nanotechnol. (2023) 1−20.

[130] J. Hayden, M.D. Hossain, Y. Xiong, K. Ferri, W. Zhu, M.V. Imperatore, N. Giebink, S. Trolier-McKinstry, I. Dabo, J.-P. Maria, Ferroelectricity in boron-substituted aluminum nitride thin films, Phys. Rev. Mater. 5 (2021) 044412.

[131] D. Wang, P. Wang, B. Wang, Z. Mi, Fully epitaxial ferroelectric ScGaN grown on GaN by molecular beam epitaxy, Appl. Phys. Lett. 119 (2021).

[132] C. Liu, Q. Wang, W. Yang, T. Cao, L. Chen, M. Li, F. Liu, D.K. Loke, J. Kang, Y. Zhu, Multiscale modeling of Al 0.7 Sc 0.3 N-based FeRAM: the steep switching, leakage and selector-free array, in: Proceedings of the 2021 IEEE International Electron Devices Meeting (IEDM), IEEE, 2021, pp. 1−8.

[133] P. Wang, D. Wang, N.M. Vu, T. Chiang, J.T. Heron, Z. Mi, Fully epitaxial ferroelectric ScAlN grown by molecular beam epitaxy, Appl. Phys. Lett. 118 (2021).

[134] J.X. Zheng, M.M.A. Fiagbenu, G. Esteves, P. Musavigharavi, A. Gunda, D. Jariwala, E.A. Stach, R.H. Olsson, Ferroelectric behavior of sputter deposited Al0. 72Sc0. 28N approaching 5 Nm thickness, Appl. Phys. Lett. 122 (2023).

[135] G. Schönweger, M.R. Islam, N. Wolff, A. Petraru, L. Kienle, H. Kohlstedt, S. Fichtner, Ultrathin Al1− x Sc x N for low-voltage-driven ferroelectric-based devices, Physica Status Solidi (RRL)−Rapid Research Letters 17 (2023), 2200312.

[136] R. Mizutani, S. Yasuoka, T. Shiraishi, T. Shimizu, M. Uehara, H. Yamada, M. Akiyama, O. Sakata, H. Funakubo, Thickness scaling of (Al0. 8Sc0. 2) N films with remanent polarization beyond 100 MC Cm− 2 around 10 Nm in thickness, APEX 14 (2021) 105501.

[137] D. Wang, P. Wang, S. Mondal, M. Hu, D. Wang, Y. Wu, T. Ma, Z. Mi, Thickness scaling down to 5 Nm of ferroelectric ScAlN on CMOS compatible molybdenum grown by molecular beam epitaxy, Appl. Phys. Lett. 122 (2023).

[138] S.-L. Tsai, T. Hoshii, H. Wakabayashi, K. Tsutsui, T.-K. Chung, E.Y. Chang, K. Kakushima, Room-temperature deposition of a poling-free ferroelectric AlScN film by reactive sputtering, Appl. Phys. Lett. 118 (2021).

[139] M.R. Islam, N. Wolff, M. Yassine, G. Schönweger, B. Christian, H. Kohlstedt, O. Ambacher, F. Lofink, L. Kienle, S. Fichtner, On the exceptional temperature stability of ferroelectric Al1-XScxN thin films, Appl. Phys. Lett. 118 (2021).

[140] N. Wolff, M.R. Islam, L. Kirste, S. Fichtner, F. Lofink, A. Žukauskaitė, L. Kienle, Al1− XScxN thin films at high temperatures: Sc-dependent instability and anomalous thermal expansion, Micromachines 13 (2022) 1282.

[141] D. Drury, K. Yazawa, A. Zakutayev, B. Hanrahan, G. Brennecka, High-temperature ferroelectric behavior of Al0. 7Sc0. 3N, Micromachines 13 (2022) 887.

[142] R. Guido, P.D. Lomenzo, M.R. Islam, N. Wolff, M. Gremmel, G. Schönweger, H. Kohlstedt, L. Kienle, T. Mikolajick, S. Fichtner, Thermal stability of the ferroelectric properties in 100 Nm-thick Al0. 72Sc0. 28N, ACS Appl. Mater. Interfaces 15 (2023) 7030−7043.

[143] W. Zhu, J. Hayden, F. He, J.-I. Yang, P. Tipsawat, M.D. Hossain, J.-P. Maria, S. Trolier-McKinstry, Strongly temperature dependent ferroelectric switching in AlN, Al1-XScxN, and Al1-XBxN thin films, Appl. Phys. Lett. 119 (2021).
[144] D. Zhou, J. Xu, Q. Li, Y. Guan, F. Cao, X. Dong, J. Müller, T. Schenk, U. Schröder, Wake-up effects in Si-doped hafnium oxide ferroelectric thin films, Appl. Phys. Lett. 103 (2013).
[145] J. Liao, S. Dai, R.-C. Peng, J. Yang, B. Zeng, M. Liao, Y. Zhou, $HfO_2$-Based ferroelectric thin film and memory device applications in the post-moore era: a review, Fundamental Res. 3 (3) (2023) 332–345.
[146] A. Muliana, Time dependent behavior of ferroelectric materials undergoing changes in their material properties with electric field and temperature, Int. J. Solid Struct. 48 (2011) 2718–2731.
[147] E.D. Grimley, T. Schenk, X. Sang, M. Pešić, U. Schroeder, T. Mikolajick, J.M. LeBeau, Structural changes underlying field-cycling phenomena in ferroelectric $HfO_2$ thin films, Adv Electron Mater 2 (2016), 1600173.
[148] F. Huang, X. Chen, X. Liang, J. Qin, Y. Zhang, T. Huang, Z. Wang, B. Peng, P. Zhou, H. Lu, Fatigue mechanism of yttrium-doped hafnium oxide ferroelectric thin films fabricated by pulsed laser deposition, Phys. Chem. Chem. Phys. 19 (2017) 3486–3497.
[149] Y. Xu, Ferroelectric Materials and Their Applications, Elsevier, 2013. ISBN 1483290956.
[150] Y.A. Genenko, J. Glaum, M.J. Hoffmann, K. Albe, Mechanisms of aging and fatigue in ferroelectrics, Mater. Sci. Eng., B 192 (2015) 52–82.
[151] V. Mikheev, E. Kondratyuk, A. Chouprik, Retention model and express retention test of ferroelectric Hf O 2-based memory, Phys. Rev. Appl. 18 (2022) 064084.
[152] Y. Zhou, H.K. Chan, C.H. Lam, F.G. Shin, Mechanisms of imprint effect on ferroelectric thin films, J. Appl. Phys. 98 (2005).
[153] K. Takada, S. Takarae, K. Shimamoto, N. Fujimura, T. Yoshimura, Time-dependent imprint in Hf0. 5Zr0. 5O2 ferroelectric thin films, Adv Electron Mater 7 (2021), 2100151.
[154] W.-H. Kim, J.Y. Son, Y.-H. Shin, H.M. Jang, Imprint control of nonvolatile shape memory with asymmetric ferroelectric multilayers, Chem. Mater. 26 (2014) 6911–6914.
[155] J.Y. Park, K. Yang, D.H. Lee, S.H. Kim, Y. Lee, P.R. Reddy, J.L. Jones, M.H. Park, A perspective on semiconductor devices based on fluorite-structured ferroelectrics from the materials–device integration perspective, J. Appl. Phys. 128 (2020).
[156] M. Materano, P.D. Lomenzo, H. Mulaosmanovic, M. Hoffmann, A. Toriumi, T. Mikolajick, U. Schroeder, Polarization switching in thin doped $HfO2$ ferroelectric layers, Appl. Phys. Lett. 117 (2020).
[157] P. Chandra, M. Dawber, P.B. Littlewood, J.F. Scott, Scaling of the coercive field with thickness in thin-film ferroelectrics, Ferroelectrics 313 (2004) 7–13.
[158] K. Florent, S. Lavizzari, L. Di Piazza, M. Popovici, J. Duan, G. Groeseneken, J. Van Houdt, Reliability study of ferroelectric Al: $HfO_2$ thin films for DRAM and NAND applications, IEEE Trans. Electron. Dev. 64 (2017) 4091–4098.
[159] T. Mimura, T. Shimizu, H. Uchida, O. Sakata, H. Funakubo, Thickness-dependent crystal structure and electric properties of epitaxial ferroelectric Y2O3-HfO2 films, Appl. Phys. Lett. 113 (2018).
[160] S. Migita, H. Ota, H. Yamada, A. Sawa, A. Toriumi, Thickness-independent behavior of coercive field in HfO2-based ferroelectrics, in: Proceedings of the 2017 IEEE Electron Devices Technology and Manufacturing Conference (EDTM), IEEE, 2017, pp. 255–256.

[161] M.C. Sulzbach, S. Estandía, X. Long, J. Lyu, N. Dix, J. Gàzquez, M.F. Chisholm, F. Sánchez, I. Fina, J. Fontcuberta, Unraveling ferroelectric polarization and ionic contributions to electroresistance in epitaxial Hf0.5Zr0.5O2 tunnel junctions, Adv Electron Mater 6 (2020), 1900852.

[162] K. Toprasertpong, K. Tahara, Y. Hikosaka, K. Nakamura, H. Saito, M. Takenaka, S. Takagi, Low operating voltage, improved breakdown tolerance, and high endurance in Hf0.5Zr0.5O2 ferroelectric capacitors achieved by thickness scaling down to 4 Nm for embedded ferroelectric memory, ACS Appl. Mater. Interfaces 14 (2022) 51137−51148.

[163] Z. Gao, Y. Luo, S. Lyu, Y. Cheng, Y. Zheng, Q. Zhong, W. Zhang, H. Lyu, Identification of ferroelectricity in a capacitor with ultra-thin (1.5-nm) Hf0.5Zr0.5O2 film, IEEE Electron. Device Lett. 42 (2021) 1303−1306, https://doi.org/10.1109/LED.2021.3097332.

# CHAPTER 11

# Thin films in semiconductor memory

**S.B. Herner**
ASM International, Portland, OR, United States

## 1. Introduction

Semiconductor memory devices are a crucial element in the continued march of ever greater computing power. Since the commercial introduction of dynamic random access memory (DRAM) cells in 1970, and flash memory cells in 1987, the regularly decreasing size and cost per bit of these two memory cells has been a foundation of the incredible amount of computing power available in 2023 [1]. DRAM cells are volatile, meaning their memory is lost when power is removed. DRAM cells are characterized by their fast access times, near infinite endurance, and high cost. Flash memory cells are commonly called NAND, which is an acronym for the logic operation Not And. NAND cells are nonvolatile, meaning their memory is retained when power is removed. NAND cells are characterized by their slow access time, limited endurance, and very low cost. There are many other kinds of memory cells, including but not only Not Or (NOR), Resistance-change random access memory (ReRAM), Magnetic RAM (MRAM), ferroelectric RAM (FeRAM), and one time programmable (OTP) memory. DRAM and NAND are dominant commercially, accounting for over 90% of nonembedded revenue for the memory sector in 2021 [2]. Static RAM (SRAM) forms an important part of the hierarchy of memory in a typical computing system. SRAM is generally located onboard logic chips, and is formed from the same transistors used in logic functions. SRAM cells therefore do not require specialized films that are different than transistors used for logic operations, and so will not be discussed in this chapter.

Since the third edition of this book was published in 2012, 3D NAND, in which the memory cells are stacked on top of another, has become the dominant architecture, replacing 2D planar NAND. Similarly, since 2012, silicon nitride has become the most commonly used charge storage film in NAND, replacing earlier floating gate architectures with polysilicon charge storage films. It is projected that DRAM cells will similarly undergo 3D stacking in commercial devices as early as 2025, replacing current 2D planar devices. In another development since the third edition, a major commercial effort in ReRAM (crosspoint memory) has been discontinued by several manufacturers. Continued innovation in architecture and thin films in both DRAM and NAND will continue increasing bit density and decreasing bit cost in the coming years.

Thin film deposition development since the third edition was published has focused on atomic layer deposition, or ALD, for critical layers. A detailed discussion of the principles of ALD is undertaken in another chapter. This chapter will focus on the practical application of ALD in semiconductor memories. The ability of ALD deposition to deposit many novel materials with subatomic thickness accuracy into structures with very high aspect ratios has been critical to continued memory scaling. Other thin film development trends have been: decreasing deposition temperatures to enable the use of temperature-sensitive materials and avoid threshold voltage ($V_t$) shifts, and the development of hard masks to enable etching of very thick films, such as in 3D NAND stacks. The film stacks that require etching are now over 5 μm tall.

This chapter will primarily review thin film usage in DRAM and NAND memory cells. The sections on each will be organized in similar fashion: Each section will have a description of present-day memory cells, a discussion of the critical thin films for each, their methods of deposition, and the characteristics needed to satisfy device requirements. Each section will conclude with speculation on the future direction of the architecture for each type of memory cell. A brief section will review chalcogenide phase memory and ferroelectric memory before concluding.

Before a detailed discussion of each type of memory cell, a review of the hierarchy of semiconductor memory is useful. There are now several kinds of memory, each optimized not only for speed but also for density and cost. A typical hierarchy of memory before data enters the logic functions of the chip is shown below in Fig. 11.1 [3].

Data moves from nonvolatile memory into volatile memory, where the data is lost when power is removed from the device. Nonvolatile memory is typically stored in either NAND solid state drives or hard drives which use magnetic memory. NAND solid state drives have increasingly been

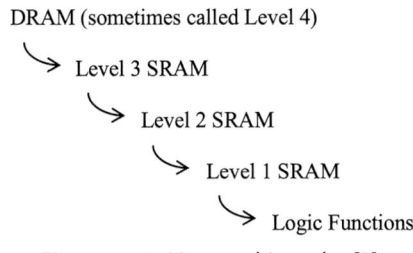

**Figure 11.1** Memory hierarchy [3].

replacing magnetic hard drives due to better access times and small form factors. Read latency is the time needed to "read" 1 bit of data from that particular memory. The read latency of typical hard drives or NAND flash is relatively slow at 85 μs [4]. It should be noted that the read latency can vary within each category (e.g., DRAM, NAND, SRAM). For example, single level cell (SLC) NAND, in which the memory cell holds 1 bit of data, can have a read latency of 16 μs read latency, while multilevel NAND cells (MLC), which hold two or more bits per cell, such as quad level cells (QLC), have longer read latencies . Present day quad level (QLC—4 bits per cell), have a 90 μs read latency [5,6]. Multilevel NAND cells are by far the dominant form commercially due to their low cost per bit.

DRAM usually draws memory from nonvolatile memory. DRAM cells are much faster than NAND cells, with current cells having a read latency typically around 100 ns [7]. Again, DRAM cells can have a range of read latencies, dependent on density and application. The data then moves from DRAM into the SRAM memory cells. Here again, there can be several different kinds of SRAM cells with different read latencies. Level 2 and 3 SRAM may be high density SRAM (HD-SRAM) cells, which is comprised of four transistors and two polysilicon resistors. HD-SRAM is denser but slower than high performance SRAM (HP-SRAM), which is comprised of six (or more) transistors and does not use resistors [8].

The hierarchy in Fig. 11.1 is characterized by increasing speed going from top-to-bottom, but also by increasing cost as shown in Table 11.1.

**Table 11.1** Memory cell performance and cost [1,9–12].

| Type | Read latency | Cost per Gb |
| --- | --- | --- |
| NAND (MLC) | 85 μs | $0.01 |
| DRAM | 100 ns | $0.20 |
| SRAM | 3 ns | $386 |

The last column showing cost per bit of each kind of memory is the system designer's optimization dilemma involving performance and cost. As of 2023, NAND single die capacity had reached 1 Tb while DRAM single die capacity had reach 32 GB. The size of SRAM varies greatly with application. For example, one of Intel's 7 nm-generation System On Chip (SOCs) had 30 MB of L3 cache (read latency 17 ns), 14 MB of L2 cache (read latency 14 ns), and 80 kB of L1 cache (read latency 1 ns) [9,10]. In general, each time data must be moved from one chip to another via wire bonds or microbumps, there is a speed penalty due to high resistance of these die-to-die interconnects [13]. Increasing use of heterogeneous integration in chip assemblies with low resistance hybrid bonds will allow better customization of the amounts of each kind of memory for particular applications, while minimizing the speed penalty for going "off-chip [13]." Heterogeneous integration describes the use of hybrid bonding to make direct chip-to-chip interconnects using copper thin films [14].

## 2. DRAM

The basic DRAM cell consists of a capacitor to store charge, serving as the state change element, and a transistor to read, program, and erase the data stored in the capacitor. In first embodiments in the 1970s, the capacitors were planar, or two dimensional (2D) devices. In the mid-1980s, three dimensional (3D) capacitors were implemented to maintain a large area for the capacitor and minimize the capacitor's footprint on the wafer. Two kinds of 3D capacitors were developed simultaneously: trench and stack. Trench capacitors are formed beneath the plane of the substrate by etching deep vias and filling them with dielectric and conductive films. Stack capacitors are formed above the plane of the substrate by etching deep vias in dielectric mold films deposited on top of the transistor. These vias are then filled with dielectric and conductive films similar to the trench capacitor. In the 2000s, the stack capacitor became the dominant architecture, although trench capacitors continue to be produced [8,15].

A typical modern DRAM cell is shown in isometric view to scale in Fig. 11.2 with its equivalent circuit diagram shown in Fig. 11.3 [16–18]. The transistor, often called an array or access transistor to distinguish from other transistors on the chip, is connected to the capacitor at the drain. The drains of other access transistors are interconnected, forming the bit line. The access transistor gate is connected to the gates of other access transistors, forming the wordline. A portion of an array of DRAM cells is shown in

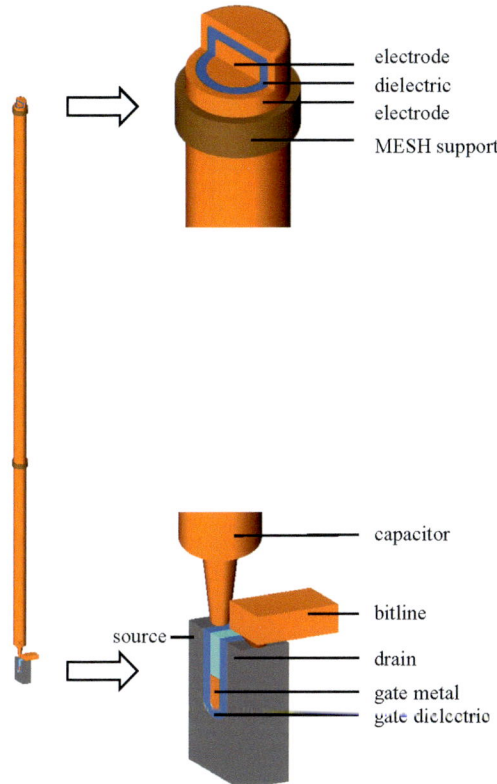

**Figure 11.2** DRAM cell with capacitor and BCAT access transistor drawn to scale, shown at left. Detail showing access transistor, shown at bottom right. Detail showing top of capacitor with cutaway and MESH support, shown at top right.

Fig. 11.4a, with the bit lines and wordlines connecting the cells in an array. This array forms a $6F^2$ cell, with one bit line pitch (3F) and one wordline pitch (2F) per cell. The honeycomb arrangement of the capacitors, shown in Fig. 11.4b, allows the tightest possible packing of the cylindrical capacitors.

## 2.1 Access transistor

The access transistor, shown in detail in Fig. 11.2b is referred to either as a Buried Channel Array Transistor (BCAT) or Recessed Channel Array Transistor (RCAT) in present day devices [19]. The peripheral transistors in DRAM chips, which are used to perform logic functions, are formed in different process steps than the those used to form the access transistors. In

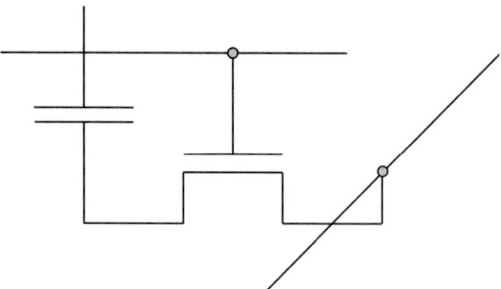

**Figure 11.3** Circuit diagram of a DRAM cell.

the BCAT, a trench is etched in epitaxial silicon between the doped source and drain. This effectively elongates the channel compared with its equivalent previous planar channel, avoiding the short channel effects in a compact two dimensional layout. Short channel effects contribute to the leakage of the cell in the OFF-state. Leakage in the OFF-state decreases the life time of the memory, requiring either more frequent refresh cycles or a larger capacitance. This architecture minimizes the footprint of the transistor while maintaining a lower leakage than would otherwise be possible.

The gate oxide of the access transistor has remained silicon oxynitride to maintain low leakage, while high speed logic transistors transitioned to high k films like hafnium oxide ($HfO_2$) many years ago [20]. These silicon oxynitride films continue to be deposited in batch furnaces at high temperatures. The gate metal is typically a combination of titanium nitride (TiN) and tungsten (W) to take advantage of the high work function of TiN (4.6 eV) and the lower resistance of W [21]. The titanium nitride and tungsten films are deposited by thermal ALD technique. The buried wordline shown in Fig. 11.4a isolates adjacent wordlines, reducing cell disturb and parasitic capacitance.

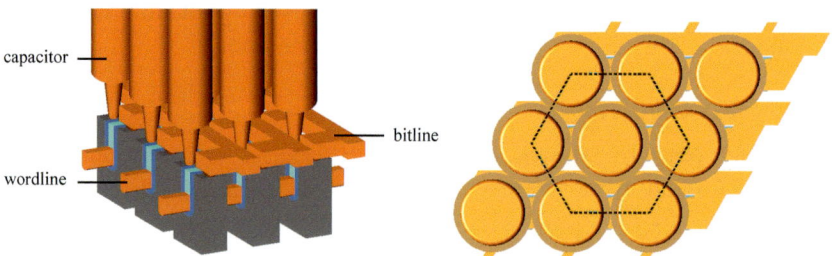

**Figure 11.4** (a) Array of DRAM cells (isometric view) and (b) top down view of capacitors in same array.

## 2.2 DRAM capacitor

The DRAM capacitor has undergone multiple evolutions in architecture, highlighted in cross sectional schematic in Fig. 11.5 [22]. These evolutions are driven by the need to maintain a high capacitance while shrinking the footprint of the capacitor. The capacitor shape fourth from the left is the roughened electrode surface produced by hemispherical silicon grains, a short-lived architecture that did not scale well [23]. Despite this capacitor architectural evolution yielding more capacitor area in a smaller footprint, the replacement of doped silicon electrodes with more conductive metals, and the adoption of capacitor dielectric materials with higher dielectric constant, the capacitance of the capacitor has steadily decreased, as shown in Fig. 11.6. This decrease is due to the reduction in area of the capacitor, driven by the need to reduce the footprint of the capacitor. The very high aspect ratio of modern capacitors has made them structurally unsound and difficult to manufacture.

The capacitance of the capacitor is determined by the equation:

$$C = \epsilon_0 k \frac{A}{t}$$

where $\epsilon_0$ is the vacuum permittivity, $k$ is the dielectric constant, $A$ is the area of the surface area between the dielectric and electrode, and $t$ is the thickness of the dielectric. The aspect ratios of present day DRAM capacitors exceed 100 as of 2023, and have become so tall that mesh support structures, shown in Figs. 11.2 and 11.5, were introduced beginning in 2004 to lend mechanical support to the capacitors before they are surrounded by passivation [18,24]. These mesh support structures bind an array of capacitors physically (not electrically).

**Figure 11.5** Evolution of DRAM capacitor shapes, shown in cross section, with early devices shown at left and later devices shown at right.

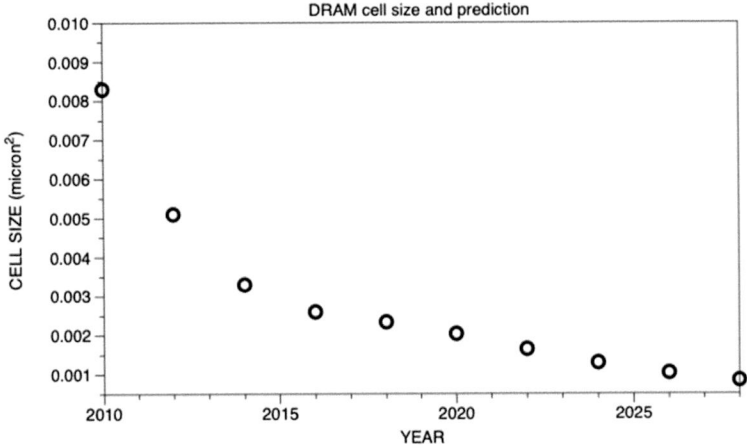

**Figure 11.6** DRAM cell size by year [22].

In present day DRAM cells the capacitance is $\sim$10 fF/cell and the leakage of the capacitor is $<10^{-7}$ A/cm$^2$. The ability to decrease the thickness of the dielectric layer is limited by leakage and the physical requirement to have a continuous film. Thickness is often expressed as equivalent oxide (SiO$_2$) thickness, or EOT.

$$\text{EOT} = \frac{3.9\, t}{k}$$

where 3.9 is the dielectric constant of SiO$_2$, $t$ is the thickness of the dielectric, and $k$ is the dielectric constant.

In 2023, EOT is generally <0.5 nm, and the favored dielectric is a stack of tetragonal ZrO$_2$/amorphous Al$_2$O$_3$/tetragonal ZrO$_2$ with $k$ values of 40/9/40, respectively, with a total thickness of <5 nm [25,26]. The ZrO$_2$ films provide the higher dielectric constant while the Al$_2$O$_3$ film maintains low leakage. These films can only be deposited by Atomic Layer Deposition (ALD) to achieve good step coverage over HAR structures. For ZrO$_2$ films, Zr(NEtMe)$_4$ and ozone are used as precursor and oxidant. For Al$_2$O$_3$ films, trimethyl aluminum (TMA) and ozone are used as precursor and oxidant. The films are deposited without the use of plasma to eliminate the possibility of plasma damage and to promote the best possible step coverage.

The electrodes for the capacitor are TiN, also deposited by ALD, typically using TiCl$_4$ or TdMat and NH$_3$. Titanium nitride has a relatively high work function, which is desirable to suppress leakage, and can be deposited as a smooth continuous film with good adhesion to dielectric film substrates.

The supporting structures, known as MESH, are made from silicon nitride, and are formed by a sequential deposition of SiN and various "mold" oxide films (BPSG, PSG, $SiO_2$). Modern capacitors are pillar structures. The pillar structure replaced the cup capacitor in 2018 to enable further footprint shrinking. As the pillar had 2× smaller area than the cup capacitor, the pillar capacitor again grew taller to mitigate capacitance reduction.

### 2.3 Future DRAM cells

Even with the introduction of FinFET cells and pillar capacitors, the scaling of DRAM cells has slowed dramatically in recent years (Fig. 11.6). The next leap in footprint reduction could be accomplished by the introduction of vertical access transistors [27]. Vertical access transistors have the channel oriented vertically, with the source and drain at the top or bottom of the channel, decreasing the footprint compared to horizontal channels. The vertical transistor can improve the drive current with a gate fully surrounding the channel. By placing a vertical epitaxial pillar underneath the capacitor, a $4F^2$ architecture is achieved (Fig. 11.7). An impediment to implementation is the floating body effect, in which reduces the threshold voltage and increases leakage in the off state. Increasing capacitance will help to reduce capacitance reduction caused by cell shrinkage.

New dielectric materials being explored for the capacitor that can increase the capacitance below 0.5 nm EOT include $TiO_2$ and $SrTiO_3$ [27]. Capacitor electrodes can also increase the capacitance by increasing the work function to suppress leakage. Ruthenium has a work function of 4.8 eV, which is higher than the work function of 4.56 eV of present day TiN electrodes. Achieving continuous Ru films of less than 5 nm thickness

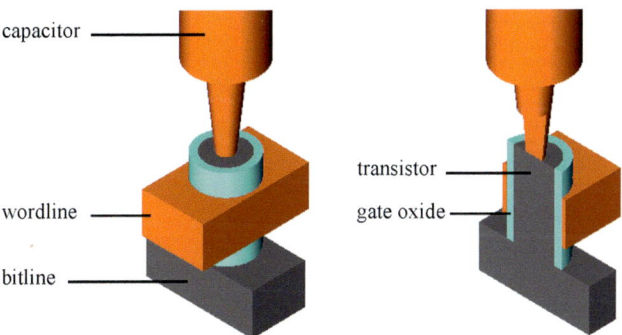

**Figure 11.7** Vertical transistor DRAM cell with cut out showing inner features at right.

has proven challenging, though, due to its high surface energy. Conductive oxides such as $RuO_2$ and $SrRuO_3$, which have even higher work functions than Ru, are also being examined. These conductive oxides are difficult to integrate because of their instability at elevated temperatures [16].

A more radical increase in the density of DRAM cells can be accomplished by stacking devices on top of one another. Device stacking is already accomplished by heterogeneously stacking DRAM die on top of one another and connecting by through silicon vias (TSVs), in an arrangement known as high bandwidth memory (HBM). As many as 12 die have been stacked and integrated into a chip assembly, as shown in Fig. 11.8 [28]. But even greater densities can be achieved by stacking DRAM cells homogeneously, such as done in NAND cells (see next section).

One proposed architecture for homogeneous DRAM cell stacking is shown in Fig. 11.9 [29]. The key to low cost 3D would be the development of multilayer patterning and formation. In this architecture, the capacitors retain their current cylindrical shape but are stacked like logs on top of one another. The access transistors are similarly stacked on top of one another. Access to the bit lines would be through a staircase pattern, similar to 3D NAND (see next section).

As the capacitors would have to be at least as long as they are tall now, the number of stacked DRAM cells most probably would number 64 or more for this architecture to be economically competitive with current 2D DRAM architecture. While 3D NAND has used thin film transistors, the

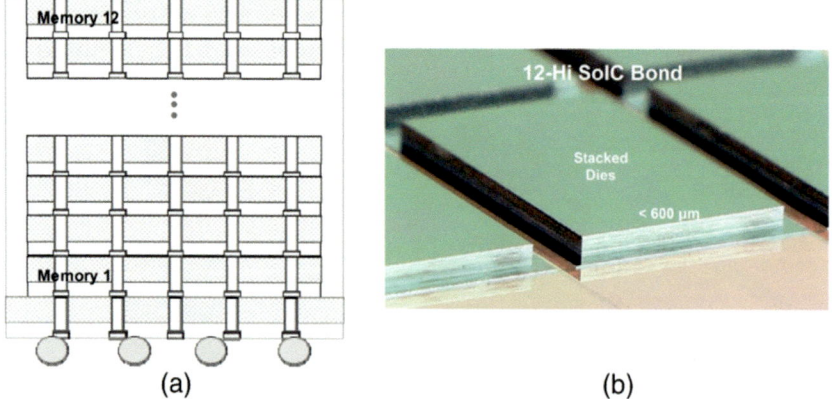

**Figure 11.8** (a) High bandwidth memory stacking of DRAM chips. Cross-sectional schematic showing 12 DRAM chips stacked and connected by TSVs and microbumps. (b) Optical isometric image of the same stack in a package [28].

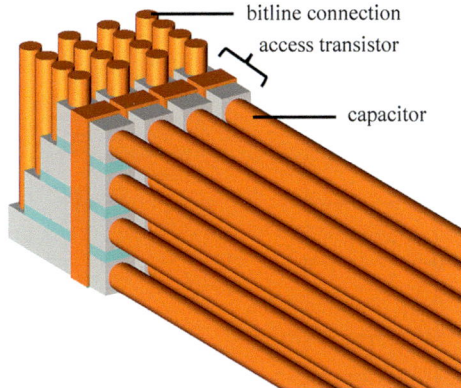

**Figure 11.9** One example of a proposed 3D DRAM architecture [29].

smaller read latency time and low leakage requirements makes single crystal transistors a more likely 3D DRAM feature. Stacking of epitaxial transistors on top of one another for 3D DRAM has been an active research topic and is expected to enter high volume manufacturing within the next 5 years.

## 3. NAND

The basic NAND cell comprises a transistor and a storage element in the gate dielectric stack of the transistor. In present day 3D NAND cells the storage element is usually a silicon nitride charge trap layer. Polysilicon ("floating gate") is the dominant storage element in 2D NAND cells, but 3D NAND volume is far greater than 2D NAND volume. NAND memory was commercially initiated by Samsung and Toshiba in 1989 as an evolution of NOR "flash" memory, invented at Toshiba in 1984 [1]. NOR memory is direct write, meaning each bit can be accessed independently. NAND memory exists as a string of connected transistors, electrically connected to one another by shared source/drains (see Fig. 11.10). NAND strings now comprise more than 100 storage transistors in addition to the select transistors at either ends of the NAND string [30]. With the long string, NAND basically requires only one contact per cell: the gate. NAND memory achieves a smaller footprint per single cell compared to NOR memory.

NAND memory began commercially in 1989 with 2D cells. The 2D NAND string is arranged horizontally using single crystal transistors. With a long evolution and attendant cell shrinkage, cell-to-cell disturb had increased to the point at the 14 nm mode where continued cell shrinkage

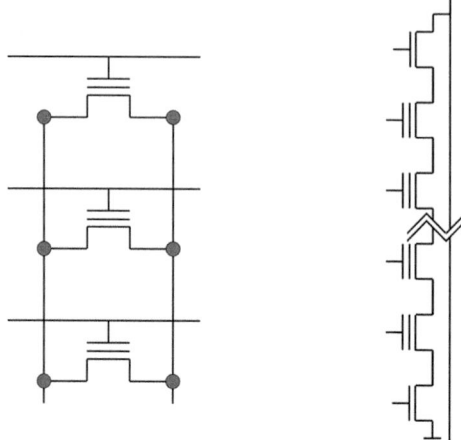

**Figure 11.10** Circuit diagram of NOR cells, left and NAND cells, at right.

was impossible, and 3D NAND was born [31]. In 2012, Samsung introduced 3D NAND cells wherein the string was vertical and comprised polysilicon thin film channels. The first commercial 3D NAND memory had 32 layers of cells. While Matrix Semiconductor pioneered 3D stacked memory devices commercially in 2003, 3D NAND improved on stacked device fabrication technology by using multi-layer patterning [32,33]. Multi-layer patterning describes fabricating many layers of devices in the same process step using a single patterning step, for example the deep etches of dozens of device layers that are a hallmark of 3D NAND fabrication. Multilayer patterning confers great cost advantage compared to patterning and fabricating each layer independently. Fewer interconnects in NAND also made for an easier path to 3D fabrication compared to NOR.

A single 3D NAND cell is shown in Fig. 11.11. The metal gates, or wordline, are arranged perpendicular to the transistor. Layers of interlayer dielectric (ILD) $SiO_2$ are on the top and bottom of the wordline. The channels, or bit lines, are perpendicular to the metal gates and are vertical relative to the wafer. Between the channel Si and the metal are tunneling SiON, storage $Si_3N_4$, blocking $SiO_2$ layers and blocking $Al_2O_3$. Inside the channel Si, described as macaroni-shaped, is filler $SiO_2$. The functions and characteristics of these films will be discussed in the next section.

An array of NAND cells is shown in Fig. 11.12 at left. To maintain some detail, the middle layers are not shown. Present day 3D NAND devices stack more than 200 layers high, with a total height exceeding

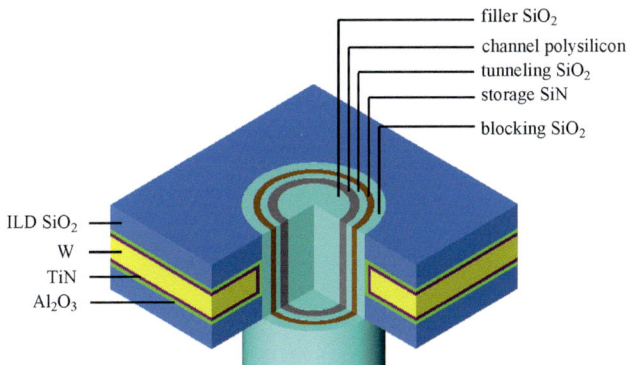

**Figure 11.11** Cut out of NAND cell.

10 μm. Much like DRAM cells, the channel vias are arranged in a hexagonal pattern, shown in top down view in Fig. 11.13 for maximum density.

The threshold voltage of the storage transistors depends on the amount of charge in the SiN charge storage film within the gate dielectric, which

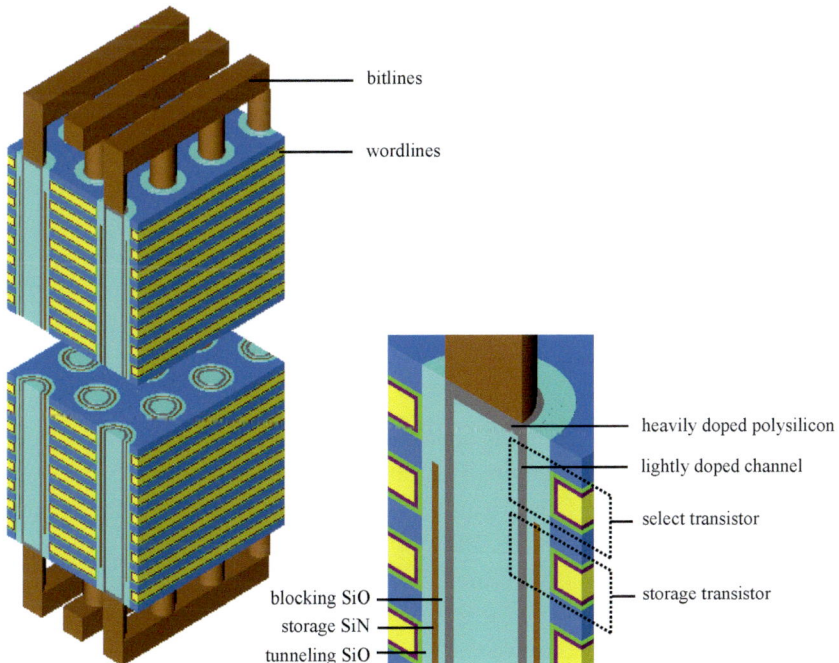

**Figure 11.12** Portion of NAND array at left. Detail of NAND array at right.

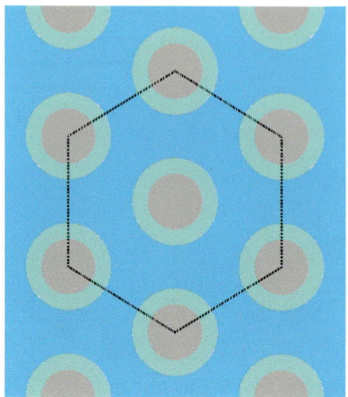

**Figure 11.13** Top down view of NAND strings showing hexagonal packing.

forms the basis of the memory. Charge is directed to the SiN film by turning on the transistor using a high gate bias, inducing Fowler-Nordheim tunneling through the tunnel oxide. The blocking $SiO_2$ and $Al_2O_3$ films largely prevent charge from leaking from the SiN film into the metal gate (TiN + W). The tunnel oxynitride composition and thickness is selected to allow sufficient charge to tunnel through to the SiN film at high gate bias (>18 V) while minimizing leakage during normal read bias (<5 V). The leakage needs to minimized to contain the charge stored on the charge trap SiN films for a period up to 10 years [4].

By varying the amount of charge in the SiN, multiple Vts can be created. Present day 3D NAND cell can hold up to 4 bits per cell. A 4 bit cell requires $4^2$ Vts, or 16 total Vts. Multilevel cells with so many different possible Vts requires not only careful engineering of the SiN charge trap film, but also careful programming and read verify functions to ensure the right Vt is achieved.

## 3.1 NAND transistor

The thin polysilicon channel is deposited into a via more than 5 μm deep and 100–200 nm in diameter. The film requires excellent step coverage and thickness control. The macaroni shape of the channel Si film reduces the volume of traps and the Vt of the transistor [24]. The channel is typically 7–11 nm thick. Thinner film cannot be deposited without occasional discontinuities. The saturation current of the transistor is partially determined by the thickness of the channel film, with thicker films having larger saturation currents [34].

Channel silicon films are typically deposited amorphous by low pressure chemical vapor deposition (LPCVD). Present day LPCVD reactors may deposit up to 150 wafers at a time, ensuring good productivity. These relatively simple deposition reactors have been a workhorse in the industry for many decades. Their simple design, using a quartz pressure vessel, quartz gas injectors, and either quartz or SiC "boats" which hold the wafers, ensure low cost [35]. For amorphous silicon deposition by LPCVD, $SiH_4$ is flowed at a pressure of 100–400 mTorr at temperature of 520°C or less [36].

With more than 100 transistors sharing the same channel that is many microns in length, it is crucial to deliver high-mobility channels. The amorphous film is crystallized slowly, with a typical crystallization temperature of 600–650°C for many hours. This crystallization anneal, referred to as solid phase crystallization (SPC), is designed to grow large polycrystalline grains. Large polycrystalline grains enable a higher carrier mobility than small polycrystalline grains, leading to larger saturation currents [34]. Depositing the film amorphous and crystallizing afterward ensures a smoother surface than depositing the film in the polycrystalline state. Smoother channel surfaces yield fewer trap sites and a higher carrier mobility.

A highly doped polycrystalline film is formed at either end of the long channel as shown in Fig. 11.12. These highly doped films allow for a low contact resistance contact to be made at either end of the string. These highly doped polycrystalline films are typically deposited by LPCVD using $SiH_4$ and a dopant-containing gas such as phosphine ($PH_3$) using similar deposition conditions for the amorphous channel silicon deposition. These are be deposited as amorphous films and converted to polycrystalline by a later anneal, similar to the channel silicon films. The last transistors at either end of the NAND string are referred to as select gates and do not store memory, unlike the other transistors in the string [4].

The bit lines and wordlines for each string of NAND cells are accessed independently. The bit lines are accessed by vertical vias that connect to horizontal metal lines (see Fig. 11.12). The wordlines for each layer are accessed by a structure called a staircase, shown below in Fig. 11.14. The staircase structure is achieved by printing a thick photoresist layer and then etching to form the first or lowest stair on the staircase [37]. After the first etch is finished, the photoresist mask is then shrunk by ashing, which shrinks all of the dimensions of the patterned photoresist by a similar amount. The etch is then repeated, forming the second lowest stair on the staircase, and so on. After staircase formation is complete, passivation oxide

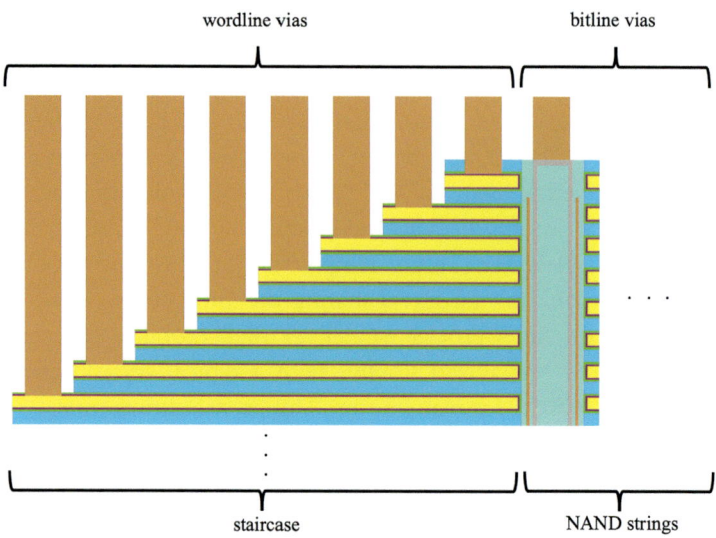

**Figure 11.14** Cross-sectional view of staircase.

is deposited over the staircase and planarized. Vias then connect each layer to interconnects above the array [33].

### 3.2 ONO storage layers

The critical films in the storage comprise the tunnel oxynitride layer, the charge trap layer, and the blocking oxide layers, often referred to collectively as the "ONO" layer. Combining multi bit storage with 3D stacking of layers has allowed for the fabrication of chips capable of storing 1 Tb of information. The ONO films are typically deposited in situ and in sequence in a single deposition tool to ensure high quality and low leakage. Historically these layers have been deposited by low pressure CVD (LPCVD) at a typical temperature of 750–800°C in a batch furnace, where 100 or more wafers can be deposited at one time. More recently the layers have been deposited at >700°C by atomic layer deposition (ALD). Atomic layer deposition can achieve better thickness and composition control in high aspect ratio (HAR) structures at lower temperature compared with LPCVD.

Present day tunneling dielectric films are SiON [38]. SiON tunnel dielectric films replaced SiO/SiN/SiO tunnel dielectrics due to improved retention and erase windows. Present day SiON tunnel dielectric films are 5–8 nm thickness with 5–15 at% N. The band diagrams of ONO layers with SiON tunnel dielectrics are shown in Fig. 11.15. The high quality needed in these layers requires relatively high deposition temperature.

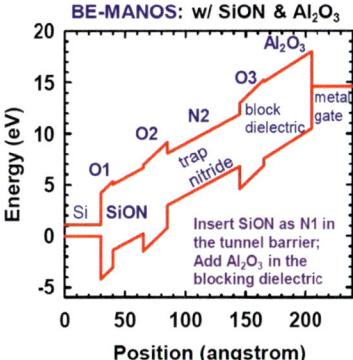

**Figure 11.15** Band diagram of SiON—SiN—SiO films for memory storage [38].

These films are deposited by thermal decomposition with no plasma. LPCVD SiON films may use source gases of dichlorosilane (DCS), nitrous oxide (NO), and ammonia ($NH_3$) as precursors, with the gas flow ratios determining the precise SiON composition.

The SiN charge storage films are Si-rich films to enable high concentrations of charge storage compared to stoichiometric SiN films [39]. Excess silicon in the SiN matrix provides charge trap sites that can be more easily detrapped when the cell needs to be erased. The SiN composition is generally measured by index of refraction, with $n = 2.0$ being stoichiometric or near-stoichiometric $Si_3N_4$. A typical index of refraction for a charge storage film is nearer $n \sim 2.2$. The SiN charge storage film may be further engineered by changing the composition at the beginning and end of the deposition to be more stoichiometric relative to the Si-rich center of the film. This graded profile of the SiN charge trap film relatively to a monolithic composition improves charge retention.

Blocking oxide films are now a combination of $SiO_2$ and $Al_2O_3$. Previously they were only $SiO_2$. The blocking $SiO_2$ film is deposited as a part of the ONO sequence inside the via, while the blocking $Al_2O_3$ film is deposited before the gate metal in the replacement sequence. The large band gap of $Al_2O_3$ allows an ultrathin layer of just a few nms to effectively block charge leakage on this side of the SiN film, as shown in the band diagram (Fig. 11.15). The $SiO_2$ blocking layer reduces the chance of metal contamination in the SiN film from the $Al_2O_3$ layer. Metal contamination within the ONO layer will reduce charge retention. These blocking oxide films can be deposited by either LPCVD or ALD techniques, with ALD now being more common.

A short discussion of the macro 3D NAND process flow follows. Many details are omitted, as there are many discussions of 3D NAND processes available to the reader. Present day 3D NAND fabrication begins with the formation of single crystal transistors in the silicon substrate. These transistors, known as circuits-under-the-array (CuA), comprise the circuitry needed to program, erase, and read the memory in the storage array. For most NAND manufacturers, these circuits will need to survive the higher temperatures needed for ONO deposition and channel Si crystallization, which means high temperature metalization like tungsten gates and larger channel lengths to allow for dopant diffusion. These CuA transistors comprise both relatively large transistors operating at high voltages (>20 V) to program and erase the NAND strings, as well as smaller transistors for other operations, such as the sense amplifiers that read the current values in the NAND strings.

After CuA formation, interconnect layers of tungsten are then formed. These tungsten interconnects are formed by a damascene process in which trenches are etched in dielectric layers, metal films are introduced into the trenches, and then metal is removed from the upper dielectric surfaces by chemo mechanical planarization (CMP). A liner/barrier film of either Ti + TiN or TiN-only is deposited before W. This liner/barrier film enables good adhesion of the W film. Titanium provides superior adhesion properties relative to TiN-only, but Ti films are very reactive and the TiN film prevents a reaction between the Ti film and the source gases for tungsten. The Ti and TiN films can be deposited by ALD using $TiCl_3$ and TdMat and $NH_3$ for Ti and TiN, respectively. In recent years, some manufacturers have been using TiN-only, despite its' poorer adhesion properties relative to Ti. Both Ti and TiN have relatively higher resistivity than W. As interconnect lines shrink laterally, a greater fraction of the damascene trench is occupied by Ti and/or TiN (there is a minimum film thickness needed to achieve a continuous film), increasing the resistivity of the interconnect. By removing the Ti adhesion layer, the increase in the interconnect resistivity with decreasing linewidth can be minimized.

Memory array formation begins with the deposition of alternating films of $SiO_2$ and $Si_3N_4$, which are deposited by plasma enhanced CVD (PECVD), with each pair forming one memory layer. These $SiO_2$ and $Si_3N_4$ films are each 20–30 nm thick. Precise control of these thicknesses, both vertically and laterally, is needed as the total layer count has exceeded 100. Variations in these thicknesses will result in variations in (later formed) interconnect resistance and in the capacitance between layers. The layers also need to be very smooth to prevent compounding errors [40].

The staircase, which allows via contacts to be formed to the gates, is formed as described previously (Fig. 11.14). After staircase formation, vias approximately 80 nm in diameter are patterned and etched through the many $SiO_2/Si_3N_4$ layers. These HAR vias are several microns deep, requiring the use of hard masks that can withstand long etch times. The vias are filled with the blocking $SiO_2$, $SiN_x$ charge storage, $SiON_x$ tunneling, channel Si, and filler $SiO_2$ layers in that order of deposition. These layers are removed from the surface of the stack by CMP.

The metal gate is formed by selectively removing the SiN layers by wet etching in hot phosphoric acid, leaving the $SiO_2$ layers. Before metal gate formation, access "slits" are formed in the $SiO_2/Si_3N_4$ stacks [41]. These access slits allow etch liquids and/or gases and subsequent film deposition to occur uniformly throughout the array. The blocking $Al_2O_3$, TiN adhesion, and tungsten gate films are deposited by ALD into the volume previously occupied by SiN. Similar to the tungsten interconnects between the CuA and memory array, the TiN thickness is minimized to allow for the maximum W fraction to occupy the interconnect. The TiN also forms the gate contacted the $Al_2O_3$ gate dielectric. To enable thinner gate films and/or reduce the resistivity of the gate films, vendors have been exploring the use of Ru or Mo films to replace tungsten word lines. Molybdenum can have decent adhesion to dielectric films without the use of liner/barrier films, unlike tungsten), and its resistivity scales relatively better than tungsten [42].

Micron, Hynix, and Kioxia have pursued a manufacturing strategy of make several "decks" of NAND memory (Fig. 11.16). Each deck is a stack of memory cells that have been fully fabricated before the next deck of cells is fabricated [30,43]. By dividing the fabrication into two or more decks, the aspect ratios which must be etched and then films deposited into are kept to a reasonable value. With high temperature formation steps in the memory array now complete, the final few lays of interconnects above the array are formed with low resistivity copper by a damascene process.

Yangtze Memory Technology company (YMTC) has pursued a different method of 3D NAND fabrication. YMTC forms the CuA and the memory array on two different wafers, and then bonds the two wafers together, shown in Fig. 11.17 [44]. The substrate of the memory array is then removed. This innovative technique has the advantage of allowing the CuA circuits to avoid high temperatures post formation, which enables smaller transistors due to limited dopant diffusion. Additionally, silicide contacts like NiSi with low silicon consumption and low temperature stability are enabled.

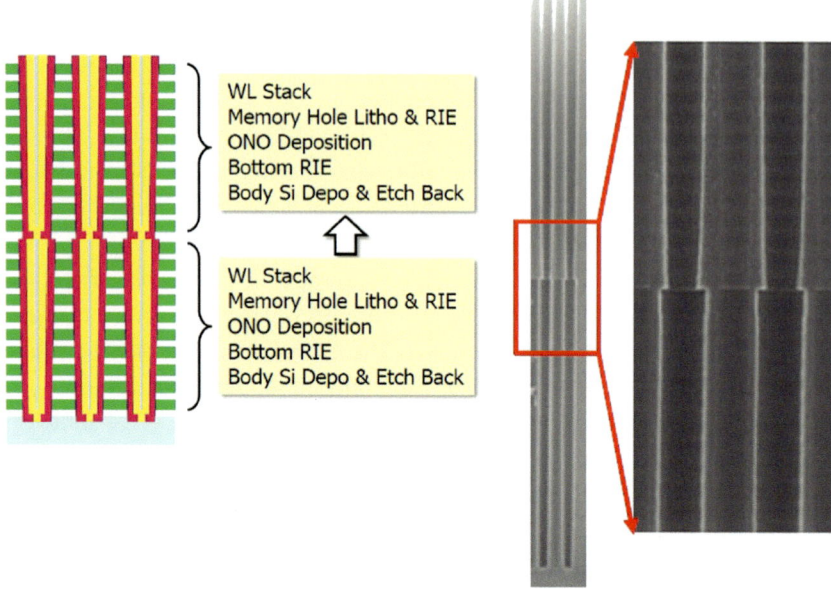

**Figure 11.16** Schematic and SEM images of deck "stacking" for 3D NAND devices [43].

A further advantage of this technique is the conservation of area needed to contact the wordlines, shown in Fig. 11.18. In the conventional staircase/CuA arrangement shown at left, lateral area is consumed by the wordline vias needing to be routed outside of the array and underneath to the CuA. In YMTC's case, the staircase vias can go directly to the CuA underneath. While the wafer bonding process adds some cost to the process, the use of smaller faster transistors in the CuA enables a reduction in the read latency

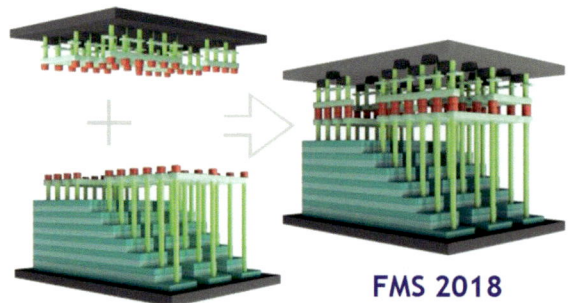

**Figure 11.17** Xtacking technique, fabricating the CuA transistors and memory array on separate wafers and then bonding together [44].

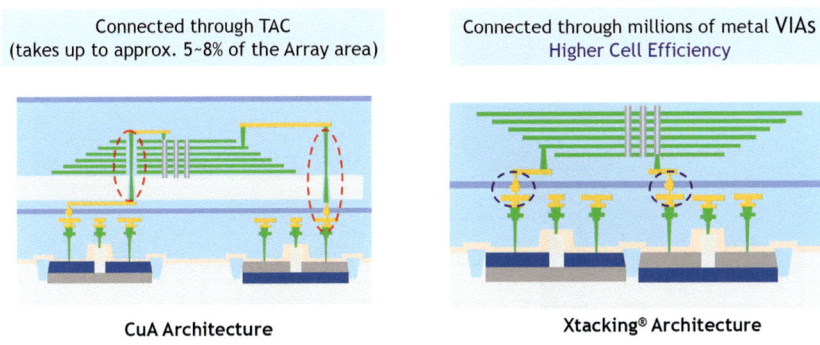

**Figure 11.18** Comparison of standard CuA NAND fabrication and wafer bonding CuA and memory arrays fabricated on separate wafers [45].

time. The reduction in area dedicated to wordline connections to the CuA results in an improved array efficiency. There is some indication that other NAND manufacturers will adopt this approach [45].

## 4. Other semiconductor memories

### 4.1 3D XPoint memory

Despite many attempts to introduce a new nonvolatile memory into the marketplace, there have been only a few successful introductions that currently occupy commercial niches. The most serious recent attempt was the introduction of chalcogenide phase change memory by Micron and Intel [46]. Chalcogenide phase change memory functions by having the memory material change from crystalline to amorphous states, and back again (set/reset operations), with the amorphous state having a higher resistivity than the crystalline state. A diode selector connected to the chalcogenide material programs and erases the material at high voltages, and reads the state of the memory at a lower voltage (Fig. 11.19).

As shown in Fig. 11.20, the memory array overlies single crystal transistor circuitry, or CuA, similar to 3D NAND. Four layers of interconnects separate the CuA and memory array, and another layer of interconnects lies above the memory array.

Phase change memory was introduced commercially by Intel and Micron as 3D XPoint Memory in 2015 after many years of research and development. Micron ended production of 3D XPoint memory in 2021, and Intel followed suit in 2022. 3D XPoint Memory was meant to fill the gap between DRAM with very fast read latency but high cost per bit, and

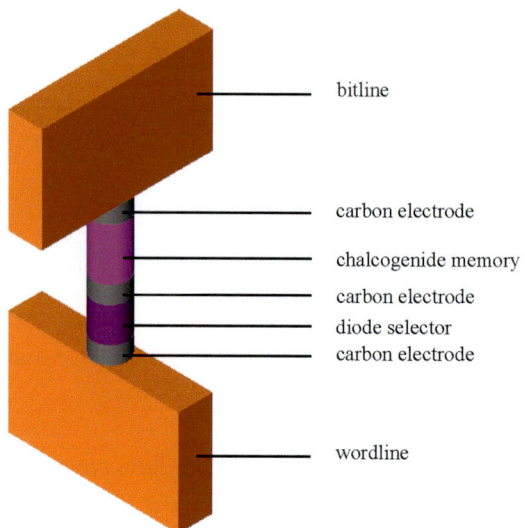

**Figure 11.19** 3D XPoint memory cell.

NAND, with slow read latency by low cost per bit (see Fig. 11.1). With a memory cell with a theoretical $(2\times)^2$ cell size, and the potential to stack the cells in a 3D array, the attributes seemed promising.

However, the speed advantage over NAND was not significant enough to overcome the high cost. The manufacturers did not adopt the multilevel patterning approach that has been so successful in 3D NAND in enabling significant cost-per-bit reduction in manufacturing. Multilevel patterning describes one mask operation that is used to pattern many layers of cells

**Figure 11.20** SEM cross sections of two layer 3D XPoint memory [46].

simultaneously in an etch or fill operation. While second generation 3D X Point, introduced in 2020, had four layers of cells compared with the initial product with two layers, that compared poorly with 2020 3D NAND where the layer count exceeded 100. With low sales volumes relative to NAND and DRAM persisting for several years, the manufacturers were not able to achieve economies-of-scale similar to DRAM and NAND. In addition, NAND continued to increase the storable number of bits-per-cell in commercial products, whereas commercially available 3D XPoint did not, although a number of research papers demonstrated multiple bits per cell in various chalcogenide memories [47].

## 4.2 Ferroelectric memory

Recent efforts to introduce a new form of nonvolatile memory cell have focused on $HfO_2$-based ferroelectric memory. Ferroelectric memory is based on switching between two stable polarization states. Electric dipoles within the grains of ferroelectric material are aligned in the presence of an electric field. As shown in Fig. 11.21, a hysteresis curve where the $P_{up}$ and $P_{down}$ states have remanent polarization values creates the memory. The FRAM memory cell is the same architecture as DRAM, namely 1T1C. The memory cell is programmed by biasing the capacitor to effect the atoms to be in the "up" or "down" state. To read the memory state, a consistent pulse is applied to the transistor with the drain set at ground. The polarization of the capacitor will cause the bit line to either remain at zero

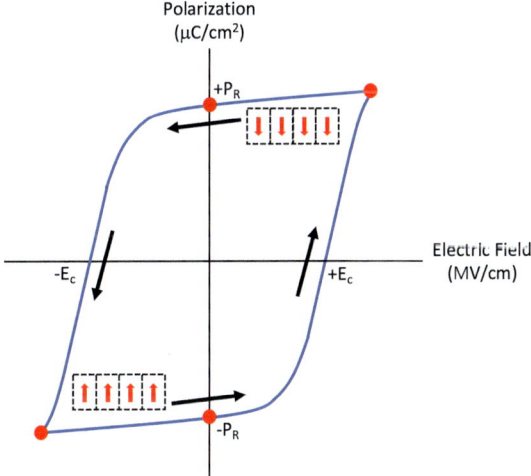

**Figure 11.21** Hysteresis curve of ferroelectric material.

or be elevated depending on whether the pulse switched the polarization or not. The ability to switch the polarization is proportional to the remanent polarization and the area of the capacitor. This effective value had to be large enough to be "read" by sense amplifier circuit, and maintaining a large enough value has limited the ability to shrink the cell. Similar to DRAM, reading the cell is destructive to the memory, and must be rewritten after a read.

Ferroelectric memory based on lead zirconate titanate (PZT) has been commercially available since 1996 [48–50]. PZT-based FRAM remains a relatively small semiconductor application (~$300 M/yr in 2022), with the largest application being smart cards. FRAM has a much faster write time than NAND, consumes less power during write, and has superior endurance. However, scalability issues, the temperature limitations of PZT after deposition, and cross contamination concerns of using PZT materials in silicon fabs have limited the application of PZT-based FeRAM to niche markets.

Interest in ferroelectric memory was renewed in 2011 with the publication of the ferroelectric properties of $HfO_2$ [51]. Not only is $HfO_2$ widely used in semiconductor fabrication plants, but good polarization characteristics can be achieved with much thinner films than is presently possible in PZT. Research into application of $HfO_2$-based FRAM has resulted in three principal cell architectures (see Fig. 11.22): (1) The 1T1C architecture previously used by PZT-based FRAM (FeRAM); (2) the 1T architecture

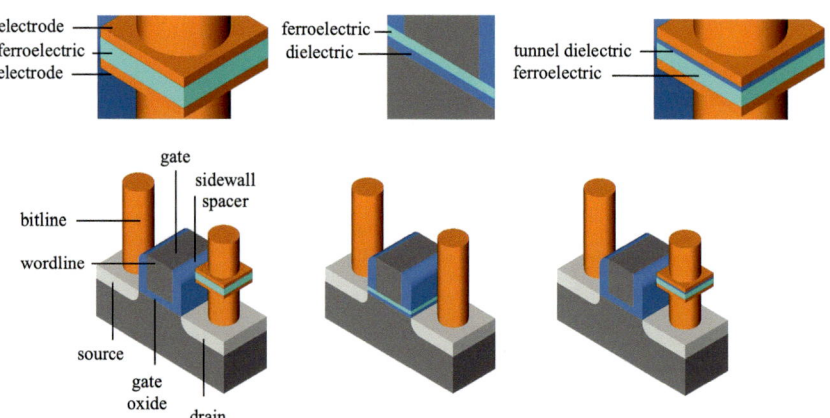

**Figure 11.22** Bottom row: Various memory devices using ferroelectric $HfO_2$ left-to-fight: 1T1C (FeFRAM), 1T FeFET, and 1T1R FTJ Top row: details of the memory state change stack.

field-effect transistor (FeFET); and (3) the 1T1R architecture ferroelectric tunnel junction (FTJ) [52,53].

In FeRAM devices, the ferroelectric $HfO_2$ is between two electrodes as a capacitor, which is attached to the drain of a transistor. In FeFET devices, the $HfO_2$ film is incorporated on top of the channel in the transistor. Another dielectric film separates the $HfO_2$ from the gate. A field is applied by biasing the gate, causing the $HfO_2$ polarization to switch. The polarization state of the $HfO_2$ affects the threshold voltage $V_t$ of the transistor. In FTJ devices, the architecture is like the FeRAM, but now the $HfO_2$ serves as a tunnel barrier between the electrodes of the capacitor. An additional tunnel dielectric film is between the $HfO_2$ and the upper electrode. The polarization state of the $HfO_2$ changes the effectiveness of the barrier, resulting in a difference in the resistance across the capacitor. In this instance, the FTJ device is a ReRAM device.

Various schemes to build 3D versions of $HfO_2$-based ferroelectric memories using multilevel patterning show a pathway to achieving very high density, and therefore cost competitiveness with 3D NAND. Several challenges remain in implementing $HfO_2$-based ferroelectric memory in high volume manufacturing: for FeRAM, the destructive reads add to the cycle count before breakdown, reducing endurance; forming the capacitors into cup-shaped 3D capacitors to maintain higher signal-to-noise ratio while scaling the cell laterally has been shown to decrease the endurance [53]; for FeFET, scaling the FeFET size reduces the signal-to-noise ratio due to size of the grains in the $HfO_2$ film; $HfO_2$ grain size variation can also cause variability in characteristics; for FTJ, the on/off ratio needs to be increased, especially important as this window decreases with scaling to smaller sizes.

## 5. Conclusion

Progress in scaling semiconductor memories has resulted in regular decreases in memory cell size and increases in capacity. DRAM chips were introduced in 1970 with a feature of 8 μm and a capacity of 1 kB. DRAM chips available in 2023 have a feature size of 12 nm and a capacity of 32 Gb. NAND chips were introduced in 1987 with a feature size of 1 μm and a capacity of 4 Mb. NAND chips available in 2023 have a feature size of 50 nm (gate pitch in vertical channel) and a capacity of 1 Tb. The cost per bit has been reduced at a similar pace as capacity scaling, enabling ubiquitous computing ability. DRAM and NAND have remained the dominant forms of semiconductor memory for more than 30 years. NAND memory cells have offered increased capacity by stacking memory cells ($z$ scaling), and DRAM is likely to take a similar path as lateral size scaling becomes increasingly difficult. Continued evolution of these memory cells, the optimization of the compute system to work with these cells, and the economies of scale afforded by the size of these industries makes it difficult for novel memory cells to emerge and take a meaningful place in the industry.

Advances in thin film deposition have been a part of this progress, enabling a growing set of material compositions, the ability to deposit on high aspect ratio features, and ever smaller thickness and composition nonuniformities across a wafer. Indeed, the number of elements from the periodic table used in high volume semiconductor manufacturing has grown each decade. The number of varied manufacturing techniques, such as replacement films and thick hard masks, has meant the possibilities of device shape and arrangement is limited only by the engineer's imagination.

## References

[1] R. Bez, P. Fantini, A. Pirovano, Historical review of semiconductor memories, in: A. Redaelli, F. Pellizer (Eds.), Semiconductor Memories and Systems, Elsevier, Cambridge, 2022, p. 2.
[2] https://www.statista.com/statistics/934579/global-DRAM-and-NAND-market-revenue (accessed 02.07.2023).
[3] S.S. Iyer, From deep trenches concerning the technology and design of embedded DRAM and three dimensional integrated circuits for memory applications, Electrochem. Soc. Trans. 31 (1) (2010) 31−41.
[4] G. Hemink, A. Goda, NAND flash technology status and perspectives, in: A. Redaelli, F. Pellizer (Eds.), Semiconductor Memories and Systems, Elsevier, Cambridge, 2022, p. 128.
[5] B. Talliz, The Samsung 983 ZET (Z-NAND) SSD Review: How Fast Can Flash Memory Get?, 19.02.2019. Available from: https://www.anandtech.com/show/13951/the-samsung-983-zet-znand-ss-reviews. (Accessed 11 June 2023).
[6] W. Cho, J. Jung, J. Kim, S. Lee, Y. Noh, D. Kim, W. Lee, K. Cho, K. Kim, J. Lee, S. Chai, E. Jo, H. Cho, J.-S. Kim, C. Kwon, C. Park, H. Nam, H. Won, T. Kim, K. Park, S. Oh, J. Ban, J. Park, J. Shin, T. Shin, J. Jang, J. Mun, J. Choi, H. Choi, S.-W. Choi, W. Park, D. Yoon, M. Kim, J. Lim, C. An, H. Shim, H. Oh, H. Park, S. Shim, H. Huh, H. Choi, S. Lee, J. Sim, K. Gwon, J. Kim, W. Jeong, J. Choi, K.-W. Jin, A. 1-Tb, 4b/cell, 176-stacked WL 3D-NAND flash memory with improved read latency and 14.8 Gb/mm$^2$ density, in: IEEE International Solid-State Circuits Conference, 2022, pp. 134−135.
[7] M. Helm, The usage of memory in current systems, in: A. Redaelli, F. Pellizer (Eds.), Semiconductor Memories and Systems, Elsevier, Cambridge, 2022, p. 44.
[8] A.K. Sharma, Advanced Semiconductor Memories, IEEE Press, Piscataway, 2003, pp. 19−20.
[9] https://www.digitaltrends.com/computing/intel-alder-lake-news-rumors-specs-release-date> (accessed 02.07.2023).

[10] https://en.www.wikichip.org/wiki/intel/microarchitectures/alder-lake> (accessed 02.07.2023).
[11] 16 Gbit DDR4 spot price $3.29; 3D TLC 256 Gbit spot price $2.50; <https://www.dramexchange.com (accessed 11.06.2023).
[12] Intel Core i5-13600k price $309 with 20MB SRAM; assumption: 20% of die area is SRAM https://www.windowscentral.com/hardware/computers-desktops/the-best-cpu-deals (accessed 11.06.2023).
[13] https://www.appliedmaterials.com/us/en/blog/blog-posts/expanding-the-ecosystem-for-hybrid-bonding-technology.html (accessed 11.06.2023).
[14] J.H. Lau, Advanced Semiconductor Packaging, Springer Link, Singapore, 2021.
[15] D. James, Recent innovations in DRAM manufacturing, in: IEEE/Semi Advanced Semiconductor Manufacturing Conference, 2010, pp. 264−269.
[16] S.K. Kim, M. Popovic, Future of dynamic random-access memory as main memory, MRS Bull. 43 (2018) 334−338.
[17] A. Spessot, H. Oh, 1T-1C dynamic random access memory status, challenges, and prospects, IEEE Trans. Electron. Dev. 67 (4) (2020) 1382−1393.
[18] D.H. Kim, J.Y. Kim, Y.S. Hwang, J.M. Park, D.H. Han, D.I. Kim, M.H. Cho, B.H. Lee, H.K. Hwang, J.W. Song, N.J. Kang, G.W. Ha, S.S. Song, M.S. Shim, S.E. Kim, J.M. Kwon, B.J. Park, H.J. Oh, H.J. Kim, D.S. Woo, M.Y. Jeong, Y.I. Kim, Y.S. Lee, H.J. Kim, J.C. Shin, J.W. Sheo, S.S. Jeong, K.H. Yoon, T.H. Ahn, J.B. Lee, Y.W. Hyung, S.J. Park, H.S. Kim, W.T. Choi, G.Y. jin, Y.G. Park, K. Kim, A mechanically enhanced storage node for virtually unlimited height (MESH) capacitor aiming at sub 70nm DRAM, in: IEEE International Electron Device Meeting, 2004, 04-69-04-72.
[19] S.-W. Park, C.S.-D. Lee, S.-A. Jang, M.-S. Yoo, K.-O. Kim, C.-O. Chung, S.Y. Cho, H.-J. Cho, L.-H. Lee, S.-H. Hwang, J.-S. Kim, B.-H. Lee, H.G. Yoon, H.-S. Park, S -J. Baek, Y.-S. Cho, N.-J. Kwak, H.-C. Sohn, S.-C. Moon, K.-D. Yoo, J.-G. Jeong, J.-W. Kim, S.-J. Hong, S.-W. Park, Highly scalable saddle-fin (S-Fin) transistor for sub-50nm DRAM technology, in: IEEE VLSI Symposium, 2006.
[20] N.-H. Lee, S. Lee, S.-H. Kim, G.-J. Kim, K.W. Lee, Y.S. Lee, Y.C. Hwang, H.S. Kim, S. Poe, Transistor reliability characterization for advanced DRAM with HK + MG & EUV process technology, in: IEEE International Reliability Physics Symposium, 2022, pp. 6A1-1−6A1-6.
[21] S. Jeon, J. Choi, H.-C. Jung, S. Kim, T. Lee, Investigation on the local variation in BCAT process for DRAM technology, in: IEEE International Reliability Physics Symposium, 2017, pp. FA4.1−FA4.2.
[22] J. Choe, DRAM Scaling Trend and Beyond, 2023. https://www.techinsights.com/blog/dram-scaling-trend-and-beyond. (Accessed 11 June 2023).
[23] H. Watanabe, T. Tatsumi, S. Ohnishi, K. Kitajima, I. Honma, T. Ikarashi, H. Ono, Hemispherical grained Si formation on in-situ phorphorus-doped amorphous-Si electrode for 256Mb DRAM's capacitor, IEEE Trans. Electron. Dev. 42 (7) (1995) 1247−1254.
[24] J.M. Park, Y.S. Hwang, S.-W Kim, S.Y. Han, J.S. Park, J. Kim, J.W. Seo, B.S. Kim, S.H. Shin, C.H. Cho, S.W. Nam, H.S. Hong, K.P. Lee, G.Y. Jin, E.S. Jung, 20nm DRAM: a new beginning of another revolution, International Electron Device Meeting (2015) 15-676−15-679.
[25] D.-S. Kil, H.-S. Song, K.-J. Lee, K. Hong, J.-H. Kim, K.-S. Park, S.-J. Yeom, J.-S. Roh, N.-J. Kwak, H.-C. Soh, J.-W. Kim, S.-W. Park, Development of new TiN/ZrO$_2$/Al$_2$O$_3$/ZrO$_2$/TiN capacitors extendable to 45nm generation DRAMs replacing HfO$_2$ based dielectrics, in: VLSI Symposium, 2016.
[26] H.J. Cho, Y.D. Kim, D.S. Park, E. Lee, C.H. Park, J.S. Jang, K.B. Lee, H.W. Kim, Y.J. Ki, I.K. Han, Y.W. Song, New TIT capacitor with ZrO$_2$/Al$_2$O$_3$/ZrO$_2$ dielectrics for 60nm and below DRAMs, Solid State Electron. 51 (2007) 1529−1533.

[27] D.-H. Kim, H.-J. Kwon, S.-J. Bae, DRAM circuit and technology, in: A. Redaelli, F. Pellizer (Eds.), Semiconductor Memories and Systems, Elsevier, Cambridge, 2022, pp. 105–106.
[28] M.F. Chen, C.H. Tsai, T. Ku, W.C. Choi, C.T. Wang, D. Yu, Low temperature SoIC bonding and stacking technology for 12-/16-Hi high bandwidth memory (HBM), IEEE Trans. Electron. Dev. 67 (12) (2020) 5343–5349.
[29] B. Meyer, Will Monolithic 3D DRAM Happen?, 2021. Available from: https://www.semiengineering.com/will-monolithic-3d-dram-happen.
[30] https://www.semiengineering.com/micron-b47r-3d-ctf-cua-nand-die-worlds-first-176l-195t/ (accessed 11.06.2023).
[31] https://www.tomshardware.com/reviews/glossary-hbm-hmb2 (accessed 11.06.2023).
[32] M. Johnson, A. Al-Shamma, D. Bosch, M. Crowley, M. Farmwald, L. Fasoli, A. Ilkbahar, B. Kleveland, T. Lee, T.-Y. Li, Q. Nguyen, R. Scheuerlein, K. So, T. Thorp, 512-Mb PROM with a three-dimensional array of diode/antifuse memory cells, IEEE J. Solid State Circ. 38 (2003) 1920–1927.
[33] Y. Fukuzumi, R. Katsumata, M. Kito, M. Kido, M. Sato, H. Tanaka, Y. Nagata, Y. Matsuoka, Y. Iwata, H. Aochi, A. Nitayama, Optimal integration and characteristics of vertical array devices for ultra high density bit cost scalable flash memory, in: IEEE International Electron Device Meeting, 2007, pp. 449–452.
[34] M. Oda, K. Sakuda, Y. Kamimuta, M. Saitoh, Carrier transport analysis of higher performance poly-Si nanowire transistor fabricated by advanced SPC with record-high electron mobility, IEEE International Electron Device Meeting (2015).
[35] https://www.asm.com/vertical-furnace/sonora-vertical-furnace> (accessed 11.06.2023).
[36] T. Kamins, Structure and properties of LPCVD Si films, J. Electrochem. Soc. 127 (1980) 686.
[37] H. Tanaka, M. Kido, K. Yahashi, M. Oomura, R. Katsumata, M. Kito, Y. Fukuzumi, M. Sato, Y. Nagata, Y. Matsuoko, Y. Iwata, H. Aochi, A. Nitayam, Bit cost scalable technology with punch and plug process for ultra high density flash memory, in: VLSI Symposium, 2007.
[38] W.-C. Chen, H.-T. Lue, S.-T. Fan, T.-H. Hsu, P.-C. Jhang, U.G. Vej-Hansen, P.A. Khomyakov, K.-H. Lin, K.-C. Wang, C.-Y. Lin, First theoretical modeling of the bandgap-engineered oxynitride tunneling dielectric for 3D flash memory devices starting from ab initio calculation of the band diagram to understand the programming, erasing, and reliability, in: International Electron Device Meeting, 2021.
[39] H.-C. Chien, C.-H. Kao, J.-W. Chang, T.-K. Tsai, Two-bit SONOS type flash using a band engineering in the nitride layer, Microelectr. Eng. 80 (17) (2005) 256–259.
[40] https://www.lamresearch.com/product/vector-product-family> (accessed 11.06.2023).
[41] G.H. Lee, S. Hwang, J. Yu, H. Kim, Architecture and process integration overview of 3D NAND flash technologies, Appl. Sci. 11 (15) (2021).
[42] A. Ajaykumar, L. Breuil, F. Schleicher, F. Sebaai, Y. Oniki, S. Ramesh, A. Arreghini, L. Nyns, J.-P. Soulie, J. Stiers, G. Van den Bosch, M. Rosmeulen, First demonstration of ruthenium and molybdenum word lines integrated into 40nm pitch 3D NAND devices, in: VLSI Symposium, 2021.
[43] S. Inaba, 3D Flash memory for data-intensive applications, in: IEEE International Memory Workshop, 2018.
[44] Z. Huo, W. Cheng, S. Yang, Unleash scaling potential of 3D NAND with innovative Xtacking architecture, in: VLSI Symposium, 2022.
[45] S.-H. Lee, The rise of memory in the ever changing AI era – from memory to more-than-memory, in: VLSI Symposium, 2022.
[46] A. Fazio, Advanced Technology and Systems of Cross Point Memory, in: International Electron Device Meeting, 2020.

[47] J. Handy, T. Coughlin, Optane's Dead — Now What? IEEE Computer Society, 2023, pp. 125—130.
[48] H. Ishiwara, Current status and prospects of ferroelectric memory, in: International Electron Device Meeting, 2001, 01-275-278.
[49] K. Kim, G. Jong, H. Jeong, S. Lee, Emerging memory technologies, in: Custom Integrated Circuits Conference, IEEE, 2005, pp. 423—426.
[50] K. Kim, 1T1C FRAM, International Symposium on VLSI technology, systems, and applications. Proceedings of Technical Papers (Cat. No. 01TH8517) (2001) 81—84.
[51] T.S. Böste, J. Müller, D. Bräuhaus, U. Schröder, U. Böttges, Ferroelectricity in hafnium oxide thin films, Appl. Phys. Lett. 99 (2011) 102903.
[52] T. Schenk, S. Mueller, A new generation of memory devices enabled by ferroelectric hafnia and zirconia, in: IEEE International Symposium on Applications of Ferroelectrics, 2021.
[53] G. Molas, L. Grenouillet, Other emerging memories, in: A. Redaelli, F. Pellizer (Eds.), Semiconductor Memories and Systems, Elsevier, Cambridge, 2022, pp. 289—296.

## CHAPTER 12

# Yield impact of defects from thin films and other processing steps

**Ishtiaq Ahsan**
Semiconductor Technology Research & Development, IBM Research, Albany, NY, United States

## 1. Introduction

Semiconductor chips are made of connected building blocks of integrated circuits. Key blocks like these are logic core, local memory, high level memory, IO circuits, etc. These blocks are shown in a pictorial schematic in Fig. 12.1a. Example schematics of a logic circuit and a memory SRAM cell are also shown in that diagram. In order to make complex semiconductor chips work, these individual blocks and circuits need to work.

In order to give a little more detail as to how semiconductor circuits are constructed, layout of a simple logic circuit is shown in Fig. 12.1b together with its schematic and various electrical nodes/pins (connection points). The electrical nodes/pins are power supply node (VDD), ground node (GND), input signal node, and output signal node. Semiconductor circuits are made with patterned features like lines and contacts formed on multiple layers. In Fig. 12.1b, top view layout and the cross-sectional view of the layout give an idea of different levels of a circuit and how they are used to physically manufacture an integrated circuit. In this figure, electrical connections (VDD, GND, input, output) are shown to be connected to Metal1 lines (boxes shown in solid gray with no border). The gate of the transistor is shown in hashed lines. Source-drain diffusion areas are shown as boxes with dotted patterns in them. Contacts between diffusion and metal 1, contacts between gate and metal 1, and contacts between two metal lines are shown as circular shapes. The metal 2 line is shown as a gray box with a solid black border line.

A fully integrated circuit in state-of-the-art technologies will have many more complex layers and many more metal lines and contacts, but diagrams in Fig. 12.1a and b give us a basic understanding of the building blocks of a

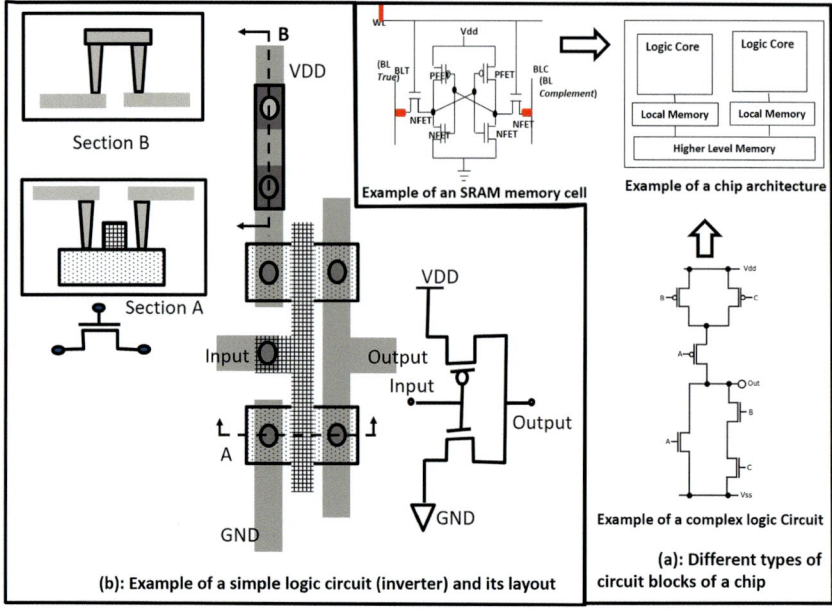

**Figure 12.1** (a) Example of a simple logic circuit (inverter) and its layout; (b) Different types of circuit blocks of a chip.

complex semiconductor chip. All these levels need to be manufactured without killer defects so that the entire chip works. The cross-sectional diagram "A" in Fig. 12.1b shows a cross section of a basic transistor. Large semiconductor microprocessor chips are made of billions of these transistors [1]. One can imagine the magnitude of challenge of manufacturing these chips in a fashion so that a high percentage of them works! These chips first need to be "functional" (e.g., no missing lines or contacts creating an open circuit or no bridged lines creating electrical shorts). They then need to operate at specific frequencies and are allowed to consume only a limited amount of power. This is so that a large number of computations can be performed in a short amount of time without consuming too much power. A semiconductor chip's computational speed and power governs how fast a computer or a phone can run and the chip's electrical power consumption rate governs how much battery life a phone can have. "Yield" of a semiconductor chip is defined as a percentage of chips on a wafer that "functions" in an allowable performance and power consumption window. Hence, yield loss of a chip can be broken up into two parts, namely.

(1) Defect limited yield loss => this is the yield loss resulting from the circuits not functioning properly due to structural integrity issues like contact open, line opens/shorts, etc.
(2) Power performance limited yield loss => this is yield loss resulting from chips not operating within a specific power-performance window. This can result from (among other things) transistor not having the right "on" current, "off" current, capacitance or metal lines not having the right resistance, capacitance, etc.

In this chapter, we will primarily discuss the first kind of yield loss.

## 2. Examples of different fail modes

In this section, examples of fail modes that can cause functional fails in electrical circuits are shown on both layout and schematics. A simple inverter circuit is used as an example in this cause, although in large chips, much more complex circuits are used. The top left image of Fig. 12.2 shows a missing pattern on the metal line causing an open in the output line. The top right example shows an unlanded contact creating a disconnection to the VDD electrical node. In the bottom left, the gate of the MOSFET is touching the contact on one side due to an overlay error. In these occasions, it will create an electrical short between the input and the ground node. On the bottom right example, we have an

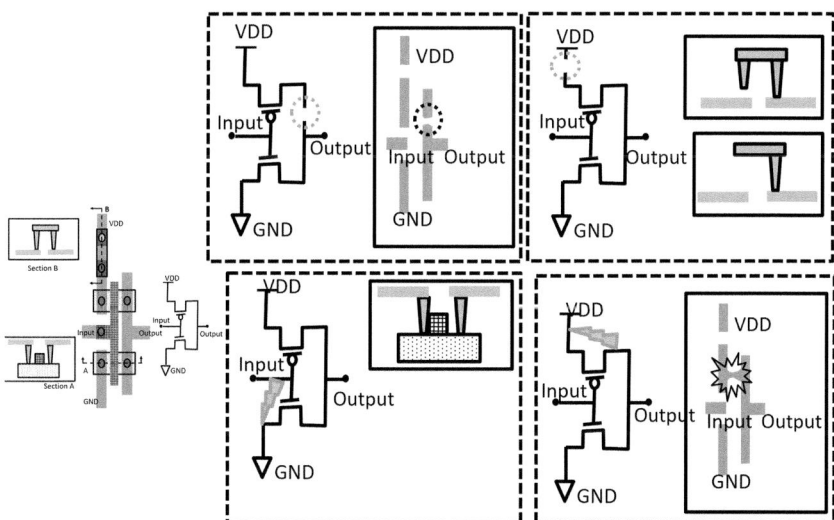

**Figure 12.2** Examples of different type of fails causing defect limited yield.

extra pattern between two metal lines causing an electrical short between VDD and the output node. All these defects will cause functional fail of the circuit.

## 3. Defect density and its impact on yield

In this section, we will discuss defect density and its impact on yield. Before we go into the details of defect density, it is first important to understand that the defect size of relevance depends on the dimensions of the levels where the defects are occurring. Let us consider Fig. 12.3. This image pictorially shows the different pitches used for different metal levels used in

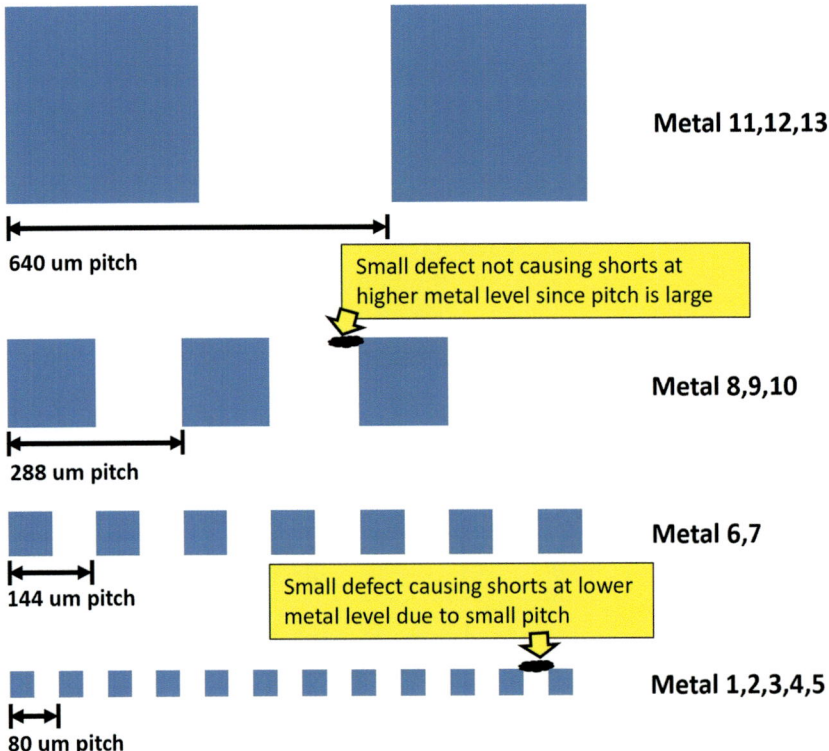

**Figure 12.3** A visual demonstration of yield impact of small defects at lower metal levels. Pitch numbers are taken from a publication on a 22 nm semiconductor technology [2]. For simplicity, lines and spaces are shown to have equal width in this figure and height and width of lines are also shown to be equal. This is not always the case in real chips.

a semiconductor logic chip. Pitch numbers in Fig. 12.3 are taken from a publication on a 22 nm semiconductor technology [2]. As shown in the figure, typically, lower metal levels have much smaller pitches. For simplicity, lines and spaces are shown to have equal width in this figure and height and width of lines are also shown to be equal. This is not always the case in real chips. Nevertheless, it is obvious that as we go down to lower levels, metals and the dimension of the levels get smaller, even small defects start having an impact on yield. The same small defects will likely case less problems at higher levels. This is why it is important to focus on defect improvement at levels that have much smaller dimensions.

For the same defect density, yield of chips of larger size go down exponentially. This can be seen visually in an example shown in Fig. 12.4. In this example, the total number of defects (the black dots) and its location on the wafer are the same between the two wafer maps. The only difference between the wafer maps is the chip size. The wafer on the right has a significantly larger chip. Assuming that existence of a single defect on the chip makes the chip nonfunctional, it becomes obvious that yield of the larger chip will be significantly lower than yield of the smaller chip. All the chips marked in gray in Fig. 12.4 are nonyielding chips. The larger chip in this case has 35% yield, whereas the smaller chip has a yield of 96%. There are many statistical models that are used to predict or calculate the yield of a given defect density [3–5]. Among one of the simplest yield models is the Poisson yield model, which is given by the following equation

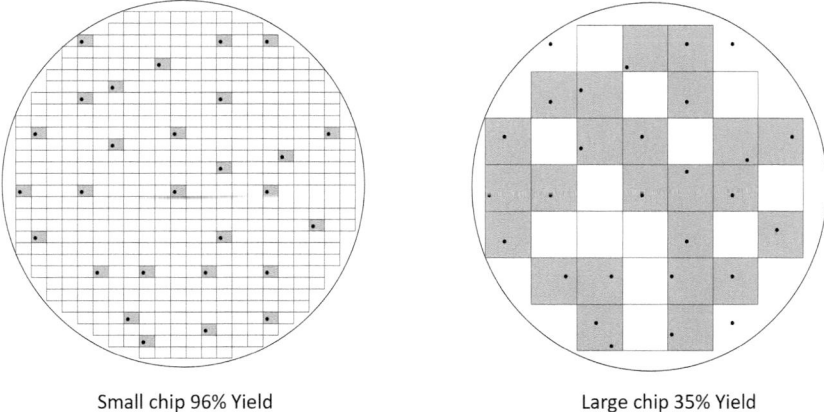

Small chip 96% Yield          Large chip 35% Yield

**Figure 12.4** How defect density impacts yield of large chips versus smaller chips?

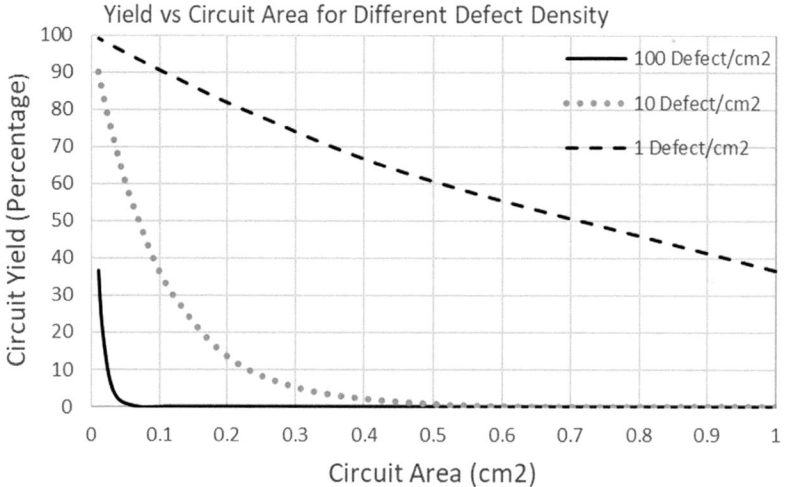

**Figure 12.5** Relationship between yield and defect density for circuits of different sizes.

$$\text{Yield} = \text{Exp}^{-(D0*A)}$$

where $D0$ is the defect density expressed in units of number defects/cm$^{-2}$, and the $A$ is the area of the circuit expressed in cm$^{-2}$.

Using this model, one can show the impact of defect density on yield. Fig. 12.5 shows the relationship of yield versus defect density of circuits of different sizes. Fig. 12.6 shows dependency of yield versus circuit size for different defect densities. From these curves, it is clear that when the defect density is even moderately high, yield of large ships can quickly approach 0%. Whereas for much smaller circuits or chips, you can be relatively insensitive to variations to defect density in this range.

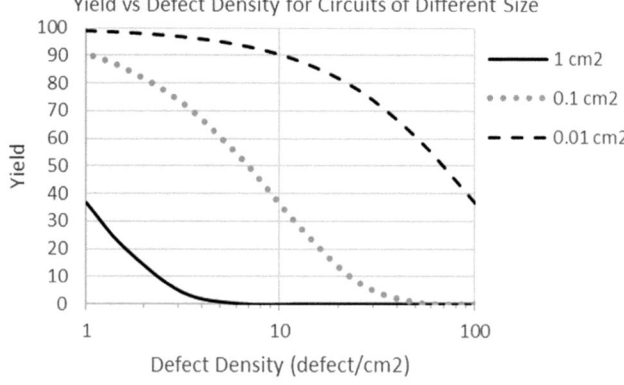

**Figure 12.6** Relationship between yield and circuit size for different defect densities.

In an earlier section of the chapter, it was mentioned that semiconductor chips are made with multiple layers of patterned material laid on top of each other, and it was shown how defects at any of these layers can cause a circuit not to function. Fig. 12.7 shows how cumulative effects of defects at multiple levels quickly add up and cause a severe loss of overall yield of a chip even though the defect density of individual layers are relatively low. Hence, it is important to have tight control on the defect density of each level or layer used in the manufacturing process of the chip. The same effect is shown with an example layout in Fig. 12.8. In this figure, it is shown that the same circuit is failing for fails caused at different levels of the chip, but each defect is making the chip nonfunctional, and the cumulative effect of all these defects is resulting in only one out of six chips being defect free and fully functional. It is also clear from the diagram that the same chip can fail from multiple defects at different levels particularly if the defect density of a level is high. For example, there is one chip shown in this diagram that fails for both metal open and missing contact. In this chapter, simple circuits and fail modes are used as examples so that the reader can have an easy understanding of the impact of defects at different layers on yield, but these impacts are observed routinely in manufacturing

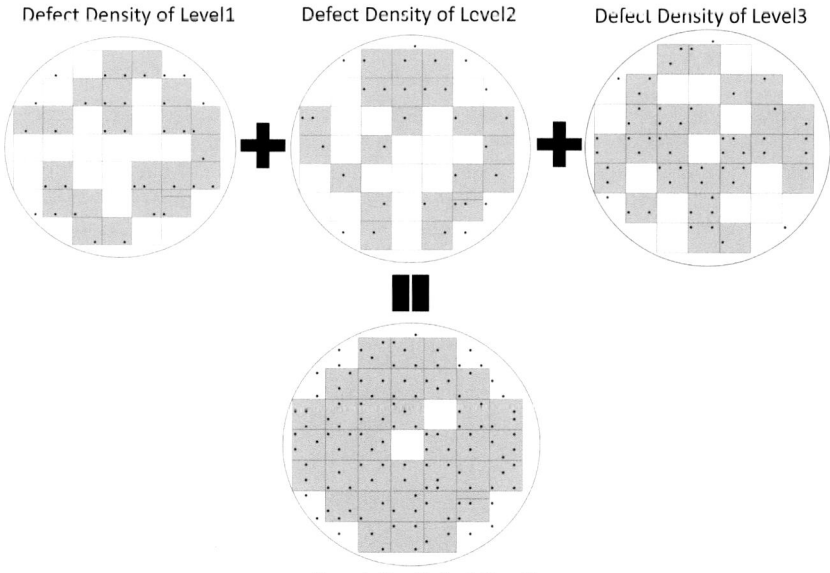

**Figure 12.7** Cumulative effects of defects at multiple levels => number of nonyielding chips (gray ones) drops significantly.

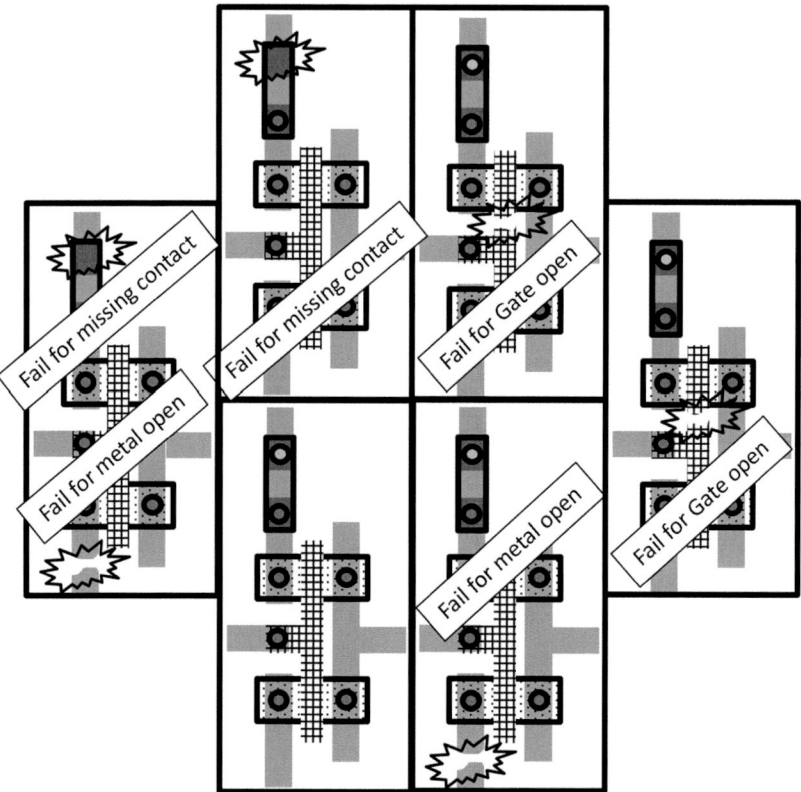

**Figure 12.8** An example for visually understanding the cumulative yield impact of defects at different levels. Different chips can fail for low-level defects generated at different levels but their cumulative effect can be big on the overall yield of the circuit on the wafer. In this example, although individual fail-modes at different levels each impact only 2 out of 6 chips, their combined effects results in a total 5 out of the 6 chips to fail.

or developing semiconductor lines of real semiconductor plants in a much greater magnitude on much more complex circuits.

Up to this point, we have discussed impact of randomly distributed defects on yield. Not all defects are distributed randomly. Sometimes, the defect distribution can be regional or semi-regional/semi-random. In these cases, yield loss on large chips can be more forgiving. As shown in Fig. 12.9, we can see that for the same number of total defects (same number of black dots) and for the same yield for small chips (76%), yield of large chips is significantly higher when the defects are confined to one specific region. Large chip yield is 56% when the defects are strictly regional and 22% when it is semi-regional. This is called clustering effect.

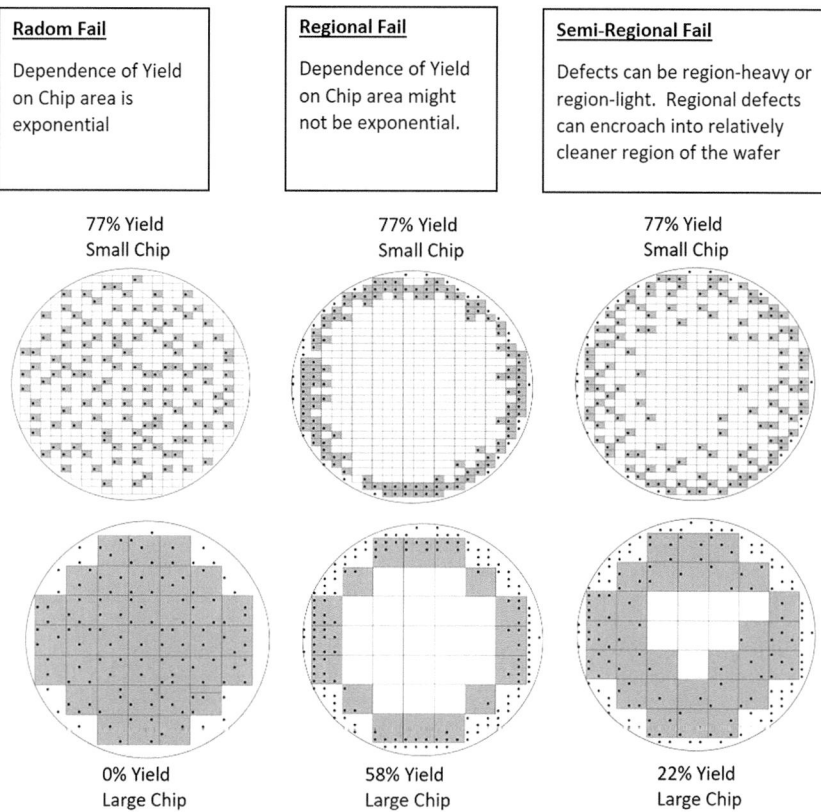

**Figure 12.9** Example of yield impact of regional and semi-regional defects on yield of large chips.

## 4. Various yield assessment structures

In this section of the chapter, a brief overview of different kinds of yield assessment structures is described. These range from simpler passive structures to complex functional circuits like the SRAM. Passive structures are the ones that do not involve transistors. These are structures like comb-serp structures madeup of interleaving lines or via chain structures made-up of links of lines and contacts between the lines. Examples of such structures are shown in Fig. 12.10a. Structures like these are used to measure line shorts or opens on contact via opens. Fig. 12.10b shows a simplified cross section of an MOSFET transistor, which is the basic building block of most of the integrated circuits manufactured today. The figure also shows some basic measurements that are typically done on the MOSFET. Examples of these

**(a)**

**Passive test Structure:**

Measures line opens/shorts.. Contact open

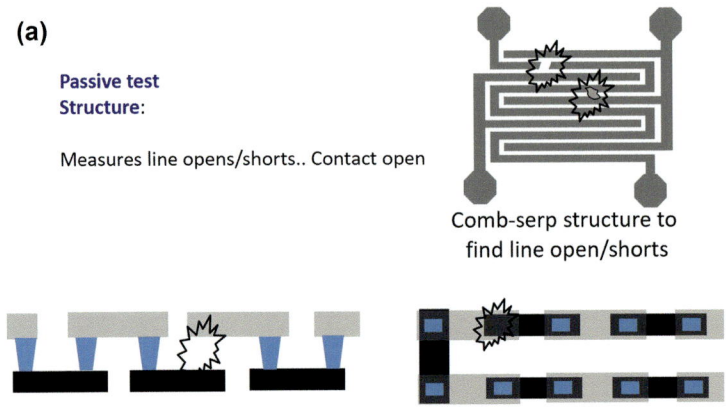

Comb-serp structure to find line open/shorts

Contact chain to find contact open

**Figure 12.10a** Examples of passive test structures like comb-serp and via-chain structures.

**(b)**

**Device test Structures:**

Measure device parameters like Ion/Ioff/Vt Ring speed

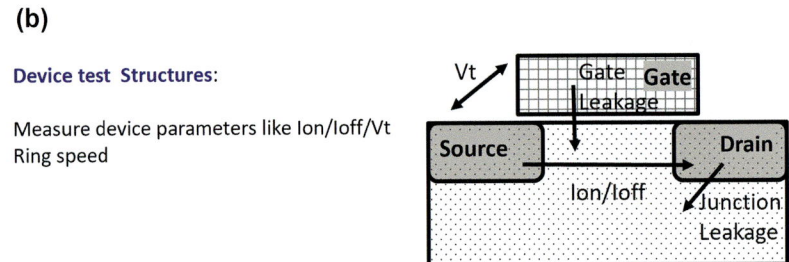

**Figure 12.10b** Example of MOSFET measurement structure.

measurements are voltage required to turn on the transistor ($V_t$), on-state current through the transistor ($I_{on}$), off state current ($I_{off}$), junction leakage, etc. These transistor characteristics are measured from individual transistors, and these parameters, together with some other capacitance and resistance measurements, play a key role in determining the circuit performance and power consumption. Hence, these play an important role in determining the power performance limited yield loss of semiconductor chips.

Fig. 12.11a shows the circuit diagram of a building block of a more complex yield assessment structure—the static random access memory (SRAM) bit cell. As the name suggests, this is a unit cell of a memory circuit. Fig. 12.11b shows how this unit cell is connected in an array of memory. SRAM memory is used heavily in logic technologies as an on–chip memory element of the chip. Circuits like this are also used as yield assessment and

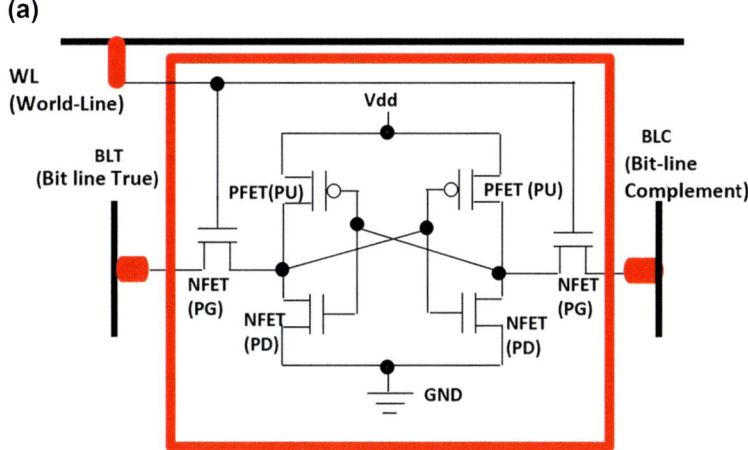

**Figure 12.11a** Schematic of SRAM bit-cell.

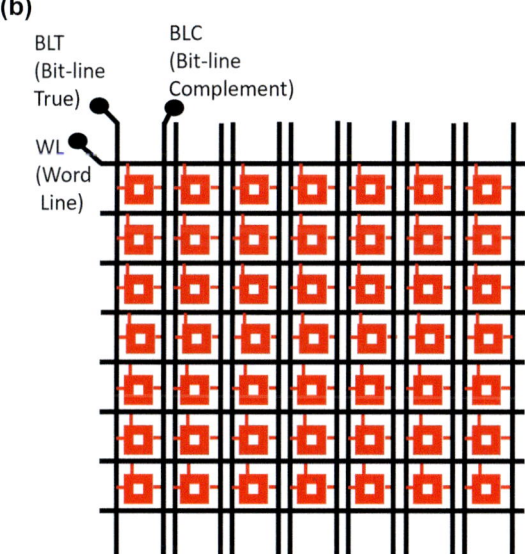

**Figure 12.11b** Array of SRAM bit-cells.

yield monitoring vehicles during semiconductor technology development and manufacturing. Fig. 12.11c shows an example layout of an SRAM array. The unit cell boundaries of each SRAM bit cell are marked on the layout. It is clear from the layout that there are many lines and contacts that are shared between adjacent cells. This is done to make the array dense so that more

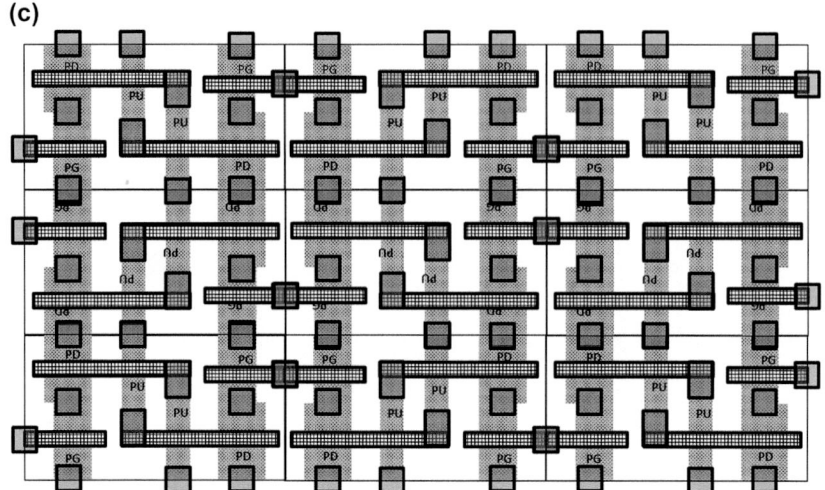

**Figure 12.11c** Example layout of SRAM array. 3 × 3 connected SRAM bits-cells are shown here.

memory bits can be placed on a given silicon area. The SRAM array is connected to power supply VDD and GND (not shown in Fig. 12.11b), and individual bits can be addressed by selecting (powering on) the corresponding word line and bit-line/bit-line—complement intersection.

The SRAM memory cells are written by flowing current through the bit-line/bit-line complement lines and the corresponding NFET PG transistor pairs of the bit cell. This puts the corresponding SRAM bit in a state of "zero" or "one" (this is how the SRAM cell is "written" to its desired state). These currents are flown using external circuitry often called SRAM periphery circuit (these circuits are not shown in this diagram). When we read back the value stored in the SRAM bit, current flows from the bit-cell into the pair of bit-lines/bit-line complement lines and are fed into a sensing circuit which can sense the voltage difference between the bit-line and bit-line complement lines. The sensing circuit thus ascertains the state of the SRAM bit. The whole SRAM array can be tested by writing different combinations of zeros and ones in the array and then reading them back and checking if the expected values were read or not. Bit-cells that do not return the expected values are called failing bits. These failing bits are marked as dark dots on a map called a "bit-fail map." These dots mark the physical location of the failing bits relative to the other bits in the array. Physical failure analysis can then be performed on these bits to determine root cause of fail.

## 5. SRAM yield learning methodology

It was mentioned in the previous section that the SRAM circuit is often used for yield learning and technology development. In this section, we will go into a little more detail of how this circuit is used. Let us use a 3 × 3 array of SRAM bit-cells as an example (Fig. 12.12a). If the entire array is healthy and passes electrical functional test, it will create a clean 3 × 3 bit map as shown in Fig. 12.12a. Let us now assume that one of the contacts, which is not shared between the bit-cells, is missing or defective as shown in the layout of Fig. 12.12b. In this case, only functionality of the cell where the contact is missing will be impacted and the failing cell will show up in the bit-fail map as a single bit fail or single cell fail. This is shown in the corresponding bit-fail map in Fig. 12.12b. Now, if the contact used for both gate and diffusion is defective as shown in Fig. 12.12c, it will also create a single cell fail since this contact is also not shared with any neighboring bit-cells. In the next example (Fig. 12.12d), it is shown that a gate contact that is shared between 2 cells is missing or defective. In this case, since this contact was needed for functionality of two horizontally

**(a)**

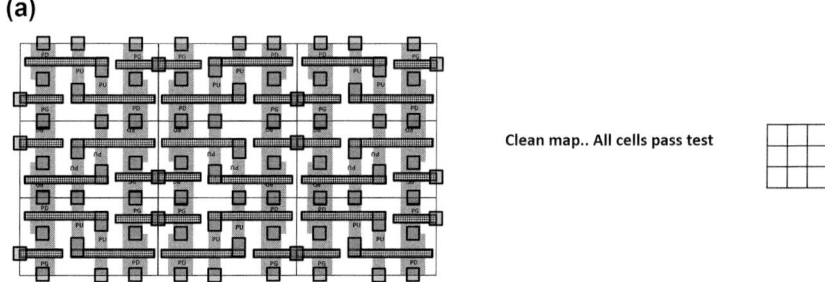

**Figure 12.12a** Top down view and bit-fail map of a fully functioning 3 × 3 array.

**(b)**

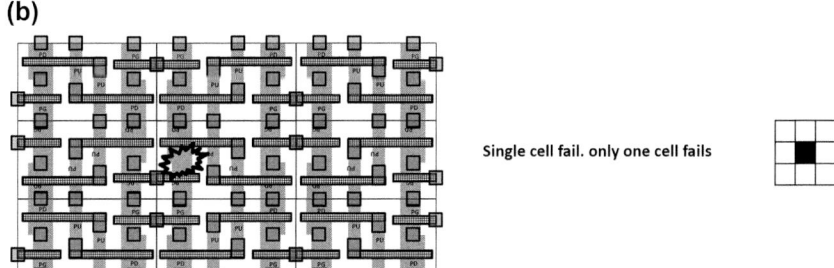

**Figure 12.12b** Top-down view and bit-fail map of a 3 × 3 array with one missing internal diffusion contact.

**(c)**

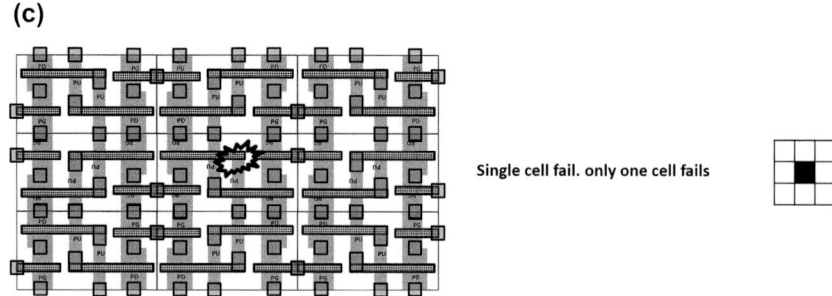

**Figure 12.12c** Top-down view and bit-fail map of a 3 × 3 array with one missing internal diffusion + gate contact.

**(d)**

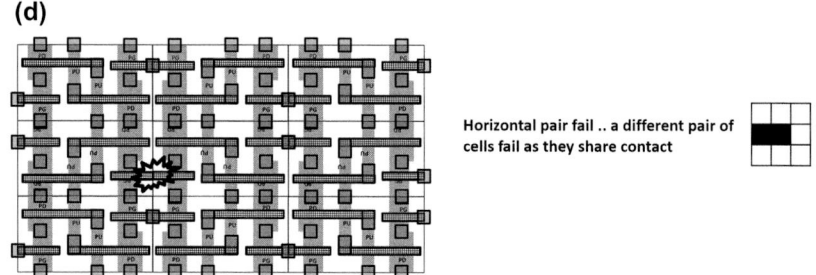

**Figure 12.12d** Top-down view and bit-fail map of a 3 × 3 array with one missing shared gate contact.

**(e)**

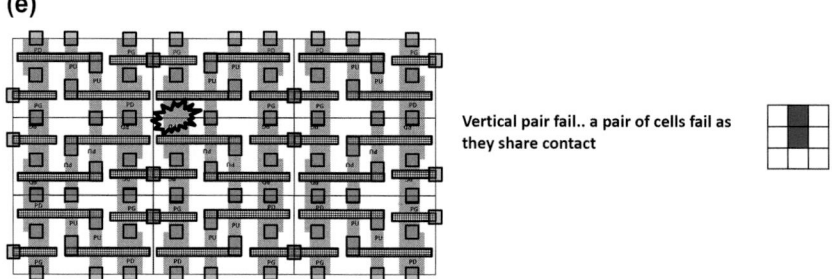

**Figure 12.12e** Top-down view and bit-fail map of a 3 × 3 array with one missing shared diffusion contact.

placed neighboring cells, it will cause both of these cells to malfunction, thereby creating what is known as a horizontal pair fail in the bitmap (shown in Fig. 12.12d). Similarly, if there is a defect on one of the diffusion contacts that is shared between two vertical neighboring cells, it will create a vertical pair fail as shown in Fig. 12.12e.

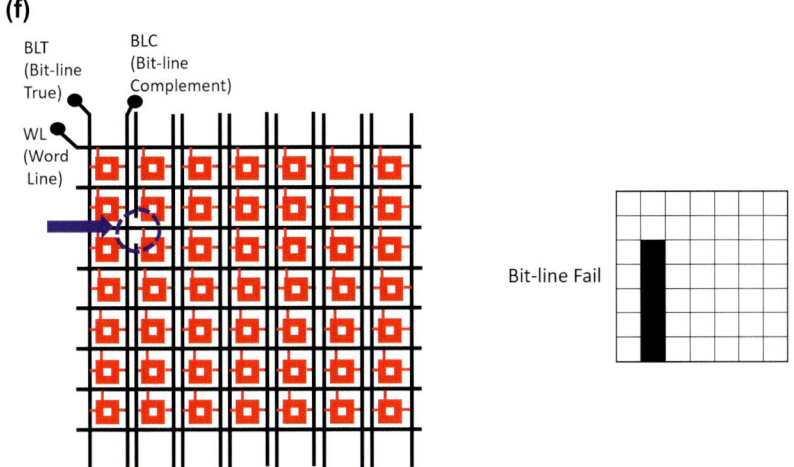

**Figure 12.12f** 7 × 7 SRAM array with a broken bit-line and the corresponding bit-fail map.

Structural fails can also occur in circuits outside of the core bit-cell itself. One such example is shown in Fig. 12.12f where there is a breakage in the bit-line (BLT) that feeds a column of bit-cells. In this case, cells that are past that breakage point will not get the signal necessary to be functional and hence all of them will fail the functional electrical test. This will manifest itself in the bit-fail map as a partial column fail also known as partial bit-line fail. Similarly, if such a fail occurs in a word-line as shown in Fig. 12.12g,

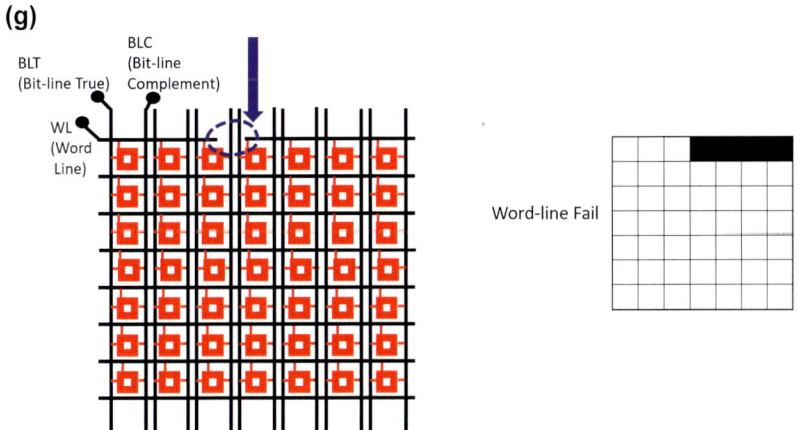

**Figure 12.12.g** 7 × 7 SRAM array with a broken word-line and the corresponding bit-fail map.

we will get a partial row fail or a partial word-line fail in the bit-fail map. If there is a catastrophic fail in the external circuit that feeds the signal or power to the entire SRAM array, it is possible to get a bit-fail map where all bits of the area are failing. Other kind of bit-fail maps are also possible to get depending on the nature of the fail and the architecture of the bit-cell and the periphery circuit. Knowledge of the design and architecture gives the yield analysis engineer tools to diagnose fail signals and gives clues about possible root cause of fail. Hence, bit-field maps are one possible method of doing electrical fault isolation where the physical locations of structural falls can be isolated. This gives the failure analysis team guidance so that the physical failure analysis (TEM/SEM imaging) can be performed at the correct physical location of the circuit.

In previous sections, we discussed various types of yield learning structures starting from simple comb-serp, via chain, MOSFET structures to complex functional SRAM circuits. A high-level comparison between these different of macros is listed in Table 12.1. Structures like comb-serp, via chain and MOSFETs are simpler; usually testable earlier in the process flow, but does not always capture all the different layout types found in a product circuit. The functional SRAM is better at capturing product type fails and is also better at pin-pointing location of fails. However, functional SRAMs often require that wafers get processed to a higher metal level for test. In technology development and manufacturing, both kind of macros are used for yield learning/yield monitoring as they often complement each other. A more comprehensive comparison of different types of yield learning test structures can be found in Ref. [6].

## 6. How to cheat defect density by adding redundancy

In this section, we will discuss methods to cheat defect density by adding redundancy in circuits. Adding "redundancy" is a method where the circuit designer add some "extra" or "additional" circuit content beyond what is absolutely required [7,8]. These "redundant" or "extra" elements can be used if the main part of the circuit is defective. An example is presented in Fig. 12.13. We can see that, in this figure, one failing bit in a memory array renders the entire memory nonuseable. This is because a sellable chip requires the entire array to be working. If the array had one redundant (spare) bit, the chip would have been usable. Redundant columns can be added to the array block as shown in Fig. 12.13b. Columns with defective bits of the

Table 12.1 Pros and cons of different types of yield assessment structures.

| Yield learning test structure | Design complexity | Testability in semiconductor process flow | Coverage of product like circuit/layout | Fault isolation capability through electrical test |
|---|---|---|---|---|
| Functional SRAM | Higher | Later | Higher | Higher |
| Comb-serp, via-chain, MOSFET Parametric test-structures | Lower | Earlier | Lower | Lower |

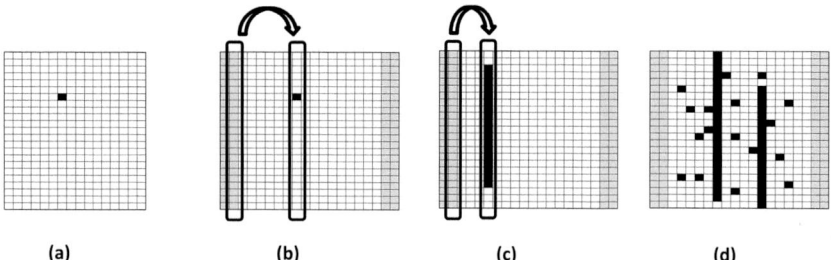

**Figure 12.13** (a) Failing chip with no redundancy; (b) Fixable chip with redundancy; (c) Fixable chip with redundancy; (d) Non-fixable chip with redundancy.

main block can be replaced by clean redundant columns in the software address space. This would make the chip a "fully functional" chip. This invocation of redundant elements is often called "repairing" of the memory array, and the yield obtained after this repair action is called "repair yield." In the example shown here in Fig. 12.13b and c, both single cell fails and bit-line fails can be "repaired" using redundant elements. If the number of failing bits become too high, as shown in Fig. 12.13d, there might not be enough redundant elements to fully "repair" all the fails and the array would be considered a "failing" array even for the repair yield criteria. Sometimes, an on-ship programmable fuse ("eFuse") is used to store the information of repair solution [9].

Fig. 12.14 gives some additional examples of what kind of fails in an SRAM are "repairable" and which are not. It is worth mentioning that the total number of redundant elements is small compared to the size of the actual array block, and hence, these redundant elements do not take up too much additional area. On the other hand, a small number of redundant elements can significantly increase the postrepair yield, thereby significantly increasing the number of SRAM array units that are usable/sellable. This is shown through an example in Figs. 12.15a,b. In this example, we see that, while not all chips can be "repaired" using redundant elements, many of the chips can be repaired and the repair action improved the yield from 19% to 73%. Adding redundant elements in SRAM arrays to boost yield are a routine feature for logic semiconductor chip designs these days.

In the previous section, it was shown how adding "spare" redundant elements in an SRAM array can significantly improve yield of an SRAM array. Similar technique can be added in other logic constructs in the design. Some of these examples are shown in Fig. 12.16. Fig. 12.16a gives an example of how "redundant (spare)" contacts can be uses in circuits to

Yield impact of defects from thin films and other processing steps 503

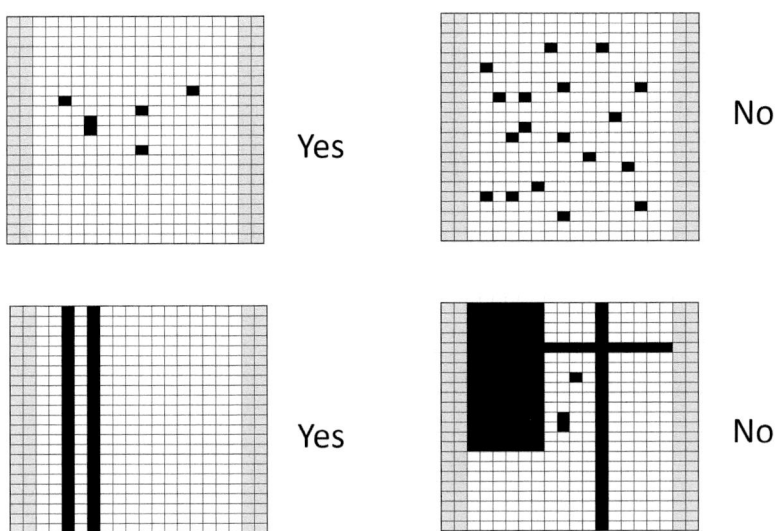

**Figure 12.14** Additional examples of fail maps that are repairable and not.

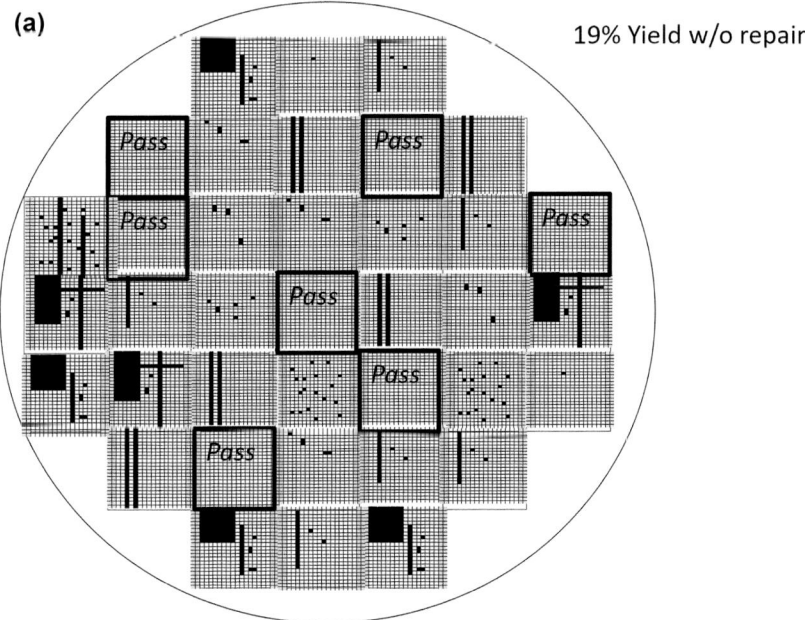

**Figure 12.15a** Cheating defect density by having redundant bits — yield without repair is very low (19%) in this example.

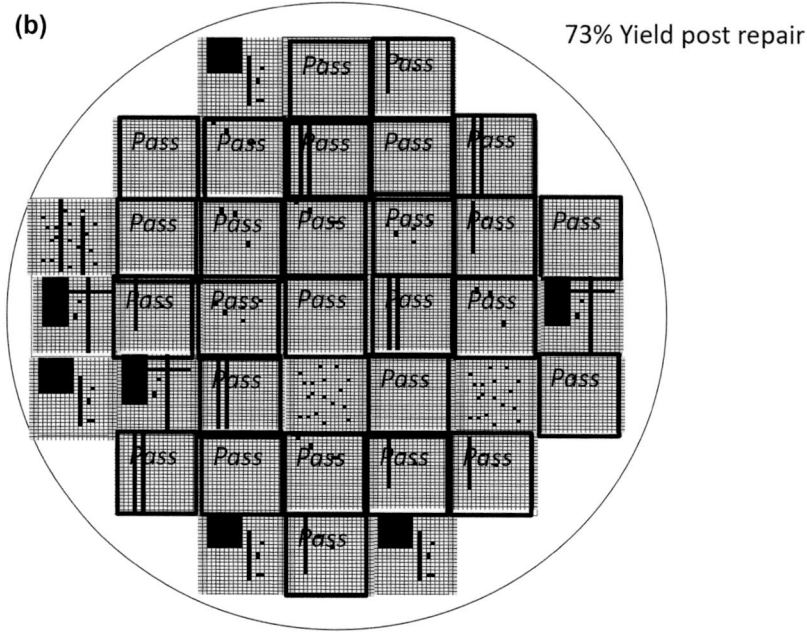

**Figure 12.15b** Cheating defect density by having redundant bits — yield with repair recovers to a high number (73%) in this example.

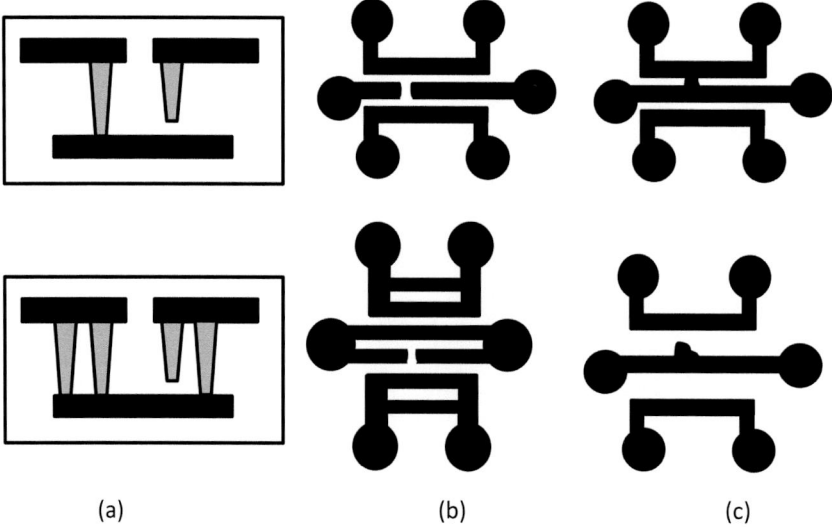

**Figure 12.16** (a) Use of redundancy in logic circuits by adding extra contacts, (b) by adding extra lines, (c) by adding extra space.

make it more robust and immune to yield issues. In this example, even if one of the contacts (vias) do not work due to defects, current can still flow through the second redundant contact, which will keep the circuit functional. Fig. 12.16b gives an example of usage of redundant lines between nodes. Here again if a single wiring line is broken due to defects, current can still be carried by the second line, and the circuit might still function. For both cases in Fig. 12.16a and b, it is true that the resistance of the current carrying path will increase when one of the contacts or lines do not work, but line and contact resistance might not be the limiting factor of these circuits anyway, and the involved circuit might still function even with higher resistance. In Fig. 12.16c, an example is shown of a layout, where lines are drawn far apart from each other to make them more immune to shorting mechanism so that even if a line is protruded due to presence of some defects, the protrusion might still not be enough to create a short to the next line. As is obvious from all these examples, all these designs that have redundancies take up additional space. Hence, a circuit designer will need to strike the correct balance between space usage and making a circuit more robust to yield issues when designing circuits.

## 7. Conclusion

In this chapter of the book, we have attempted to introduce the reader to some basic principles, methodologies, and test structures associated with yield learning in a semiconductor technology. Although semiconductor technologies in recent years have become very complex, some basic aspects of yield learning remain the same. An integral part of yield learning is finding root cause of the electrical fail signal. Defect inspection and failure analysis techniques are an integral part of this root cause finding work. Details of those techniques are out of scope for this book chapter. The result of these defect inspection and failure analysis work is often an image of the defect, like some of the examples shown in Fig. 12.2. Corrective process actions can then be taken to address these yield fails in subsequent lots/wafers. Other chapters of this book discuss details of various processing steps of the semiconductor chip manufacturing flow. Hopefully, this chapter on yield will give the reader better appreciation for the need for semiconductor processes that are less prone to defectivity and are more "yield friendly."

## Acknowledgments

The author of this chapter would like to thank Rama Divakaruni, Steve Crown, and Ronald Newhart for their input and help. The author would also like to thank the management chain in the IBM Semiconductor Research division for their support.

## References

[1] S.K. Moore, D. Schneider, The state of the transistor: in 75 years, it's become tiny, mighty, ubiquitous, and just plain weird, IEEE Spectr. 59 (12) (2022) 30−31.
[2] S. Narasimha, et al., 22nm high-performance SOI technology featuring dual-embedded stressors, Epi-Plate high-K deep-trench embedded DRAM and self-aligned via 15LM BEOL, in: Internationl Electron Devices Meeting, 2012, https://doi.org/10.1109/IEDM.2012.6478971.
[3] C.H. Stapper, LSI yield modelling and process modelling, IBM J. Res. Dev. 20 (3) (1976).
[4] C.H. Stapper, P.P. Castrucci, R.A. Maeder, W.E. Rowe, R.A. Verhelst, IBM J. Res. Dev. 26 (5) (1982).
[5] J.A. Cunningham, The use and evaluation of yield models in integrated circuit manufacturing, IEEE Trans. Semicond. Manuf. 3 (2) (1990).
[6] I. Ahsan, C. Schiller, F. Towler, Z. Song, R. Wong, D. Clark, S. Lucarini, F. Beaudoin, Tristate inverter array: a new technology development yield learning vehicle complementing traditional SRAM arrays, IEEE Trans. Semicond. Manuf. 28 (4) (2015).
[7] S.E. Schuster, Multiple word/bit line redundancy for semiconductor memories, IEEE J. Solid State Circ. SC-13 (5) (1978).
[8] I. Kim, Y. Zorian, K. Goh, H. Pham, F.P. Higgins, J.L. Lewandowski, Built in self repair for embedded high density SRAM, in: Paper 43.2, International Test Conference, 1998, pp. 1112−1119.
[9] N. Robson, J. Safran, K. Chandrasekharan, A. Cestero, X. Chen, R. Rajeevakum, A. Leslie, D. Moy, T. Kirihata, I. Subramanian, Electrically programmable fuse (eFUSE): from memory redundancy to autonomic chips, in: IEEE Custom Integrated Circuits Conference, 16−19 September, 2007.

# Summary

## Dominic J. Schepis

We have seen throughout this publication that many authors have described detailed information containing state-of-the-art materials and methods for a broad variety of thin films used in various nanotechnology applications.

As the reader has progressed through each chapter of this handbook, we hope that certain themes have surfaced. One of the most apparent is the theme of scaling and increased 3-dimensional requirements, which have driven new developments for thin films to retain the integrity, defectively, and reliability, over large topographies.

An overall theme highlighted in the introduction was the many challenges and potential future directions for semiconductors as the world's needs for microelectronics are expanding almost exponentially. New prospects such as artificial intelligence are taxing the computing power to today's applications. Several new initiatives on quantum computing have sprung up in response to this need for growing computing power. It also demonstrates the importance of our continual development and understanding of the equipment and materials science that create these new chips.

*New materials.* As we saw, over many process nodes, semiconductors have employed many known materials which were used for gate oxides, back-end-of-the-line dielectrics, spacer materials, etc. However, in recent years, many new innovations in materials science have come forth providing high-k dielectrics, porous dielectrics with low-k, and metallurgical contacts. The earliest aluminum wiring which gave way to copper has even come under scrutiny as metal contacts, resistance, and wiring that can be directly patterned are in demand. First-generation high-k materials have given way to material such as HfOxN and HfLaOxN and bilayers. Silicon substrates have also in some cases been replaced by SiGe substrates or channel materials. New materials have been introduced for innovative LED technologies as well. New developments suggest that cubic InGeN/GaN LEDs may show promise for next-generation devices. Finally, a discussion of new developments in ferroelectrics was reviewed with not only the known BST materials but new candidates such as AlScN which can be deposited with ALD, PLD, and MBE techniques.

*Topography.* Again, as with recent generations of semiconductor devices, the challenge of depositing homogenous contiguous films over topography has grown. Semiconductor devices, once planar in fashion, have given way to stacks of materials with great aspect ratios. Finfet fin heights grew, followed by new nanowire or gate-all-around (GAA) architectures which drove 3D structures with extremely high aspect ratios. Spaces between horizontal nanosheets also became a challenge to fill even with atomically grown ALD methods. The aggressive scaling and stacking of devices present challenges to chemical−mechanical planarization (CMP) process technologies, including the need for more CMP steps and polishing of new materials. The major CMP performance metrics, such as planarity, uniformity, and defectivity, are being stretched to meet specifications at nanometer scales and for 3D stacks. Novel endpoint methods and AI/APC are mandatory. Even silicon memory technologies are pushing the limits of topography by which thin films are required to bridge with atomic repeatability. Also challenging, is the metrology to characterize these films due to spot sizes for many methods being too large for the reduced dimensions. These trends are likely to continue even as dimensions are pushed to their limits.

*Interconnects.* Interconnects have also advanced along with the newer device structures. We have seen that Cu and various low-k dielectrics were materials of choice for several generations of on-chip interconnects. However, challenges with integration including patterning, liner coverage, chemical mechanical polishing (CMP), and packaging remain. These processes took careful optimization to ensure adequate reliability for TDDB, electromigration, stress-induced voiding, and packaging stresses. Current and future directions suggest new materials, as well as more complex homogeneous integration where interconnects can be placed on the backside of the device wafer through the use of through-silicon-vias (TSVs), or off-chip, on silicon or organic interposers.

Regardless of the architectures which follow in the future, we anticipate that multiple technologies will use many of the developments discussed in this book for a variety of new technologies including microprocessors, sensors, memory, ferroelectric applications, and perhaps even superconducting thin films for quantum computing. We look with anticipation as the technology of thin films continues to enable innovations in the foreseeable future.

# Index

Note: 'Page numbers followed by *f* indicate figures and *t* indicate tables'

## A

Ab-initio modeling, 311–320
  effective work function (EWF) engineering, 316–320, 318f, 319t
  higher K dielectric, 311–316, 315t–316t
Accelerated life testing, 364
Aluminum scandium nitride (AlScN), 438–440
Artificial intelligence (AI), 10, 266
Atmospheric pressure CVD (APCVD), 224–226
Atomic force microscopy, 213–215
Atomic layer deposition (ALD), 10, 218–221, 219f

## B

Barium–strontium–titanate (BST), 436–438
Barrier layer deposition, 37
Benzotriazole (BTA), 269–270
Bias temperature instability, 361
Bilayer gate dielectrics, 305–306

## C

Carbon nanotubes (CNTs), 70–71
Cell performance and cost, 457t
Chemical-mechanical planarization (CMP)
  challenges, 259–264
  defectivity, 262–263
  equipment, 264–267, 265f–266f
  fine patterns, post CMP metal loss in, 264
  hybrid wafer bonding, 259f
  pads and conditioners, 270–281, 271f–272f
    pads with engineered asperities (EA pads), 274–277, 276f–277f
    post clean, 277–281, 278f–281f
  planarity and surface roughness, 262
  slurry, 267–270, 268f–269f
  steps, 259–260
  sub-nm wafer-level uniformity, 261, 261f–262f
Chemical mechanical polishing (CMP), 42–46, 43f, 44t, 45f–46f
Chemical vapor deposition (CVD) technology
  applications, 205–211
    dielectrics, 207–210
    metals, 205–207
    semiconductors, 210–211
  atmospheric pressure CVD (APCVD), 224–226
  atomic layer deposition (ALD), 218–221, 219f
  basic principles, 194–197, 195f
  contamination, 211–213
  history of, 198–205, 199f
  low-pressure CVD (LPCVD), 222–224
  metal-organic CVD (MOCVD), 226–230
  metrology, 213–218
    atomic force microscopy, 213–215
    electrical, 217–218
    film thickness, 213–215
    Fourier transform infrared, 215–216
    mass spectroscopy, 215–216
    mechanical, 216–217
    nanoindentation, 216–217
    Nomarski, 213–215
    optical microscopy, 213–215
    sample imaging, 213–215
    scanning electron microscopy, 213–215
    secondary mass spectrometry, 215–216
    stud pull, 216–217

509

Chemical vapor deposition (CVD) technology (*Continued*)
   transmission electron microscopy, 213–215
   wafer bowing, 216–217
   XRD, 215–216
  plasma-enhanced CVD (PECVD), 230–232
  subatmospheric CVD (SACVD), 221–222
  tool selection for research and manufacturing, 233–236, 236f
  trends and projection, 236–242
  vertical diffusion furnace, 200–201, 201f
Clustered vacuum tools, 106
CMP. *See* Chemical mechanical polishing (CMP)
Constant electric-field scaling, 290–291, 291f
Contamination control/prevention, sputter processing, 130–132
  extrinsic contamination control, 131–132
  residual gas analyzer (RGA) monitoring, 131–132
  tooling and shielding considerations, 130–131
Copper interconnect processing, 21–46
  chemical mechanical polishing (CMP), 42–46, 43f, 44t, 45f–46f
  dielectric patterning, 26–32, 28f–30f, 32f–33f
  low-k dielectrics, 23–26, 23f, 25f
  metallization, 32–41, 35f–36f, 38f–39f, 40t, 41f
  process flow, 21–23, 22f
Corona-Kelvin technique, 426, 427f
Corrosion inhibitor, 269–270
Cosine sputtering law, 97
Cubic GaN and InGaN/GaN, 379–382, 381f
Cubic GaN crystal, 379–380
Cubic GaN on U-grooved Si (100), 392–398, 395f, 397f
Cubic InGaN/GaN, computation-based design of, 382–392, 385f–388f, 390f–391f

## D

Defect density, 488–492, 488f–491f
  redundancy, 500–505, 504f
Defectivity, even lower tolerance for, 262–263
Dennard scaling theory
  bilayer gate dielectrics, 305–306
  constant electric-field scaling, 290–291, 291f
Device junction formation, dopant diffusion control for, 166–167, 167f
Device scaling, 18–21, 18f–20f
Dielectrics, 207–210
  patterning, 26–32, 28f–30f, 32f–33f
  reliability, 58
Direct deposition on cubic substrates, 380
3D XPoint memory, 475–477, 476f
Dynamic random access memory (DRAM), 21, 458–465, 459f
  access transistor, 459–460
  capacitor, 461–463
  future cells, 463–465

## E

Effective conditioning, 275–276
Effective work function (EWF) engineering, 316–320, 318f, 319t
Efficiency droop, 376–378
Electrical metrology, 217–218
Electrical results, 339–341, 340f–342f, 344f–345f
Electromigration, 47–52, 47f, 49f–51f, 361
Electron blocking layer (EBL), 391–392
E-model, 59–60
End-pointing, 265
Energy dependence, sputter processing, 95–97
  cosine sputtering law, 97
Energy dispersive X-ray spectroscopy (EDS), 9
Engineered asperities (EA) pads, 274–277, 276f–277f
Epitaxial process
  FinFET device, Si:As SDE formation on, 169

Index 511

gate-all-around (GAA) pFET device, performance enhancement techniques for, 173–182
  (110) versus (001) channel orientation, 178–182
  Ge fraction/SiGe thickness, 178
  highly compressive strained SiGe channel application, 174–178
  nFET devices fabrication, 179
  SiGe cladded nanosheet structure formation, 174–177, 175f–176f
  stacked Si NS pFET, 179
  TSi for device performance, electrical sensitivities to, 180–182
nanoscale strained SiGe FinFET, 148–165
  local strain, 157–165
  SiGe FinFET, Ge content on, 152–155
  strained SiGe pFinFET formation on SRB, 150–152
  strain in SiGe channel FinFET, 155–157, 155f
scalded devices, advanced source drain extension formation for, 165–173
  device junction formation, dopant diffusion control for, 166–167, 167f
  source drain extension formation, advanced epitaxial growth technique for, 168–173, 168f
  SDE material and lateral recess for device performance, electrical sensitivities to, 171–173
strained Si technology with, 143–148
  mobility enhancement technology, 143–147, 144f–145f
  three-dimensional structure, new device structure with, 148
Equivalent oxide thickness (EOT) scaling, 292–300
  HfO2/TiO2 higher K for, 305–306
  interfacial layer, 307–311
  through interfacial layer, 307–311
    interfacial layer scavenging, 308–311, 309f–310f
    nitrided interfacial layer (SiON), 307

Error correction codes (ECC), 364
Extrinsic contamination control, 131–132

F

Fail modes, examples of, 487–488
Failure mode and effects analysis (FMEA), 364
Ferroelectric devices, 429–431
  ferroelectric tunnel junctions (FTJ), 431
  metal ferroelectric metal (MFM), 430
  metal ferroelectric semiconductor (MFS), 430
  variable capacitors, 430
Ferroelectric memory, 477–479, 477f–478f
Ferroelectric models, 409–412, 411f–412f
Ferroelectric thin films
  aluminum scandium nitride (AlScN), 438–440
  applications, 428–434
  barium–strontium–titanate (BST), 436–438
  characterization of, 420–428, 423t
  corona-Kelvin technique, 426, 427f
  ferroelectric devices, 429–431
    ferroelectric tunnel junctions (FTJ), 431
    metal ferroelectric metal (MFM), 430
    metal ferroelectric semiconductor (MFS), 430
    variable capacitors, 430
  ferroelectric models, 409–412, 411f–412f
  hafnium oxide based ferroelectric films, 434–436
  history, 405–407
  hysteresis P-E loop, 423–425
  in-memory and neuromorphic computing, 432f, 433–434
  multiple properties of, 406f
  patterning of, 420, 421t–422t
  piezoresponse force microscopy (PFM), 425
  principle, 408–412, 408f

Ferroelectric thin films (*Continued*)
  pyroelectric, piezoelectric and energy harvesting, 433, 433f
  reliability of ferroelectric films, 440–443, 440t
    fatigue, 442
    field cycling endurance, 442
    imprint effect, 442–443
    retention, 442
    scaling, 443
    wake-up, 441
  thin film deposition processes, 414–418, 414f–415f
    atomic layer deposition (ALD), 416–417, 417f
    chemical solution deposition (CSD), 414
    metal-organic chemical vapor deposition (MOCVD), 418
    molecular beam epitaxy (MBE), 418
    pulsed laser deposition (PLD), 415, 416f
    sputter deposition, 414–415
  thin films, 412, 413t
  three exemplary ferroelectric films, 434–440
  tunable microwave devices, 433–434
Ferroelectric tunnel junctions (FTJ), 431
Field cycling endurance, 442
Field Programmable Gate Arrays (FPGAs), 67
Film
  deposition, 229
  thickness, 213–215
Fine patterns, post CMP metal loss in, 264
FinFET device, Si:As SDE formation on, 169
Flip-chip attach, 64
Fourier transform infrared, 215–216

# G

Gas-phase interactions, 229
Gate-all-around (GAA) pFET device, performance enhancement techniques for, 173–182
  (110) *versus* (001) channel orientation, 178–182
  Ge fraction/SiGe thickness, 178
  highly compressive strained SiGe channel application, 174–178
  nFET devices fabrication, 179
  SiGe cladded nanosheet structure formation, 174–177, 175f–176f
  stacked Si NS pFET, 179
  TSi for device performance, electrical sensitivities to, 180–182
Gate dielectrics scaling
  Ab-initio modeling, 311–320
    effective work function (EWF) engineering, 316–320, 318f, 319t
    higher K dielectric, 311–316, 315t–316t
  Dennard scaling theory
    bilayer gate dielectrics, 305–306
    constant electric-field scaling, 290–291, 291f
  electrical results and discussions, 339–341, 340f–342f, 344f–345f
  EOT scaling, HfO2/TiO2 higher K for, 305–306
  EOT scaling through interfacial layer, 307–311
    interfacial layer scavenging, 308–311, 309f–310f
    nitrided interfacial layer (SiON), 307
  gate oxide, 292–300
  gate oxides, FinFET era, 320–322, 320f, 322f
  gate stack characteristics, 342–347, 347f, 351f
  generalized scaling, 292, 293t
  hafnium based ternary, quaternary and bilayer oxides for EOT scaling, 300–306
    hafnium dioxide, physical structure of, 300, 300f
    hafnium lanthanum oxynitride (HfLaOxN), 304–305, 304f–305f
    hafnium oxynitride (HfOxN), 301–304, 301f–303f
  heterostructure device structure results, 337–338

Index 513

high mobility (high atomic % Ge, SiGe) channel, 347–350
high voltage (HV) input/output (I/O) gate oxides, 322–327
I/O gate oxide reliability, 346–347
logic IL and I/O gate oxide research, 347–350
low temperature, Si/SiGe heterostructure-based I/O devices with, 336–347
  nano-sheet gate-all-around device, integrate with, 336–337
SiGe as a pFET channel (cSiGe), 327–329, 329f
  co-integration with logic transistor, 334–335
  FinFET vs. nano-sheet gate-all-around structure comparison, 330–331
  logic transistor interfacial layer, 333–334
  nano-sheet devices, 333–334
  Nano-sheet (NS) gate-all-around (GAA) transistor technology, 329–335
  nano-sheet high voltage I/O transistor gate dielectrics, 334–335
  nano-sheet transistor structure process integration, 331–333
Gate oxides, 292–300
  breakdown, 361
  FinFET era, 320–322, 320f, 322f
Gate stack characteristics, 342–347, 347f, 351f
Ge fraction/SiGe thickness, 178
Generalized scaling, 292, 293t
Green-emitting hexagonal InGaN/GaN LEDs, 373–379, 375f

# H

Hafnium based ternary, quaternary and bilayer oxides for EOT scaling, 300–306
  hafnium dioxide, physical structure of, 300, 300f

hafnium lanthanum oxynitride (HfLaOxN), 304–305, 304f–305f
hafnium oxynitride (HfOxN), 301–304, 301f–303f
Hafnium dioxide, physical structure of, 300, 300f
Hafnium lanthanum oxynitride (HfLaOxN), 304–305, 304f–305f
Hafnium oxide based ferroelectric films, 434–436
Hafnium oxynitride (HfOxN), 301–304, 301f–303f
Hard pads, 271
Heterogeneous integration, 242
Heterojunction bipolar transistors (HBTs), 359–360
Heterostructure device structure results, 337–338
HfO2/TiO2 higher K, 305–306
Higher K dielectric, 311–316, 315t–316t
Highly compressive strained SiGe channel application, 174–178
High mobility (high atomic % Ge, SiGe) channel, 347–350
High-performance processes, 6
High voltage (HV) input/output (I/O) gate oxides, 322–327
Hot carriers, 361
Hybrid wafer bonding, 259f
Hysteresis P-E loop, 423–425

# I

Imprint effect, 442–443
In-line chemical monitoring, 267
In-memory and neuromorphic computing, 432f, 433–434
Interfacial layer, 307–311
Interfacial layer scavenging, 308–311, 309f–310f
International Technology Roadmap for Semiconductors (ITRS), 6
I/O gate oxide reliability, 346–347

# K

Knock-on sputtering, 95–96

## L

Light emitting diodes (LEDs)
    technologies, thin film
    development for
  cubic GaN and InGaN/GaN, 379—382, 381f
  cubic GaN on U-grooved Si (100), 392—398, 395f, 397f
  cubic InGaN/GaN, computation-based design of, 382—392, 385f—388f, 390f—391f
  future work, 398—399
  green-emitting hexagonal InGaN/GaN, 373—379, 375f
Line reliability
  back end of, 362—364
  front end of, 362—364
  middle of, 362—364
Logic IL and I/O gate oxide research, 347—350
Logic transistor interfacial layer, 333—334
Low-energy (subthreshold) sputtering, 95
Low-k dielectrics, 23—26, 23f, 25f
Low-pressure CVD (LPCVD), 222—224
Low temperature, Si/SiGe heterostructure-based I/O devices with, 336—347
  nano-sheet gate-all-around device, integrate with, 336—337

## M

Machine learning (ML), 266
Mass spectroscopy, 215—216
Metal ferroelectric metal (MFM), 430
Metal ferroelectric semiconductor (MFS), 430
Metallization, 32—41, 35f—36f, 38f—39f, 40t, 41f
Metal-organic CVD (MOCVD), 226—230
Metal-oxide-semiconductor field-effect transistor (MOSFET), 18, 141—143
Metals, 205—207

Metrology, 265—266
Metrology, sputtered films, 126—130
  non-destructive thickness measurements, 127—130
    time-resolved picosecond ultrasound, 129—130
    X-ray fluorescence (XRF), 128
    X-ray reflectance (XRR), 129
  resistance/four-point probe measurement, 126—127
Microtrenching, 28
Mitigation strategies, 364—365
Mobility enhancement technology, 143—147, 144f—145f
Multiplaten configuration, 264—265

## N

NAND, 465—475, 467f
  ONO storage layers, 470—475, 471f
  transistor, 468—470, 470f
  Yangtze Memory Technology company (YMTC), 473
Nanoindentation, 216—217
Nanoscale strained SiGe FinFET, 148—165
  local strain, 157—165
  SiGe FinFET, Ge content on, 152—155
  strained SiGe pFinFET formation on SRB, 150—152
  strain in SiGe channel FinFET, 155—157, 155f
Nano-sheet devices, 333—334
Nano-sheet (NS) gate-all-around (GAA) transistor technology, 329—335
Nano-sheet high voltage I/O transistor gate dielectrics, 334—335
Nano-sheet transistor structure process integration, 331—333
nFET devices fabrication, 179
Nitrided interfacial layer (SiON), 307
Non-destructive thickness measurements, 127—130
  time-resolved picosecond ultrasound, 129—130
  X-ray fluorescence (XRF), 128
  X-ray reflectance (XRR), 129

Index 515

**O**
On-chip interconnects
  copper interconnect processing, 21—46
    chemical mechanical polishing
      (CMP), 42—46, 43f, 44t, 45f—46f
    dielectric patterning, 26—32, 28f—30f,
      32f—33f
    low-k dielectrics, 23—26, 23f, 25f
    metallization, 32—41, 35f—36f,
      38f—39f, 40t, 41f
    process flow, 21—23, 22f
  device scaling, 18—21, 18f—20f
  future directions, 66—71, 66f, 68f—70f
  reliability, 46—66
    electromigration, 47—52, 47f,
      49f—51f
    package reliability, 62—66, 63f, 63t,
      65f
    stress-induced voiding, 52—57,
      52f—54f, 56f
    time dependent dielectric breakdown
      (TDDB), 57—62, 57f, 60f—61f
ONO storage layers, 470—475, 471f
Optical microscopy, 213—215
Oxygen-containing plasmas, 29

**P**
Package reliability, 62—66, 63f, 63t, 65f
Packaging processes, 63
Piezoresponse force microscopy (PFM),
  425
Planarity and surface roughness, 262
Plasma-enhanced CVD (PECVD),
  230—232
Plasma-induced damage, 361
Polarization, 376
Post chemical-mechanical planarization
  (CMP) clean, 277—281,
  278f—281f
Process control, 364
Process flow, 21—23, 22f
Process variability, 361
Pyroelectric, piezoelectric and energy
  harvesting, 433, 433f

**Q**
Quantum Computing, 10

**R**
Reactive ion etching (RIE), 30—31
Reactive sputter cleaning, 34
Redundancy, 364
Reliability, 46—66, 360
  assessment, test structures and
    methodologies for, 364—366
  design for, 365
  electromigration, 47—52, 47f, 49f—51f
  ferroelectric films, 440—443, 440t
    fatigue, 442
    field cycling endurance, 442
    imprint effect, 442—443
    retention, 442
    scaling, 443
    wake-up, 441
  package reliability, 62—66, 63f, 63t, 65f
  physics analysis, 364
  semiconductor reliability balancing act,
    360
  stress-induced voiding, 52—57, 52f—54f,
    56f
  time dependent dielectric breakdown
    (TDDB), 57—62, 57f, 60f—61f
Residual gas analyzer (RGA)
  monitoring, 131—132
Resistance/four-point probe
  measurement, 126—127
Retention, 442

**S**
Sample imaging, 213—215
Scalded devices, advanced source drain
  extension formation for,
  165—173
  device junction formation, dopant
    diffusion control for, 166—167,
    167f
  source drain extension formation,
    advanced epitaxial growth
    technique for, 168—173, 168f
Scaling, 443
Scanning electron microscopy (SEM),
  213—215
SDE material and lateral recess for device
  performance, electrical
  sensitivities to, 171—173

## Index

Secondary ion mass spectroscopy (SIMS), 9
Secondary mass spectrometry, 215–216
Semiconductor memory
  cell performance and cost, 457t
  3D XPoint memory, 475–477, 476f
  dynamic random access memory (DRAM), 458–465, 459f
    access transistor, 459–460
    capacitor, 461–463
    future cells, 463–465
  ferroelectric memory, 477–479, 477f–478f
  hierarchy, 457f
  NAND, 465–475, 467f
    ONO storage layers, 470–475, 471f
    transistor, 468–470, 470f
    Yangtze Memory Technology company (YMTC), 473
Semiconductors, 210–211
  reliability, 360
    challenges and principal degradation mechanisms, 361–362
    line reliability
      back end of, 362–364
      front end of, 362–364
      middle of, 362–364
    mitigation strategies, 364–365
    semiconductor reliability balancing act, 360
  reliability assessment
    test structures and methodologies, 364–366
SiGe as a pFET channel (cSiGe), 327–329, 329f
  co-integration with logic transistor, 334–335
  FinFET vs. nano-sheet gate-all-around structure comparison, 330–331
  logic transistor interfacial layer, 333–334
  nano-sheet devices, 333–334
  Nano-sheet (NS) gate-all-around (GAA) transistor technology, 329–335
  nano-sheet high voltage I/O transistor gate dielectrics, 334–335
  nano-sheet transistor structure process integration, 331–333
SiGe channel FinFET strain, 155–157, 155f
SiGe cladded nanosheet structure formation, 174–177, 175f–176f
SiGe FinFET, Ge content on, 152–155
Slurry, 267–270, 268f–269f
Soft error rate, 362
Soft pads, 271
Sonication, 281
Source drain extension formation, advanced epitaxial growth technique for, 168–173, 168f
Sputter deposition, 35
Sputter processing
  contamination control and prevention, 130–132
    extrinsic contamination control, 131–132
    residual gas analyzer (RGA) monitoring, 131–132
    tooling and shielding considerations, 130–131
  energy and kinematics of, 94–95
  energy dependence, 95–97
    cosine sputtering law, 97
  metrology of sputtered films, 126–130
    non-destructive thickness measurements, 127–130
    resistance/four-point probe measurement, 126–127
  physical vapor deposition (PVD)
    advanced-groundrule interconnects copper fill using reflow, 121–124, 123f
    nanometer-scale engineering, 115–126, 116f
    self-capping layers for Cu interconnects, copper-alloy seedlayers, 117–121, 118f–119f
  plasmas
    DC diode plasmas, 98–100
    magnetron designs, 102–103
    magnetron sputtering, 101–103, 101f
    RF plasmas, 100
  reactive sputter deposition, 103–105
    current–voltage hysteresis, 103–105, 104f
  tool design and applications for semiconductor technology

batch/planetary systems, 105–106
clustered sputter-tool layout, 107–108
collimation, 109–115
directional sputter deposition, 108–115
ionized and self-ionized sputtering, 111–114, 112f
self-sustained sputtering, 114–115
single-wafer systems, 106–108
SRAM yield learning methodology, 497–500, 497f–499f, 501t
Stacked Si NS pFET, 179
Strained SiGe pFinFET formation, SRB, 150–152
Strained Si technology, 143–148
mobility enhancement technology, 143–147, 144f–145f
three-dimensional structure, new device structure with, 148
Stress-induced voiding (SIV), 51t, 52–57, 52f–54f, 56f
Stud pull, 216–217
Subatmospheric CVD (SACVD), 221–222
Sub-nm wafer-level uniformity, 261, 261f–262f
Subtractive metalization process, 241
Superfilling, 38–39

# T

Temperature management, 364
Thermal stress, 361
Thin films, 412, 413t
deposition processes, 414–418, 414f–415f
atomic layer deposition (ALD), 416–417, 417f
chemical solution deposition (CSD), 414
metal-organic chemical vapor deposition (MOCVD), 418
molecular beam epitaxy (MBE), 418
pulsed laser deposition (PLD), 415, 416f
sputter deposition, 414–415
nanotechnology

challenges, 10
device scaling, 4–9
characterization, 8–9, 10f
device roadmap, 6–8, 7f
handbook organization, 11–14
Three-dimensional structure, new device structure with, 148
Three exemplary ferroelectric films, 434–440
Time dependent breakdown test (TDDB), 58
Time dependent dielectric breakdown (TDDB), 57–62, 57f, 60f–61f
Time-resolved picosecond ultrasound, 129–130
Titanium nitride (TiN), 103
Tool ambient control, 267
Tooling/shielding considerations, 130–131
Transistor, 468–470, 470f
Transmission electron microscopy, 213–215
TSi for device performance, electrical sensitivities to, 180–182
Tunable microwave devices, 433–434

# V

Variable capacitors, 430
Vertical diffusion furnace, 200–201, 201f

# W

Wafer bowing, 216–217
Wafer handling, 267
Wake-up, 441

# X

X-ray diffraction (XRD), 215–216
X-ray fluorescence (XRF), 128
X-ray reflectance (XRR), 129

# Y

Yangtze Memory Technology company (YMTC), 473
Yield assessment structures, 493–496, 494f, 496f

Printed in the United States
by Baker & Taylor Publisher Services